MEASURING
&
MONITORING
Plant Populations

AUTHORS:

Caryl L. Elzinga Ph.D.
Alderspring Ecological Consulting
P.O. Box 64
Tendoy, ID 83468

Daniel W. Salzer
Coordinator of Research and Monitoring
The Nature Conservancy of Oregon
821 S.E. 14th Avenue
Portland, OR 97214

John W. Willoughby
State Botanist
Bureau of Land Management
California State Office
2135 Butano Drive
Sacramento, CA 95825

This technical reference represents a team effort by the three authors. The order
of authors is alphabetical and does not represent the level of contribution.

Though this document was produced through an interagency effort, the following BLM numbers
have been assigned for tracking and administrative purposes:

BLM Technical Reference 1730-1

BLM/RS/ST-98/005+1730

ACKNOWLEDGEMENTS

The production of this document would not have been possible without the help of many individuals. Phil Dittberner of the Bureau of Land Management's National Applied Resource Sciences Center (NARSC) coordinated the effort for BLM. Ken Berg, former BLM National Botanist, provided support and funding for the project.

The content of many chapters in this Technical Reference has benefited from the review of lecture outlines included in "Vegetation Monitoring in a Management Context," a week-long monitoring workshop offered jointly by The Nature Conservancy and the U.S. Forest Service.

The authors would also like to acknowledge those persons who reviewed the document and provided valuable comments, including Jim Alegria of the BLM Oregon State Office; Paul Sawyer of the BLM Arizona State Office; Rita Beard, Andrew Kratz, Will Moir, and David Wheeler of the Forest Service; Peggy Olwell of the National Park Service; and Gary White of Colorado State University.

We'd like to thank Sherry Smith of Indexing Services for the many hours she donated to this project in developing the index to this TR.

We extend a special thank you to Janine Koselak (Visual Information Specialist) of NARSC for doing a masterful job in layout, design, and production of the final document.

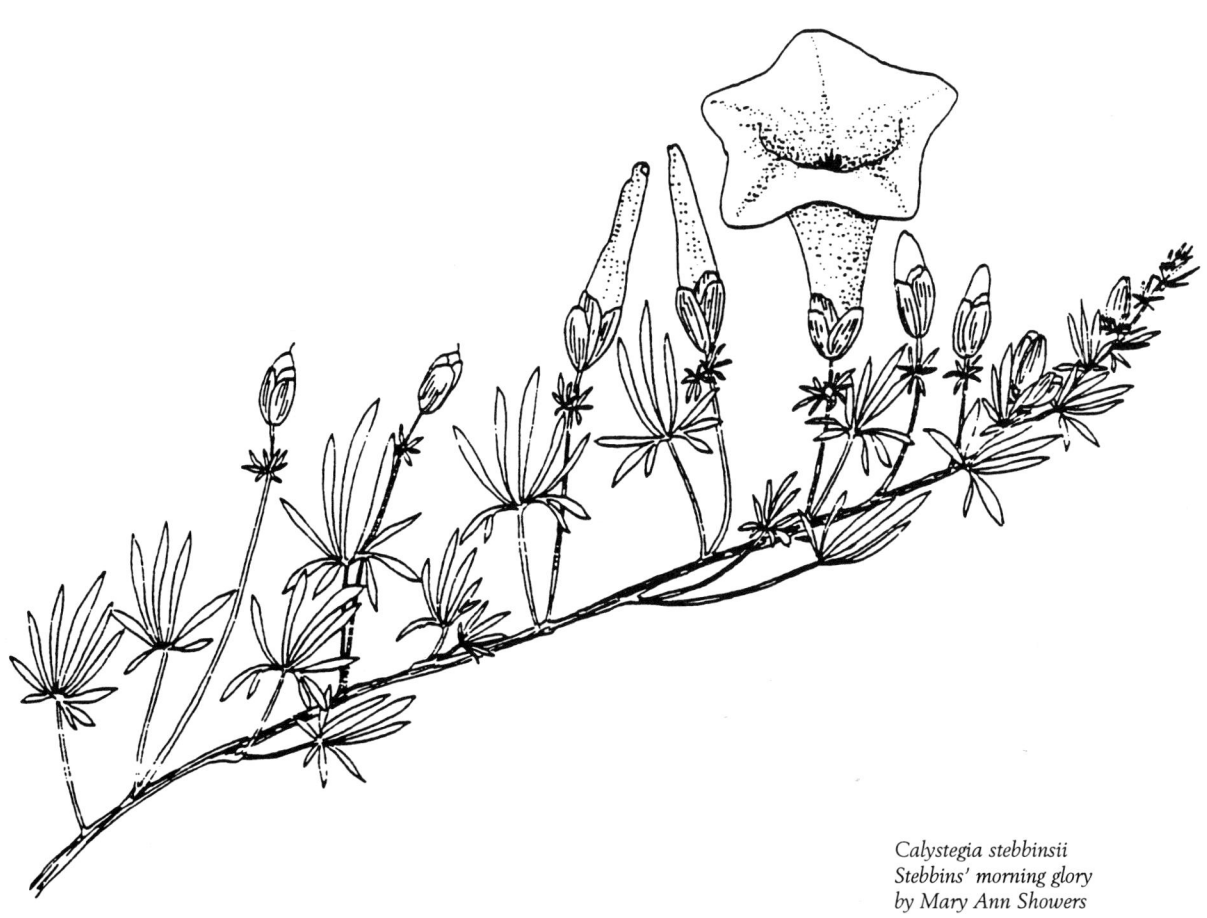

Calystegia stebbinsii
Stebbins' morning glory
by Mary Ann Showers

PREFACE

This technical reference applies to monitoring situations involving a single plant species, such as an indicator species, key species, or weed. It was originally developed for monitoring special status plants, which have some recognized status at the Federal, State, or agency level because of their rarity or vulnerability. Most examples and discussions in this technical reference focus on these special status species, but the methods described are also applicable to any single-species monitoring and even some community monitoring situations. We thus hope wildlife biologists, range conservationists, botanists, and ecologists will all find this technical reference helpful.

Monitoring is not a new activity for land management agencies, but there is a renewed interest and a new national emphasis on improving the quality of monitoring. Monitoring designed and executed effectively is a powerful tool for better management of resources. Good monitoring, while initially expensive to implement, is eventually cost-effective because management problems can be detected at an early stage, when solutions may yet be relatively inexpensive. Good monitoring can demonstrate that management is effective and successful, can silence critics, and can encourage the widespread adoption of an effective management technique.

Often, however, the results from monitoring are inconclusive and fail to provide the information needed to evaluate the success of management. Inconclusive or ambiguous monitoring results are expensive, both in terms of the resources wasted on the monitoring project and the potential costs of incorrect action. These costs are often difficult to measure because they are exacted from the environment in the form of environmental damage, or from industry in the form of unnecessary controls. Reduced public confidence and litigation expenses are additional hidden costs of poor monitoring.

Many monitoring projects suffer one of five unfortunate fates: (1) they are never completely implemented; (2) the data are collected but not analyzed; (3) the data are analyzed but results are inconclusive; (4) the data are analyzed and are interesting, but are not presented to decision makers; (5) the data are analyzed and presented, but are not used for decision-making because of internal or external factors (see Appendix 1 for some typical scenarios). The problem is rarely the collection of data. Agency personnel are often avid collectors of field data because it is one of the most enjoyable parts of their jobs. Data collection, however, is a small part of successful monitoring.

Because of the difficulty and importance of effective monitoring, agencies developed standard monitoring approaches in the 1960s through 1980s. While these techniques effectively met the challenges of that time, they are inadequate now for several reasons:

◆ The resources and management effects of interest today are more variable and complex. It is difficult for standard designs to keep pace with the rapid changes in issues. Monitoring data from standard techniques are sometimes inconclusive because the studies are not specifically designed for the issue in question.

◆ Many standard techniques do not address issues of statistical precision and power during design; thus, standard monitoring techniques that involve sampling may provide estimates that are too imprecise for confident management decisions.

◆ Commodity and environmental groups have become more sophisticated in resource measurement and are increasingly skeptical of data from standard agency techniques.

◆ Funding reductions are restricting resources available for monitoring projects. Concurrently, agencies are being required to more clearly demonstrate through monitoring that funds are being used to effectively manage public lands. This situation requires the design of efficient monitoring projects that provide data specific to the current issues.

The challenges of successful monitoring involve efficient and specific design, and a commitment to implementation of the monitoring project, from data collection to reporting and using results. We have designed this technical reference with these challenges in mind. Our approach differs radically from the development of standard techniques for field offices to apply. We instead provide technical guidance that assists field personnel in thinking through the many decisions that they must make to specifically design monitoring projects for the site, resources, and issues. We base this approach on the belief that local resource managers and specialists understand their issues and their resources best and, therefore, are best able to design monitoring to meet their specific needs. With this technical reference, local personnel can design much of the monitoring done at the local level, and recognize when they need additional specialized skills for a successful project.

We encourage you to treat this technical reference not as a step-by-step guide on how to implement a monitoring study, but as a collection of pieces that you need to choose among and put together for your particular situation and species. We have organized this technical reference to follow a logical progression of planning and objective setting, designing the methodology, taking the measurements in the field, analyzing and presenting the data, and making the necessary management responses. Many of these steps, however, occur simultaneously, or provide feedback to others. Decisions made at each step of the monitoring process can affect the whole project, and those made at later stages sometimes require the reassessment of previous decisions. A listing and short content description of each chapter should make it clear that those chapters we have placed in the latter part of the reference are also important in the conceptual stage if the monitoring is to be efficient and effective:

Chapter 1. Introduction—Describes the role of monitoring in adaptive management. Contrasts monitoring with other data-collection activities, such as inventory and long-term ecological studies.

Chapter 2. Monitoring Overview—Provides a step-by-step overview of the entire monitoring process, and references chapters where information on each step can be found in more detail. Flow charts are included to illustrate feedback loops and interrelationships among the steps.

Chapter 3. Setting Priorities and Selecting Scale—Presents criteria and techniques for setting priorities among species or populations and choosing the most appropriate scale and intensity for monitoring.

Chapter 4. Management Objectives—Illustrates the foundational nature of management objectives and describes their components, types, and development.

Chapter 5. Basic Principles of Sampling—Describes basic terms and concepts relevant to sampling using simple examples. This chapter provides background information critical to understanding material presented in Chapters 6, 7, and 11.

Chapter 6. Sampling Objectives—Describes objectives that complement management objectives whenever the monitoring includes sampling procedures. A sampling objective sets a specific goal for the level of precision or acceptable error rates associated with the sampling process.

Chapter 7. Sampling Design—Describes how to make the six basic decisions that must be made in designing a sample-based monitoring study: (1) What is the population of interest? (2) What is an appropriate sampling unit? (3) What is an appropriate sampling unit size and shape? (4) How should sampling units be positioned? (5) Should sampling units be permanent or temporary? (6) How many sampling units should be sampled?

Chapter 8. Field Techniques for Measuring Vegetation—Discusses selecting an appropriate vegetation attribute to measure when monitoring (e.g., cover, density, frequency, biomass, etc.) in terms of the biology and morphology of the species, and the practical limitations involved in each type of measurement. Field techniques for measuring each vegetation attribute and advice on field techniques and tools are provided.

Chapter 9. Data Management—Covers different ways of recording monitoring data in the field and describes means for entering and managing field monitoring data sets with computers.

Chapter 10. Communication and Monitoring Plans—Encourages the use of monitoring plans to solicit involvement in the development of a monitoring project, and to document the accepted monitoring protocol. Describes parties whose support may be critical for a successful monitoring project.

Chapter 11. Statistical Analysis—Describes the methods used to analyze monitoring data collected using sampling procedures, the use of graphs to examine data prior to analysis and to display the results of analysis, and the interpretation of monitoring data following analysis.

Chapter 12. Demography—Describes techniques for demographic analysis of populations and provides cautions and suggestions for their use.

Chapter 13. Completing Monitoring and Reporting Results—Summarizes the final stages of a monitoring project and describes methods for reporting results.

Effective monitoring is not easy; it requires a commitment of time and a willingness to think through alternatives during planning and design. We believe you will find that increasing time spent in design reduces total monitoring costs by making monitoring more efficient and effective. Above all, we hope to help you avoid wasting time on a monitoring project that fails to yield results useful for management decisions.

Because this is a somewhat novel approach, and because we intend to eventually update this handbook, we are especially interested in receiving your comments and opinions. You can send comments to:

Dr. Phil Dittberner
National Applied Resource Sciences Center, RS-140
Denver Federal Center, Building 50
P.O. Box 25047
Denver, CO 80225-0047

CONTENTS

APPENDICES

CHAPTER 1
Introduction

Arctomecon humilis
Dwarf bear-claw poppy
by Kaye H. Thorne

CHAPTER 1. Introduction

The root of the word monitoring means "to warn," and an essential purpose of monitoring is to raise a warning flag that the current course of action is not working. Monitoring is a powerful tool for identifying problems in the early stages, before they become dramatically obvious or crises. If identified early, problems can be addressed while cost-effective solutions are still available. For example, an invasive species that threatens a rare plant population is much easier to control at the initial stages of invasion, compared to eradicating it once it is well established at a site. Monitoring is also critical for measuring management success. Good monitoring can demonstrate that the current management approach is working and provide evidence supporting the continuation of current management.

In order for monitoring to function as a warning system or a measure of success, we must understand what monitoring is and the close relationship between monitoring and improved natural resource management decision-making. This chapter describes that relationship and defines the concept of monitoring used throughout this technical reference. The distinction between monitoring and many similar data-collecting activities is also discussed.

A. Definition of Monitoring

In this technical reference, we define monitoring as the collection and analysis of repeated observations or measurements to evaluate changes in condition and progress toward meeting a management objective. While the focus is plant populations, this concept is applicable to management of any natural resource. Monitoring is a key part of what has been termed "adaptive management," in which monitoring measures progress toward or success at meeting an objective and provides the evidence for management change or continuation (Holling 1978; Ringold et al. 1996). At its most rigorous, monitoring incorporates a research design so effects may be attributed to management causes (see Section C.3. Research, below). In practice, most monitoring measures the change or condition of the resource; if objectives are being met, management is considered effective (more on this in Chapter 13).

The adaptive management cycle is illustrated in Figure 1.1: (1) objectives are developed to describe the desired condition; (2) management is designed to meet the objectives, or existing management is continued; (3) the response of the resource is monitored to determine if the objective has been met; and (4) management is adapted (changed) if objectives are not reached. Figure 1.2 illustrates the fate of the adaptive management cycle if monitoring produces inconclusive data.

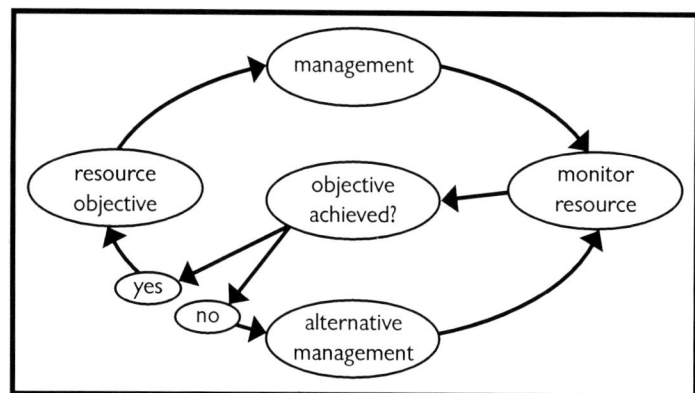

FIGURE 1.1. Diagram of a successful adaptive management cycle. Note that monitoring provides the critical link between objectives and adaptive (alternative) management.

Inherent in defining monitoring as part of the adaptive management cycle are two key concepts. The first is that monitoring is *driven by objectives*. What is measured, how well it is measured, and how often it is measured are design features that are defined by how an objective is articulated. The objective describes the desired condition. Management is designed to meet the objective. Monitoring is designed to determine if the objective is met. Objectives form the

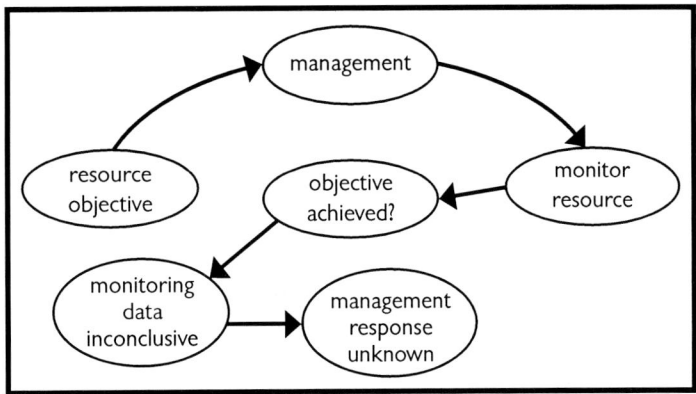

FIGURE 1.2 Diagram of monitoring that fails to close the adaptive management cycle. Because monitoring data is inconclusive, the management response is unknown and the cycle is unsuccessful.

foundation of the entire monitoring project.

The second concept is that monitoring is only initiated if opportunities for management change exist. If no alternative management options are available, expending resources measuring a trend in a species population is futile. What can you do if a population is declining other than document its demise? Because monitoring resources are limited, they should be directed toward species for which management solutions are available.

B. Resource Monitoring and Habitat Monitoring

Two types of monitoring are included in this technical reference. *Resource monitoring* focuses on the plant resource itself and monitors some aspect of that resource such as population size, average density, cover, or frequency. *Habitat monitoring* describes how well an activity meets the objectives or management standards for the habitat. Establishing residual plant height standards in riparian areas and measuring how well a grazing system meets those standards is an example of habitat monitoring. Another example is to set a threshold level of the percentage of habitat that may be disturbed by a particular activity, such as a logging project. Neither of these examples measures the response of the resource. The objective instead describes a habitat standard against which the effectiveness of management is monitored.

In resource monitoring, the actual causes of population condition and trend are unknown. Changes or condition could be the result of management, but could also result from weather patterns, insect infestations, changes in herbivory, etc. This problem also exists when monitoring a habitat factor. An additional problem with habitat monitoring is that you do not know whether assumed relationships between the habitat factor and the species are true. Habitat monitoring is most effective when research has demonstrated a relationship between a habitat parameter and the condition of a species. For most rare plants, these data are lacking, and the relationship between a habitat parameter and a species must be inferred from hypothesized and known ecological relationships.

In spite of these difficulties, the use of habitat monitoring should be expanded in rare plant management. Habitat monitoring is often more easily implemented and evaluated than resource monitoring (MacDonald and Smart 1993). For some species, such as annuals that fluctuate dramatically from year to year or long-lived perennials that change very little, habitat monitoring may be more sensitive to detecting undesirable change than monitoring the plant species directly.

C. Related Activities

The term "monitoring" has been applied to a variety of data-gathering activities. We have defined monitoring in this technical reference as driven by objectives and implemented within a management context. This differs from many activities described below that are often implemented under the general term "monitoring." We believe that many of these activities will benefit from applying the concepts described, but throughout this technical reference we will maintain our narrower definition of monitoring as objective based.

1. Inventory

Inventory can be described as a point-in-time measurement of the resource to determine location or condition. For rare plants, inventories may be designed to:

◆ Locate populations of a species.
◆ Determine total number of individuals of a species.
◆ Locate all populations of rare species within a specific area (often a project area).
◆ Locate all rare species occurring within a specified habitat type.
◆ Assess and describe the habitat of a rare species (e.g., associated species, soils, aspect, elevation).
◆ Assess existing and potential threats to a population.

Data collected during an inventory may be similar to those collected during monitoring. For example, the number of individuals in each population may be counted during an inventory. Similarly, a monitoring project may require counts of a single or several populations every year for several years.

Information collected during an inventory may provide a baseline, or the first measurement, for a monitoring study. Often, however, the necessary type and intensity of monitoring will not be known until the inventory is completed. The information collected during inventory, while useful for the development of a monitoring study, may not be useful for monitoring itself. Here is a typical example.

> During inventory for *Physaria didymocarpa* var. *lyrata*, qualitative estimates of population size and various habitat parameters were noted. Exhaustive inventory, however, only identified four populations. All are over 500 individuals, but all are restricted to small areas of extremely steep scree slopes. Management conflicts are severe at all four sites. Because of the demonstrated rarity of the species, public and agency concern over management is intense. Qualitative estimates of population size are considered inadequate for monitoring this species, and quantitative objectives and monitoring are recommended.

2. Natural history study

We do not consider investigations into basic ecological questions (pollination ecology, life history, seed viability, seedbank longevity) to be monitoring. These questions often must be answered before effective monitoring can be designed, but such studies are not monitoring. Unfortunately, in the past, agency cost coding provided no incentive for basic ecological investigation, and so creative specialists used the term "monitoring" for activities that were really natural history studies. With changes in cost coding and increased emphasis on ecosystem management, there is more flexibility to do natural history studies and recognize them for what they are.

3. Research

Although natural history studies can be considered a type of research, the term here refers to a study designed to determine the cause(s) of some observed ecological phenomenon. Many monitoring projects intend to determine the response of a plant population to a particular management activity, but in reality few monitoring projects conclusively identify the cause of the response. Monitoring data are usually of limited value in determining causes of change, and you must be careful to not misrepresent monitoring data as information on cause and effect. For example, simply noting a decline in a species population after logging supports the hypothesis that logging negatively impacts the species but does not prove that logging is the cause of the decline. The decline has to be consistently found at several logging sites and not found in unlogged areas to confidently determine logging activities as the cause of the decline.

A continuum can be identified from monitoring to research as shown in Figure 1.3. The confidence of attributing a change to a particular cause increases along the continuum, but the

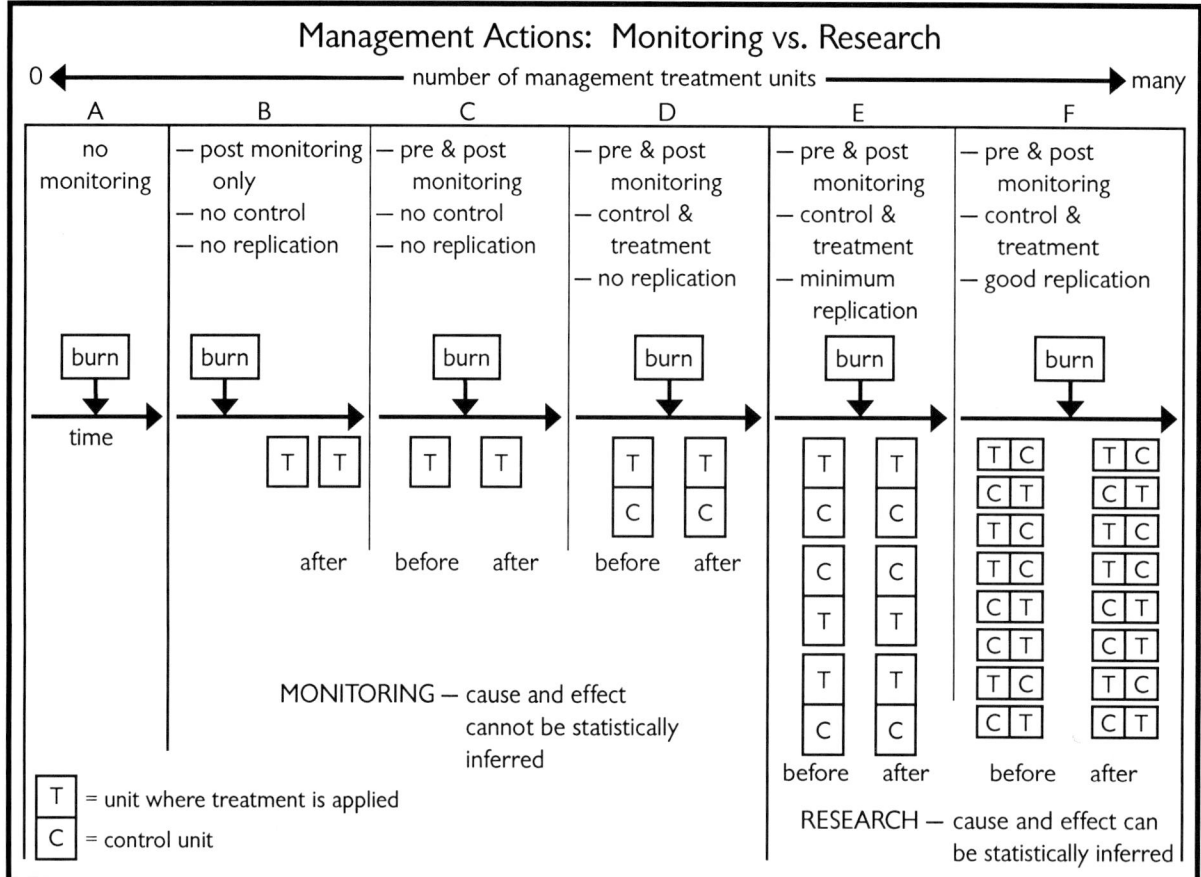

FIGURE 1.3. A comparison of monitoring and research approaches for detecting a treatment effect from a prescribed burn. For each of the scenarios shown in columns B-F above, statistical comparisons can be made between different time periods and a decision can be made as to whether or not a statistically significant difference occurred. However, the interpretation of that difference can be confounded by factors that are independent of the treatment itself. The diagram and the following examples illustrate a continuum of increasing confidence in determining likely causation as you move from left to right in the diagram. In column B, there is no pre-treatment measurement but you may see differences between years one and two after the burn. There is no way of knowing the conditions prior to treatment, and changes may be due to the burn, or they may be the result of some other factor such as lower precipitation. In column C, where data was gathered both before and after the burn, you still don't know if changes were due to the burn or some other factor that differed between the two time periods. In column D, there is a single treatment unit and a single control unit. Perhaps you see a change occur in the burned area but not the control area. The change could be caused by the burn or there may be some other factor that differentially affects the treatment area compared to the control. The burn unit, for example, could have a slightly lower water table than the control unit, a factor independent of the burn but not apparent to the naked eye. Other factors such as disease, insect infestation, and herbivory often occur in concentrations, heavily affecting one area but not adjacent areas. Any of these factors could be the cause of observed differences. In the last two columns, the treatment and control are replicated in space; thus there is a possibility of attributing differences to the treatment. Since ecological systems are variable, the example in column E with three replicates may have inadequate statistical power to detect differences. The differences due to the treatment may be hidden by differences that occur due to other factors. The larger number of replicates in column F greatly increases the likelihood of detecting treatment differences due to the higher statistical power associated with 8 replicates as compared to 3 replicates.

* the term "significant" means that a statistical test was carried out and the difference was significant according to the test.

cost of acquiring the needed data also increases. Statistical significance is often erroneously equated with cause. In Figure 1.3, statistically significant differences were found in several scenarios, but only for the last two scenarios (Columns E and F) can you attribute significant differences to a cause. Only with several replications of treatment and control can you confidently attribute changes to a treatment or cause.

Natural resource managers must decide during the development of a monitoring project if proving causal relationships is important. If demonstrating causality is required, the cost of obtaining that information must be evaluated. In many cases of resource management, a research approach may not be feasible. Some typical problems are the complexity of the system, the nonlinear response of organisms to causal mechanisms, and the lack of available replicates because only one "treatment" area is available (Thomas et al. 1981).

This technical reference is not designed as a guide for developing research projects. Good design is imperative to the success of a research project, and often requires specialized skills. If you intend to embark on a research project, resources spent acquiring the skills of a statistician and perhaps an academic researcher are a good investment, especially if the treatments are expensive (such as construction of exclosures or prescribed burns). Hairston (1989) and Manly (1992) provide excellent introduction to good, effective research designs in ecology.

4. Implementation monitoring

Implementation monitoring assesses whether the activities are carried out as designed. For example: Was the fence built in the right spot to specifications to protect the plant population from deer? Was the off-highway vehicle (OHV) closure maintained? Were the cows moved on the right date to allow the rare plant to successfully produce seed? While such monitoring does not measure the plant population, it does provide critical feedback on whether the planned management is being implemented. Implementation monitoring can also identify which variables are most likely to be causing a change in the resource, and help eliminate from consideration some potential causes of change. This type of monitoring, although critical to successful management, is not discussed further in this technical reference.

5. Measuring change

Measuring change over time is a main characteristic of monitoring, but simply measuring change does not meet the definition of monitoring in this technical reference. Studies that measure change can be implemented in the absence of an identified need for decision-making. In contrast, monitoring is characterized primarily by objectives and by being part of an adaptive management cycle. Monitoring uses change data to evaluate management and make decisions (Perry et al. 1987).

Studies measuring change in the absence of a management context have been collectively termed "surveillance" (Perry et al. 1987), but three types are recognized and described here: trend studies, baseline studies, and long-term ecological studies. The distinction among the types is blurred, and the terms have frequently been used interchangeably by resource managers.

a. Measuring trend

Much of the work done as monitoring is really designed to gather basic information about the species. The most common approach is measuring to learn how the resource is changing over time—measuring trend (some authors call this "baseline monitoring"- see next section).

Here is an example of a study objective for measuring trend: "Determine if the density of *Primula alcalina* is increasing, decreasing, or remaining stable at the Texas Creek Population over the next 5 years." While the trend of the population may be important, you could more rigorously develop this study objective into a management objective:

Management objective: Maintain at least the current density of *Primula alcalina* at the Texas Creek population (within 10% of the first measurement) over the next 5 years.

Management response (if cause is unknown): If *Primula* density declines by more than 10% over the 5-year period, more intensive monitoring or research will be initiated to determine the cause of the decline; or

Management response (if cause is suspected): If *Primula* density declines by more than 10% over the next 5 years, grazing at the site will be limited to late fall to allow seed set and dissemination.

A fundamental difference exists between monitoring for trend and monitoring for management, even though the actual measurements and analysis may be the same. The second approach places the measurements within the adaptive management cycle and identifies the changes in management that will occur if the monitoring has a certain result. At the time the study begins we do not know whether the population is stable, declining, or increasing. By conducting the study within the framework of an objective and a management response, the course of action at the end of 5 years is known before monitoring begins. If monitoring shows the population is increasing or stable, current management may continue. If populations are declining, an alternative management approach is outlined. If the study is done simply to detect change, the course of action at the end of the 5 years will be unclear. What will likely occur is that existing management and the trend study will simply be continued to determine if the decline of the rare plant continues.

Specialists are often hesitant to develop objectives and management responses because of a lack of information on the desired condition of the plant population and the relationship of management to that condition. At a minimum, however, an objective to maintain the current condition can be established and a commitment made to respond with more extensive monitoring, study, or research if a decline is measured.

b. Baseline studies and long-term ecological studies

Another type of activity implemented as monitoring is called "baseline monitoring." This is the assessment of existing conditions to provide a standard, or "baseline," against which future change is measured. Commonly, a large number of variables is measured in hopes of capturing within the baseline data set the ones that turn out to be important later. Baseline monitoring is sometimes termed "inventory monitoring" (MacDonald et al. 1991) because it often involves the collection of data to describe the current condition of a resource. Remeasurement at a later date may be intended, but a commitment or plan for periodic remeasurement is lacking. Periodic remeasurement is integral to a monitoring study. The problem with baseline studies, and using inventory data as baseline data, is that the design of the study may be inadequate to detect changes. This inadequacy usually results from including too many variables and using too small of a sample size.

If the measurements are taken with a scheduled remeasurement, a baseline study may be termed a long-term ecological study. The most common goal of these studies is to learn about the natural range of temporal variability of the resource by documenting the rates and types of changes that occur in response to natural processes such as succession and disturbance. The term "long-term ecological monitoring" usually is used to describe the measurement of community variables to determine change over the long term, 50-200 years or more. In most studies, many variables are measured on a few large permanent plots (usually greater than 0.1 hectare). Commonly measured variables include cover or density of all plant species, demographic parameters of important species, soil surface conditions, fuel loads, and animal signs (Greene 1984, 1993; Dennis 1993; Jensen et al. 1994). The term "baseline monitoring" is also sometimes used for this activity.

Two key differences exist between baseline and long-term ecological studies and the monitoring described in this technical reference:

♦ Baseline and long-term ecological studies do not specifically evaluate current management or result in a management decision, although they may provide management direction in the future by describing system functions and fluctuations (Perry et al. 1987). In monitoring, the application of the data to management is identified before the measurements are taken because monitoring is part of the adaptive management cycle.

♦ These studies often attempt to maximize the number of characteristics measured because those most sensitive for measuring change are not known. In contrast, in this technical reference, we advocate the explicit selection of one or a few measurable variables to be monitored.

One type of monitoring explicitly involves the measurement of a "baseline" and is sometimes termed "baseline monitoring." In this monitoring design a series of measurements are taken prior to the initiation of a management activity and used for comparison (a "baseline") with the series of measurements taken afterward (Green 1979; MacDonald et al. 1991). This type of situation is common in water quality monitoring. For example, measurements of water column sediment in a river may be taken for 5 years prior to the construction of a power plant, and then for 5 years afterward to determine the background, or baseline level of sediment, and whether the pollution controls of the plant are adequate to prevent elevated sediment levels. When measurements are made at both treatment and control areas, this type of monitoring design is termed the before-after, control-impact (BACI) design (Bernstein and Zalinski 1983; Faith et al. 1995; Long et al. 1996). It is rare in land management agencies and rare plant management to have several years notice before initiating an activity during which a baseline can be measured, but if the opportunity arose, such a monitoring design can be very effective.

D. Mandates for Monitoring

1. Congressional mandates

Monitoring is so important to successful management that Congress gives specific direction for monitoring in several pieces of legislation. Pertinent sections are reprinted in Appendix 2 and are summarized below.

Endangered Species Act (ESA) of 1973 (as amended in 1988): Directs all Federal agencies to carry out programs for the conservation of threatened and endangered (T/E) species. Agencies must also aid the Secretary of the Interior in implementing a system to effectively monitor the status of all listed species, as well as those species for which it has been determined that listing is "warranted, but precluded." These latter species were previously classified as Candidates (C1) and more recently as "Species of Concern."

Federal Land Policy and Management Act of 1976: Requires that "...the public lands be periodically and systematically inventoried" (section 102(a)). This inventory "...shall be kept current so as to reflect changes in conditions and to identify new and emerging issues and their values" (Section 201(a)). While the term "inventory" is used, the requirements for a continuing process and collection of data that measure change suggest monitoring more than inventory.

National Environmental Policy Act (1969): Directs Federal agencies to use ecological information to examine direct, indirect, and future consequences when planning and developing projects. Such examination requires the availability of monitoring data to describe how resources responded to past similar actions and the implementation of monitoring to determine the accuracy of the assessment and prediction of the impacts from the proposed project.

2. Bureau of Land Management regulations

Based on congressional direction and Executive Orders, BLM has developed agency regulations that all its offices are required to follow. These regulations are codified in the BLM manual. Sections 1622, 6500, 6600, and 6840 clearly show that monitoring is recognized at national levels as an extremely important activity that should be given high priority by each local office.

Section 1622: Requires that priority species and habitats be identified, management objectives developed for them, and the objectives monitored. The manual section especially encourages the setting of "threshold levels which indicate when modifications in management direction will be made."

Section 6500: Requires an inventory of special status species be maintained on a "continuing basis," which suggests monitoring more than inventory. It also directs the BLM to prepare site-specific objectives for special status species in all activity plans (such as allotment management plans) and monitor them to evaluate success.

Section 6600: Requires that objectives for special status plants be developed and monitored. These objectives are to be clearly defined, site-specific, and measurable, with a timetable for accomplishment.

Section 6840: Directs the BLM to "...evaluate ongoing management activities to ensure T/E conservation objectives are being met." Special status plants are to be conserved by "...monitoring populations and habitats...to determine whether management objectives are being met."

Literature Cited

Bernstein, B. B.; Zalinski, J. 1983. An optimal sampling design and power tests for environmental biologists. Journal of Environmental Sciences 16: 35-43.

Dennis, A. 1993. Sampling methods for vegetation description and monitoring. Berkeley, CA: unpublished draft on file at USDA Forest Service, Pacific Southwest Station/Region 5, Research Natural Areas Program.

Faith, D. P.; Dostine, P. L.; Humphrey, C. L. 1995. Detection of mining impacts on aquatic macroinvertebrate communities: results of a disturbance experiment and the design of a multi-variate BACIP monitoring program at Coronation Hill, Northern Territory. Australian Journal of Ecology 20: 167-180.

Green, R. H. 1979. Sampling design and statistical methods for environmental biologists. New York, NY: John Wiley & Sons.

Greene, S. 1984. Botanical baseline monitoring in research natural areas in Oregon and Washington. In: Johnson, J. L.; Franklin, J. F.; Krebill, R. G., coordinators. Research natural areas: baseline monitoring and management: proceedings of a symposium; 1984 March 21; Missoula, MT. General Technical Report INT-173. Ogden, UT: USDA Forest Service, Intermountain Research Station: 6-10.

Greene, S. 1993. RNA Monitoring protocols installation instructions. Corvallis, OR: unpublished draft (July, 1993) on file at: USDA Forest Service, Pacific Northwest Station, Research Natural Areas Program.

Hairston, N. G. 1989. Ecological experiments: purpose, design and execution. Cambridge, England: Cambridge University Press.

Holling, C. S., ed. 1978. Adaptive environmental assessment and management. New York, NY: John Wiley & Sons.

Jensen, M. E.; Hann, W.; Keane, R. E.; Caratti, J.; Bourgeron, P. S. 1994. ECODATA—A multire-source database and analysis system for ecosystem description and analysis. In: Jensen, M. E.; Bourgeron, P. S., eds. Eastside forest ecosystem health assessment, volume II: Ecosystem management: principles and applications. General Technical Report GTR-PNW-318. Portland, OR: U.S. Department of Agriculture, Forest Service: 203-216.

Long, B. G.; Dennis, D. M.; Skewes, T. D.; Poiner, I. R. 1996. Detecting an environmental impact of dredging on seagrass beds with a BACIR sampling design. Aquatic Botany 53: 235-243.

MacDonald, L. H.; Smart, A. W. 1993. Beyond the guidelines: practical lessons for monitoring. Environmental Monitoring and Assessment 26: 203-218.

MacDonald, L. H.; Smart, A. W.; Wissmar, R. C. 1991. Monitoring guidelines to evaluate effects of forestry activities on streams in the Pacific Northwest and Alaska. EPA/910/9-91/001. Seattle, WA: U.S. Environmental Protection Agency.

Manly B. F. J. 1992. The design and analysis of research studies. Cambridge, England: Cambridge University Press.

Perry, J. A.; Schaeffer; D. J.; Herricks, E. E. 1987. Innovative designs for water quality monitoring: are we asking the questions before the data are collected? In: Boyle, T. P., ed. New approaches to monitoring aquatic ecosystems, ASTM STP 940. Philadelphia, PA: American Society for Testing and Materials: 28-39.

Ringold, P. L.; Alegria, J.; Czaplewski, R. L.; Mulder, B. S.; Tolle, T.; Burnett, K. 1996. Adaptive monitoring design for ecosystem management. Ecological Applications 6(3): 745-747.

Thomas, J. M.; McKenzie, D. H.; Eberhardt, L. L. 1981. Some limitations of biological monitoring. Environment International 5: 3-10.

CHAPTER 2
Monitoring Overview

Atriplex canescens
Four wing saltbush
by Jennifer Shoemaker

CHAPTER 2. Monitoring Overview

The steps described below and illustrated in the flow diagrams in Figure 2.1-2.5 provide an overview of the development of an adaptive management cycle (Figure 1.1). This chapter focuses on the development of objectives and monitoring methods and briefly addresses the development of management strategies.

The following steps should not be considered sequential. Feedback loops and reviews are many, as shown by the multidirectional arrows in the flow diagrams. At nearly any point in the process of developing a monitoring project, earlier decisions may have to be revisited and changes made.

A. Complete Background Tasks
(Figure 2.2)

1. Compile and review existing information

Compile relevant information on the species and/or populations. For those monitoring projects where the target species and/or population is predetermined, you will only need the information specific to the species. For rare plant management programs that are just beginning, you'll likely want to assemble the information needed to set priorities among all the sensitive species occurring in your administrative unit. If you manage many species, you may wish to start with a short list of species that are high priority, perhaps because of legal reasons, such as federally listed species and Species of Concern (Chapter 3).

2. Review upper level planning documents

Consistent local land management depends on following upper level planning documents. These documents describe to the public the agency's planned activities. Because managers are accountable for implementing these plans, specific management activities for rare plant populations should demonstrate progress toward meeting

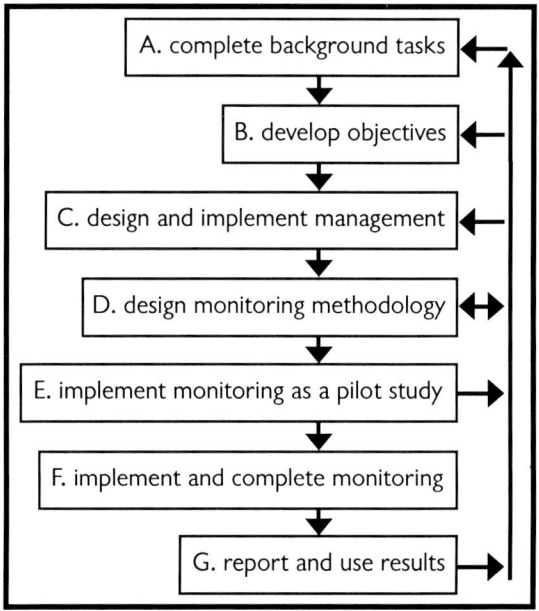

FIGURE 2.1. These seven major steps are broken into sub-steps and illustrated in figures 2.2 - 2.5.

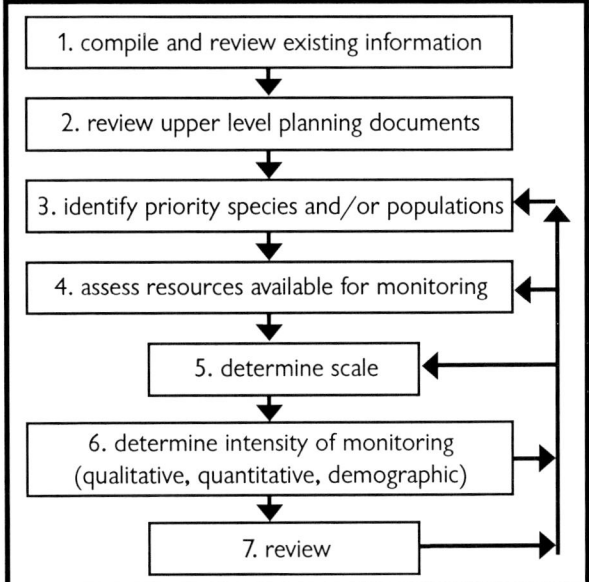

FIGURE 2.2. Flow diagram of the monitoring process, continued. Steps associated with completing background tasks are illustrated in detail.

goals and objectives described in them. Even if you believe your agency's land use plan provides little specific direction for rare plant management (many of the older ones don't), you will increase support for your specific project if you can show a clear relationship between it and the general directives outlined in planning documents (Chapter 3).

3. Identify priority species and/or populations

Prioritize the species for monitoring, and document your thought process. This documentation will be useful to you and your successor if managers and other parties question your priorities. For priority species select priority populations. These priorities may periodically require reassessment due to changes in threats, management, conflicts, and the interest of outside parties (Chapter 3).

4. Assess the resources available for monitoring

Resources for monitoring depend on management support, priorities, and the people and equipment available. Has management placed a priority on this monitoring project, or is support and funding limited? You may need to promote the importance of the project before you begin working on it. Are qualified personnel available to do the work? Do you have the necessary field equipment such as vehicles and measuring tapes? Is any high-tech equipment available (e.g., geographic information systems, global positioning systems, survey or forestry equipment)? Are people willing to give reviews and help sharpen your thinking? Do you have access to people with specialized skills? The types and amounts of resources will limit the extent and complexity of a monitoring project (Chapter 3).

5. Determine scale

Identify the scale of interest for monitoring (e.g., the range of the species, the populations within a certain watershed, populations in certain types of management units, a single population, a portion of a single population such as a key area or macroplot). Decide the scale of interest early in the monitoring process because it will influence later decisions and design. If, for example, the scale of interest is the species across its entire range, you will need to coordinate with various administrative units to develop a network of monitoring studies (Chapter 3).

6. Determine intensity of monitoring

Will qualitative monitoring be adequate? Do you need quantitative data? Does the rarity of the species, the degree of threats, or the political sensitivity of potential decisions warrant the use of an intensive demographic approach? You may need to reevaluate the selected intensity of monitoring as you work through the remaining monitoring decisions (Chapter 3).

7. Review

At this point, management should be briefed, and opinions and review solicited. For small projects, you could complete these steps on your own and then solicit internal and possibly external review. For larger programs or highly controversial species and populations, you may need to assemble a team (Chapter 10).

B. Develop Objectives (Figure 2.3)

1. Develop an ecological model

In this technical reference, we promote the use of narrative or diagrammatic summaries (models) of the ecological and management interrelationships of the species of interest (Chapter 4 gives examples). Completing a model will help develop objectives, focus your monitoring, and improve interpretation and application of the data.

2. Identify general management goals

Using your ecological model, try to refine conservation goals. Should the population size of the species be increased? Maintained? Recruitment increased? Mortality decreased? Describing these general management goals is the first step toward developing specific objectives.

B. DEVELOP OBJECTIVES

1. develop an ecological model
2. identify general management goals
3. select indicator
4. identify sensitive attributes
5. specify direction and quantity of change
6. specify time frame
7. develop management objective
8. specify management response
9. review management objective

FIGURE 2.3. Flow diagram of the monitoring process, continued. Tasks associated with developing objectives are illustrated in detail.

3. Select indicator

You may choose to monitor the species itself or some aspect of the habitat that serves as an indicator of species success. Monitoring threats as indicators can form an effective basis for management changes. Indicators are also useful for species that are difficult to measure or monitor (e.g., very small species, annuals, long-lived species).

4. Identify sensitive attribute

Common measurable vegetation attributes include density of individuals, cover, frequency, and production. Attributes also include qualitative and semi-quantitative measures such as presence or absence of the species, estimates of cover by cover class, and visual estimates of population size. The attribute most sensitive for measuring progress toward the described goal will vary by species and situation. For example, individuals of some species such as rhizomatous grasses are difficult to count. Instead of density, you would need to select another measure of success or improvement such as cover or frequency. Examples of attributes of habitat indicators include density of weeds, the percentage of the site impacted by some activity (such as disturbance from recreational vehicles), and the height of stubble remaining after livestock grazing. The attribute most sensitive and useful for monitoring depends on the life history and morphology of the species and the resources available to measure the attribute (Chapter 8). Some species are so poorly known you may have difficulty identifying a sensitive parameter. Make the best choice you can or postpone monitoring until you know more about the natural history of the species.

5. Specify direction and quantity of change

Will you monitor for a percentage change or an absolute change, a target or threshold value (Chapter 4)? What increase do you want to see, or what decrease will you tolerate? Can you specify a target population size? The quantity has to be measurable (confidently measuring a 1% change in average density is extremely difficult) and biologically meaningful (a 10% change in density of an annual species is probably not important). Again, you may be limited by lack of information. You may also be limited by the amount of change you can detect in a sampling situation (Chapters 5, 6, 7, and 11).

6. Specify time frame

How soon will management be implemented? How quickly do you expect the species to respond? How long do you want this monitoring program to continue if some threshold is not reached? The time frame should be biologically meaningful for the change you are anticipating. A 50% increase in the density of a long-lived woody plant, for example, is unlikely to occur over the next 3 years (although a decline of that magnitude may be possible and alarming).

7. Develop management objective

The priority species or population, selected scale (location), sensitive attribute, quantity and direction of change, and time frame of change (Steps A5, A6, B3, B4, B5, and B6) are the critical components of the objective. Combine them into a simple, measurable, understandable objective (Chapter 4).

8. Specify management response

Given the potential alternative results of monitoring, what management changes would be implemented in response to each alternative (Chapters 4, 10, and 13)? These management responses should be clarified before monitoring begins so all parties know the implications of monitoring results.

9. Review management objective

Preferably, several of these steps would be completed by a team of specialists and management, but often the rare plant specialist will work alone through these steps. Before proceeding to the design of monitoring, solicit internal and external review, especially from parties that may be affected by management changes made in response to monitoring data (Chapter 10). Do others have information about the biology or ecology of the species that you should incorporate into the model? Do all agree on the management objective? Do all agree with the proposed management response?

C. Design and Implement Management

Depending on the situation, current management may be continued or new management proposed. Often current management is continued and monitored because little is known about the ecology and management requirements of a particular rare species. In some cases, however, previous monitoring data or natural history observations may suggest a need for management change. The ecological model may provide insight on needed changes as well. If new management is required, it must be completely described so it can be implemented effectively.

The design of conservation management strategies for rare plant species involves consideration of the ecology of the plant, funding, management options, conflicting uses and activities, and communication and coordination with public and user groups. This complex and difficult step is unique to each situation, and is a subject beyond the scope of this technical reference.

D. Design the Monitoring Methodology (Figure 2.4)

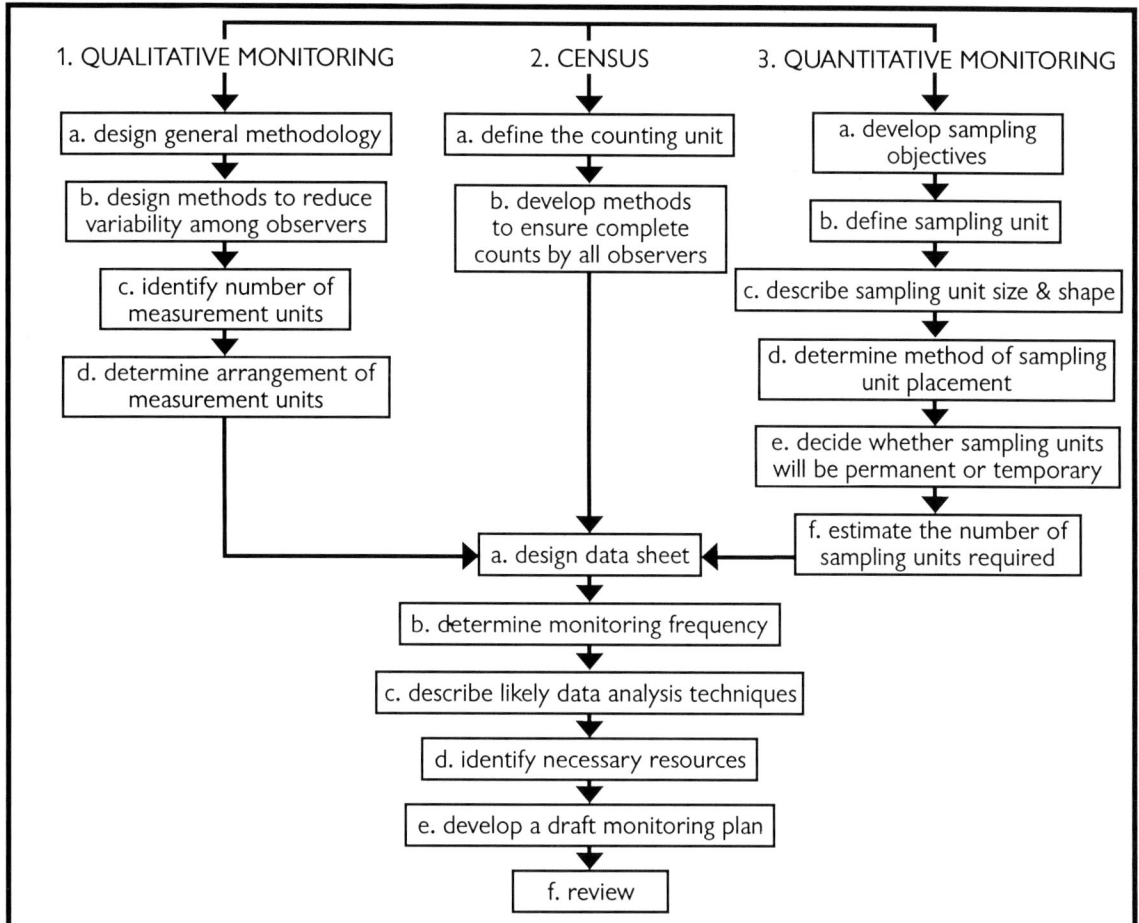

FIGURE 2.4. Flow diagram of monitoring process illustrating the sub-steps of designing a monitoring methodology. The decisions required for each of the three types of monitoring—qualitative, census, and quantitative (sampling)—are summarized.

1. Qualitative monitoring

a. Design general methodology

Methods for qualitative monitoring include estimating quantity (e.g., ranked abundance, cover class) and quality (e.g., population stage class distribution, habitat condition), and using a permanent recording method, such as a photopoint or a video sequence (Chapter 8).

b. Design methods to reduce variability among observers

The biggest drawback of using qualitative techniques is that estimates among observers can vary significantly. Between-observer variability can be reduced by several strategies described in Chapter 8.

c. Identify number of measurement units

Some qualitative monitoring situations may require several to many measurement units, such as macroplots or photoplots. These are not sampling units, since they will not be combined and analyzed as a sample. Many design decisions, however, are similar to those required for sampling units and include selecting size, shape, and permanence.

d. Determine arrangement of the measurement units

How will these measuring units be distributed in the population or across the landscape? Will you selectively place them based on some criteria such as threat or ease of access? Will you distribute these units evenly across the population to enhance dispersion and avoid bias?

2. Census

a. Define the counting unit

Will you count individuals (genets), stems, clumps, or some other unit? Will you count all individuals or only certain classes (such as flowering)? These questions must be clearly addressed in the design to ensure different observers conduct counts using the same criteria (Chapter 8).

b. Develop methods to ensure complete counts

Will you have standardized methods (transects, plots, or grids)? Counts that are intended to be a complete census are often incomplete. What strategies will you use to ensure small individuals are not overlooked (Chapter 8)?

3. Quantitative Studies with Sampling

a. Develop sampling objectives

If you are using sampling to estimate population sizes or mean values (such as density, cover, or frequency), you must also identify an acceptable level of precision of the estimate. If you are sampling and determining the statistical significance of changes over time, you must identify the size of the minimum detectable change (previously specified in your management objective), the acceptable false-change error rate, and missed-change error rate (or statistical power level). What is the risk to the species if your monitoring fails to detect a real change (missed-change error), and how confident must you be of detecting a change over time (statistical power)? What is the risk to alternative uses/activities if your monitoring detects a change that is not real (false-change error)? (Chapters 5, 6, and 7)

b. Define the sampling unit

Will sampling units be individually placed plots, plots or points placed along a line, a line of points, individual plants, seedpods, or some other unit? The sampling unit must be explicitly identified to ensure the selected units are random and independent (Chapters 7 and 11).

c. Describe unit size and shape

The most efficient size and shape of the sampling unit depends on the spatial distribution of the species you are sampling. Most plants grow in clumps. Unless careful consideration is made of plot size and shape, most plots will rarely intersect clumps of the target species. Many plots will be required in such a design to meet the specified precision and power of the sampling objective. Efficient sampling design using plots of appropriate size and shape can dramatically reduce the number of sampling units that must be measured, reducing the time and resources required for the field work and data entry. The size and shape of the sampling unit may be the most important decision affecting the success of projects where sampling is used (Chapters 7 and 11).

d. Determine sampling unit placement

Sampling units must be positioned without bias. There are several methods described in Chapter 7.

e. Decide whether sampling units will be permanent or temporary

Permanent sampling units are suitable for some situations, while temporary ones are more suitable for others (Chapter 7). If the sampling units are permanent, monumenting or another method of relocation becomes critical and will require additional field time for plot establishment during the first year of the monitoring project (Chapter 8).

f. Estimate the number of sampling units required

Data from a pilot study are the most reliable means to estimate the number of sampling units required to meet the targets of precision and power established in the sampling objective (see section E, below). Chapter 7 and Appendices 7, 16, 17, and 18 describe estimation of sample size based on pilot data as well as some alternative methods.

4. Design issues common to all three types

a. Design data sheet

While some studies may use electronic tools to record data, in most studies the researcher will record measurements on a data sheet. A well-designed data sheet can simplify rapid and accurate data recording and later computer data entry (Chapter 9).

b. Determine monitoring frequency

How often should the parameter be measured? Will you be monitoring annually? Every 3 years? The frequency varies with the life form of the plant and the expected rate of change (long-lived plants may require infrequent measurement), the rarity and trend of the species (the risk of loss for very rare or very threatened species is higher), and the resources available for monitoring.

c. Describe the likely data analysis techniques

For all projects, describe how the data will be evaluated and analyzed. If you are using quantitative sampling, identify the statistical tests appropriate for the data you're planning to collect so the assumptions of the tests can be considered in the design stage (Chapter 11). Don't assume you can collect data, give it to an "expert" and expect meaningful results. Useful data analysis starts with good field design and data collection. This is also a good point to check whether the data will actually address the objective, given the analyses you plan to use.

d. Identify necessary resources

Now that you have specifically designed the monitoring project, estimate the projected annual and total costs, analyze the resources needed, and compare to resources available. Reevaluate equipment and personnel required to successfully implement your project and ensure they are available. Document the responsible individual/team for implementation of the monitoring, the source and amount of the funding for monitoring (annually and over the life of the project), and the necessary equipment and personnel.

e. Develop a draft monitoring plan

If all these steps have been documented and reviewed, many components of your monitoring plan have been completed. The draft monitoring plan provides four important benefits: (1) it focuses the thinking of the author by forcing articulation; (2) it provides a vehicle for communication and review; (3) it documents approval and acceptance when finalized; and (4) it provides a history of the project and guards against the untimely end of the monitoring project if the primary advocate leaves (Chapter 10). For those monitoring projects requiring minimal review from people outside the agency , the monitoring plan may be postponed until after data from the pilot stage have been analyzed.

f. Review plan

Use the monitoring plan to solicit review of your proposed project (Chapter 10). Do all reviewers agree with the methodology? Does the proposed methodology really monitor the objective? It may be necessary to revise either the methodology, or the objective, or both. For example, your objective may involve increasing cover of the target species, but as you design the monitoring you may realize that measuring cover of this particular species will be difficult. Treat development of objectives and design as an interactive process; the objective drives the design of the monitoring, but the practical constraints of the morphology of the plant, the characteristics of the site, or the availability of monitoring resources may require reevaluation of the objective.

E. Implement Monitoring as a Pilot Study (Figure 2.5)

1. Collect field data and evaluate field methods

The first trial of a monitoring method in the field often exposes problems with the methodology (e.g., plots cannot be positioned due to dense vegetation; the proposed counting unit cannot be applied consistently; lacy vegetation proves a problem for measuring shrubs along a line intercept). This is why the pilot period is important for testing the feasibility of the proposed monitoring approach and identifying improvements. You may find at this stage that the project cannot be implemented as planned and requires substantial revision, or even abandonment, in spite of all the work done to this point.

2. Analyze pilot study data

Analyze data from the pilot study. Do assumptions of the ecological model still appear correct? Are sampling objectives of precision and power met? If not, you may need to alter your monitoring design (add more sampling units or improve the efficiency), the sampling objective (accept lower precision and/or power), or perhaps abandon the entire project. Is the level of change or difference you've specified seem realistic? Do changes due to weather seem larger than you anticipated, thus swamping the quantity specified in your objective, or do the plants appear so slow-growing that the proposed change is unrealistic? You may need to reassess the quantity or time frame component of your objective.

3. Reassess time/resource requirements

The pilot project should provide a better estimate of the resources required for monitoring. Your estimate of costs should include the amount of time it has taken to develop the monitoring to this point as well as how much time it will take to continue the monitoring annually and complete final data analysis and reporting.

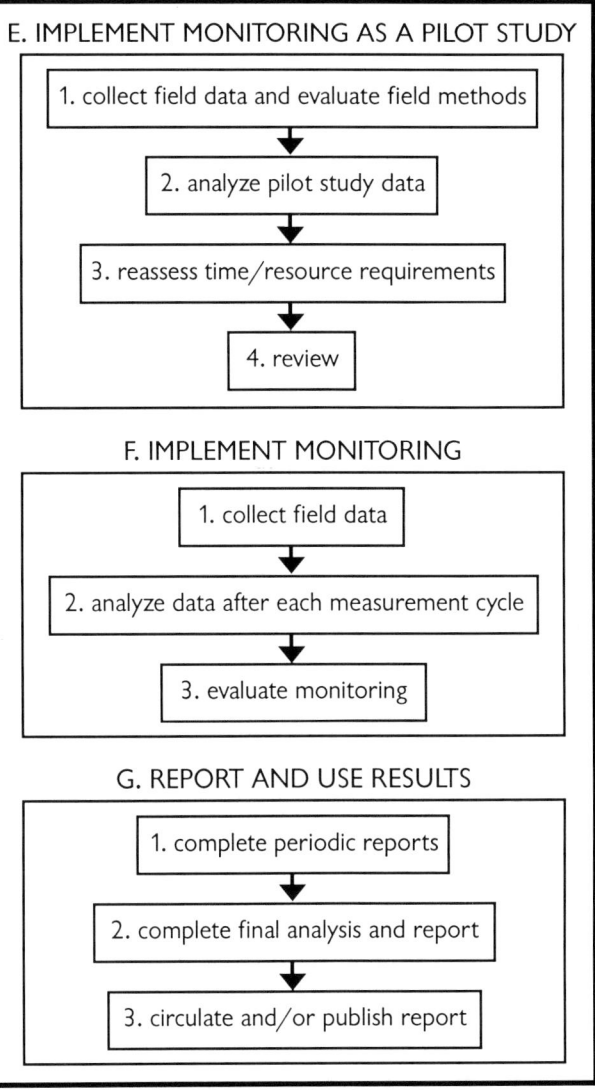

FIGURE 2.5. Flow diagram of the monitoring process, continued. Tasks associated with implementing monitoring as a pilot study, continuing monitoring, reporting and using results are illustrated.

4. Review

Solicit review of the results of the pilot period. Do all parties still agree to continue the monitoring and abide by the results? Are the resources available to implement monitoring throughout its life span? Make necessary changes to the monitoring design and the monitoring plan and solicit final review.

F. **Implement Monitoring** (Figure 2.5)

1. Collect field data

Complete data collection at specified intervals. Ensure data sheets are completely filled out, duplicated, and stored in a safe place.

2. Analyze data after each measurement cycle

Complete data analysis soon after data collection. Data should not be stored over several years before analysis for a final report. Timely analysis identifies problems early, reduces the work associated with the final report, and ensures that questions requiring additional field visits can be addressed. In addition, questions that occur as field data sheets are entered into the computer can often be answered because the field work is still fresh in your memory.

3. Evaluate monitoring

Evaluate field methods, costs, sample size, and relevancy of the monitoring project after each data collection. Recognize that at any time in the process a problem may arise that causes you to change or abandon your monitoring effort. All the steps preceding this one reduce that risk, but do not eliminate it.

G. Report and Use Results (Figure 2.5)

1. Complete periodic reports

Completing a summary report each time data are collected will yield the following benefits: (1) display the importance and usefulness of the monitoring to management, thus increasing continued support; (2) provide a summary for successors in the event of your departure; and (3) provide a document that can be circulated to other interested parties.

2. Complete final analysis and report

At the end of the specified time frame (or earlier if objectives are achieved), prepare a final monitoring report and distribute to all interested parties (Chapter 13). This final report presents and summarizes the data, analyses, and results, and provides recommendations. If the monitoring project has been designed and documented as described above and data have been analyzed periodically, this final report should be easy to complete and not contain major surprises.

3. Circulate and/or publish report

Sharing the results of your monitoring increases the credibility of the agency, assists others in the design of their monitoring projects, enhances partnerships, and reduces redundancy. Sharing the results in a technical forum such as a symposium or a journal article is also good professional development for you.

CHAPTER 3
Setting Priorities, Selecting Scale and Intensity

Fremontodron decumbens
Pine Hill flannelbush
by Mary Ann Showers

CHAPTER 3. Setting Priorities, Selecting Scale and Intensity

Resources and funding for monitoring are limited. You will not likely be able to develop objectives and monitor achievement of objectives for all the species and populations for which you are responsible. Priorities must be set, and the scale and intensity of monitoring these priorities must be determined. Scale describes the spatial extent, and intensity describes the complexity and cost of the monitoring. The scale of monitoring can range from a macroplot subjectively placed within a population to all populations of a species across its range. Intensity can vary from a single photopoint that is revisited every 5 years to a labor-intensive demographic technique that requires annual assessment of every individual in a population.

Clearly, as you increase the scale and intensity you will know more about the species and its trend and status, but the monitoring will be more expensive. With limited funds, you can monitor one or a few species at a large scale and high intensity, or more species at a more limited scale and lower intensity. The setting of priorities is the first step in determining the importance and number of species and/or populations that require attention, the monitoring resources that should be allocated to each, and the complexity of the objective for each species or population that can be monitored.

In the absence of priorities, species that are in need of monitoring because of their rarity and sensitivity may be ignored while more common species may be addressed due to their relationship with an urgent or high profile issue (such as plants found in riparian areas or old growth), or because they are public favorites (such as orchids). The narrow margin of existence of some species and the crisis rate of decline in others leave little room for misallocation of management and monitoring resources. Although you would expect that the rarest species are monitored most intensively, a review of monitoring in the United States found that according to priority classifications used by The Nature Conservancy and the U.S. Fish and Wildlife Service, nearly half of the species monitored were of low priority ranking (Palmer 1986, 1987). Surprisingly, nearly a third of the studies reviewed used a demographic approach, the most intensive method of monitoring, and a choice that likely meant ignoring other species. Explicit setting of priorities would alleviate this problem.

A. Assembling Background Information

1. Upper-level planning documents and guidance

Priority species or populations may have already been identified in an accepted land use plan or activity plan. These documents provide overall management direction for large areas of land (e.g., a District or Resource Area). Your office and priority species may also be addressed in a recent interagency regional plan that encompasses portions of several States occurring within an ecosystem boundary. In the absence of compelling reasons, such as new information or the appearance of new threats, the priorities identified in these plans should be accepted and used.

Many land use plans, especially older ones, provide little direction for management of rare plants. If the plan lacks goals directly related to rare plants, look for supporting goals, such as "maintain a full complement of flora and fauna" or "maintain viable populations of native species." Occasionally, directions for conservation of rare species may be found in lists of standard operating procedures.

Another source for setting priorities may be State or regional lists of rare species. In some States, priorities are recommended at an annual meeting of representatives from Federal and State agencies, universities, and private firms. Some BLM State Offices issue a list of priority species, as do some Forest Service Regional Offices. The U.S. Fish and Wildlife Service assigns a listing priority to species based on threats and taxonomic status (Figure 3.1). The Nature Conservancy and its associated Natural Heritage Programs rank all rare species with a State and Global rarity ranking based on the number of occurrences (Figure 3.1).

LISTING PRIORITY, U.S. FISH AND WILDLIFE SERVICE

Priority*	Threat Magnitude	Threat Immediacy	Taxonomic Status
1	high	imminent	monotypic genus
2	high	imminent	species
3	high	imminent	subspecies/variety
4	high	not imminent	monotypic genus
5	high	not imminent	species
6	high	not imminent	subspecies/variety

* Priority 7-12 uses the same approach, but for species with low magnitude of threat

THE NATURE CONSERVANCY/NATURAL HERITAGE PROGRAM RATING SYSTEM

1. Critically imperiled (5 or fewer occurrences or very few [<1,000] individuals or few acres).
2. Imperiled (6 to 20 occurrences or few [<3,000] remaining individuals, or few acres).
3. Very rare and local, found locally in a restricted range, or vulnerable to extinction or extirpation by outside factors (21-100 occurrences or <10,000 individuals).
4. Apparently secure, though it may be rare in parts of its range.
5. Demonstrably secure, though it may be rare in parts of its range.
6. Status uncertain, with the need for more information; possibly in peril.

These rankings can be used either at the State scale (within the State only) or at a global scale (the entire range of the species), and are often presented, for example, as "S1" or "G1", for critically imperiled at the state or global level.

STATE LISTING CRITERIA*

State Priority 1. A taxon in danger of becoming extinct or extirpated from the State in the foreseeable future if identifiable factors contributing to its decline continue to operate; these are taxa whose populations are present only at critically low levels or whose habitats have been degraded or depleted to a significant degree.

State Priority 2. A taxon likely to be classified as Priority 1 within the foreseeable future if factors contributing to its population decline or habitat degradation or loss continue.

Sensitive. A taxon with small populations or localized distributions within the State that presently do not meet the criteria for classification as Priority 1 or 2, but whose populations and habitats may be jeopardized without active management or removal of threats.

Monitor. Taxa that are common within a limited range, as well as those taxa which are uncommon, but have no identifiable threats (for example, certain alpine taxa).

Review. Taxa which may be of conservation concern, but for which we have insufficient data to recommend an appropriate classification.

Possibly extirpated. Taxa which are known in the State only from historical (pre-1920) records or are considered extirpated from the State.

* Used for prioritizing State-listed species in several States.

FIGURE 3.1. Three widely used systems for ranking species. Such approaches may be useful for setting monitoring priorities.

2. Existing species information

a. Applications of existing information

Reviewing and assembling existing information serves five important functions in the development of a monitoring project:

◆ The compilation and comparison of existing information are important for setting priorities. You must know about the relative rarity and threats to all the species you manage in order to allocate monitoring resources among them.

◆ The response to a management approach may have already been monitored elsewhere. An initial review may identify the need for immediate changes in management, and thereby avoid monitoring a decline before the management action is initiated.

◆ Some measurement techniques may have been tried previously on your species (or a similar one) with minimal success. Knowing the monitoring history may help you avoid repeating mistakes.

◆ This information will be used in developing the ecological model and setting objectives.

◆ A compilation of existing information will identify parties that should be included in the development of the monitoring project. For example, an assessment of distribution of a species across its range might identify the need to coordinate monitoring of populations on adjacent Federal lands managed by another office or agency. An assessment of threats may identify user groups who should be involved in the development of a monitoring project, since their resource use may affect or be affected by the results (Chapter 10).

b. Documenting existing information

All existing information should be documented and stored in a single place (you should duplicate and archive one copy to protect from loss). A summary of the information that should be included is given in Figure 3.2. For many species, little is known, and many of the information items must be filled in with hypotheses. Avoid simply leaving the information out. Your hypotheses are likely better than nothing, and, by forcing yourself to try to describe all of the species' biology and threats, you will identify those information items that are critical to your ecological model (Chapter 4) and to the monitoring design. These may require additional study before initiating monitoring.

The sources of the information in your summary should be documented. Cite published sources and personal communication, and comment about the reliability of the information. Hypotheses and your observations should be clearly identified.

These summaries are time-consuming, but they have benefits in addition to improving the quality of monitoring projects. The summaries can be referenced or included in biological evaluations and assessments. They can be helpful in training technicians or other specialists. They also communicate your observations and knowledge of the species to your successors. Once completed, the summaries are easily updated, incorporating new information as it becomes available.

Species Biology	Habitat
life history (annual, perennial)	physical features
life expectancy (long or short lived)	soil
reproductive ecology	elevation
pollinators	aspect
flowering period	slope
annual variability in flowering	moisture
seed maturation period	biotic features
seed production	community
seed viability	potential importance of competition
seedling ecology	animal use
regularity of establishment	natural disturbance
germination requirements	fire
establishment requirements	slope movement
Population	small scale (e.g., animal diggings)
population size (range, average)	**Threats**
annual variation	natural
number and distribution of populations	herbivory
	disease
	predators
	succession
	weed invasion
	anthropogenic
	on-site (grazing, logging, etc.)
	off-site (changes in hydrology, pollinators)
	Trend
	Causes of Trend
	Management Options

FIGURE 3.2. Components of information that may be useful in a review of a species. Summarizing all that is known or hypothesized about each of these components is not only helpful is setting priorities among species and populations for monitoring, it is also critical for developing ecological models, designing studies, and ensuring that anecdotal information about a species is not lost during changes in personnel.

c. Locating information

Sources of information are varied, and are rarely in an accessible published form. Much of the knowledge about a species resides in the experience of individuals, and may be difficult to extract.

Natural Heritage Programs and Conservation Database Centers associated with a State agency or The Nature Conservancy maintain databases on location and condition of rare plant populations. They also provide access to that information in adjacent States. State Native Plant Societies and environmental groups may have information on the species, and may also be able to put you in contact with amateur and professional botanists who know about the species.

Academic experts who have worked with the species or related species may sometimes be found at universities or colleges. Herbaria may be a source of information on additional populations. Specimen labels often contain habitat notes, and some herbaria have computerized these to facilitate searching and summarizing. Many Heritage Programs/Databases have completed herbarium searches for rare species and may have the information in an accessible form. Be cautious about using herbarium records, however. Specimens may not be accurately identified (those that have been annotated as part of a recent study are the most reliable), may be misfiled, or may be poor representations of a species. Place names may provide only general location information or even be incorrect.

People familiar with the species may be found within your local office (other specialists), in local Federal offices administering adjacent lands (e.g., BLM, Forest Service, National Park Service, military), in regional and State federal offices, in research units, and associated with private consulting firms. You may also be able to find knowledgeable individuals using Internet/World Wide Web resources such as discussion groups. (The Ecological Society of America administers a discussion group that may be helpful.[1])

Published information on rare species is most often found in symposium proceedings, technical reports, and project reports. This information can be difficult to locate through conventional computerized searches and is often best found through contact with reputable sources. Often State Natural Heritage Programs maintain extensive collections of unpublished and published literature on sensitive species.

B. Setting Priorities

1. Involving managers, using teams

Because establishing priorities in land management is a subjective process, different people will list the same species in different priority order. For a manager, the highest priority species may be the one that conflicts with the dominant commodity activity. For the botanist, the highest priority species may be the rarest one. Legal direction, existing plans, and pet projects may all conflict with priorities that would result from a strict following of biological criteria.

Because the setting of priorities is subjective, we recommend they be set by a team, or at a minimum, with input from management and other specialists. Solicit input from others outside of the agency as well, such as from Native Plant Society members and commodity groups. Setting priorities is a situation-specific activity. The lists of criteria that follow are not meant to be exhaustive; there may be other criteria important to your specific situation.

2. Criteria for species comparisons

Rank. Some approaches have utilized the conservation status or rank assigned a species, such as from one of the systems illustrated in Figure 3.1. Note that in many systems, this rank is already a composite of criteria. For example, the ranking used by the U.S. Fish and Wildlife Service combines taxonomic distinctness with the magnitude and imminence of the threat.

Rarity. Rarity relates to population size, number of populations, and distribution of populations across the landscape. In comparing species, perhaps the most useful aspect of rarity for monitoring is the number of populations. A species restricted to a single large population is

---

[1] To subscribe, send an e-mail message to LISTSERV@UMDD.UMD.EDU. Leaving the subject line blank, include the following message: SUBSCRIBE ECOLOG-L [your name], typing your name where the brackets are. Put nothing else in the message. The listserver will send additional information.

at more risk than one with fewer total numbers distributed in several populations. Similarly, populations clustered in a small area may all be affected by the same threat, while populations that are widely distributed are less likely to be affected by a single impact.

Taxonomic distinctness. A species that is the only representative in its family would rank above one that is the only representative in its genus, which would rank above a species that occurs in a genus with many species. The concept behind this ranking is that the taxonomic distinctness of a single-species family correlates with high genetic uniqueness. A subspecies or variety would for the same reasons be considered of less value. A drawback to this approach is that taxonomic divisions are largely based on morphological differences and may not directly relate to genetic diversity.

Sensitivity to threats. Species vary in their sensitivity to threats depending on their biology and ecology. Species with a long-lived seed bank are buffered from population declines because a single good germination year can function as a rescue. Species with populations that vary widely from year to year but lack a seed bank are more prone to local extinction. Species that are limited to midsuccessional stages are vulnerable to both disturbance and succession caused by lack of disturbance. Species dependent on other species for pollination or seed dissemination are more sensitive than those that are wind-pollinated and dispersed. Species that have exacting habitat requirements are more sensitive than those that are more cosmopolitan in habitat.

Known declines. Species with known declines based on monitoring or observation are more important for monitoring and management than species that are considered stable.

Extent of threats. Threats can be evaluated in terms of scale and intensity. Scale describes the percentage of the populations affected and the distribution of threats across the landscape. Intensity describes the degree to which populations are affected by threats (e.g., extirpation of the population, mortality of a few individuals).

Immediacy of threats. The rate at which threats may occur and populations decline is another important consideration. Species or populations with ongoing or immediate threats would rank higher than those with potential threats.

Conflict. The degree of management conflict between potential conservation actions and existing or alternative uses (usually commercial) may be an important consideration in prioritizing populations. The degree of conflict may also dictate the form of monitoring (high conflict situations may require quantitative monitoring or even research into cause and effect).

Monitoring difficulty. Monitoring some types of plants, such as annuals and geophytes, can be nearly impossible due to temporal and spatial variability. Some species, such as those found on cliffs, are difficult to access. Monitoring species growing on fragile sites, such as erosive slopes or semi-aquatic habitats, may cause unacceptable investigator impacts.

Availability of management actions. If no management options are available, resources should be directed toward other species with management alternatives.

Recovery potential. Some species will only recover with a large expenditure of resources, while others have high recovery potential. You may choose to focus on the species with the highest potential, especially if several species could be managed for recovery with the same resources required for one.

Public interest. The fact that birds and mammals (e.g., bald eagles and grizzly bears) corner vastly more recovery resources than amphibians or invertebrates is largely a result of this factor. A common orchid will have more public support and interest than a rare algae.

Potential for crisis. Crisis can be defined in biological terms (potential for extinction) and in management terms (potential for politically heated conflict).

3. Criteria for population comparison

Population size. Investing in larger populations may be a better conservation strategy than salvaging small populations. Larger populations are better able to weather annual variability, and they provide a larger buffer for decline. Conversely, it may be more important in some situations to monitor small populations because they are more prone to extinction, and assume the larger ones are at less risk.

Population viability. A population with individuals distributed among all age classes is more demographically "healthy" than one with obviously skewed stage distributions (e.g., all old or dying individuals). Monitoring may be concentrated on those populations with the best potential for long-term survival or on those that are obviously in trouble.

Population location. Selecting populations on the fringe of the distribution of the species usually increases the range of genetic variability conserved. These populations may also occupy fringe habitats that are marginal and stressful, and may express response to rangewide stresses, such as climate changes, before more central populations.

Habitat quality. Depending on the situation, higher priority for monitoring may be applied to populations found on degraded or disturbed habitat (because they are more at risk) or on stable or pristine habitat (because protection is a better conservation investment than restoration).

Unique habitat. Populations located on unique habitat likely contain unique genetic combinations and are important for conserving the range of genetic diversity of the species.

Previous information/monitoring/research. Populations with previous monitoring or natural history studies may be a higher priority if data suggest a decline or problem, or a lower priority if data suggest the population is stable or increasing.

Special management area. Specially designated areas such as Research Natural Areas (RNAs) and purchased preserves represent a significant investment of resources. If rare plant populations are an important factor for establishment or purchase, maintaining the population is a management priority. Monitoring of these populations would be a higher priority than populations in non-designated areas. Conversely, it may be assumed that the protection afforded by designation reduces threats, as well as the need for monitoring.

Other. Most of the criteria applicable to prioritizing among species are also applicable to prioritizing among populations (e.g., sensitivity to threats, extent of threats monitoring difficulty, availability of management actions, recovery potential, public interest, and potential for crisis). Using these criteria to establish priorities among populations, however, differs from their application to comparisons of species. For example, characteristics such as high sensitivity to threat, known declines, and extensive immediate threats would usually result in a high priority for a species. For species, the goal is to protect and maintain as many species as possible (to combat extinction) so species most at risk are those most important to manage. These same characteristics may result in either a low or high priority for a population, depending on the specific situation. When comparing populations, efforts may further conservation goals by concentrating on populations with few or minor threats, low sensitivity to potential threats, and known stable population size. Conversely, populations with minimal threats may appear so secure that monitoring can focus instead on those populations that are threatened or have recovery needs.

4. Using criteria matrices to set priorities

Several methods for setting priorities have been developed that use various criteria. The most widely applied systems are those developed by The Nature Conservancy and the U.S. Fish and Wildlife Service (Figure 3.1). These systems combine criteria of rarity and threat. Because each situation is different, however, a better approach allows you to design your own system, identifying criteria that are important to the specific situation. A matrix approach can be used when a large number of criteria are to be incorporated, and you wish to weight each criterion individually. In the example given in Figure 3.3, biological criteria are given higher emphasis than management criteria. Figure 3.4 and Figure 3.5 provide blank work sheets for comparing species and populations.

C. Assess Available and Needed Resources

Management must be committed to the monitoring project and willing to expend the resources required for a successful project. Priorities and allocation of time and dollars are the responsibility of management. Managers are also the ones who will make decisions based on the monitoring. Be wary of your inclination

SPECIES		rarity	taxonomic status	sensitivity	known decline	extent of threats	immediacy of threats	existing conflict	monitoring difficulty*	availability of management actions	recovery potential	public interest	potential for crisis	Total
		BIOLOGICAL CRITERIA							MANAGEMENT CRITERIA					
	WEIGHTING	4	2	5	5	5	5	2	1	5	5	1	1	
species A (a rare variety)	rating for species	3	1	3	3	2	3	1	3	1	2	1	1	24
	rating × weight	12	2	15	15	10	15	2	3	5	10	1	1	91
species B	rating for species	2	2	1	3	3	3	3	1	3	3	3	3	30
	rating × weight	8	4	5	15	15	15	6	1	15	15	3	3	105
species C	rating for species	1	3	3	2	2	1	1	3	3	3	1	3	26
	rating × weight	4	6	15	10	10	5	2	3	15	15	1	3	89
species D	rating for species	1	1	1	2	1	1	1	3	1	1	1	1	15
	rating × weight	4	2	5	10	5	5	2	3	5	5	1	1	48
species E	rating for species	3	1	3	3	3	3	1	1	3	3	1	1	26
	rating × weight	12	2	15	15	15	15	2	1	15	15	1	1	109

* note that all weights range from 1-5 and species ratings range from 1-5 and species ratings range from 1-3, with the lowest number having the lowest importance. For monitoring difficulty, a low number means it is a difficult species to monitor (the more difficult species receive a lower importance for monitoring).

FIGURE 3.3. Completed matrix for setting priorities among five species.

to do self-driven monitoring, where you choose to devote what resources you can toward your pet monitoring project. Although the monitoring may be implemented as long as you're there to do it, if you leave, your pet project may die. A monitoring project needs other advocates besides the specialist(s), preferably in management.

Once management is supportive, you should consider three limiting factors when designing a monitoring project: (1) the skill level of those planning and implementing the project; (2) the equipment available; and (3) the time and money available for field work and analysis.

The project may require special skills at the planning level. Depending on the complexity of the project and your knowledge, you may need a statistician or someone with expertise in sampling design. State offices and regional offices may have people who can help. You may be able to solicit or contract advice from specialists associated with universities, private consulting firms, and conservation groups. Rare plant experts associated with State agencies and those with the U.S. Fish and Wildlife Service may also provide advice. Use as many resource people as possible for review.

Special skills may also be needed at the implementation level. Field work that will be completed mostly by summer technicians may need to be designed differently than that done by experienced botanists.

FIGURE 3.4. Blank matrix worksheet for setting priorities among species.

Most plant monitoring projects require inexpensive equipment such as measuring tapes, pin flags, and a camera (a list of standard field equipment is provided in Appendix 12). Some projects may require specialized equipment, such as Global Positioning Systems, survey equipment, and video equipment. These are becoming more commonly available at agency offices. Other specialists in your forestry and range programs may have ideas about useful equipment that will reduce your field time. Many of these people also have experience in sampling vegetation and can provide ideas and help sharpen your thinking through discussion.

Finally, the time required must be compared to the time allocated for a monitoring project. Most botanists and specialists are fairly good at estimating field time for gathering data.

CRITERIA	criteria weighting	Population 1		Population 2		Population 3		Population 4		Population 5		Population 6		Population 7	
		rating	w × r	rating	w × r	rating	w × r	rating	w × r	rating	w × r	rating	w × r	rating	w × r
population size															
population viability															
population location															
habitat quality															
unique habitat															
previous information/ monitoring/research															
special management area															
conflict															
monitoring difficulty															
availability of management actions															
recovery potential															
public interest															
potential for crisis															
sensitivity															
known decline															
extent of threats															
immediacy of threats															
rating total															

FIGURE 3.5. Blank matrix worksheet for setting priorities among populations.

Estimating the office time required is more difficult. Estimate at least one work week to develop and document the objectives and design the monitoring. Complex projects requiring consideration of various points of view, and extensive review will take much longer. To estimate analysis and reporting time, multiply field time by 2-5 times, depending on the complexity of the data gathered. Qualitative data will take less time to analyze and report than a detailed, data-intensive method that requires statistical analysis.

It is important that the time required for monitoring be estimated liberally. Many field data sets have not been analyzed because time needed for analysis was not included in the budget. Managers must know and support the total time required for completion of the monitoring project.

D. Selecting Scale

Monitoring scales can vary from a single small local population of a few individuals (local scale), to many large populations and the range of the species (landscape scale). The scale should be decided explicitly, because scale has important implications for monitoring design (Chapter 7).

1. Landscape scale

The selection of scale will be guided by management considerations and priorities and limited by resources available for monitoring. Landscape scale can be defined in a number of ways:

◆ all known populations of the species
◆ populations on Federal and State lands
◆ populations within an administrative boundary (e.g., BLM District or Resource Area, National Forest)
◆ populations within a watershed
◆ populations within a vegetation type
◆ populations within a management unit (e.g., an allotment, a wilderness area)
◆ populations within a treatment area
◆ populations with a specific management treatment (e.g., grazed populations)

Establishing a system of monitoring populations of a species across its entire range provides the most accurate measure of the overall trend and condition of a species. Because of the required coordination efforts for species that cross administrative boundaries, however, such rangewide approaches are unfortunately rarely attempted. If you share a species of limited distribution with only one or two other agencies, consider trying to coordinate monitoring efforts. For species that cross several administrative boundaries, the new efforts at interagency regional planning and ecosystem management provide hope that coordination of rangewide monitoring of species may become easier in the future.

Once you've identified the landscape scale and the pool of populations that you will consider monitoring, you need to decide if all populations at that scale will be monitored or only a portion of them (perhaps because of limited monitoring resources). If monitoring only a portion, you must decide if you want to draw a *sample* of populations from all those that occur at that scale, or *select* specific populations. If you wish to draw conclusions about all of the other populations at that scale from the portion monitored, you will need to draw a random sample of monitored populations from the entire set of populations. For example, if you monitor only populations that are easily accessible along roads, your sample would be biased (not random) and only represent roadside populations. You would be unable to draw any conclusions about populations in native habitat. You may, however, decide that you will select only roadside populations because those are the ones about which you have conservation concerns. This is a perfectly valid approach, as long as you recognize that you are limited to conclusions only about those selected roadside populations.

In statistical terms, when you identify the set of all populations that are of interest, you define a "sampling universe" from which you will randomly draw "sampling units" (in this example, individual populations). You must carefully consider both the sampling universe and sampling units if you want to be able to draw conclusions about several populations. These concepts are described in more detail in Chapter 7, and also apply to consideration of scale at the single population and macroplot level (below).

2. Local scale

You may be constrained by limited monitoring resources to selecting one or a few populations from all of those known. You may select this population over others based on some criteria such as previous information, ease of access, degree of threat or lack of it, size, and inclusion in a special management area such as an RNA. Be aware that monitoring a population does not allow you to draw conclusions about all of the populations. Even if you select a population at random, it is not appropriate to assume it represents the range of conditions and trends occurring within other populations. Common sense and biological experience suggest just the opposite. You may, however, be able to use qualitative monitoring at other populations to support conclusions that trend or condition is similar to the site you are monitoring quantitatively.

Unless your selected population is very small, you will face the same sample versus selection issue previously described for populations at the landscape scale. If you wish to draw conclusions about the entire population you are monitoring, you must sample the population randomly. Sometimes this is not possible. For example, a population comprising individuals dispersed over a very large area may be difficult and time-consuming to sample randomly, or some portions of the population may be physically inaccessible.

One option is to select a portion of the population as a key area or macroplot, monitor only within that area, and agree among interested parties that the results will be applied to management of the entire population (Chapter 7). This approach is common in range studies. The drawback is that you must assume the key area functions as an indicator for the entire population. Inferences cannot be made to the entire population based on data. Changes measured on the macroplot may or may not represent those occurring outside of the macroplot. This problem can be partially addressed by supplementing the quantitative studies within a macroplot with qualitative studies dispersed throughout the population. While you will still be unable to conclusively state that the changes observed within the macroplot represent those outside the macroplot, the supporting evidence may be sufficiently strong for management decisions.

Situating a macroplot requires some decisions. Will the plot(s) be located in the area most likely to be affected by adverse management? Will you attempt to locate the plot(s) in a representative area of the population, and if so, how will you define what is representative? Will your main criteria be ease of access? Chapter 7 discusses these issues in more detail.

E. Selecting Intensity

Intensity of monitoring can be defined as the complexity of methods used to collect information. Monitoring intensity roughly equates to time, but also relates to the skills required to collect information. Monitoring can be generally classified into qualitative and quantitative techniques. Qualitative techniques are usually less intensive than quantitative. Within each class, levels of intensity also vary.

1. Examples of qualitative monitoring

Presence or absence. Noting whether the species of interest is still at the site may be an effective way to monitor a species with many roadside populations. Populations located along roads can be noted by a "windshield check" by other specialists in the course of their work.

Site condition assessment. Site condition assessments provide a repeated evaluation of the quality of the habitat. The monitoring is designed to detect obvious and dramatic changes that can be recorded photographically, with video, or in written descriptions aided by a standard form (see Appendix 10).

Estimates of population size. Visual evaluation of population size, often in classes such as 0-10, 11-100, 101-1000, 1001-5000, etc., provides more information than simply noting presence or absence.

Estimation of demographic distribution. A population's demographic distribution is the percentage of the population or number of individuals within classes such as seedling, non-reproductive adult, reproductive, and senescent.

Assessment of population condition. In this approach, the observer evaluates the condition of the population by noting occurrence and extent of utilization, disease, predation, and other factors.

Photopoints. Photopoints are pictures that are retaken from the same position of the same frame at each observation (see Chapter 8).

Photoplots. Photoplots straddle the division between qualitative and quantitative monitoring. These are usually close-up photographs of a bird's-eye view of a plot within the frame. Plot size varies with camera height and lens type, but commonly ranges from 50cm x 30cm to 1m x 1m. Photoplots can provide a qualitative record of a small portion of the population, or they can be used as a plot to measure cover and/or density (see Chapter 8).

Boundary mapping. Mapping the perimeter of a plant population monitors change in the area occupied by the population.

2. Types of quantitative monitoring

By definition, in quantitative studies some attribute is measured or counted. Three basic types of quantitative approaches can be described:

Census. A census of the population counts or measures every individual. The main advantage of this approach is that the measure is a count and not an estimate based on sampling. No statistics are required. The changes measured from year to year are real, and the only significance of concern is biological.

Sample. A sample measures only a portion of the plant population. No sample is an identical representation of the population as a whole. It is an estimate of that population; thus, some error is associated with the sample (the difference between the sample estimate and the real value of the population). Statistics is the tool used to assess that error (see Chapters 5, 6, 7, and 11). A sample of quantitative data should only be taken if the results are to be analyzed statistically, because the error associated with that sample can be quite large, and the monitoring useless for detecting change. Only through statistical analysis can the magnitude of sampling error be assessed.

Some monitoring designs avoid statistics by doing a complete census or full counts in a small portion of the populations in a representative plot. For example, height may be used to measure plant vigor annually. Rather than sampling, a single representative plot is established in the middle of the population and the height of all individuals within that plot is measured. No statistics are necessary, because you know the true average height *of all the plants in the plot*. If the decision has been made to base management changes on the changes within the plot, this is an acceptable approach, but be aware that the average height of the plants in the plot is not an estimate of the average height of the plants in the population.

Demographic monitoring. Demographic monitoring involves marking and monitoring the fate of individuals through time (Chapter 12). It is extremely labor-intensive, and represents the most intense level of monitoring that can be used.

F. Priorities, Resources, Scale, Intensity

Priorities, available resources, selected scale, and selected intensity are closely related and must be considered together when developing a monitoring project. Given limited monitoring resources, scale and intensity are inversely related. You can choose to monitor many populations (large scale) with low intensity or devote all your monitoring resources toward monitoring a single population intensely. If you have many high priority species, limited monitoring resources may allow you to monitor only a single population of each species at a low intensity.

This explicit consideration of the interplay of priorities, resources, scale, and intensity is critical to the effective allocation of monitoring resources. In the absence of this analysis, we tend to ignore inexpensive monitoring solutions and focus on intensive data-collecting techniques. Other techniques, such as qualitative methods and photographs, are generally less time-consuming to design and implement, but can be effective for many situations. Low-intensity monitoring may be designed as a warning system that triggers more intensive monitoring or research if a problem appears. In other situations, low-intensity techniques may provide the data needed for making decisions. Most changes monitored by these techniques must be fairly large or obvious before they are detected; thus, it is often appropriate to take immediate management action based on these measures. Implementing a high-intensity study to quantify a problem that is obvious only delays remedial action.

In summary, allocating monitoring resources is a critical initial stage in the development of a monitoring project. Ranking priorities and selecting scale and intensity are not trivial activities, but are fundamental to the effective design of good monitoring. Using teams and soliciting review will help focus decisions about allocation, and avoid premature sidetracks into selecting methods.

Literature Cited

Palmer, M. E. 1986. A survey of rare plant monitoring: programs, regions and species priority. Natural Areas Journal 6: 27-42.

Palmer, M. E. 1987. A critical look at rare plant monitoring in the United States. Biological Conservation 39: 113-127.

CHAPTER 4
Management Objectives

Sorghastrum nutans
Indian grass
by Jennifer Shoemaker

CHAPTER 4. Management Objectives

A. Introduction

In this technical reference we are promoting objective-based monitoring whose success depends upon developing specific management objectives. Objectives are clearly articulated descriptions of a measurable standard, desired state, threshold value, amount of change, or trend that you are striving to achieve for a particular plant population or habitat characteristic. Objectives may also set a limit on the extent of an undesirable change.

As part of the adaptive management cycle, management objectives:

♦ Focus and sharpen thinking about the desired state or condition of the resource.
♦ Describe to others the desired condition of the resource.
♦ Determine the management that will be implemented, and set the stage for alternative management if the objectives are not met.
♦ Provide direction for the appropriate type of monitoring.
♦ Provide a measure of management success.

As the foundation for all of the management and monitoring activity that follows, developing good management objectives is probably the most critical stage in the monitoring process (MacDonald et al. 1991). Objectives must be realistic, specific, and measurable. Objectives should be written clearly, without any ambiguity.

B. Components of an Objective

Six components are required for a complete management objective:

♦ **Species or Habitat Indicator:** identifies what will be monitored
♦ **Location:** geographical area
♦ **Attribute:** aspect of the species or indicator (e.g., size, density, cover)
♦ **Action:** the verb of your objective (e.g., increase, decrease, maintain)
♦ **Quantity/Status:** measurable state or degree of change for the attribute
♦ **Time frame:** the time needed for management to prove itself effective

Management objectives lacking one or more of these components are unclear. Figure 4.1 gives examples of typical incomplete objectives and identifies their missing components.

1. Species or habitat indicator

Monitoring may involve measuring the change or condition of some aspect of the species itself. If you are monitoring the species, the objective should include its scientific name. If the objective will address a subset of the species (e.g., only flowering individuals, fruits, or seedlings), this should be specified.

What's Missing?

1. Increase <u>Physaria</u> <u>didymocarpa</u> var. <u>lyrata</u> at the Williams Creek Shale Pit by 1999.

2. Exclude livestock from the Summit Creek <u>Primula</u> <u>alcalina</u> population.

3. Exclude livestock from the Summit Creek if cattle are impacting <u>Primula</u> <u>alcalina</u>.

4. Increase percent cover of the Lime Creek population of <u>Astragalus</u> <u>aquilonius</u> by 50%.

5. Decrease the percent of <u>Astragalus</u> <u>aquilonius</u> individuals trampled by livestock at the Grandview site by the 1996 grazing season.

6. Maintain a population of at least 400 individuals of <u>Astragalus</u> <u>diversifolius</u> at the Birch Creek site between 1994 and 2000.

7. Allow no more than 30% herbivory of inflorescences in any two years in a row between 1997 and 2000.

8. Increase <u>Astragalus</u> <u>aquilonius</u> at the Wood Creek site by 30% between 1997 and 2003.

9. Increase the habitat occupied by <u>Gymnosteris</u> <u>nudicaulis</u> by 300 hectares.

10. Increase the viability of the onion.

11. Maintain, at a minimum, 300 <u>Happlopappus</u> <u>radiatus</u>.

12. Increase the number of hectares of <u>Primula</u> <u>alcalina</u> habitat under protective management by 240 hectares by 2003.

---

1. Increase what attribute of <u>P. didymocarpa</u>? Increase from what level or from which time?

2. This is a management action, not an objective.

3. Not an objective, more similar to a management response. The term "impacting" is ambiguous. Need to identify some measurable parameters.

4. Increase by 50% over current value? By when?

5. How large a decrease in percent? From when?

6. Looks OK.

7. What plant? What site or population? Is an inflorescence included in the 30% if it is only partially eaten, or does it have to be completely consumed? (this would be addressed in the methodology)?

8. What attribute of <u>A. aquilonius</u>? Cover? Density? Something else?

9. Where? In a certain population or watershed or throughout the resource area? By when should this increase occur?

10. What is viability? How much increase? What onion? What population? By when?

11. Where? Time frame? Maintain 300 of what attribute (individuals, stems, flowering plants)?

12. What is protective management? Where should this increase occur?

FIGURE 4.1. Examples of objectives missing one of the six components of a management objective: species or habitat indicator, location, attribute, action, quantity/status, and time.

Measurement attributes can also focus on aspects of the habitat of a species rather than direct measurements of the plant population itself. Attributes may be selected that serve as indicators or surrogates for the condition of a particular species. Useful indicators may focus directly on known or perceived threats to a particular population. Here are three examples that illustrate the use of indicator measurement attributes based on threats assessments:

◆ **Off-road vehicle impacts.** A rare plant population exists in a remote area where the only known threat is disturbance from off-road vehicles. Monitoring the presence, number, or spatial extent of tire tracks may provide the most sensitive feedback information needed to adjust management activities on the site (e.g., installing new signs or fences).

◆ **Non-native species impacts.** A particular site contains populations of several rare plant populations. The only known threat to the site is the potential invasion of non-native species which presently do not occur on the site but do occur along a nearby road. Tracking the weed-free condition of the site may provide the most critical information needed to prescribe management activities (e.g., organizing a volunteer work party to remove all non-native species from the site).

◆ **Woody species encroachment.** An open meadow site supports several rare plant populations. Historical records indicate that lightening-ignited fires burned through the site on a regular basis. Fire suppression activities in the nearby forest have dramatically reduced the fire frequency and the principal threat to the rare species population is shading from woody species encroachment. Tracking the abundance of woody species in the meadow will alert you to the need for management action (e.g., prescribed burning).

Other potential indicators or surrogates for directly measuring attributes of a plant population include abiotic variables (e.g., water quality parameters), or other plant or animal species.

Using habitat indicators to indirectly measure species' success or condition is a common practice in resource management, but is not without problems. Your chosen indicator may have a weaker relationship with the species than you hypothesize. Another factor may have important effects on the species, but have little relationship to the selected indicator (e.g., cattle grazing of a wetland species and a selected indicator of soil water levels). For these reasons, when threats-based attributes, indicator species, or abiotic variables are used as surrogates for tracking individual plant populations, it is advisable to periodically assess the plant population itself to ensure the validity of the surrogate relationship.

2. Location

Clear delineation of the specific entity or geographic area of management concern allows all interested parties to know the limits to which management and monitoring results will be applied. The spatial bounds of interest defined in a management objective will vary depending on land management responsibilities (e.g., you may only have access to a portion of a particular population due to multiple land ownership patterns) or particular management activities (e.g., you may only be interested in plants located within a fenced macroplot that is located within a larger population). The location is related to the selected scale of monitoring (Chapters 3 and 7), which is affected by conservation goals and responsibilities, the biology of the species, and the realities of limited monitoring resources.

3. Attribute

Five major classes of quantitative vegetation measures are available. A brief description and comparison is given here; Chapter 8 presents a more thorough comparison.

◆ **Density.** Density is the number of individuals or stems (or another counting unit) per unit area. It can only be used when a consistent counting unit can be recognized.

◆ **Cover.** The amount of ground covered by the vertical projection of plant matter can be visualized by considering a bird's-eye photograph of the vegetation. The percentage of the ground obscured by vegetation is canopy (or aerial) cover. Basal cover is the percentage of the ground covered by the base or trunk of the plant.

◆ **Frequency.** If a population is visualized overlaid with a grid defining sampling units, the percentage of those units occupied by the species is the frequency. As quadrat size changes, frequency also changes; thus, frequency is a measure that is dependent on the size of the sampling unit.

◆ **Vigor.** Measures include biomass production, the number of new shoots produced, the number of reproductive shoots produced, the number of seeds produced, plant height, plant volume, and many others.

◆ **Demography.** Demographic approaches use rates of reproduction and mortality to model population dynamics. These techniques have recently become widely applied in rare plant management and are described in Chapter 12.

Other possible attributes include population size, qualitative estimates of abundance, presence/absence, and areal extent. Attributes of habitat factors may be similar to quantitative vegetation measures (e.g., density of tire tracks or cover of woody species) or peculiar to the factor (e.g., level of a trace contaminant expressed in parts per million).

When selecting an attribute, first narrow the list of potential attributes given constraints of species morphology and site characteristics (e.g., density is not an option if your species lacks a recognizable counting unit). Then narrow the list further by considering the following criteria:

◆ The measure should be sensitive to change (preferably the measure should differentiate between human-caused change and "natural" fluctuation).
◆ Biologically meaningful interpretations of the changes exist that will lead to a logical management response.
◆ The cost of measurement is reasonable.
◆ The technical capabilities for measuring the attribute are available.
◆ The potential for error among observers is acceptable.

4. Action

There are three basic actions: increase, decrease, and maintain. There is a tendency when managing rare things to want to have them increase. Some populations, however, may already be at the maximum potential for their habitat or suffer from no apparent threats. For these, a more realistic objective would be to maintain current condition. For other populations you may wish to set a threshold that will trigger a management action if the population falls below the threshold. Some questions to consider:

◆ Are current populations viable or have recovery needs such as increased population size, improved vigor, or change in demographic distribution been identified? Species with potential for rapid declines or existing significant degradation of habitat may deserve a more aggressive approach than simply maintaining the current condition.

◆ Are management options available that you believe will increase the abundance or improve the condition of the species?

◆ Will increases occur with removal of threats, or will more active management efforts be necessary (e.g., prescribed fire, augmentation by transplants, control of competing exotics).

The following is a list of common action verbs used in management objectives and guidelines describing when each is appropriate:

◆ **Maintain:** use when you believe the current condition is acceptable or when you want to set a threshold desired condition (e.g., maintain a population of 200 individuals; maintain a population of at least 4000 individuals).

◆ **Limit:** use when you wish to set a threshold on an undesirable condition or state of the species or habitat (e.g., limit cheatgrass cover to 10%; limit mortality to 10% per year).

◆ **Increase:** use when you want to improve some aspect of the species or habitat factor (e.g., increase the average density by 20%; increase the number of populations to 16).

◆ **Decrease:** use when you want to reduce some negative aspect of the species or habitat (e.g., decrease livestock utilization of inflorescences to 50% or less; decrease cheatgrass cover by 20%).

5. Quantity/state

The condition or change must be described with a measurable value. This can be a quantity (e.g., 500 individuals, 20% cover, 30% change), or a qualitative state (e.g., all life stages present at the site, cover class 4).

Determining these quantities or states requires consideration of a number of factors:

◆ How much can the species respond? Populations of long-lived plants (like trees or some cacti) may be very slow to respond to management changes. Changes may be small and difficult to detect, or take many years to express.

◆ What is necessary to ensure species or population viability (e.g., how much change, what population size, what qualitative state)?

◆ How much change is biologically meaningful? Some species (such as annuals) can have tremendous annual variability, and an objective that specifies, for example, a 10% increase in density is meaningless.

◆ What is the intensity of management? Will you continue existing management, remove current threats, or implement a radical alternative?

◆ What is the implementation schedule of management? If the monitoring project is scheduled to last 5 years, but new management will not be implemented until the second year of the study, the change results from only 3 years of management.

◆ What are the costs and problems associated with measuring the amount of change specified? Small changes are often difficult and expensive to detect (Chapters 5 and 7).

The task of specifying a measurable quantity or state is usually a challenging one. The ecology of many plants, especially rare ones, is poorly understood. Predicting the response of a plant to particular management activity is often difficult. Many plant populations undergo natural fluctuations as they respond to varying climatic conditions or to the fluctuating populations of pollinators or herbivores. Most plant populations have been subject to impacts from human activities and there may be little or no knowledge of historical conditions or natural population levels. Few species have been studied in enough detail to reliably determine minimum viable population levels. These challenges should not serve as obstacles to articulating measurable objectives. Use the tools described in Section D and do the best that can be done. If you do not articulate a measurable management objective, you have no way to assess if current management is beneficial or deleterious to the species of interest.

6. Time frame

The time required to meet a management objective is affected by the biology of the species, the intensity of management, and the amount of change specified. Populations of short-lived plants that reproduce annually can probably respond fairly quickly, but long-lived plants and those with episodic reproduction may require more time. Intense management will result in more rapid changes than low intensity or no special management. Large changes will require more time than smaller ones, unless a management action will have immediate, large impacts (e.g., timber harvest).

It is recommended that time frames be as short as possible for several reasons:

◆ Changes in agency budgets and personnel often doom long-term monitoring projects.

◆ Short-term objectives promote regular reassessment of management and implementation of management changes.

◆ Monitoring often uncovers unexpected information; short-term objectives encourage modification of objectives and monitoring based on this information.

◆ Short-term objectives circumvent the trap of monitoring *ad infinitum* while avoiding difficult decisions.

Objectives with time frames as short as several months to a year may be appropriate in some situations. The adaptive management cycle must occur within a short enough period that opportunities for species recovery or alternative management are not lost.

C. Types and Examples of Management Objectives

Objectives can be described in one of two ways:

◆ A **condition** (e.g., increase the population size of Species A to 5000 individuals; maintain a population of Species B with at least 2500 individuals; maintain Site B free of noxious weeds X and Y). We will call these *target/threshold management objectives*.

◆ A **change** relative to the existing situation (e.g., increase mean density by 20%; decrease the frequency of noxious weed Z by 30%). We will call these *change/trend management objectives*.

For target/threshold objectives, you assess your success in meeting your objective by comparing the current state of the measurement attribute to the desired state or to an undesirable state that operates as a red flag or threshold. With a change/trend objective you measure the trend over time. The two objectives are obviously related. Consider the following change/trend objective:

◆ Increase mean density of *Primula alcalina* at Texas Creek by 20% between 1998 and 2005.

You could sample your population, and estimate the current density (say 50 plants/m²). Once the current density is estimated, you could write your objective as follows:

◆ Increase mean density of *Primula alcalina* at Texas Creek to 60 plants/m² by 2005 (a target/threshold objective).

In spite of this relationship, the two types of objectives are appropriate for different situations. You may choose a change/trend objective when you have insufficient information to describe a realistic future condition. You would also use a change/trend objective when you believe the current state is less important than the trend over time. For example, whether a population has 8000 individuals or 6000 individuals may not matter; a decline from 8000 individuals to 6000 individuals (a 25% decline), may be very important to detect. Usually change objectives are more appropriate than target/threshold types of objectives when management has changed and you want to monitor the response (trend) of the selected attribute.

The two types of objectives also require different considerations in designing the monitoring methodology and analyzing the results, especially when the monitoring of the objective requires sampling. Chapter 6 describes these issues in detail.

Management objectives can be written to describe either desirable or undesirable conditions and trends. You would frame your objective in desirable terms if you believe improvement of the plant population or habitat is necessary and you have implemented management you believe will result in improvement. These objectives are sometimes referred to as "desired condition objectives" because they describe the target condition or trend of the resources (e.g., increase to 2000 individuals, decrease cover of a noxious weed by 40%).

If you believe that the current condition is acceptable, and that a continuation of current management will likely maintain that condition, you could frame your objective using undesirable thresholds of condition or trend. These are sometimes referred to as "red flag objectives" because they state the level of an undesirable condition or change that will be tolerated (e.g., no fewer than 200 individuals; no more than 20% cover of the noxious weed; no more than a 20% decrease in density). These objectives act as a warning signal that management must change when the threshold is exceeded. Red flag objectives can be written to identify an unacceptable decline in a rare species or a surrogate habitat variable, or an unacceptable increase in a negative factor (e.g., an exotic species, encroaching shrub cover, the percentage of habitat disturbed by recreational vehicle traffic, etc.).

Different types of management objectives require varying intensities of monitoring (Chapter 3). Qualitative objectives can be monitored using techniques that assess condition or state without using quantitative estimators. Simply finding if the plant still occurs at a site is a type of monitoring that can be very effective for some situations. Another approach is to use estimates of abundance such as "rare," "occasional," "common," and "abundant," or to map the areal extent of the population. Objectives may also be written so they can be monitored by complete counts. Complex objectives may require more intensive monitoring involving quantitative sampling or demographic techniques.

The following examples are arranged in order approximating increasing intensity and include desired condition and red flag types. More examples are provided in Appendix 3.

1. Examples of Target/Threshold Objectives

◆ Maintain the presence of *Penstemon lemhiensis* in the 12 photoplots located in the Agency Creek drainage over the 10-year time span of the Agency Creek Allotment Management Plan (1998-2008).

◆ Maintain the current knapweed-free condition of the *Penstemon lemhiensis* population in the Iron Creek drainage from 1998 to 2008.

◆ Increase the number of population areas of *Penstemon lemhiensis* within the Kenney Creek Watershed from 8 in 1998 to 15 by 2010.

◆ Maintain a population of *Thelypodium repandum* containing individuals in all stage classes (seedling, rosette, reproductive) at the Lime Creek site from 1998 to 2008.

◆ Allow no more than 2 of the 25 presence/absence photoplots at the Lake Creek population of *Physaria didymocarpa* var. *lyrata* to show a loss of the presence of the species between 1998 and 2002.

◆ Increase the Basin Creek population of *Physaria didymocarpa* var. *lyrata* to 120 individuals by 2005.

◆ Maintain at least 100 individuals of *Penstemon lemhiensis* at the Iron Creek site over the life of the Iron Creek Allotment Management Plan (1998 to 2010).

◆ Increase the number of individuals of *Penstemon lemhiensis* in the Iron Creek population to 4500 individuals by the year 2000.

◆ Maintain at least 2000 individuals of *Thelypodium repandum* at the Malm Gulch site over the 10-year period (1998-2008) of the special use permit (current estimated population size: 3000).

2. Examples of Change Objectives

◆ Increase the ranked abundance of *Penstemon lemhiensis* in each of the 10 permanently marked macroplots at the Grizzly Ridge population by one rank class by 2005.

◆ Double the population area occupied by *Penstemon lemhiensis* at the Williams Creek site by 2010.

◆ Allow a decrease in the ranked abundance of *Penstemon lemhiensis* in each of the 10 permanently marked macroplots at the Grizzly Ridge population of no more than one rank class between 1998 and 2005.

◆ Decrease the frequency of *Bromus tectorum* by 30% at the Iron Creek population of *Penstemon lemhiensis* between 1997 and 2005.

◆ Increase the mean density of *Penstemon lemhiensis* at the Warm Springs population by 20% between 1997 and 2000.

◆ Increase the population size of *Penstemon lemhiensis* at the Iron Creek site by 50% by 2005.

◆ Allow a decrease of no more than 20% from the 1998 cover of *Astragalus diversifolius* at the Texas Creek site by 2005.

◆ Allow a decrease of no more than 30% in the population size of *Primula alcalina* in the first 5 years after cattle are reintroduced to the Birch Creek site.

D. Resources and Tools for Setting Objectives

1. Existing plans

General goals for a particular plant species may be described in other planning documents such as land use plans, forest plans, or activity plans. Linking a monitoring project to these higher level planning documents may increase management support and funding for the project. The goals in these plans may also serve as a useful starting point for developing more complete and specific objectives.

2. Ecological models

Ecological models are simply conceptual visual or narrative summaries that describe important ecological components and their relationships. Constructing a model stimulates thinking about the ecology and biology of the target species. You don't have to be mathematically inclined to develop and use a model; the type of model described here rarely involves complicated formulas or difficult mathematics.

Ecological models have three important benefits. First, they provide a summary of your knowledge of the species, enabling you to see the complete picture of the ecology of the species. For example, because livestock grazing affects a species negatively by direct herbivory, you may consider that relationship first. Grazing may, however, also affect the species positively through indirect effects on community composition by reducing competition. Trampling by livestock may positively affect the population by exhuming seeds from the seed bank and increasing germination. During the development of an ecological model, you will have to think about these indirect and sometimes hidden relationships. The model will often identify several factors that can cause the change you hope to detect by monitoring, and perhaps help isolate the most important and interesting mechanism.

Second, ecological models identify the gaps in your knowledge and understanding of the species. Your model may suggest that these gaps are not important, in which case you may choose to ignore these unknowns. Conversely, the model may suggest an unknown relationship is extremely important for understanding the total ecological and management scenario. You may need additional studies before effective monitoring can begin.

Third, ecological models help identify mechanisms and potential management options. If the ecological model suggests, for example, that seedling establishment appears rare, that successional processes of canopy closure may be occurring, and that litter buildup on the ground provides few germination sites, you may be inclined to think about prescribed fire, or some other management strategy that induces germination or reverses succession. Lacking an ecological model, you may have focused on only a single attribute, such as the lack of seedling establishment, which can result from a multitude of causes.

An ecological model can be as simple or complex as you wish. You can focus on a single management activity, as shown in Figure 4.2, or you can attempt to summarize all the interactions, as shown in Figure 4.3.

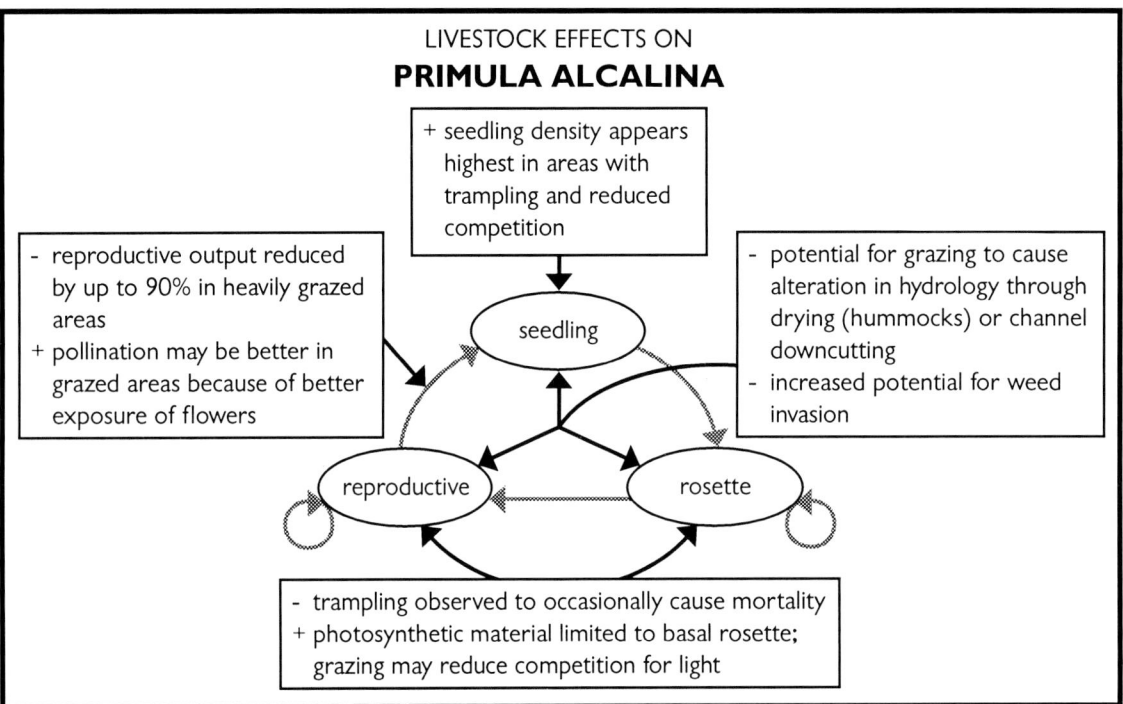

FIGURE 4.2. An ecological model showing positive and negative effects of grazing on an Idaho endemic species, Primula alcalina.

3. Reference sites

The goal in rare plant management is to ensure species are viable over the long-term. For most rare plant managers, this translates into maintaining several to many viable populations within the range of their administrative boundaries. Defining and measuring a "viable" population, however, is difficult (Chapter 12 describes some techniques). This creates a problem in identifying quantities in objectives: How big should the population be? What vigor condition equals "healthy" plants? What percentage of the population should be reproductive? Defining the desired condition of the habitat can be equally difficult.

Reference sites can serve as comparison areas to help set quantitative targets in objectives. These are areas with minimal human impact, such as Research Natural Areas (RNAs). Reference sites may also be an undesignated area with populations that appear thriving and healthy.

Reference sites can be valuable, but use them with caution. Simply because a population is located in a protected area does not ensure that it is viable or healthy. Lack of management activities within protected areas may be allowing successional processes to occur that are detrimental to the plant. In addition, populations that appear "healthy and thriving" to casual observation may actually be declining.

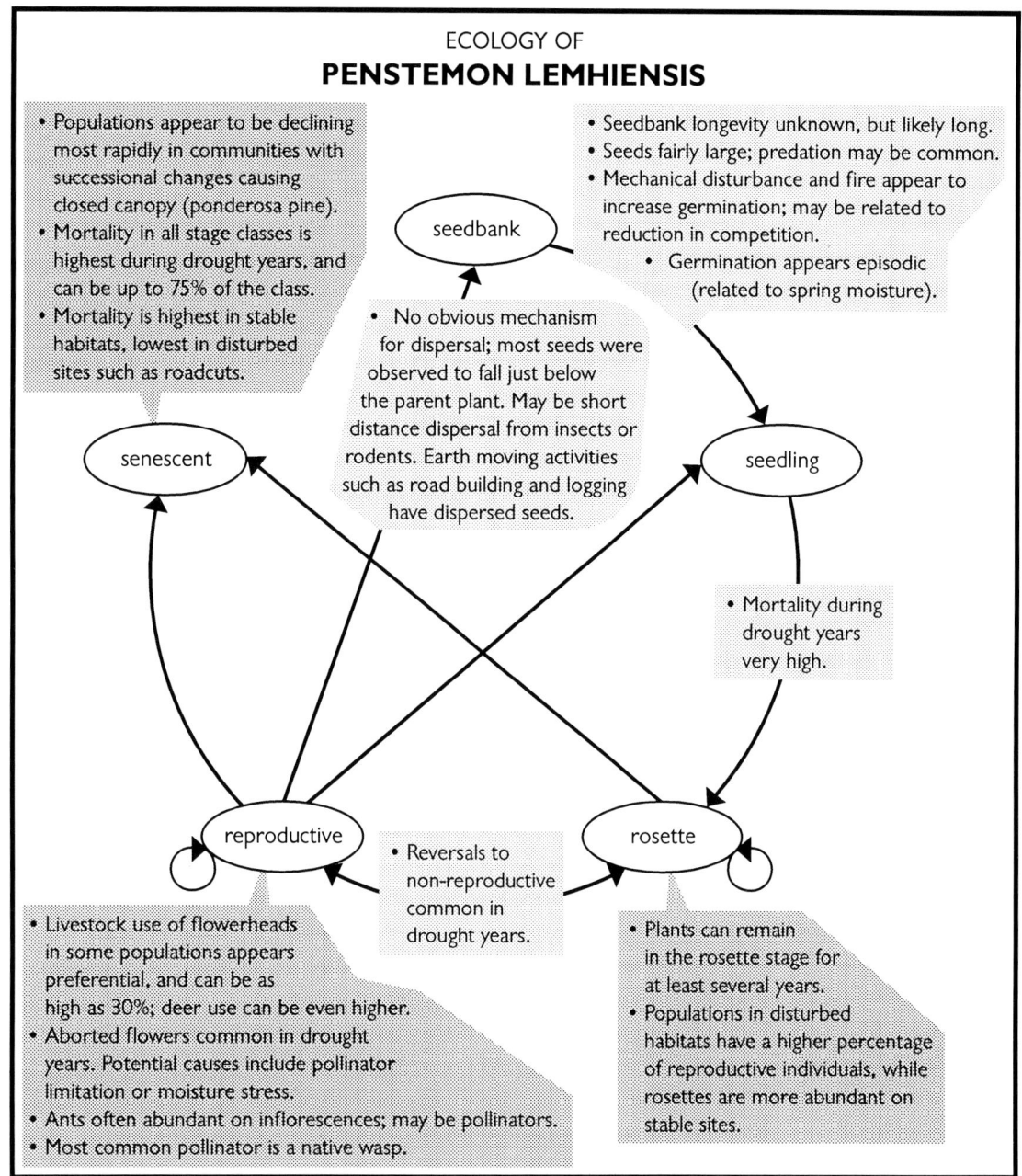

ECOLOGY OF
PENSTEMON LEMHIENSIS

- Populations appear to be declining most rapidly in communities with successional changes causing closed canopy (ponderosa pine).
- Mortality in all stage classes is highest during drought years, and can be up to 75% of the class.
- Mortality is highest in stable habitats, lowest in disturbed sites such as roadcuts.

- Seedbank longevity unknown, but likely long.
- Seeds fairly large; predation may be common.
- Mechanical disturbance and fire appear to increase germination; may be related to reduction in competition.
  - Germination appears episodic (related to spring moisture).

seedbank

- No obvious mechanism for dispersal; most seeds were observed to fall just below the parent plant. May be short distance dispersal from insects or rodents. Earth moving activities such as road building and logging have dispersed seeds.

senescent

seedling

- Mortality during drought years very high.

reproductive

- Reversals to non-reproductive common in drought years.

rosette

- Livestock use of flowerheads in some populations appears preferential, and can be as high as 30%; deer use can be even higher.
- Aborted flowers common in drought years. Potential causes include pollinator limitation or moisture stress.
- Ants often abundant on inflorescences; may be pollinators.
- Most common pollinator is a native wasp.

- Plants can remain in the rosette stage for at least several years.
- Populations in disturbed habitats have a higher percentage of reproductive individuals, while rosettes are more abundant on stable sites.

FIGURE 4.3. An ecological model of all known or suspected interactions for a rare <u>Penstemon</u> species.

4. Related or similar species

Comparisons with more "successful" related species or with species that appear ecologically similar may help set objective quantities that are biologically reasonable (Pavlik 1993). For example, Pavlik (1988) compared nutlet production in an endangered borage, *Amsinckia grandiflora*, with a weedy *Amsinckia*. In another series of studies, the demography of the rare *Plantago cordata*, which grows in freshwater tidal wetlands along the East Coast and along non-tidal streams in Indiana and Illinois, was compared to the widespread *P. major* (Meagher et al. 1978). This approach has obvious limitations. Rare species are often rare because they do not have the reproductive capacity, dispersal potential, or growth potential of more common species.

5. Experts

Experts can provide additional information and opinions on the assumptions within the ecological model. In-house experts include both regional and State ecologists and botanists, as well as specialists in other disciplines such as forestry, range management, wildlife ecology, and riparian management. External specialists include academic, professional, and amateur ecologists and botanists who may know about the species of interest, or a closely related one, or may be knowledgeable about the ecological system in which the species resides. These people can help set realistic, achievable objectives.

6. Historical records and photos

Historical conditions at a site may have been captured in old aerial photos or in historical photos or other historical records housed in museums or maintained by local historical societies. Human disturbances such as roads, trails, and buildings may be visible. Woody species density and/or cover may also be visible. Early survey records by the General Land Office often contained descriptions of general vegetation and habitat characteristics during the mid to late 1800s. Long-term elderly residents can be a fascinating source of information on historical conditions.

E. Developing Management Objectives - An Example

Developing management objectives is a challenging task (Box 4.1). The following is an example illustrating the types of assumptions and decisions that accompany each component.

Collomia debilis var. *camporum* is a long-lived, mat-forming perennial that occurs in 12 discrete locations (occurrences) along a 7-mile stretch of the North Fork of the Salmon River. Occurrences occupy stable slopes of blocky talus. Plants grow in soil pockets among the talus. Size of each occurrence ranges from 0.5 to 3 acres, each with 50 to 500+ pockets of plants. The number of plants cannot be determined because mats grow into each other and are difficult to separate into individuals. A two-lane state highway runs along the base of the slope for the entire 7 miles. Any expansion of the highway (wider shoulders or more lanes) would severely impact all

---

Box 4.1 CHALLENGES TO DEVELOPING GOOD MANAGEMENT OBJECTIVES FOR PLANT SPECIES

1. The ecology of many plants, especially rare ones, is poorly understood. Predicting the response of the plant to management strategies, selecting the attribute or expression of the plant (e.g., cover, density, etc.) most sensitive to management and likely to change, and estimating the expected rate of change in a plant attribute is often difficult.

   Ecological models and comparisons to similar species can help set realistic objectives.

2. There is a human tendency, especially when little is known about the plant, to try to measure many attributes.

   Developing objectives requires choosing a single attribute or a limited suite of attributes to represent overall improvement in the condition of the plant population or habitat. Ecological models may help identify the critical habitat factors and the most sensitive life stages of the species. If you have extensive monitoring resources and/or a complex situation you may need to identify more than one objective.

3. Some species with high annual variability, such as annuals, or small plants that are difficult to even see, present monitoring challenges.

   You may need to admit that some species simply cannot be measured directly. For these, a more effective approach focuses on a related habitat factor or a significant threat.

4. Management goals such as increasing the number of livestock grazing a meadow and increasing the density of a plant species may be in conflict.

   These conflicts must be resolved before specific objectives can be developed. You may need to assemble a team and develop a set of complementary objectives for a particular site together.

5. Goals in land use plans and activity plans for rare plant species may not exist or may be too general to provide direction for specific management objectives. Other specialists or managers may offer little direction or assistance because of their lack of experience with setting objectives for rare plant species.

   Outside experts can provide important insight and suggestions, but a team of agency specialists representing different disciplines may be required to develop a common vision for a site and plant population.

Collomia occurrences. Cheatgrass (*Bromus tectorum*) and knapweed (*Centaurea repens*) occur along the highway right-of-way. Some *Collomia* occurrences have sparse knapweed and cheatgrass; the effects of weeds on the rare species is not known.

1. Review upper-level direction

We first evaluate goals and objectives pertinent to *Collomia* in upper-level plans. The existing Land Use Plan (LUP) does not even recognize the occurrence of *Collomia* on BLM lands because the populations were discovered after the LUP was finalized. The only direction provided by the LUP is a standard operating procedure that states the effects of all projects on sensitive plant species will be evaluated through a field examination. An Allotment Management Plan (AMP) is in place for the area containing *Collomia*. It contains no references to sensitive plants. It is scheduled for evaluation and revision in 2005.

2. Identify the species or habitat factor

An objective could focus on some aspect of *Collomia* or on the most immediate threat, weed infestation. You select the species itself for the following reasons:

♦ Although weeds are a concern, they currently are not very extensive at the site, nor do you know how they affect *Collomia*. Weeds would, therefore, not serve as a reliable indicator for population health.

♦ You have no data on trends or current condition of the *Collomia* occurrences except estimates of areal extent and number of clumps of plants for each of the 12 occurrences. Although plants appear to be long-lived (many mat-forming species are), you noted in your field surveys that there seemed to be many dead individuals and no seedlings. Because of the lack of information on trend or health of the occurrences you prefer to monitor the species directly.

You may also wish to monitor the weed infestation. If so, it is better to develop a separate objective for that issue, rather than trying to combine the species and the indicator into a single complex objective.

Draft objective: *Collomia debilis* var. *camporum*

3. Specify the location

You decide to address all 12 occurrences because of the following reasons:

♦ All of the occurrences are administered by BLM.

♦ All 12 occurrences are important to the viability of the *Collomia* because this variety is so rare, and limited to such a small total area.

♦ This species is your top priority for monitoring, and will receive about half of your monitoring resources.

Draft objective: all 12 occurrences of *Collomia debilis* var. *camporum* along the North Fork.

4. Describe the attribute

Because of the high conservation priority of *Collomia*, you plan to quantitatively monitor this species at each occurrence. You select cover as an appropriate attribute for mat-forming perennials that cannot be separated into individuals (Chapter 8; Appendix 11).

Draft objective: Cover of all 12 occurrences *Collomia debilis* var. *camporum* along the North Fork.

5. Specify action

Because you know so little about the species, you are unable to design management actions that would increase any aspect of this species. The current habitat exhibits no obvious impacts from humans (except for sparse weeds); thus, you assume that current levels are "natural." You decide that maintaining the current population would be acceptable.

Draft objective: Maintain cover of all 12 occurrences *Collomia debilis* var. *camporum* along the North Fork.

6. Specify quantity

You want to maintain the current cover of *Collomia*, but you expect some natural fluctuation around a mean cover value even if *Collomia* populations are healthy and stable. You must specify the level of change that you will allow before you implement alternative management. You have no data suggesting an acceptable level of fluctuation. Because the species is so rare, you don't want to specify a level that masks real and worrisome change, but you also don't want your allowable limits of fluctuation so narrow that you are implementing new management unnecessarily. You decide to allow a decrease of 20% from current cover before you will implement alternative management. You base this value on your knowledge of natural fluctuations in unrelated perennial mat-forming species measured in a nearby range monitoring study.

Draft objective: At each of the 12 occurrences along the North Fork, limit any decrease in cover of *Collomia debilis* var. *camporum* to no more than 20%.

7. Specify time frame

Your objective is still unclear. As currently written, it suggests that an annual decrease of 15% from the previous year would be acceptable. You must identify the starting point from which you will measure the threshold decline of 20%. You also need to specify the time period for which your objective is effective. Most objectives should include a final date that triggers a complete evaluation and final report.

You decide you want to measure the population for several years before writing a final report. You select the year 2004 because the AMP is scheduled for re-evaluation in 2005. You also decide that the baseline cover value will be the 1998 measured cover, and that a decrease of more than 20% from that level would be unacceptable.

Final objective: At each of the 12 occurrences along the North Fork, limit any decrease from current (1998) cover of *Collomia debilis* var. *camporum* to no more than 20% between 1998 and 2004.

F. Difficult Situations

Three types of plants pose special difficulties for developing objectives and monitoring: annuals, long-lived perennials, and species that act as metapopulations.

1. Annuals with a long-lived seed bank

One of the most difficult situations for monitoring is an annual species that only appears above ground once every few years, or even once every few decades. Measuring above-ground expression, such as density, may provide some insight on the weather patterns that create a "good" year, but little information on long-term trends of the population. Most of the population is out of sight, below ground, expressing itself only occasionally.

The study of seed banks is a fairly new discipline (Leck et al. 1989). The biggest problem with studying seed banks is that their distribution underground is usually quite clustered, and the small soil cores used to sample the seed bank result in a large number of cores with none of the target species, and a few cores with many. This creates a serious problem for determining seed bank density with any reasonable precision (Benoit et al. 1989). A second problem with studying seed banks is the labor expense. Extracting cores is time consuming, but estimating the number of seeds in each core is even more so. Two methods are generally used: (1) growing out the cores in a greenhouse and counting the number of germinants, and (2) extracting seeds from the soil core by flotation and physically counting the seeds extracted (Gross 1990). Both are obviously labor-intensive. Both are also fraught with problems. The grow-out method is sometimes unsuccessful because dormancy-breaking and germination requirements are not met. The flotation method may extract dead seeds as well as live ones (Gross 1990).

An alternative that may be more successful than monitoring the species itself is to focus on the habitat. Habitat features such as level of human activity, invasion of exotics, and changes in community composition caused by succession may identify problems for an annual species. Note that for many annuals, some level of disturbance is necessary for exposure of the seed bank and germination; thus, change in disturbance level may be a sensitive attribute to monitor.

2. Extremely long-lived plants

Long-lived species pose the opposite problem as annuals: variability is so slight over time that there is no sensitive measure of change. For some of these species, habitat parameters can change significantly before mortality occurs. Reproduction and/or seedling establishment may be an extremely rare event, although for some long-lived species, the seedling class is the most dynamic and the most sensitive to adverse changes. Monitoring changes in habitat condition or the condition of individual plants may be a more appropriate measurement attribute for these long-lived species than measuring the plants themselves.

3. Plants that act as metapopulations

Species that exhibit metapopulation behavior occur on the landscape with both temporal and spatial variability. These plants may be viable as a metapopulation over the entire land-scape, but individual populations may be short-lived. Dispersal of seeds or propagules and available colonization sites are the two most important factors in the success of a metapopulation. A good example of a plant metapopulation is the Furbish's lousewort, found along a major river system in Maine. Populations of the species are eliminated by ice scouring and spring flooding, but new populations appear on suitable sites left bare by receding floods (Menges 1986, 1990). Because the plant has no seed bank, colonization is dependent on the dispersal of the fall seed crop to new sites. Metapopulation dynamics depend on a dispersal mechanism so that available habitat can be colonized as existing populations become extinct.

Many in the conservation community contend that consideration of metapopulation dynamics is crucial to any conservation strategy (Hanski 1989), while others argue that the importance of metapopulations has been overstated (Doak and Mills 1994). While a few empirical studies have shown the importance of metapopulation dynamics for some invertebrate and animal species, plant studies are much rarer. A review of the literature found only nine plant studies in which a parameter important to the theory of metapopulation dynamics—migration, extinction, or colonization—was actually measured (Husband and Barrett 1996).

While there are exceptions, most plant species disperse propagules locally (Harper 1977; Silvertown and Lovett-Doust 1993). In the absence of an obvious long-distance dispersal mechanism (such as the river in the Furbish's lousewort example), it is difficult to hypothesize how a species could function as a metapopulation, and how to design management to allow that function to occur. It is also questionable whether the dispersal mechanisms important to metapopulation dynamics that may have operated in the past can still operate in today's fractured and fragmented landscape.

In the absence of obvious potential for metapopulation dynamics, the most conservative strategy is to maintain both existing populations and some potential habitat areas. The latter can then provide opportunities for both natural colonization and deliberate re-introductions.

G. Management Implications

Management implications of monitoring must be identified before monitoring begins. If there are no management implications or options, monitoring resources are better spent on another species or population. Usually, however, there *are* options, but some of them may be expensive, or politically difficult to implement. There is a tendency in resource management agencies to continue monitoring, even when objectives are not met, rather than make the difficult decisions associated with changes in management. Because of this hesitancy, we recommend that management implications be an integral part of pre-monitoring planning. Management implications of monitoring are more likely to be applied if they are identified before the monitoring begins, and if all parties agree to the objectives, monitoring methods, and response to monitoring data (see more on this in Chapter 13).

Identifying management implications is difficult, because in some rare plant monitoring situations, the needed management changes are unknown. At a minimum, a management commitment can be made before monitoring begins that additional, more intensive investigation into the management needs of the species will begin if objectives are not achieved. For examples of management objectives paired with management implications, see Appendix 3.

Literature Cited

Benoit, D. L.; Kenkel, N. C.; Cavers, P. B. 1989. Factors influencing the precision of soil seed bank estimates. Canadian Journal of Botany 67: 2833-2840.

Doak, D. F.; Mills, L. S. 1994. A useful role for theory in conservation. Ecology 75(3): 615-626.

Gross, K. L. 1990. A comparison of methods for estimating seed numbers in the soil. Journal of Ecology 78: 1079-1093.

Hanski, I. 1989. Metapopulation dynamics: does it help to have more of the same? Trends in Ecology and Evolution 4:113-114.

Harper, J. L. 1977. Population biology of plants. London: Academic Press, Inc.

Husband, B. C.; Barrett, S. C. H. 1996. A metapopulation perspective in plant population biology. Journal of Ecology 84: 461-469.

Leck, M. A.; Parker, V. T.; Simpson, R. L. 1989. Ecology of soil seed banks. New York: Academic Press, Inc.

MacDonald, L. H., Smart, A. W.; Wissmar, R. C. 1991. Monitoring guidelines to evaluate effects of forestry activities on streams in the Pacific Northwest and Alaska. EPA/910/9-91-001. Seattle, Washington: United States Environmental Protection Agency, Water Division.

Meagher, T. R.; Antonovics, J.; Primack, R. 1978. Experimental ecological genetics in *Plantago*. III. Genetic variation and demography in relation to survival of *Plantago cordata*, a rare species. Biological Conservation 14: 243-257.

Menges, E. S. 1986. Predicting the future of rare plant populations: demographic monitoring and modeling. Natural Areas Journal 6: 13-25.

Menges, E. S. 1990. Population viability analysis for an endangered plant. Conservation Biology 4: 52-62.

Pavlik, B. M. 1988. Nutlet production and germination of *Amsinckia grandiflora*. I. Measurements from cultivated populations. Sacramento, CA: California Department of Fish and Game, Endangered Plants Program.

Pavlik, B. M. 1993. Demographic monitoring and recovery of endangered plants. In Bowles, M. and Whelen, C. (eds.). Recovery and Restoration of Endangered Species. Cambridge: Cambridge University Press.

Silvertown, J. W.; Lovett-Doust, J. 1993. Introduction to plant population biology. Oxford: Blackwell Scientific.

CHAPTER 5
Basic Principles of Sampling

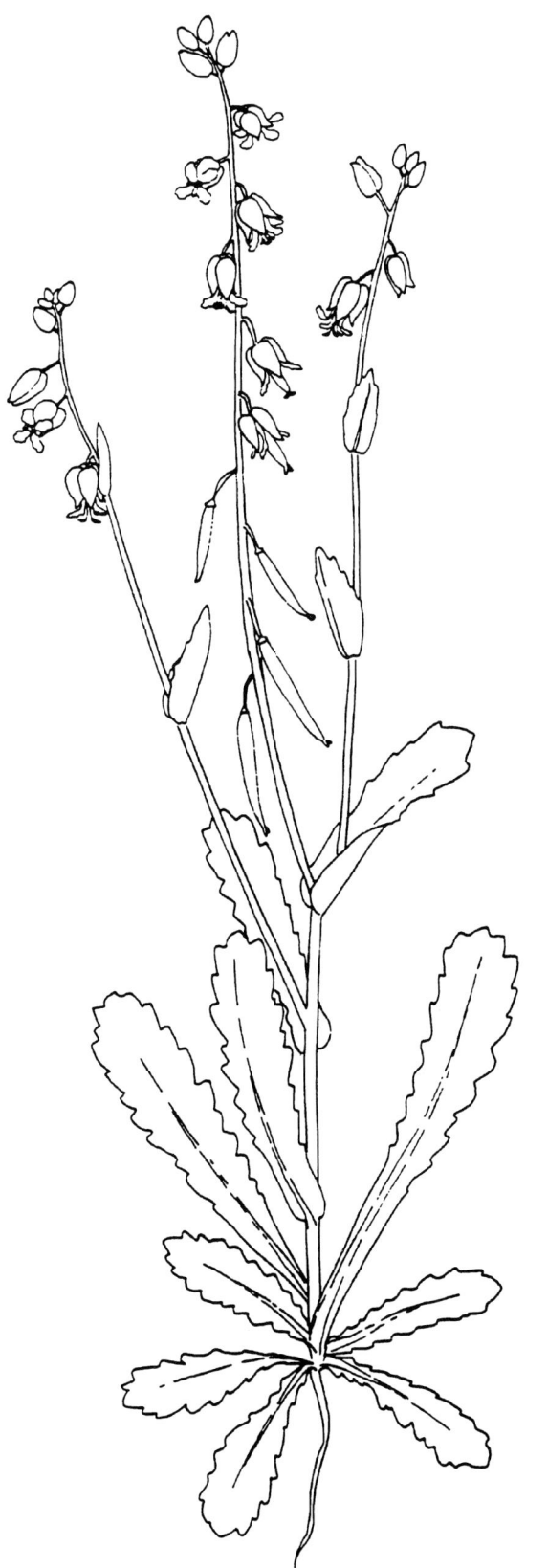

Caulanthus californicus
California jewelflower
by Mary Ann Showers

CHAPTER 5. Basic Principles of Sampling

A. Introduction

What is sampling? A review of several dictionary definitions led to the following composite definition:

> The act or process of selecting a part of something with the intent of showing the quality, style, or nature of the whole.

Monitoring does not always involve sampling techniques. Sometimes you can count or measure all individuals within a population of interest. Other times you may select qualitative techniques that are not intended to show the quality, style, or nature of the whole population (e.g., subjectively positioned photoplots).

What about those situations where you have an interest in learning something about the entire population, but where counting or measuring all individuals is not practical? This situation calls for sampling. The role of sampling is to provide information about the population in such a way that inferences about the total population can be made. This inference is the process of generalizing to the population from the sample, usually with the inclusion of some measure of the "goodness" of the generalization (McCall 1982).

Sampling will not only reduce the amount of work and cost associated with characterizing a population, but sampling can also increase the accuracy of the data gathered. Some kinds of errors are inherent in all data collection procedures, and by focusing on a smaller fraction of the population, more attention can be directed toward improving the accuracy of the data collected.

This chapter includes information on basic principles of sampling. Commonly used sampling terminology is defined and the principal concepts of sampling are described and illustrated. Even though the examples used in this chapter are based on counts of plants in quadrats (density measurements), most of the concepts apply to all kinds of sampling.

B. Populations and Samples

The term "population" has both a biological definition and a statistical definition. In this chapter and in Chapter 7, we will be using the term "population" to refer to the statistical population or the "sampling universe" in which monitoring takes place. The statistical population will sometimes include the entire biological population, and other times, some portion of the biological population. The *population* consists of the complete set of individual objects about which you want to make inferences. We will refer to these individual objects as *sampling units*. The sampling units can be individual plants or they may be quadrats (plots), points, or transects. The *sample* is simply part of the population, a subset of the total possible number of sampling units. These terms can be clarified in reference to an artificial plant population shown in Figure 5.1. There are a total of 400 plants in this population, distributed in 20 patches of 20 plants each. All the plants are contained within the boundaries of a 20m x 20m "macroplot." The collection of plants in this macroplot population will be referred to as the "400-plant population." A random arrangement of ten 2m x 2m quadrats positioned within the 400-plant population is shown in

Figure 5.1. Counts of plants within the individual quadrats are directed at the objective of estimating the total number of plants in the 20m x 20m macroplot. The sampling unit in this case is the 2m x 2m quadrat. The sample shown in Figure 5.1 is a set of 10 randomly selected quadrats. The population in this case is the total collection of all possible 2m x 2m quadrats that could be placed in the macroplot (N=100).

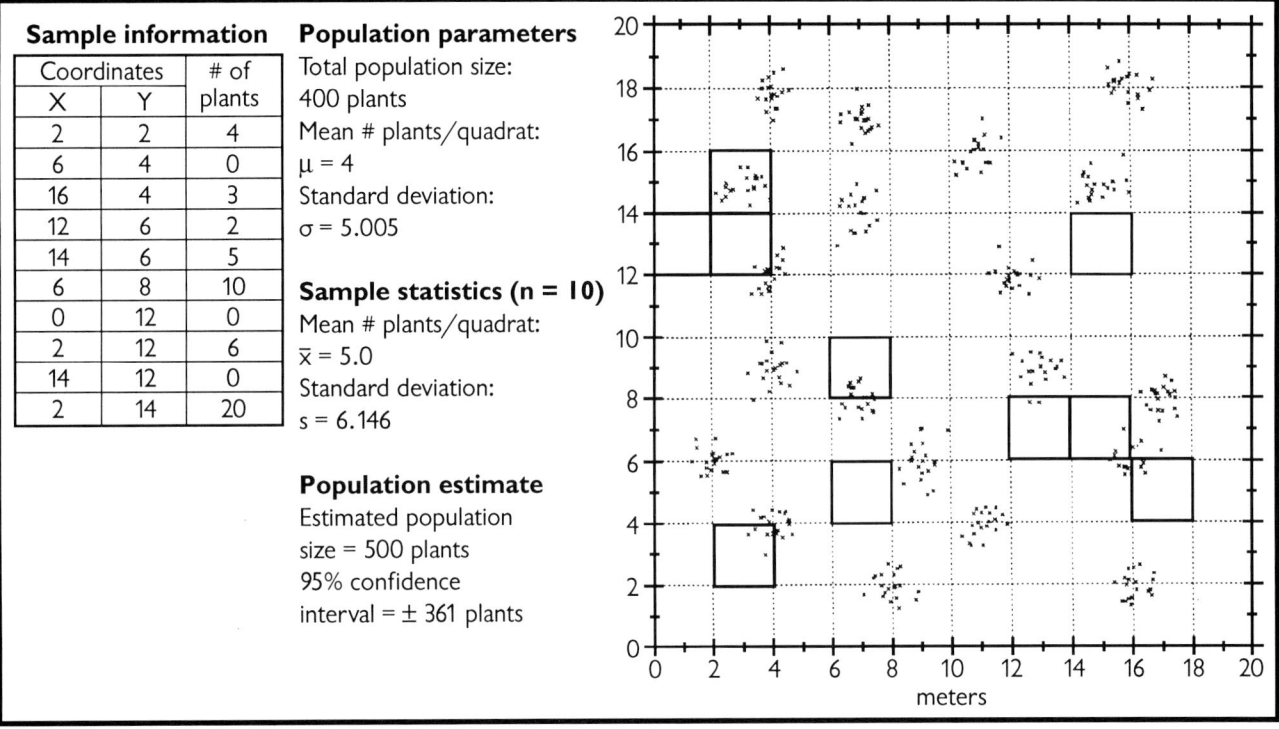

Sample information

Coordinates		# of
X	Y	plants
2	2	4
6	4	0
16	4	3
12	6	2
14	6	5
6	8	10
0	12	0
2	12	6
14	12	0
2	14	20

Population parameters

Total population size: 400 plants

Mean # plants/quadrat: $\mu = 4$

Standard deviation: $\sigma = 5.005$

Sample statistics (n = 10)

Mean # plants/quadrat: $\bar{x} = 5.0$

Standard deviation: $s = 6.146$

Population estimate

Estimated population size = 500 plants

95% confidence interval = ± 361 plants

FIGURE 5.1. Population of 400 plants distributed in 20 clumps of 20 plants. This figure shows a simple random sample of ten 2m x 2m quadrats, along with sample statistics and true population parameters.

C. Population Parameters vs. Sample Statistics

Population parameters are descriptive measures which characterize the population and are assumed to be fixed but unknown quantities that change only if the population changes. Greek letters such as μ and σ are often used to denote parameters. If we count all the plants in all the quadrats that make up the 400-plant population shown in Figure 5.1 (400 plants), and divide by the total number of possible 2m x 2m quadrat locations in the macroplot (100 quadrats), we obtain the true average number of plants per quadrat (4 plants/quadrat). This, assuming we have made no errors, is the *true population mean (μ)*. If we know how much each individual quadrat differs from the true population mean, we can calculate another important population parameter, the *true population standard deviation (σ)*. The standard deviation is a measure of how similar each individual observation is to the overall mean and is the most common measure of variability used in statistics. Populations with a large amount of variation among possible sampling units will have a larger standard deviation than populations with sampling units that are more similar to one another.

Sample statistics are descriptive measures derived from a sample (e.g., 10 of the possible 100 2m x 2m quadrats). Sample statistics provide estimates of population parameters. Sample statistics will vary from sample to sample in addition to changing whenever the underlying population changes. Roman letters such as $\bar{X}$ and s are usually used for sample statistics. Consider the following simple example where a sample of three sampling units yields values of 9, 10, and 14 plants/quadrat:

The *sample mean* ($\overline{X}$) = (9+10+14)/3 = 11 plants/quadrat

We could also calculate from this sample a *sample standard deviation (s)*. The sample standard deviation describes how similar each individual observation is to the sample mean. The derivation of a standard deviation (in case you want to calculate one by hand) is provided in Appendix 8. The standard deviation is easily calculated with a simple hand calculator using the "s" or "σ_{n-1}" key.

The standard deviation (s) for the simple example above is 2.65 plants/quadrat.

Consider another simple example with sampling unit values of 2, 10, and 21 plants/quadrat.

The mean ($\overline{X}$) = (2+10+21)/3 = 11 plants/quadrat

The standard deviation (s) for this example is 9.54 plants/quadrat.

Thus, both examples have a sample mean of 11 plants/quadrat, but the second one has a higher standard deviation (9.54 plants/quadrat) than the first (2.65 plants/quadrat), because the individual quadrat values differ more from one another in the second example.

In the example shown in Figure 5.1, the true population mean is 4.00 plants/quadrat, whereas the sample mean is 5.00 plants/quadrat. The true population standard deviation is 5.005 plants/quadrat, whereas the sample standard deviation is 6.146 plants/quadrat.

D. Accuracy vs. Precision

Accuracy is the closeness of a measured or computed value to its true value. *Precision* is the closeness of repeated measurements of the same quantity. A simple example will help illustrate the difference between these two terms. Two quartz clocks, equally capable of tracking time, are sitting side-by-side on a table. Someone comes by and advances one of the clocks by 1 hour. Both clocks will be equally "precise" at tracking time, but one of them will not be "accurate."

Efficient sampling designs try to achieve high precision. When we sample to estimate some population parameter, our sample standard deviation gives us a measure of the repeatability, or precision of our sample; it does not allow us to assess the accuracy of our sample. If counts of plants within different quadrats of a sample are similar to one another (e.g., the example above with a mean of 11 and a standard deviation = 2.65) then it is likely that different independent samples from the same population will yield similar sample means and give us high precision. When quadrat counts within a sample are highly variable (e.g., the example above with a mean of 11 and a standard deviation of 9.54), individual sample means from separate independent samples may be very different from one another giving us low precision. In either case, if the counting process is biased (perhaps certain color morphs or growth forms of individuals are overlooked), results may be inaccurate.

E. Sampling vs. Nonsampling errors

In any monitoring study errors should be minimized. Two categories of errors are described next.

1. Sampling errors

Sampling errors result from chance; they occur when sample information does not reflect the true population information. These errors are introduced by measuring only a subset of all the sampling units in a population.

Sampling errors are illustrated in Figure 5.2, in which two separate, completely random samples (2A and 2B) are taken from the 400-plant population shown in Figure 5.1. In each case, ten 2m x 2m quadrats are sampled and an estimate is made of the total number of plants within the population. The sample shown in Figure 5.2A produces a population estimate of only 80 plants, whereas the sample shown in Figure 5.2B yields an estimate of 960 plants. Both estimates are poor because of sampling error (chance placement of the quadrats resulted in severe under- or overestimates of the true population total).

You can imagine the problems that can arise if you monitor the same population two years in a row and get sample information that indicates that the population shifted from 960 plants to 80 plants when it really didn't change at all. Sampling errors can lead to two kinds of mistakes: (1) missing real changes (missed-change errors), and (2) detecting apparent changes that don't really exist (false-change errors).

Sampling errors can be estimated from the sampling data. Some of the basic sampling design tools covered in Chapter 7, enable you to evaluate the effectiveness of your monitoring study by taking a closer look at the sampling data. This can be especially helpful when setting up new projects; an evaluation of pilot sampling data can point out potential sampling error problems, enabling an investigator to fix them at an early stage of the project. Good sampling designs can reduce sampling errors without increasing the cost of sampling.

2. Nonsampling errors

Nonsampling errors are errors associated with human, rather than chance, mistakes. Examples of nonsampling errors include:

◆ Using biased selection rules, such as selecting "representative samples" by subjectively locating sampling units, or by substituting sampling units that are "easier" to measure.

◆ Using vegetation measurement or counting techniques within sampling units in which attributes cannot be accurately counted or measured. For example, counts of grass stems within a quadrat with counts in the hundreds may lead to numerous counting errors.

◆ Inconsistent field sampling effort. Nonsampling errors can be introduced if different investigators use different levels of effort (e.g., one investigator makes counts from "eye-level," whereas another counts by kneeling next to the quadrat).

◆ Transcription and recording errors. Nonsampling errors can be introduced if the data recorder's "7's" look like "1's" to the person entering the data.

◆ Incorrect or inconsistent species identification. This category also includes biases introduced by missing certain size classes or color morphs.

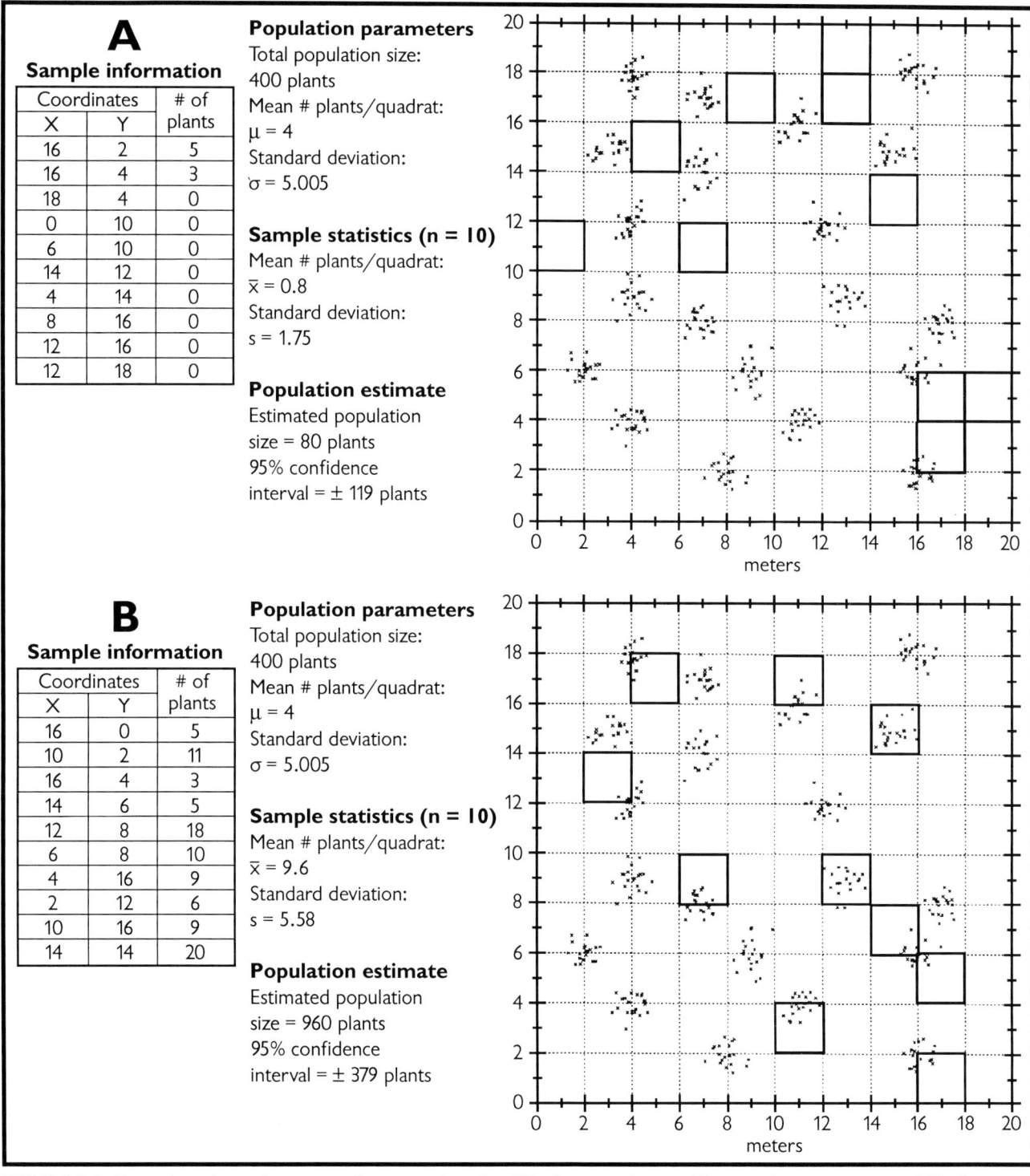

A

Sample information

| Coordinates | | # of |
X	Y	plants
16	2	5
16	4	3
18	4	0
0	10	0
6	10	0
14	12	0
4	14	0
8	16	0
12	16	0
12	18	0

Population parameters
Total population size:
400 plants
Mean # plants/quadrat:
$\mu = 4$
Standard deviation:
$\sigma = 5.005$

Sample statistics (n = 10)
Mean # plants/quadrat:
$\bar{x} = 0.8$
Standard deviation:
$s = 1.75$

Population estimate
Estimated population
size = 80 plants
95% confidence
interval = ± 119 plants

B

Sample information

| Coordinates | | # of |
X	Y	plants
16	0	5
10	2	11
16	4	3
14	6	5
12	8	18
6	8	10
4	16	9
2	12	6
10	16	9
14	14	20

Population parameters
Total population size:
400 plants
Mean # plants/quadrat:
$\mu = 4$
Standard deviation:
$\sigma = 5.005$

Sample statistics (n = 10)
Mean # plants/quadrat:
$\bar{x} = 9.6$
Standard deviation:
$s = 5.58$

Population estimate
Estimated population
size = 960 plants
95% confidence
interval = ± 379 plants

FIGURE 5.2. Examples of sampling errors from sampling the 400-plant population. The population estimates of 80 plants and 960 plants are far from the true population of 400 plants.

Because sampling designs are based on the assumption that nonsampling errors are zero, the number of nonsampling errors needs to be minimized. Ensure that your sampling unit makes sense for the type of vegetation measurement technique you have selected. When different personnel are used in the same monitoring study, conduct rigorous training and testing to ensure consistency in counts or measurements. Design field data forms (Chapter 9) that are easy to use and easy for data transcribers to interpret. Proof all data entered into computer programs to ensure that entered numbers are correct. In contrast to sampling errors, the probability of nonsampling errors occurring cannot be assessed from pilot sample data.

F. Sampling Distributions

One way of evaluating the risk of obtaining a sample value that is vastly different than the true value (such as population estimates of 80 or 960 plants when the true population is 400 plants) is to sample a population repeatedly and look at the differences among the repeated population estimates. If almost all the separate, independently derived population estimates are similar, then you know you have a good sampling design with high precision. If many of the independent population estimates are not similar, then you know your precision is low.

The 400-plant population can be resampled by erasing the 10 quadrats (as shown in either Figure 5.1 or Figure 5.2) and putting 10 more down in new random positions. We can keep repeating this procedure, each time writing down the sample mean. Plotting the results of a large number of individual sample means in a simple histogram graph yields a *sampling distribution*. A sampling distribution is a distribution of many independently gathered sample statistics (most often a distribution of sample means). Under most circumstances, this distribution of sample means fits a normal or bell-shaped curve.

A distribution of population size estimates from 10,000 separate random samples using ten 2m x 2m quadrats from the 400-plant population is shown in Figure 5.3A. The x-axis shows the range of different population estimates, and the y-axis shows the relative and actual frequency of the different population estimates. Think of this as the results of 10,000 different people sampling the same population on the same day, each one setting out 10 randomly positioned 2m x 2m quadrats (somehow without negatively impacting the population) and coming up with their own independent population estimate. The highest population estimate out of the 10,000 separate samples was 960 plants and the lowest population estimate was zero (four of the 10,000 samples yielded a population estimate of zero). The shape of this distribution indicates the magnitude of likely sampling errors. Wide distributions could yield population estimates that are "far" from the true population value. A sampling design that led to the type of sampling distribution depicted in Figure 5.3A would not be useful since few of the estimates approach the true population size of 400 plants. *One of the principal objectives in sampling design is to make the shape of sampling distributions as narrow as possible.*

Fortunately, you do not have to repeatedly sample your population and see how wide your sampling distribution is to determine if you need to change anything. There are some simple statistical tools that provide a convenient shortcut for evaluating the precision of your sampling effort from a single sample. These tools involve calculating standard errors and confidence intervals to estimate sampling precision levels.

1. Standard error

A *standard error* is the standard deviation of a large number of independent sample means. It is a measure of precision that you derive from a single sample. The formula for calculating a standard error is as follows:

> **Formula for standard error:**
>
> $$SE = \frac{s}{\sqrt{n}}$$
>
> Where: SE = Standard error
> s = Standard deviation
> n = Sample size

To paraphrase the earlier statement regarding an important objective of sampling design, *one of the principal objectives in sampling design is to reduce the size of the standard error.*

This formula demonstrates that there are only two ways of minimizing standard errors, either: (1) increase sample size (n), or (2) decrease the standard deviation (s).

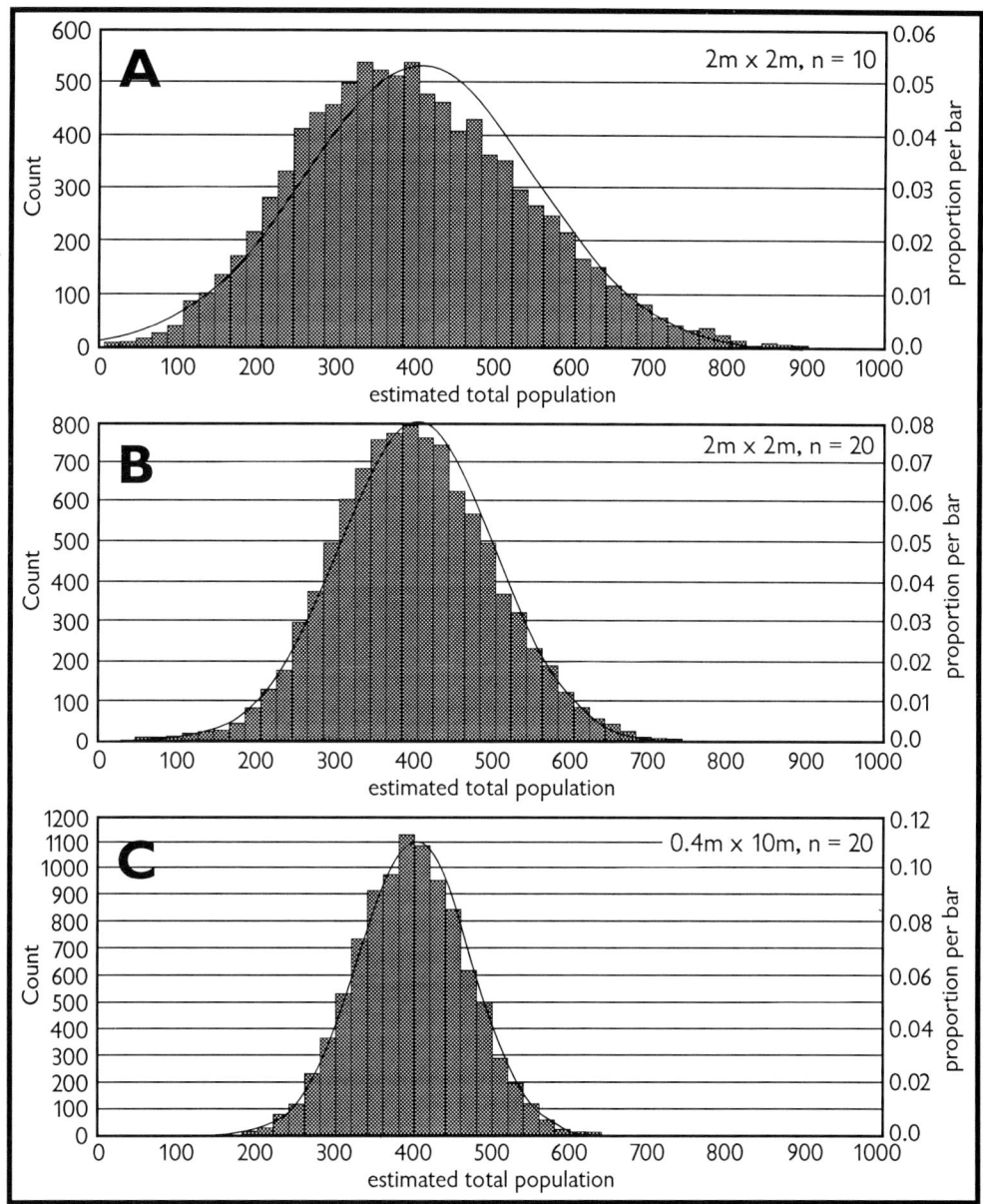

FIGURE 5.3. Sampling distributions from three separate sampling designs used on the 400-plant population. All distributions were created by sampling the population 10,000 separate times. The smooth lines show a normal bell-shaped curve fit to the data. Figure 3A shows a sampling distribution where ten 2m x 2m quadrats were used. Figure 3B shows a sampling distribution where twenty 2m x 2m quadrats were used. Figure 3C shows a sampling distribution where twenty 0.4m x 10m quadrats were used.

◆ **Increase sample size.** A new sampling distribution of 10,000 separate random samples drawn from our example population is shown in Figure 5.3B. This distribution came from randomly drawing samples of *twenty* 2m x 2m quadrats instead of the *ten* quadrats used to create the sampling distribution in Figure 5.3A. This increase in sample size from 10 to 20 provides a 29.3% improvement in precision (as measured by the reduced size of the standard error).

◆ **Decrease sample standard deviation.** Another sampling distribution of 10,000 separate random samples drawn from our 400-plant population is shown in Figure 5.3C. The sampling design used to create this distribution of population estimates is similar to the one used to create the sampling distribution in Figure 5.3B. The only difference between the two designs is in quadrat shape. The sampling distribution shown in Figure 5.3B came from using twenty 2m x 2m quadrats; the sampling distribution shown in Figure 5.3C came from using twenty 0.4m x 10m quadrats. This change in quadrat shape reduced the true population standard deviation from 5.005 plants to 3.551 plants. This change in quadrat shape led to a 29.0% improvement in precision over the 2m x 2m design shown in Figure 5.3B (as measured by the reduced size of the standard error). This 29.0% improvement in precision came without changing the sampling unit size ($4m^2$) or the number of quadrats sampled (n=20); only the quadrat shape (from square to rectangular) changed. When compared to the original sampling design of ten 2m x 2m quadrats, the twenty 0.4m x 10m quadrat design led to a 49.8% improvement in precision. Details of this method and other methods of reducing sample standard deviation are covered in Chapter 7.

How is the standard error most often used to report the precision level of sampling data? Sometimes the standard error is reported directly. You may see tables with standard errors reported or graphs that include error bars that show ± 1 standard error. Often, however, the standard error is multiplied by a coefficient that converts the number into something called a confidence interval.

2. Confidence intervals

A confidence interval provides an estimate of precision around a sample mean, a sample proportion, or an estimate of total population size that specifies the likelihood that the interval includes the true value. The vertical lines marked with the "95%" in Figure 5.4 indicate that 95% of all the samples (9,500 out of the 10,000) fit between these two lines. Five percent of the samples (2.5% in each tail of the distribution) fall outside the vertical lines. These lines

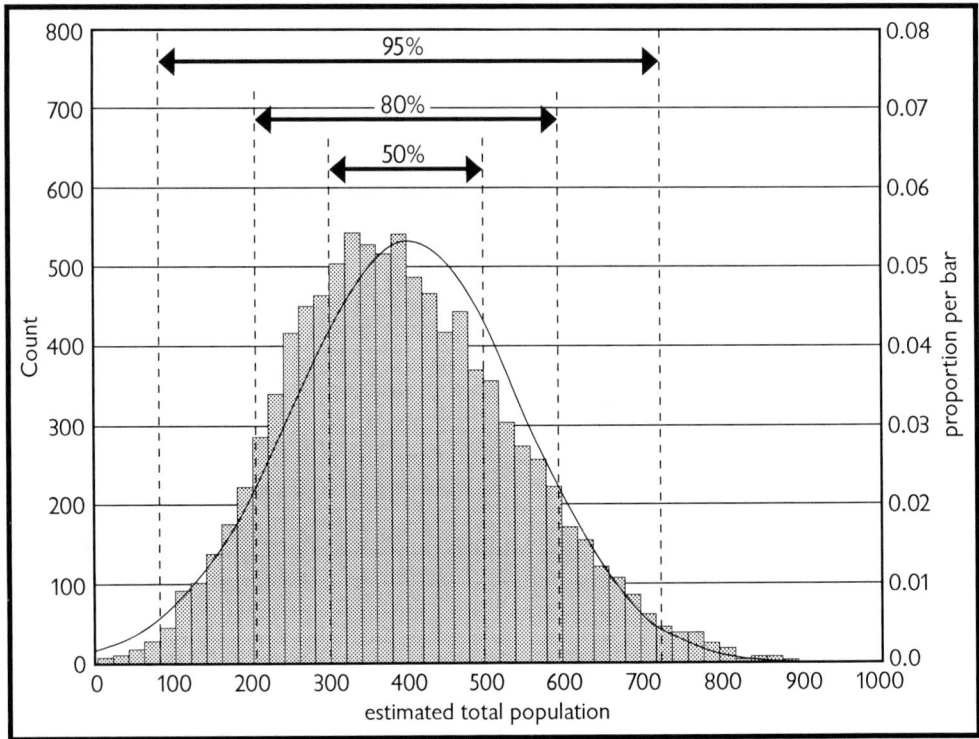

FIGURE 5.4. Distribution from sampling the 400-plant population 10,000 times using ten samples of 2m x 2m quadrats. The 95%, 80%, and 50% confidence intervals around the true population of 400 plants are shown. The smooth line shows a normal, bell-shape curve fit to the data.

are positioned equally from the center of the sampling distribution, approximately 320 plants away from the center of 400 plants. Thus, 95% of all samples are within ± 320 plants of the true population size.

A confidence interval includes two components: (1) the *confidence interval width* (e.g., ± 320 plants); and (2) the *confidence level* (e.g., 90%, 95%). The confidence level indicates the probability that the interval includes the true value. Confidence interval width decreases as the confidence level decreases. This relationship is shown in Figure 5.4 where three different confidence levels are graphed on the sampling distribution obtained by sampling the 400-plant population with ten 2m x 2m quadrats. These same three confidence intervals are shown again in Figure 5.5A, where they are graphed in a format commonly used to report confidence intervals. There is no gain in precision associated with the narrowing of confidence interval width as you go from left to right in Figure 5.5A (i.e., from 95% confidence, to 80% confidence, to 50% confidence); only the probability that the confidence interval includes the true value is altered. Another set of three confidence intervals is shown in Figure 5.5B. Like Figure 5.5A, confidence intervals get narrower as we move from left to right in the graph, but this time the confidence level is the same (95%) and the narrower widths came from using different sampling designs. There is a gain in precision associated with the narrowing of confidence interval width as you go from left to right in Figure 5.5B (i.e., from the ten 2m x 2m design to the twenty 2m x 2m design to the twenty 0.4m x 10m design) because we have reduced the uncertainty of our population estimate by tightening the confidence interval width at the same confidence level.

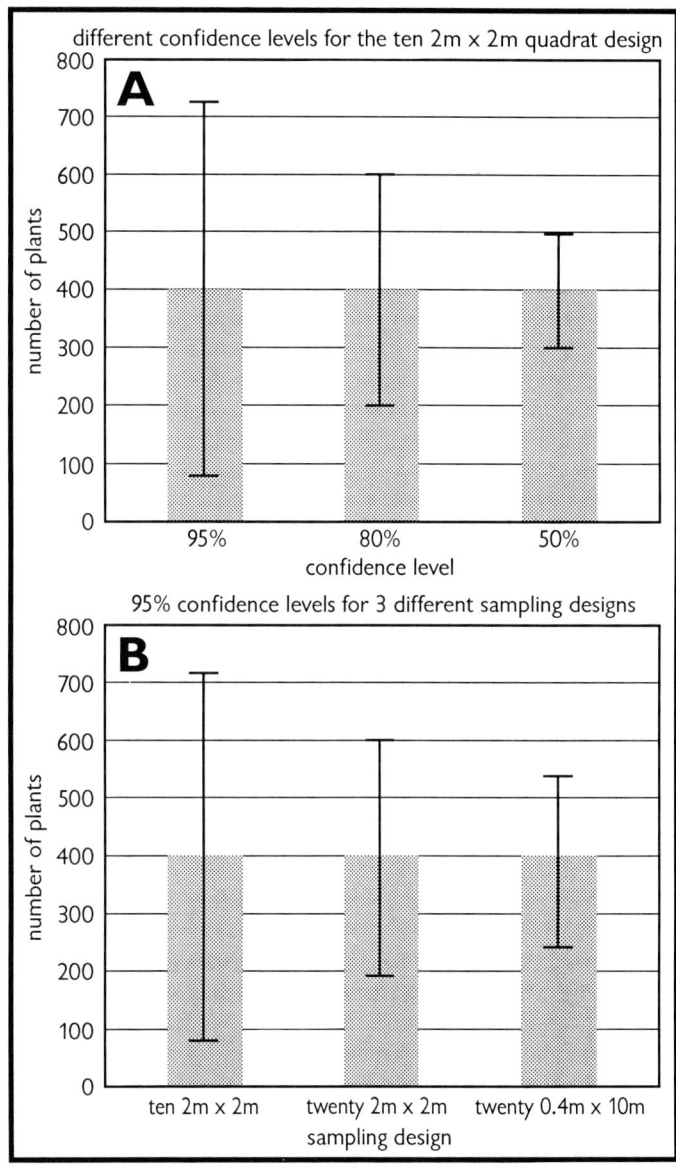

FIGURE 5.5. Comparison of confidence intervals and confidence levels for different sampling designs from the 400-plant population. Figure 5A shows three different confidence levels (95%, 80%, and 50%) for the same data set based upon sampling ten 2m x 2m quadrats. Figure 5B shows 95% confidence intervals for three different sampling designs that differ in the level of precision of the population estimates.

In order to calculate confidence intervals for sample means, we need two values: (1) the standard error calculated according to the above formula (SE = $s/\sqrt{n}$), and (2) the corresponding value from a table of critical values of the *t* distribution (see Appendix 8 for instructions on calculating confidence intervals around proportions). The confidence interval half-width, extending an equal distance on both sides of the mean, is the standard error × the critical *t* value (except when sampling from finite populations, see below). The appropriate critical value of t depends on the level of confidence desired and the number of sampling units (n) in the sample. A table of critical values for the t distribution (Zar 1996) is found in Appendix 5. To use this table, you must first select the appropriate confidence level column. If you want to be 95% confident that your confidence interval includes the true mean, use the column headed $\alpha(2) = 0.05$. For 90% confidence, use the column headed $\alpha(2) = 0.10$. You use $\alpha(2)$ because you are interested in a confidence interval on both sides of the mean. You then use the row indicating the number of degrees of freedom (v), which is the number of sampling units minus one $(n\text{-}1)$.

For example, if we sample 20 quadrats and come up with a mean of 4.0 plants and a standard deviation of 5.005, here are the steps for calculating a 95% confidence interval around our sample mean:

The standard error (SE = $s/\sqrt{n}$) = $\dfrac{5.005}{\sqrt{20}}$ = 5.005/4.472 = 1.119.

The appropriate t value from Appendix 5 for 19 degrees of freedom (v) is 2.093.

One-half of our confidence interval width is then SE $\times$ t-value = 1.119 $\times$ 2.093 = 2.342.

Our 95% confidence interval can then be reported as 4.0 ± 2.34 plants/quadrat or we can report the entire confidence interval width from 1.66 to 6.34 plants/quadrat. This indicates a 95% chance that our interval from 1.66 plants/quadrat to 6.34 plants/quadrat includes the true value.

Another way to think of 95% confidence intervals calculated from sampling data is that the interval specifies a range that should include the true value 95% of the time. If you are calculating 95% confidence intervals and independently randomly sample a population 100 different times, you should see that approximately 95 of the intervals will include the true mean and 5 will miss it. This relationship is shown in Figure 5.6 where 100 independent population estimates are graphed with 95% confidence intervals from the 400-plant populations using samples of twenty 0.4m x 10m quadrats. You will notice that the solid dots, used to show each of the 100 population estimates, fluctuate around the true population value of 400 plants. You will also notice that 96 out of 100 confidence intervals shown in Figure 5.6 include the true value. If the confidence level had been set at 80%, then approximately 20 of the intervals would have failed to include the true value. A 99% confidence level would have led to approximately only one interval out of the 100 that did not include the true population size (in order to capture the true value more often, the individual confidence interval widths for a 99% confidence level are wider than the confidence interval widths for a 95% confidence level).

G. Finite vs. Infinite Populations

If we are sampling with quadrats, there is a finite number of quadrats that can be placed in the area to be sampled, assuming that no two quadrats overlap (this is called sampling without replacement). If the sampled area is large, then the number of quadrats placed in the area may be very large as well, but nonetheless finite. On the other hand, an infinite number of points or lines could be placed in the area to be sampled. This is because points, at least theoretically, are dimensionless, and lines are dimensionless in one direction. This means, at least for all practical purposes, that a population of points or of lines is infinite.

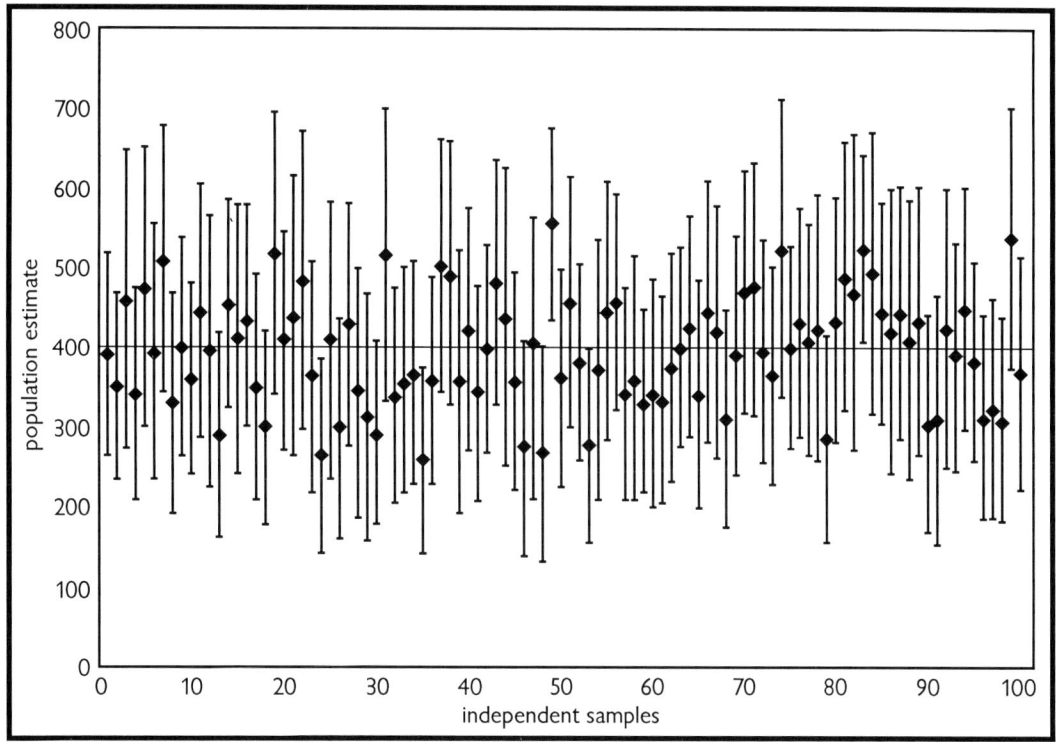

FIGURE 5.6. Population estimates from 100 separate random samples from the 400-plant population. Each sample represents the population estimate from sampling twenty 0.4m x 10m quadrats. The horizontal line through the graph indicates the true population of 400 plants. Vertical bars represent 95% confidence intervals. Four of the intervals miss the true population size.

If the area to be sampled is large relative to the area that is actually sampled, the distinction between finite and infinite is of only theoretical interest. When, however, the area sampled makes up a significant portion of the area to be sampled, we can apply the *finite population correction factor*, which reduces the size of the standard error. The most commonly used finite population correction factor is shown to the right:

Formula for the finite population correction:

$$FPC = \sqrt{\frac{N-n}{N}}$$

Where: N = total number of potential quadrat positions
n = number of quadrats sampled

When n is small relative to N, the equation is close to 1, whereas when n is large relative to N, the value approaches zero. The standard error $(s/\sqrt{n})$ is multiplied by the finite population correction factor to yield a corrected standard error for the finite population.

Consider the following example. The density of plant species X is estimated within a 20m x 50m macroplot (total area = 1000m²). This estimate is obtained by collecting data from randomly selected 1m x 10m quadrats (10m²). Sampling without replacement, there are 100 possible quadrat positions. $\frac{\text{total area}}{\text{quadrat area}} = \frac{1000m^2}{10m^2}$

Thus, our population, N, is 100. Let's say we take a random sample, n, of 30 of these quadrats and calculate a mean of eight plants per quadrat and a standard deviation of four plants per quadrat. Our standard error is thus: $s/\sqrt{n} = 4/\sqrt{30} = 0.73$. Although our sample mean is an unbiased estimator of the true population mean and needs no correction, the standard error should be corrected by the finite population correction factor shown on the top of page 72:

Example of applying the finite population correction factor:

$$SE' = (SE)\sqrt{\frac{N-n}{N}} \qquad SE' = (0.73)\sqrt{\frac{100-30}{100}} = 0.61$$

Where: SE' = corrected standard error
SE = uncorrected standard error
N = total number of potential quadrat positions
n = number of quadrats sampled

Since the standard error is one of the factors used to calculate confidence intervals (the other is the appropriate value of t from a t table), correcting the standard error with the finite population correction factor makes the resulting confidence interval narrower. It does this, however, *only* if n is sufficiently large relative to N. A rule of thumb is that unless the ratio n/N is greater than .05 (i.e., you are sampling more than 5% of the population area), there is little to be gained by applying the finite population correction factor to your standard error.

The finite population correction factor is also important in sample size determination (Chapter 7) and in adjusting test statistics (Chapter 11). The finite population correction factor works, however, only with finite populations, which we will have when using quadrats, but will not have when using points or lines.

H. False-Change Errors and Statistical Power Considerations

These terms relate to situations where two or more sample means or proportions are being compared with some statistical test. This comparison may be between two or more places or the same place between two or more time periods. These terms are pertinent to both planning and interpretation stages of a monitoring study. Consider a simple example where you have sampled a population in two different years and now you want to determine whether a change took place between the two years. You usually start with the assumption, called the null hypothesis, that no change has taken place. There are two types of decisions that you can make when interpreting the results of a monitoring study: (1) you can decide that a change took place, or (2) you can decide that no change took place. In either case, you can either be right, or you can be wrong (Figure 5.7).

monitoring for change — possible errors		
	no change has taken place	there has been a real change
monitoring system detects a change	false-change error (Type I) α	no error (Power) $1 - \beta$
monitoring system detects no change	no error $(1 - \alpha)$	missed-change error (Type II) β

FIGURE 5.7. Four possible outcomes for a statistical test of some null hypothesis, depending on the true state of nature.

1. The change decision and false-change errors

The conclusion that a change took place may lead to some kind of action. For example, if a population of a rare plant is thought to have declined, a change in management may be needed. If a change did not actually occur, this constitutes a false-change error, a sort of false alarm. Controlling this type of error is important because taking action unnecessarily can be expensive (e.g., a range permittee is not going to want to change the grazing intensity if a decline in a rare plant population really didn't take place). There will be a certain probability of concluding that a change took place even if no difference actually occurred. The probability of this occurring is usually labeled the *P*-value, and it is one of the types of information that comes out of a statistical analysis of the data. The *P*-value reports the likelihood that the observed difference was due to chance alone. For example, if a statistical test comparing two sample means yields a *P*-value of 0.24 this indicates that there is a 24% chance of obtaining the observed result even if there is no true difference between the two sample means.

Some threshold value for this false-change error rate should be set in advance so that the *P*-value from a statistical test can be evaluated relative to the threshold. *P*-values from a statistical test that are smaller than or equal to the threshold are considered statistically "significant," whereas *P*-values that are larger than the threshold are considered statistically "nonsignificant." Statistically significant differences may or may not be ecologically significant depending upon the magnitude of difference between the two values. The most commonly cited threshold for false-change errors is the 0.05 level; however, there is no reason to arbitrarily adopt the 0.05 level as the appropriate threshold. The decision of what false-change error threshold to set depends upon the relative costs of making this type of mistake and the impact of this error level on the other type of mistake, a missed-change error (see below).

2. The no-change decision, missed-change errors, and statistical power

The conclusion that no change took place usually does not lead to changes in management practices. Failing to detect a true change constitutes a *missed-change error*. Controlling this type of error is important because failing to take action when a true change actually occurred may lead to the serious decline of a rare plant population.

Statistical power is the complement of the missed-change error rate (e.g., a missed-change error rate of 0.25 gives you a power of 0.75; a missed-change error rate of 0.05 gives you a power of 0.95). High power (a value close to 1), is desirable and corresponds to a low risk of a missed-change error. Low power (a value close to 0) is not desirable because it corresponds to a high risk of a missed-change error.

Since power levels are directly related to missed-change error levels, either level can be reported and the other level can be easily calculated. Power levels are often reported instead of missed-change error levels, because it seems easier to convey this concept in terms of the certainty of detecting real changes. For example, the statement "I want to be at least 90% certain of detecting a real change of five plants/quadrat" (power is 0.90) is simpler to understand than the statement "I want the probability of missing a real change of five plants/quadrat to be 10% or less" (missed-change error rate is 0.10).

An assessment of statistical power or missed-change errors has been virtually ignored in the field of environmental monitoring. A survey of over 400 papers in fisheries and aquatic sciences found that 98% of the articles that reported nonsignificant results failed to report any power results (Peterman 1990). A separate survey, reviewing toxicology literature, found high power in only 19 out of 668 reports that failed to reject the null hypothesis (Hayes 1987). Similar surveys in other fields such as psychology or education have turned up "depressingly low" levels of power (Brewer 1972; Cohen 1988).

3. Minimum detectable change

Another sampling design concept that is directly related to statistical power and false-change error rates is the size of the change that you want to be able to detect. This will be referred to as the *minimum detectable change or MDC*. The MDC greatly influences power levels. A particular sampling design will be more likely to detect a true large change (i.e., with high power) than to detect a true small change (i.e., with low power).

Setting the MDC requires the consideration of ecological information for the species being monitored. How large of a change should be considered biologically meaningful? With a large enough sample size, statistically significant changes can be detected for changes that have no biological significance. For example, if an intensive monitoring design leads to the conclusion that the mean density of a plant population increased from 10.0 plants/m^2 to 10.1 plants/m^2, does this represent some biologically meaningful change in population density? Probably not.

Setting a reasonable MDC can be difficult when little is known about the natural history of a particular plant species. Should a 30% change in the mean density of a rare plant population be cause for alarm? What about a 20% change or a 10% change? The MDC considerations are likely to vary when assessing vegetation attributes other than density, such as cover or frequency (Chapter 8). The initial MDC, set during the design of a new monitoring study, can be modified once monitoring information demonstrates the size and rate of population fluctuations.

4. How to achieve high statistical power

Statistical power is related to four separate sampling design components by the following function equation:

Power = a function of (s, n, MDC, and α)

where: s = standard deviation
    n = number of sampling units
    MDC = minimum detectable change
    α = false-change error rate

Power can be increased in the following four ways:

1. **Reducing standard deviation.** This means altering the sampling design to reduce the amount of variation among sampling units (see Chapter 7).

2. **Increasing the number of sampling units sampled.** This method of increasing power is straightforward, but keep in mind that increasing n has less of an effect than decreasing s since the square root of sample size is used in the standard error equation (SE = $s/\sqrt{n}$).

3. **Increasing the acceptable level of false-change errors (α).**

4. **Increasing the MDC.**

Note that the first two ways of increasing power are related to making changes in the sampling design, whereas the other two ways are related to making changes in the sampling objective (see Chapter 6).

5. Graphical comparisons

As stated, power is driven by four different factors: standard deviation, sample size, minimum detectable change size, and false-change error rate. In this section we take a graphical look at how altering these factors changes power. The comparisons in this section are based upon sampling a fictitious plant population where we are interested in assessing plant density relative to an established threshold value of 25 plants/m^2. Any true population densities less than 25 plants/m^2 will trigger management action. We are only concerned with the question of whether the density is lower than 25 plants/m^2 and not whether the density is higher. In this example, our null hypothesis (H$_O$) is that the population density equals 25 plants/m^2 and our alternative hypothesis is that density is less than 25 plants/m^2. The density value of 25 plants/m^2 is the most critical single density value since it defines the lower limit of acceptable plant density.

The figures in this section are all based upon sampling distributions where we happen to know the true plant density. Recall that a sampling distribution is a bell-shaped curve that depicts the distribution of a large number of independently gathered sample statistics. A sampling distribution defines the range and relative probability of any possible sample mean. You are more likely to obtain sample means near the middle of the distribution than you are to obtain sample means near either tail of the distribution.

A sampling distribution based on sampling our fictitious population with a true mean density of 25 plants/m^2 is shown in Figure 5.8A. This distribution is based on a sampling design using thirty 1m x 1m quadrats where the true standard deviation is ± 20 plants/quadrat. If 1,000 different people randomly sample and calculate a sample mean based upon their 30 quadrat values, approximately half the individually drawn sample means will be less than 25 plants/m^2 and half will be greater than 25 plants/m^2. Approximately 40% of the samples will yield sample means less than or equal to 24 plants/m^2. A few of our 1,000 individuals will obtain estimates of the mean density that deviate from the true value by a large margin. One of the individuals will likely stand up and say, "my estimate of the mean density is 13 plants/m^2", even though the true density is actually 25 plants/m^2. As interpreters of the monitoring information, we would conclude that since 999 of the 1,000 people obtained estimates of the density that were greater than 13, the true density is probably not 13. Our best estimate of the true mean density will be the average of the 1,000 separate estimates (this average is likely to be extremely close to the actual true value).

Now that we have the benefit of 1,000 independent estimates of the true mean density, we can return to the population at a later time, take a single random sample of thirty 1m x 1m quadrats, calculate the sample mean, and then ask the question, "what is the probability of obtaining our sample mean value if the true population is still 25 plants/m^2?" If our sample mean density turns out to be 24 plants/m^2, would this lead to the conclusion that the population has crossed our threshold value? Seeing that our sample mean is lower than our target value might raise some concerns, but we have no objective basis to conclude that the true population is not, in fact, still actually 25 plants/m^2. We learned in the previous paragraph that a full 40% of possible samples are likely to yield mean densities of 24 plants/m^2 or less if the true mean is 25 plants/m^2. Thus, the probability of obtaining a single sample mean of 24 plants/m^2 or less when the true density is actually 25 plants/m^2 is approximately 0.40. Obtaining a sample mean of 24 plants/m^2 is consistent with the hypothesis that the true population density is actually 25 plants/m^2.

How small a sample mean do we need to obtain to feel confident that the population has indeed dropped below 25 plants/m^2? What will our interpretation be if we obtained a sample mean of 22 plants/m^2? Based upon our sampling distribution from the 1,000 people, the probability of obtaining an estimate of 22 plants/m^2 or less is around 20%, which represents a 1-in-5 chance that the true mean is still actually 25 plants/m^2. Based upon the sampling distribution from our 1,000 separate samplers, we can look at the likelihood of obtaining other different sample means. The probability of obtaining a sample of 20 plants/m^2 is 8.5%,

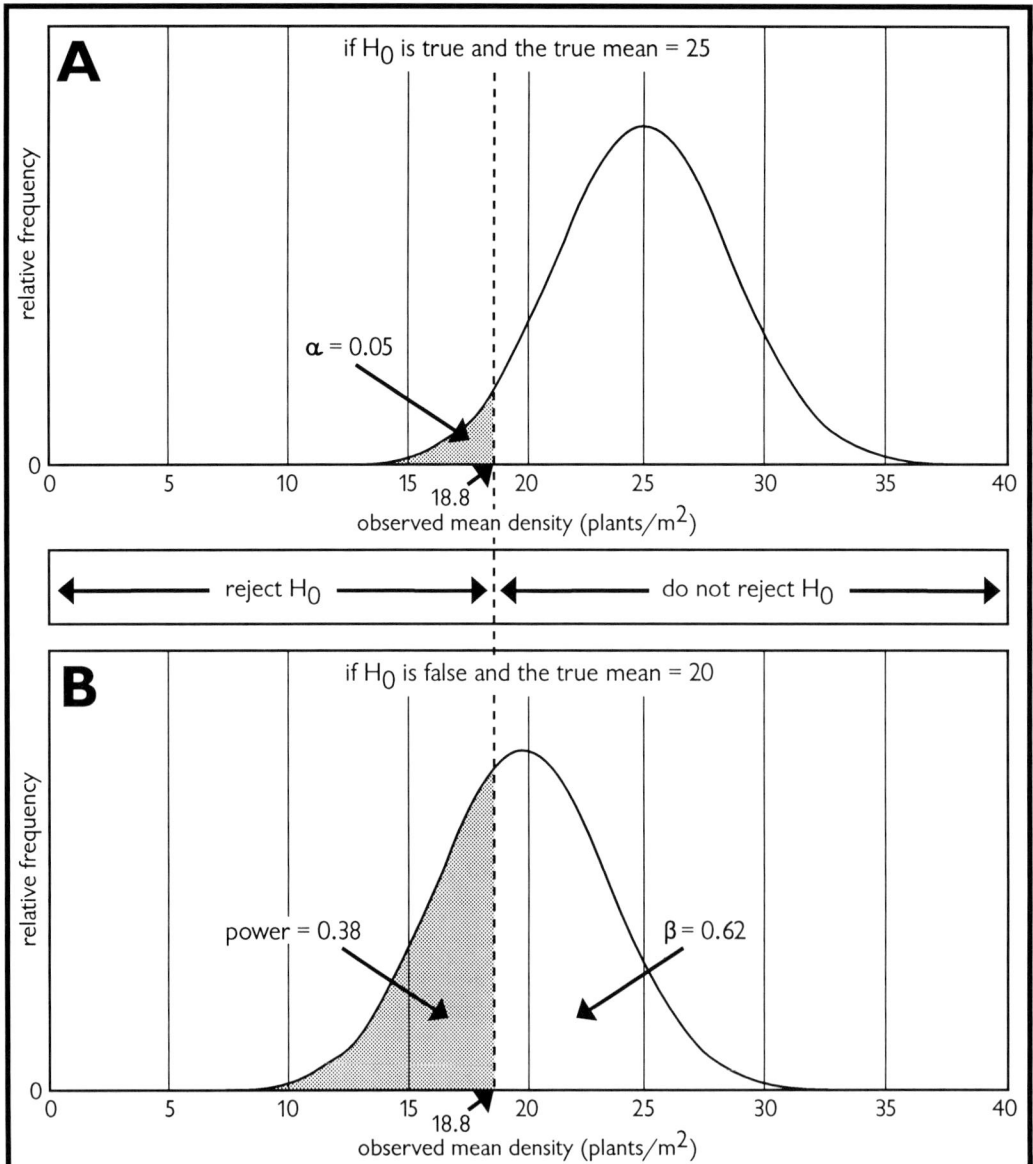

FIGURE 5.8. Example of sampling distributions for mean plant density in samples of 30 permanent quadrats where the among-quadrat standard deviation is 20 plants/m². Part A is the sampling distribution for the case in which the null hypothesis, H_0, is true and the true population mean density is 25 plants/m². The shaded area in part A is the critical region for $\alpha = 0.05$ and the vertical dashed line is at the critical sample mean value, 18.8. Part B is the sampling distribution for the case in which the H_0 is false and the true mean is 20 plants/m². In both distributions, a sample mean to the left of the vertical dashed line would reject H_0, and to the right of it, would not reject H_0. Power and β values in part B, in which H_0 is false and the true mean = 20, are the proportion of sample means that would occur in the region in which H_0 was rejected or not rejected, respectively (adapted from Peterman 1990).

and the probability of obtaining a sample of 18 plants/m² is 2.9% if the true mean density is 25 plants/m².

Since in most circumstances we will only have the results from a single sample (and not the benefit of 1,000 independently gathered sample means), another technique must be used to determine whether the population density has dropped below 25 plants/m². One method is to run a statistical test that compares our sample mean to our density threshold value (25 plants/m^2). The statistical test will yield a P-value that defines the probability of obtaining our sample mean if the true population density is actually 25 plants/m^2. As interpreters of our monitoring information, we will need to set some probability threshold P-value to guide our interpretation of the results from the statistical test. This P-value threshold defines our acceptable false-change error rate. If we run a statistical test that compares our sample mean to our density threshold value (25 plants/m^2) and the P-value from the test is lower than our threshold value, then we conclude that the population density has, in fact, declined below 25 plants/m^2. Thus, if we set our P-value threshold to 0.05 and the statistical test yields a P-value of 0.40, then we *fail to reject* the null hypothesis that the true population density is 25 plants/m^2. If, however, the statistical test yields a P-value of 0.022, this is lower than our threshold P-value of 0.05, and we would reject the null hypothesis that the population is 25 plants/m^2 in favor of our alternative hypothesis that the density is lower than 25 plants/m^2.

The relationship between the P-value threshold of 0.05 and our sampling distribution based upon sampling thirty 1m x 1m quadrats is shown in Figure 5.8A. The threshold density value corresponding to our P-value threshold of 0.05 is 18.8 plants/m^2, which is indicated on the sampling distribution by the dashed vertical line. Thus, if we obtain a mean density of 18 plants/m^2, which is to the left of the vertical line, we reject the null hypothesis that the population density is 25 plants/m^2 in favor of an alternative hypothesis that density is lower than 25 plants/m^2. If we obtain a mean density of 21 plants/m^2, which is to the right of the vertical line, then we fail to reject the null hypothesis that the population density is really 25 plants/m^2.

So far we have been discussing the situation where the true population density is right at the threshold density of 25 plants/m^2. Let's look now at a situation where we know the true density has declined to 20 plants/m^2. What is the likelihood of our detecting this true density difference of 5 plants/m^2? Figure 5.8B shows a new sampling distribution based upon the true density of 20 plants/m^2 (standard deviation is still ± 20 plants/m^2). We know from our previous discussion that sample means to the right of the vertical line in Figure 5.8A lead to the conclusion that we can't reject the null hypothesis that our density is 25 plants/m^2. If our new sample mean turns out to exactly match the new true population mean (i.e., 20 plants/m^2), will we reject the idea that the sample actually came from a population with a true mean of 25 plants/m^2? No, at least not at our stated P-value (false-change error) threshold of 0.05. A sample mean value of 20 plants/m^2 falls to the right of our dashed threshold line in the "do not reject H$_O$" portion of the graph and we would have failed to detect the true difference that actually occurred. Thus, we would have committed a missed-change error.

What is the probability of missing the true difference of five plants/m^2 show in Figure 5.8B? This probability represents the missed-change error rate (β) and it is defined by the nonshaded area under the sampling distribution in Figure 5.8B, which represents 62% of the possible sample mean values. Recall that the area under the whole curve defines the entire range of possible values that you could obtain by sampling the population with the true mean = 20 plants/m^2. If we bring back our 1,000 sampling people and have each of them sample thirty 1m x 1m quadrats in our new population, we will find that approximately 620 of them will obtain estimates of the mean density that are greater than the 18.8 plants/m^2 threshold value that is shown by the vertical dashed line. What about the other 380 people? They will obtain population estimates fewer than the critical threshold of 18.8 plants/m^2 and they will reject the null hypothesis that the population equals 25 plants per quadrat. This proportion of 0.38 (380 people out of 1,000 people sampling) represents the *statistical power* of our sampling design and it is represented by the shaded area under the curve in Figure 5.8B. If the true population mean is indeed 20 plants/m^2 instead of 25 plants/m^2, then we can be 38%

(power = 0.38) sure that we will detect this true difference of five plants/m². With this particular sampling design (thirty 1m x 1m quadrats) and a false-change error rate of α=0.05, we run a 62% chance (β=0.62) that we will commit a missed-change error (i.e., fail to detect the true difference of five plants/m²). If the difference of five plants/m² is biologically important, a power of only 0.38 would not be satisfactory.

We can improve the low-power situation in four different ways: (1) increase the acceptable false-change error rate; (2) increase the acceptable MDC; (3) increase sample size; or (4) decrease the standard deviation. New paired sampling distributions illustrate the influence of making each of these changes.

a. Increasing the acceptable false-change error rate

In Figure 5.8B, a false-change error rate of α=0.05 resulted in a missed-change error rate of β=0.62 to detect a difference of five plants/m². Given these error rates, we are more than 12 times more likely to commit a missed-change error than we are to commit a false-change error. What happens to our missed-change error rate if we specify a new, higher false-change error rate? Shifting our false-change error rate from α=0.05 to α=0.10 is illustrated in Figure 5.9

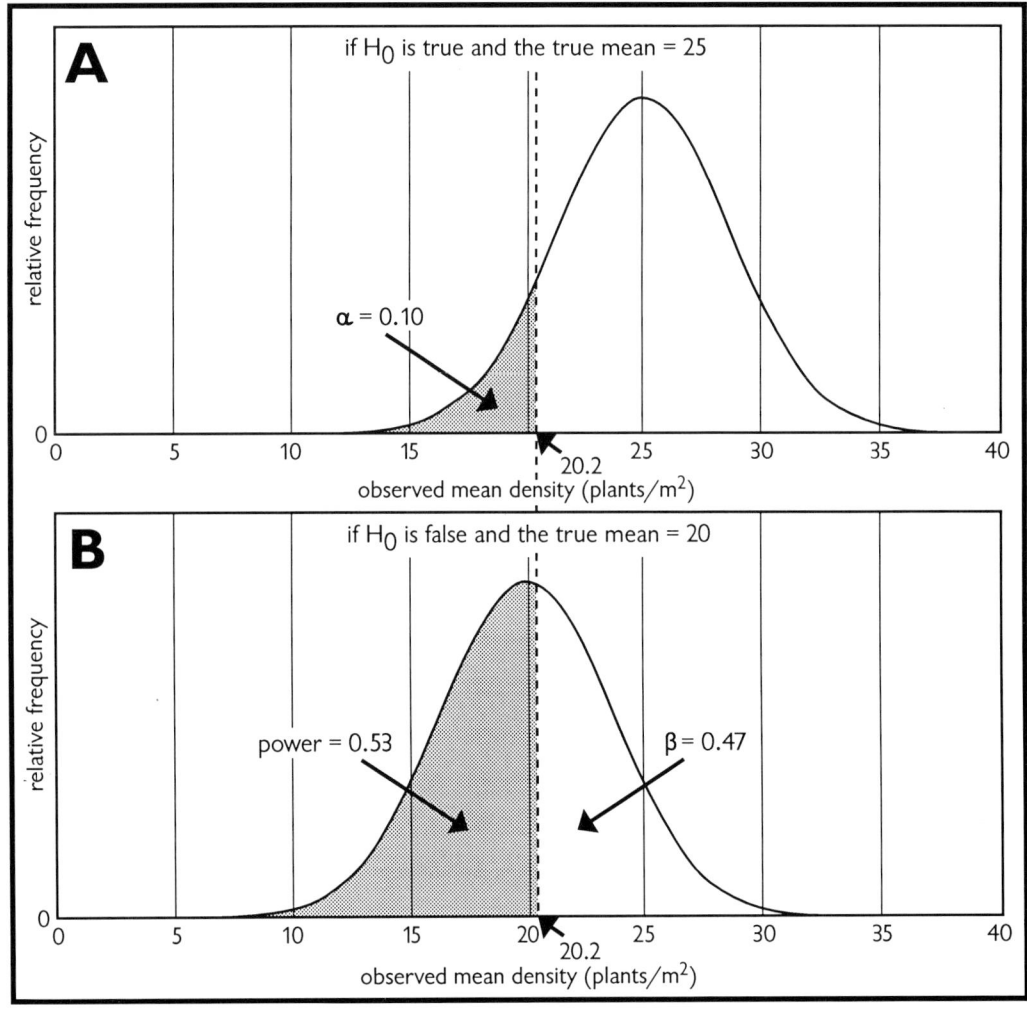

FIGURE 5.9. The critical region for the false-change error in the sampling distributions from Figure 5.8 has been increased from α = 0.05 to α = 0.10. Part B, in which the H₀ is false and the true mean = 20, shows that power is larger for α = 0.10 than for Figure 5.8 where α = 0.05 (adapted from Peterman 1990).

for the same sampling distributions shown in Figure 5.8. Our critical density threshold at the $P=0.10$ level is now 20.21 plants/m^2, and our missed-change error rate has dropped from $\beta=0.62$ down to $\beta=0.47$ (i.e., the power to detect a true five plant/m^2 difference went from 0.38 to 0.53). A sample mean of 20 plants/m^2 will now lead to the correct conclusion that a difference of five plants/m^2 between the populations does exist. Of course, the penalty we pay for increasing our false-change error rate is that we are now twice as likely to conclude that a difference exists in situations when there is no true difference and our population mean is actually 25 plants/m^2. Changing the false-change error rate even more, to $\alpha=0.20$, (Figure 5.10) reduces the probability of making a missed-change error down to $\beta=0.29$ (i.e., giving us a power of 0.71 to detect a true difference of five plants/m^2).

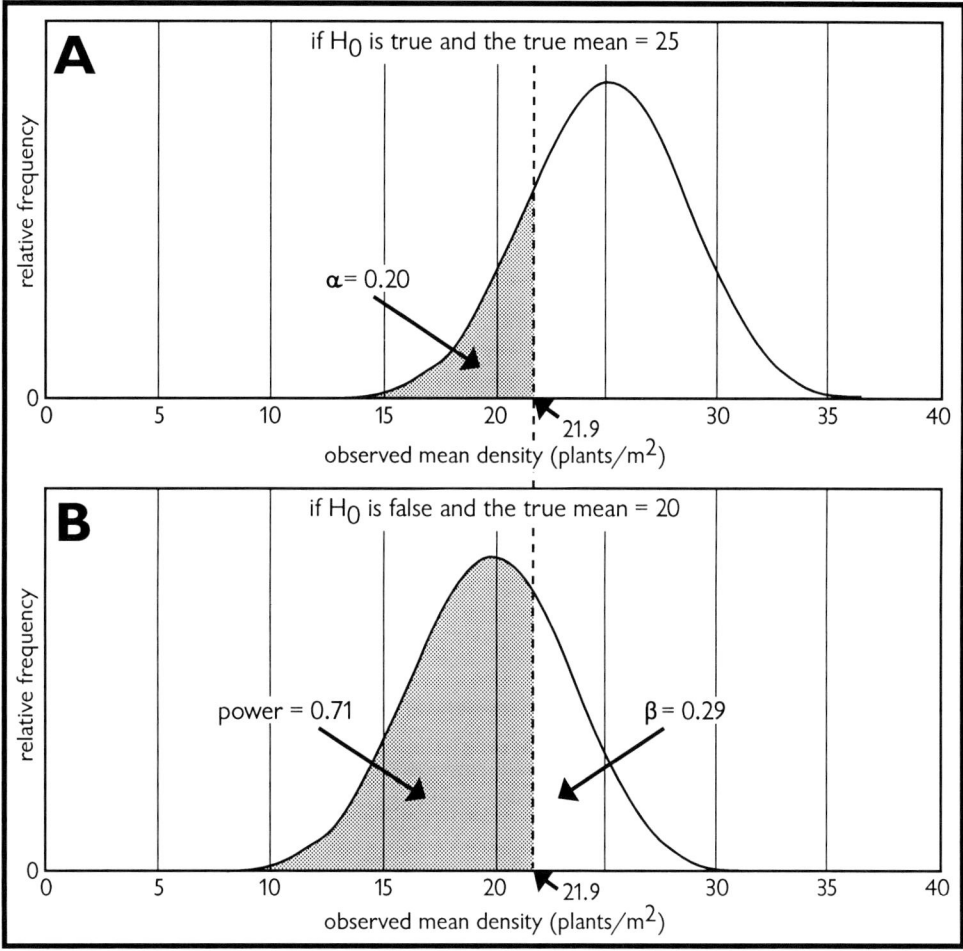

FIGURE 5.10. The critical region for the false-change error in the sampling distributions from Figure 5.8 has been increased from $\alpha=0.05$ to $\alpha=0.20$. Part B, in which the H_0 is false and the true mean = 20, shows that power is larger for $\alpha=0.20$ than for Figure 5.8 where $\alpha=0.05$ or Figure 5.9 where $\alpha=0.10$. Again, a sample mean to the left of the vertical dashed line would reject H_0, while one to the right of it would not reject H_0 (adapted from Peterman 1990).

b. Increasing the acceptable MDC

Any sampling design is more likely to detect a true large difference than a true small difference. As the magnitude of the difference increases, we will see an increase in the power to detect the difference. This relationship is shown in Figure 5.11B, where we see a sampling distribution with a true mean density of 15 plants/m^2, which is 10 plants/m^2 below our threshold density of 25 plants/m^2. The false-change error rate is set at $\alpha=0.05$ in this example.

This figure shows that the statistical power to detect this larger difference of 10 plants/m² (25 plants/m² to 15 plants/m²) is 0.85 compared with the original power value of 0.38 to detect the difference of five plants/m² (25 plants/m² to 20 plants/m²). Thus, with a false-change error rate of 0.05, we can be 85% certain of detecting a difference from our 25 plants/m² threshold of 10 plants/m² or greater. If we raised our false-change error from $\alpha=0.05$ to $\alpha=0.10$ (not shown in Figure 5.11), our power value would rise to 0.92, which creates a sampling situation where our two error rates are nearly equal ($\alpha=0.10$, $\beta=0.08$).

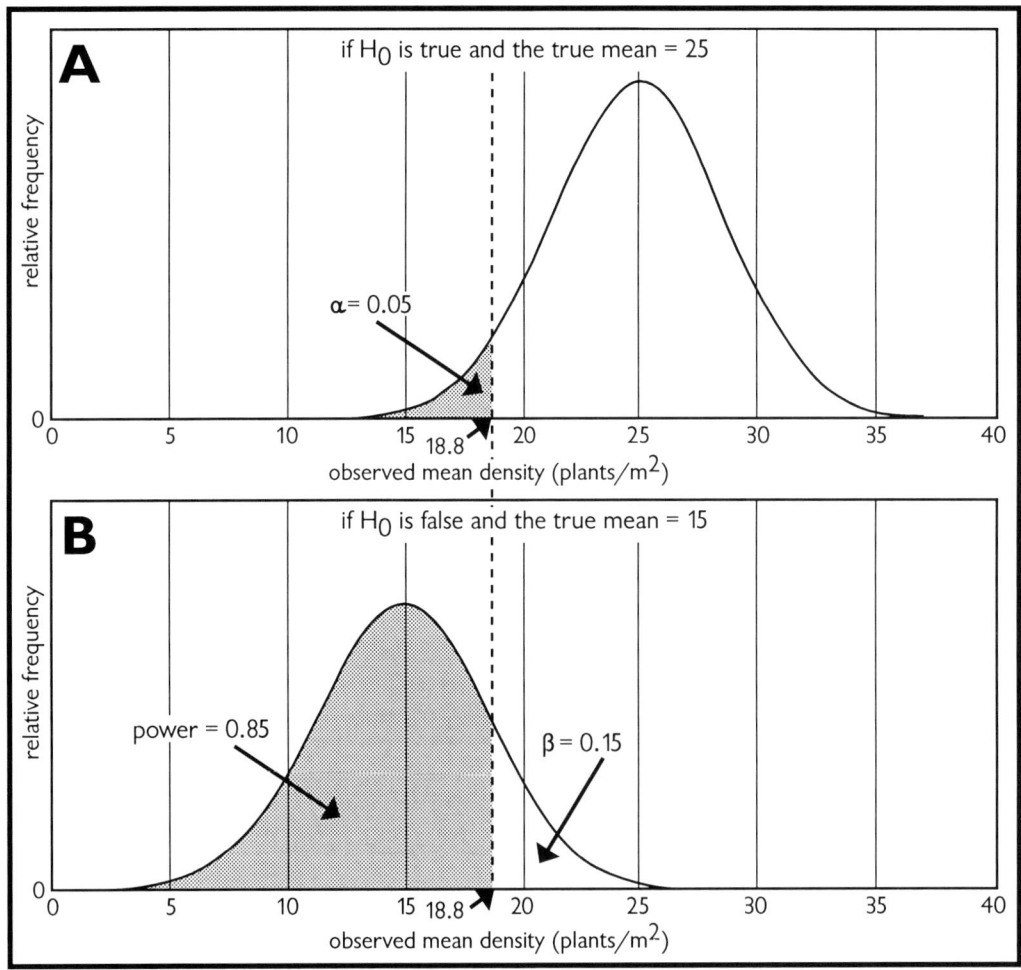

FIGURE 5.11. Part A is the same as Figure 5.8; in part B, the true population mean is 15 plants/m² instead of the 20 plants/m² shown in Figure 5.8. Note that power increases (and β decreases) when the new true population mean gets further from the original true mean of 25 plants/m². Again, a sample mean to the left of the vertical dashed line would reject H_0, while one to the right of it would not reject H_0 (adapted from Peterman 1990).

c. Increasing the sample size

The sampling distributions shown in Figures 5.8 to 5.11 were all created by sampling the populations with n=30 1m x 1m quadrats. Any increase in sample size will lead to a subsequent increase in power to detect some specified minimum detectable difference. This increase in power results from the sampling distributions becoming narrower. Sampling distributions based on samples of n=50 are shown in Figure 5.12 where the true difference between the two populations is once again five plants/m² with a false-change error rate threshold of $\alpha=0.05$. The increase in sample size led to an increase in power from power=0.38 with n=30 to power=0.54 with n=50. Note that the critical threshold density associated with an $\alpha = 0.05$ is now 20.3 plants/m² as compared to the 18.8 plants/m² threshold when n = 30.

d. Decreasing the standard deviation

The sampling distributions shown in Figures 5.8 to 5.12 all are based on sampling distributions with a standard deviation of ±20 plants/m². The quadrat size used in the sampling was a square 1m x 1m quadrat. If individuals in the plant population are clumped in distribution, then it is likely that a rectangular shaped quadrat will result in a lower standard deviation (See Chapter 7 for a detailed description of the relationship between standard deviation and sampling unit size and shape). Figure 5.13 shows sampling distributions where the standard deviation was reduced from ±20 plants/m² to ±10 plants/m². Note that the critical threshold density associated with an $\alpha = 0.05$ is now 21.9 plants/m² as compared to the 18.8 plants/m² threshold when the standard deviation was ± 20 plants/m². This reduction in the true standard deviation came from a change in quadrat shape from the 1m x 1m square shape to a 0.2m x 5m rectangular shape. Note that quadrat area (1m²) stayed the same so that the mean densities are consistent with the previous sampling distributions shown in Figures 5.8

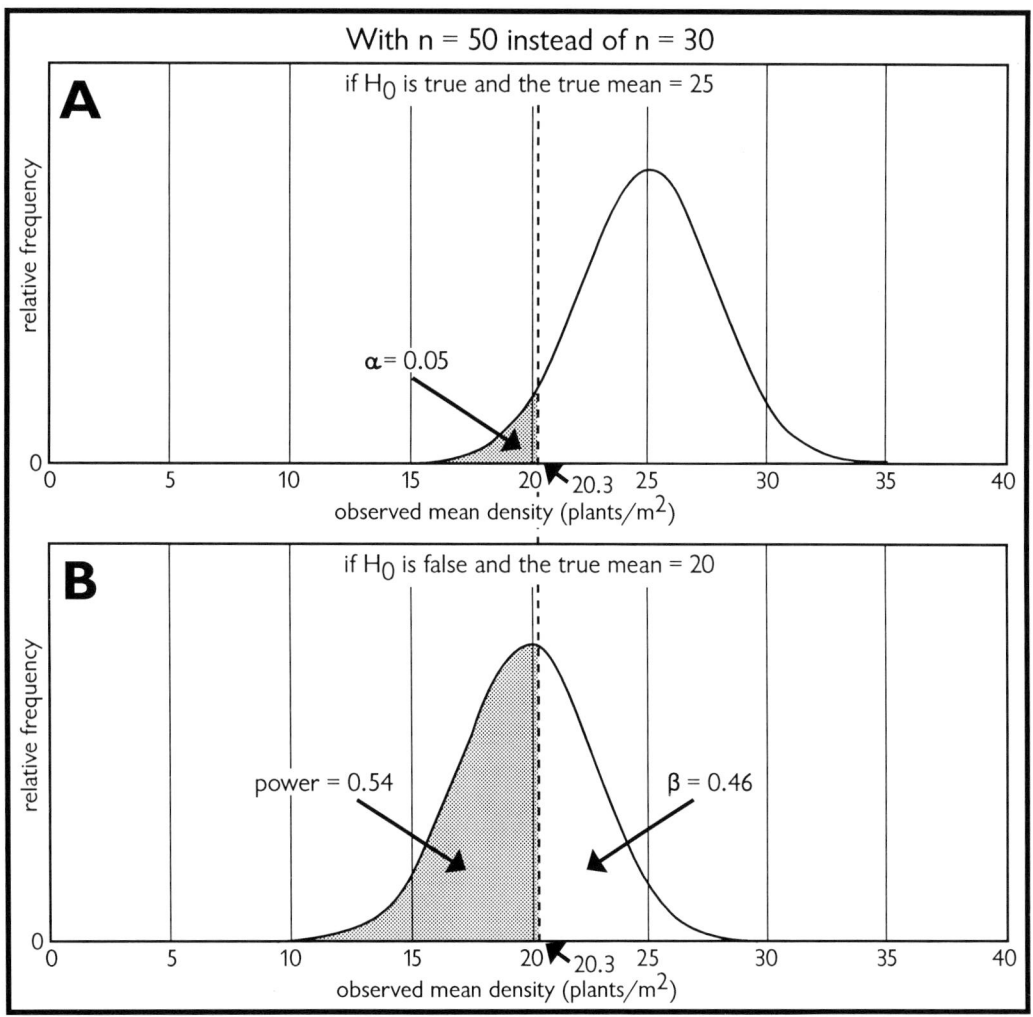

FIGURE 5.12. The sample size was increased to n = 50 quadrats from the n = 30 quadrats shown in Figure 5.8. Note that power increases (and β decreases) at larger sample sizes. Again, a sample mean to the left of the vertical dashed line would reject H_0, while one to the right of it would not reject H_0 (adapted from Peterman 1990).

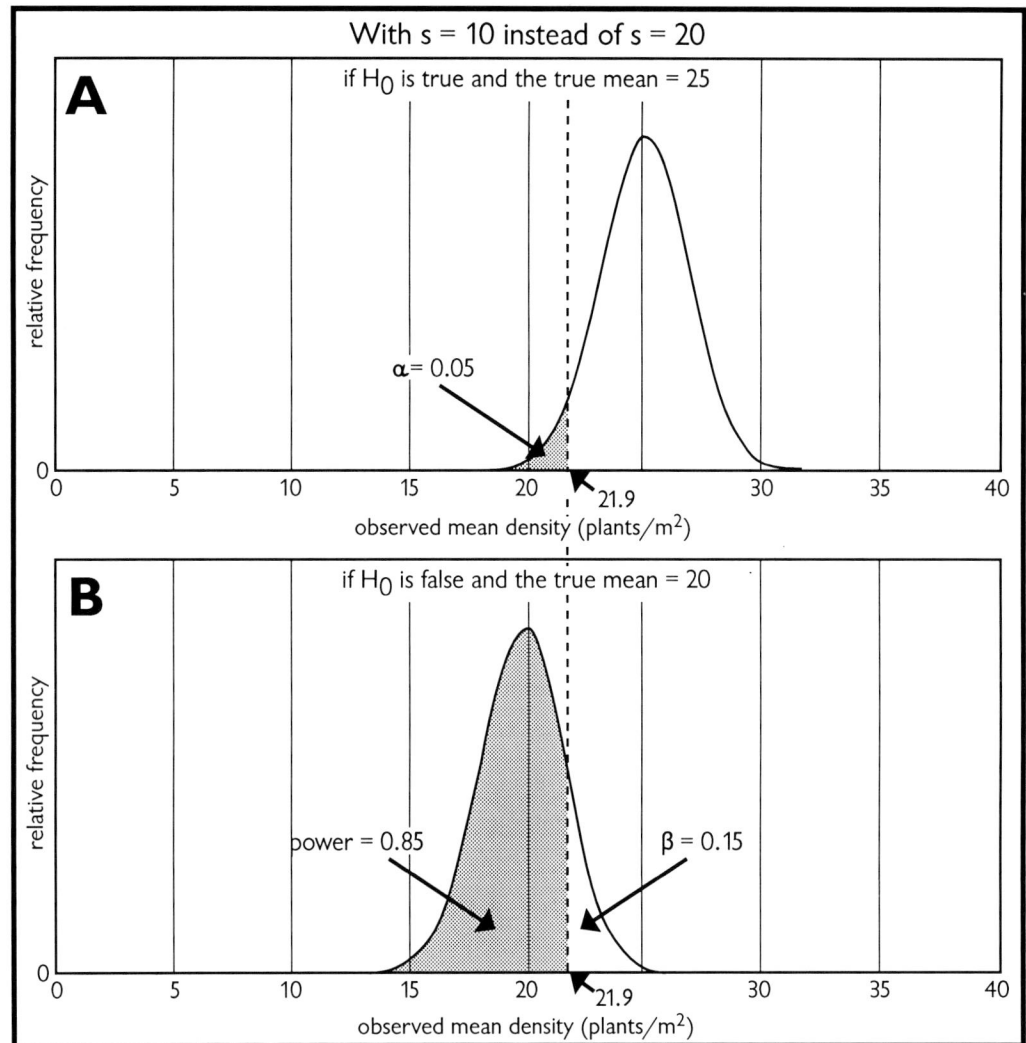

FIGURE 5.13. The standard deviation (s) of 20 plants/m² shown in Figure 5.8 is reduced to ten plants/m². Note that power increases (and β decreases), as the standard deviation decreases. Again, a sample mean to the left of the vertical dashed line would reject H_0, while one to the right of it would not reject H_0 (adapted from Peterman 1990).

through 5.12. This reduction in standard deviation led to a dramatic improvement in power, from 0.38 (with sd = 20 plants/m²) to 0.85 (with sd = 10 plants/m²). Reducing the standard deviation has a more direct impact on increasing power than increasing sample size, because the sample size is reduced by taking its square root in the standard error equation ($SE = s/\sqrt{n}$). Recall that the standard error provides an estimate of sampling precision from a single sample without having to enlist the support of 1,000 people who gather 1,000 independent sample means.

e. Power curves

The relationship between power and the different sampling design components that influence power can also be displayed in power curve graphs. These graphs typically show power values on the y-axis and either sample size, MDC, or standard deviation values on the x-axis. Figure 5.14A shows statistical power graphed against different magnitudes of change for the same hypothetical data set described above and shown in Figures 5.8 to 5.11. Four different power curve lines are shown, one for each of the following four different false-change (α) error rates: 0.01, 0.05, 0.10, and 0.20. The power curves are based on sampling with a sample size of 30 quadrats and a standard deviation of 20 plants/m². For any particular false-change error rate, power increases as the magnitude of the minimum detectable change increases. When α=0.05, the power to detect small changes is very low (Figure 5.14A). For example, we have only a 13% chance of detecting a difference of 2 plants/m² (i.e., a density of 23 plants/m² which is 2 plants/m² below our threshold value of 25 plants/m²). In contrast, we can be 90% sure of detecting a minimum difference of 11 plants/m². We can also attain higher power by increasing

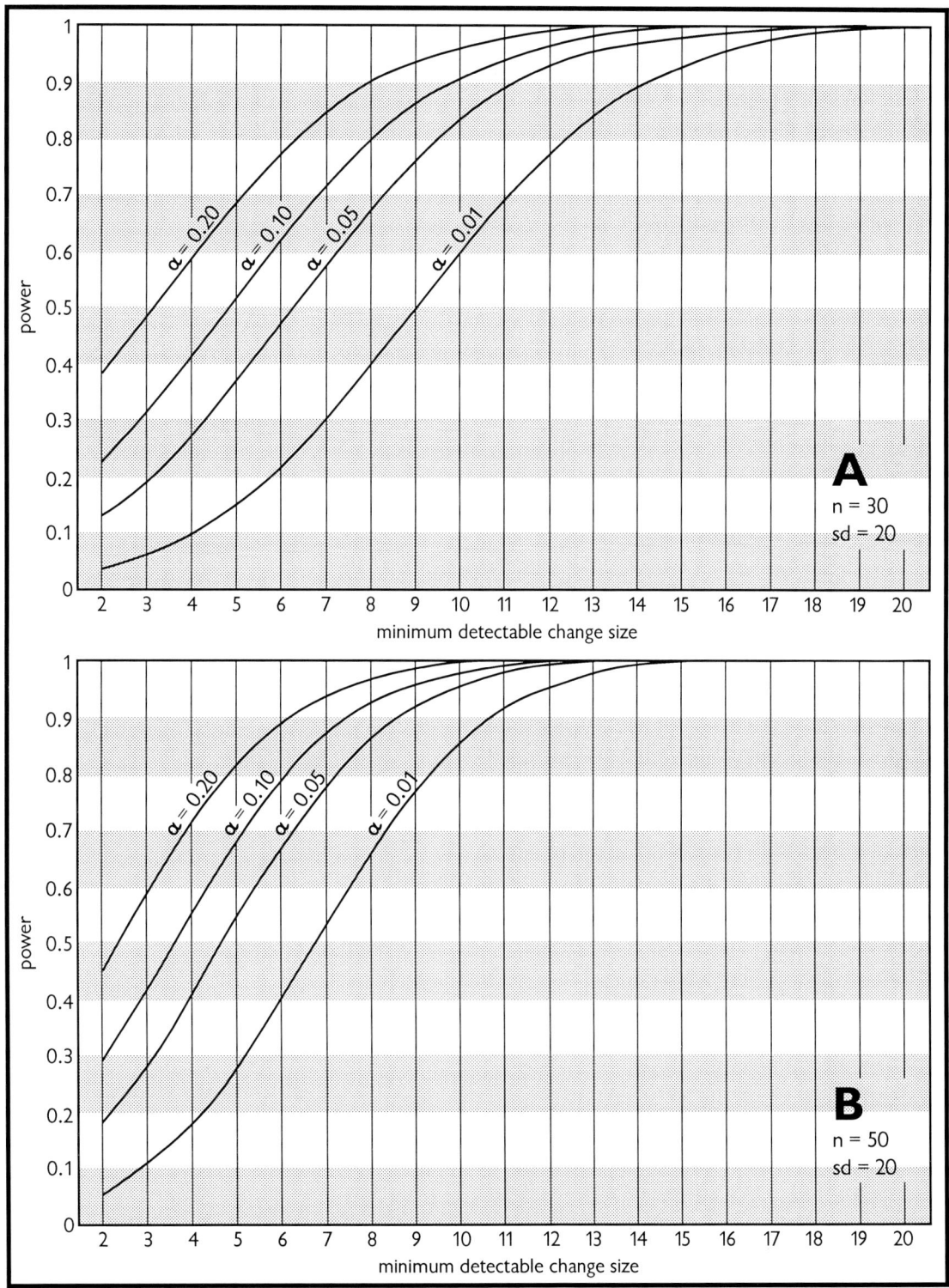

FIGURE 5.14. Power curves showing power values for various magnitudes of minimum detectable change and false-change error rates when the standard deviation is 20. Part A shows power curves with a sample size of 30. Part B shows power curves with a sample size of 50.

the false-change error rate. The power to detect a change of eight plants/m² is only 0.41 when α=0.01, but it increases to 0.69 at α=0.05, to 0.81 at α=0.10, and to 0.91 at α=0.20.

A different set of power curves are shown in Figure 5.14B where the sample size is n=50 instead of the n=30 shown in Figure 5.14A. This larger smaller sample size shifts all of the power curves to the left, making it more likely that smaller changes will be detected. For example, with a false-change error rate of α=0.10, the power to detect a seven plant/m² difference is 0.88 with a sample size of n=50 quadrats compared to the power of 0.73 with a sample size of n=30 quadrats.

Power curves that show the effect of reducing the standard deviation are shown in Figure 5.15. Figure 5.15A is the same as Figure 5.14A where the standard deviation is 20 plants/m². Figure 5.15B shows the same power curves except they are based on a standard deviation of 10 plants/m². The smaller standard deviation shifts all of the power curves to the left and results in much steeper slopes. The smaller standard deviation leads to substantially higher power levels for any particular MDC value. For example, the power to detect a change of five plants/m² with a false change error rate of α=0.10 is only 0.53 in Figure 5.15A as compared to the power of 0.92 in Figure 5.15B.

6. Setting false-change and missed-change error rates

Both false-change and missed-change error rates can be reduced by sampling design changes that increase sample size or decrease sample standard deviations. Missed-change and false-change error rates are inversely related, which means that reducing one will increase the other (but not proportionately) if no other changes are made. The decision of which type of error is more important should be based on the nature of the changes you are trying to determine, and the consequences of making either kind of mistake.

Because false-change and missed-change error rates are inversely related to each other, and because these errors have different consequences to different interest groups, there are different opinions as to what the "acceptable" error rates should be. The following examples demonstrate the conflict between false-change and missed-change errors.

◆ **Testing for a lethal disease.** When screening a patient for some disease that is lethal without treatment, a physician is less concerned about making a false diagnosis error (analogous to a false-change error) of concluding that the person has the disease when they do not than failing to detect the disease (analogous to a missed-change error) and concluding that the person does not have the disease when in fact they do.

◆ **Testing for guilt in our judicial system.** In the United States, the null hypothesis is that the accused person is innocent. Different standards for making judgement errors are used depending upon whether the case is a criminal or a civil case. In criminal cases, proof must be "beyond a reasonable doubt." In these situations it is less likely that an innocent person will be convicted (analogous to a false-change error), but it is more likely that a guilty person will go free (analogous to a missed-change error). In civil cases, proof only needs to be "on the balance of probabilities." In these situations, there is a greater likelihood of making a false conviction (analogous to a false-change error), but a lower likelihood of making a missed conviction (analogous to a missed-change) error.

◆ **Testing for pollution problems.** In pollution monitoring situations, the industry has an interest in minimizing false-change errors and may desire a very low false-change error

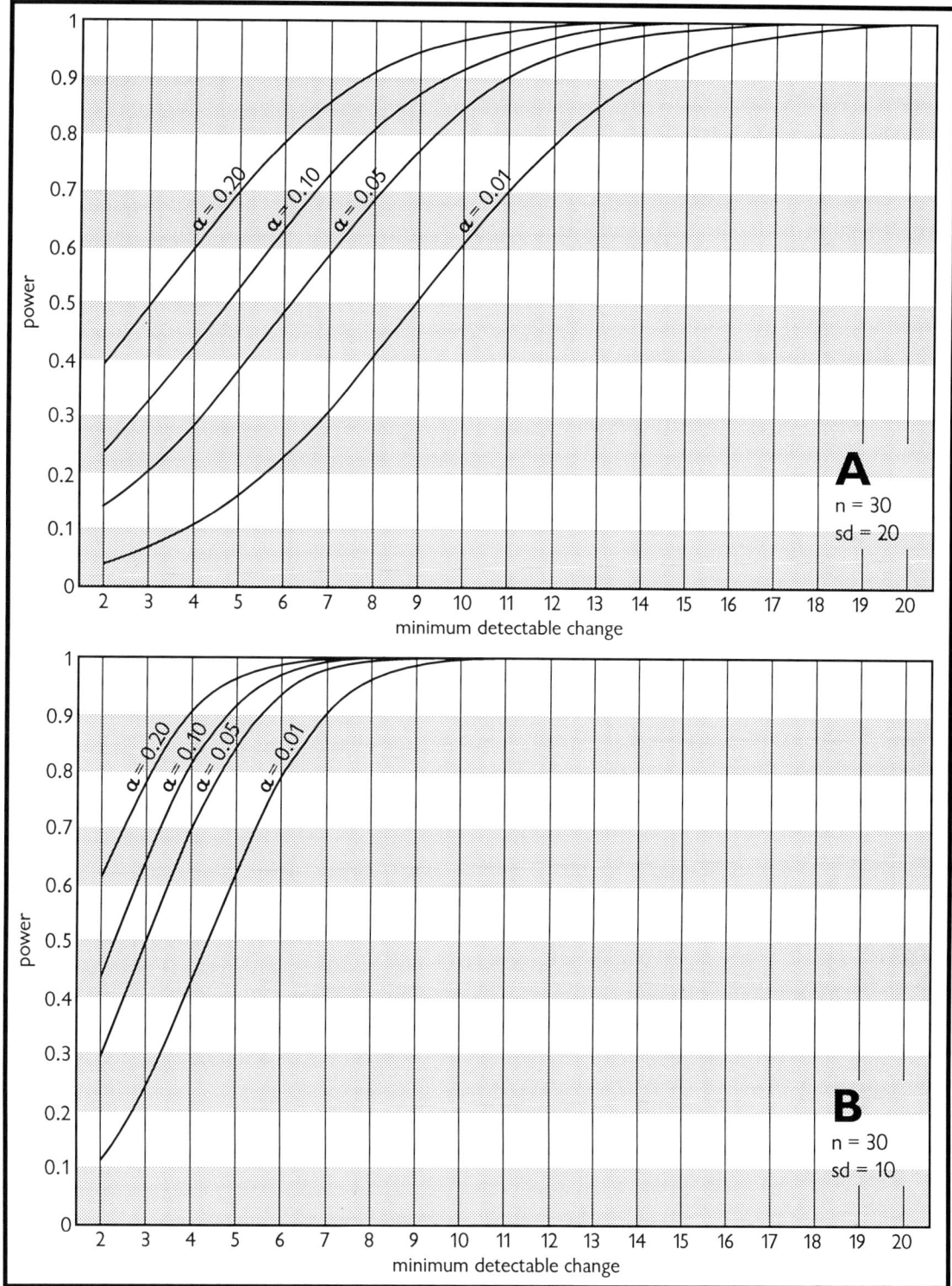

FIGURE 5.15. Power curves showing power values for various magnitudes of minimum detectable change and false-change error rates when the sample size = 30. Part A shows power curves with a standard deviation of 20 plants/m². Part B shows power curves with a standard deviation of 10 plants/m².

rate (e.g., $\alpha = 0.01$ or 0.001). Companies do not want to be shut down or implement expensive pollution control procedures if a real impact has not occurred. In contrast, an organization concerned with the environmental impacts of some pollution activity will likely want to have high power (low missed-change error rate) so that they do not miss any real changes that take place. They may not be as concerned about occasional false-change errors (which would result in additional pollution control efforts even though real changes did not take place).

Missed-change errors may be as costly or more costly than false-change errors in environmental monitoring studies (Toft and Shea 1983; Peterman 1990; Fairweather 1991). A false-change error may lead to the commitment of more time, energy, and people, but probably only for the short period of time until the mistake is discovered (Simberloff 1990). In contrast, a missed-change error, as a result of a poor study design, may lead to a false sense of security until the extent of the damages are so extreme that they show up in spite of a poor study design (Fairweather 1991). In this case, rectifying the situation and returning the system to its preimpact condition could be costly. For this reason, you may want to set equal false-change and missed-change error rates or even consider setting the missed-change error rate lower than the false-change error rate (Peterman 1990; Fairweather 1991).

There are many historical examples of costly missed-change errors in environmental monitoring. For example, many fish population monitoring studies have had low power to detect biologically meaningful declines so that declines were not detected until it was too late and entire populations crashed (Peterman 1990). Some authors advocate the use of something they call the "precautionary principle" (Peterman and M'Gonigle 1992). They argue that, in situations where there is low power to detect biologically meaningful declines in some environmental parameter, management actions should be prescribed as if the parameter had actually declined. Similarly, some authors prefer to shift the burden of proof in situations where there might be an environmental impact from environmental protection interests to industry/development interests (Peterman 1990; Fairweather 1991). They argue that a conservative management strategy of "assume the worst until proven otherwise" should be adopted. Under this strategy, developments that may negatively impact the environment should not proceed until the proponents can demonstrate, with high power, a lack of impact on the environment.

7. Why has statistical power been ignored for so long?

It is not clear why missed-change errors, power, and minimum detectable change size have traditionally been ignored. Perhaps researchers have not been sufficiently exposed to the idea of missed-change errors. Most introductory texts and statistics courses deal with the material only briefly. Computer packages for power analysis have only recently become available. Perhaps people have not realized the potentially high costs associated with making missed-change errors. Perhaps researchers have not understood how understanding power can improve their work.

The issue of power and missed-change errors has gained a lot of attention in recent years. A literature review in the 1980's would not have turned up many articles dealing with statistical power issues. A literature review today would turn up dozens of articles in many disciplines from journals all over the world (see Peterman 1990 and Fairweather 1991 for good review papers on statistical power). Journal editors may soon start requiring that power analysis information be reported for all nonsignificant results (Peterman 1990). There may also be some departure from the strict adherence to the 0.05 significance level (Peterman 1990; Fairweather 1991).

8. Use of prior power analysis during study design

Power analysis can be useful during both the design of monitoring studies and in the interpretation of monitoring results. The former is sometimes called "prior power analysis," whereas the latter is sometimes called "post-hoc power analysis" (Fairweather 1991). Post-hoc power analysis is covered in Chapter 11.

The use of power analysis during the design and planning of monitoring studies provides valuable information that can help avoid monitoring failures. Once some preliminary or pilot data have been gathered, or if some previous years' monitoring data are available, power analysis can be used to evaluate the adequacy of the sampling design. Prior power analysis can be done in several different ways. All are based upon the power function described earlier:

Power = a function of (s, n, MDC, and α)

The power of a particular sampling design can be evaluated by plugging sample standard deviation, sample size, the desired MDC, and an acceptable false-change error rate, into equations or computer programs (Appendix 16) and solving for power. If the power to detect a biologically important change turns out to be quite low (high probability of a missed-change error), then the sampling design can be modified to try to achieve higher power.

Alternatively, a desired power level can be specified and the terms in the power function can be rearranged to solve for sample size. This will give you assurance that your study design will succeed in being able to detect a certain magnitude of change at the specified power and false-change error rate. This is the format for the sample size equations that are discussed in Chapter 7 and presented in Appendix 7.

Still another way to do prior power analysis is to specify a desired power level and a particular sample size and then rearrange the terms in the power function to solve for the MDC (Rotenberry and Wiens 1985; Cohen 1988). If the MDC is unacceptably large, then attempts should be made to improve the sampling design. If these efforts fail, then the decision must be made to either live with the large MDC or to reject the sampling design and perhaps consider an alternative monitoring approach.

The main advantage of prior power analysis is that it allows the adequacy of the sampling design to be evaluated at an early stage in the monitoring process. It is much better to learn that a particular design has a low power at a time when modifications can easily be made than it is to learn of low power after many years of data have already been gathered. The importance of specifying acceptable levels of false-change and missed-change errors along with the magnitude of change that you want to be able to detect is covered in Chapter 6 the next chapter, which introduces sampling objectives.

Literature Cited

Brewer, J. K. 1972. On the power of statistical tests in the American Educational Research Journal. American Educational Research Journal 9: 391-401.

Cohen, J. 1988. Statistical power analysis for the behavioral sciences. 2nd edition. Hillsdale, N. J. Lawrence Erlbaum Associates.

Fairweather, P. G. 1991. Statistical power and design requirements for environmental monitoring. Australian Journal of Marine and Freshwater Research 42: 555-567.

Hayes, J. P. 1987. The positive approach to negative results in toxicology studies. Ecotoxicology and Environmental Safety 14: 73-77.

McCall, C. H. 1982. Sampling and statistics handbook for research. Ames, IA: The Iowa State University Press.

Peterman, R. M. 1990. Statistical power analysis can improve fisheries research and management. Canadian Journal of Fisheries and Aquatic Sciences. 47: 2-15.

Peterman, R. M.; M'Gognigle. M. 1992. Statistical power analysis and the precautionary principle. Marine Pollution Bulletin. 24(5): 231-234.

Rotenberry, J. T.; Wiens J. A. 1985. Statistical power analysis and community-wide patterns. American Naturalist. 125: 164-168.

Simberloff, D. 1990. Hypotheses, errors, and statistical assumptions. Herpetelogica. 46: 351-357.

Toft, C. A.; P. J. Shea. 1983. Detecting community-wide patterns: estimating power strengthens statistical inference. American Naturalist. 122: 618-625.

Zar, J. H. 1996. Biostatistical Analysis. Englewood Cliffs, N.J. Jersey: Prentice-Hall, Inc.

CHAPTER 6
Sampling Objectives

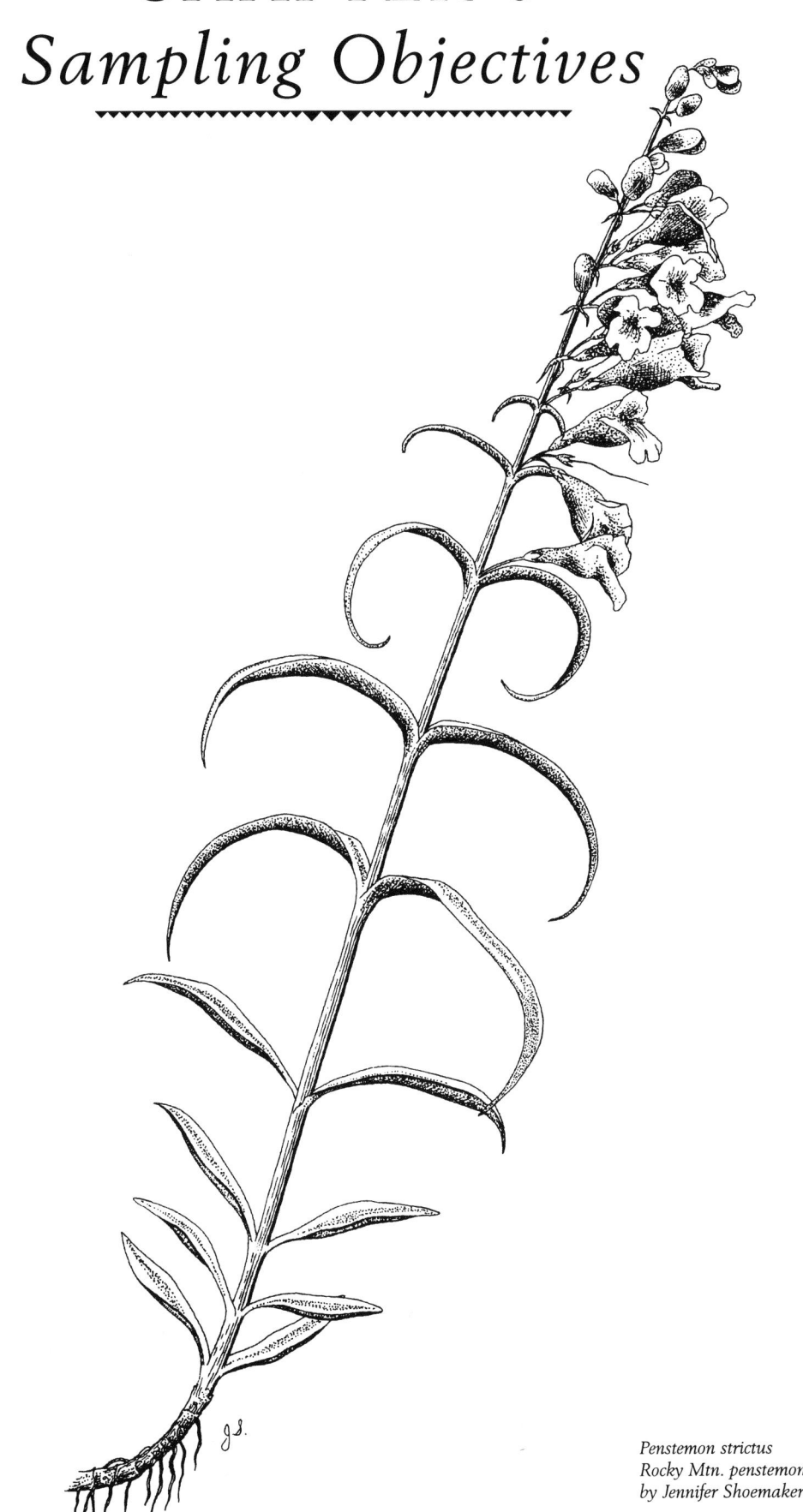

Penstemon strictus
Rocky Mtn. penstemon
by Jennifer Shoemaker

CHAPTER 6. Sampling Objectives

A. Introduction

Sampling objectives should be written as companion objectives to management objectives (Chapter 4) whenever monitoring includes sampling procedures. As described in Chapter 5, sampling involves assessing a portion of a population with the intent of making inferences to the population as a whole. Sampling objectives specify information such as target levels of precision, power, acceptable false-change error rate, and the magnitude of change you are hoping to detect. Unlike a management objective, which sets a specific goal for attaining some ecological condition or change value, a sampling objective sets a specific goal for the measurement of that value. For example, considering the following examples of management objectives, with corresponding sampling objectives:

◆ **Management objective**: We want to maintain a population of *Lomatium bradshawii* at the Willow Creek Preserve with at least 2,000 individuals from 1998 to 2008.

◆ **Sampling objective**: We want to be 95% confident that estimates are within ± 25% of the estimated true value.

◆ **Management objective**: We want to see a 20% increase in the average density of *Lomatium bradshawii* at the Willow Creek Preserve between 1997 and 1999.

◆ **Sampling objective**: We want to be 90% sure of detecting a 20% change in the density and we are willing to accept a 1 in 10 chance that we'll say a change took place when it really didn't.

The principal reason to add sampling objectives to management objectives is to ensure that you end up with useful monitoring information. If this additional information is not specified, you risk ending up with an inadequate sampling design that makes it difficult or almost impossible to assess whether you've achieved your management objective. For example, without setting sampling targets, you may end up with an estimate of population size with confidence intervals nearly as wide as the estimate itself (e.g., 1000 plants ± 950 plants) or you may find that you have low power to detect some biologically meaningful change (e.g., only a 15% chance of detecting the change you were hoping to achieve). The information specified in a sampling objective is also necessary to determine adequate sample sizes using the procedures described in Chapter 7.

For monitoring that does not involve sampling, your ability to assess success at meeting your management objective should be obvious from the management objective itself without the need to specify additional information. Consider the following management objectives that involve monitoring without sampling:

◆ Maintain the current knapweed-free condition of the *Penstemon lemhiensis* population in the Iron Creek drainage for the next 10 years.

◆ Maintain at least 100 individuals of *Penstemon lemhiensis* at the Iron Creek site over the life of the Iron Creek Allotment Management Plan.

To determine success at meeting the first objective, you simply need to visit the site at some specified interval and search for the presence of knapweed. To assess success for the second objective, you will likely be able to count all the plants in the population (or at least the first 100 that you find). Thus, the management objectives for these non-sampling types of monitoring do not require the additional components that are discussed in this chapter.

Developing sampling objectives is being covered following Chapter 5, because these objectives include terms and concepts related to sampling procedures. If some of the terms included in this chapter are not familiar to you, refer to Chapter 5 for more information.

There are two categories of sampling objectives that correspond to the two major categories of management objectives: (1) target/threshold management objectives; and (2) change/trend management objectives.

B. Target/Threshold Management Objectives

The sampling objective in this case is to estimate some parameter in the population (e.g., mean density per unit area, mean percent cover, or mean plant height), to estimate a proportion (e.g., the frequency of a particular species within a set of quadrats placed within a sampled area), or to estimate total population size (total number of plants within a sampled area). These estimates are then compared to the target/threshold value to determine if the management objective is met. Sampling objectives for this type of management objective need to include two components related to the precision of the estimate:

◆ **The confidence level**. How confident do you want to be that your confidence interval will include the true value? Is 80% confidence high enough or do you want 90%, 95%, or even 99% confidence?

◆ **The confidence interval width**. How wide a range are you willing to accept around your estimated value? Is ± 20% of the estimated mean or total value adequate or do you want to be within ± 10%?

Following is an example of a target/threshold management objective with a corresponding sampling objective:

◆ **Management objective**: Increase the number of individuals of *Penstemon lemhiensis* in the Iron Creek Population to 1,000 individuals by the year 2000.

◆ **Sampling objective**: We want to be 95% confident that population estimates are within 20% of the estimated true value.

This sampling objective specifies a relative confidence interval width (± 20% of the estimated true value) so the targeted confidence interval width in absolute units will depend upon the estimated population size. For example, if the first year of monitoring yields a population estimate of 500 plants, the targeted confidence interval width is 500 plants x 20% = ± 100 plants. Information from pilot sampling can be used to determine how many sampling units need to be sampled to achieve a confidence interval width of ± 100 plants. See Appendix 3 for additional examples of sampling objectives paired with target/threshold management objectives.

C. Change/Trend Management Objectives

The sampling objective in this case is to determine whether there has been a change in some population parameter such as a mean value (e.g., mean density per unit area of a particular species, mean percent cover, mean plant height), a proportion (e.g., the frequency of a particular species within a set of quadrats placed within the sampled area), or the total population (total number of plants within a sampled area) between two or more time periods. This category of sampling objective must include the following three components:

◆ **The acceptable level of power (or the acceptable level of the missed-change error [Type II error] rate).** How certain do you want to be that, if a particular change does occur, you will be able to detect it? If you want to be 90% certain of detecting a particular magnitude of change, then you are specifying a desired power of "90%" (power and missed-change error rates are complementary, so in this example, the missed-change error rate is 0.10).

◆ **The acceptable false-change error (Type I error) rate.** What is the acceptable threshold value for determining whether an observed difference actually occurred or if the observed difference resulted from a chance event? This represents the chance of concluding that a change took place when it really did not. The $\alpha = 0.05$ level is frequently used, but you should carefully consider the impact of this decision on the probability of making missed-change errors.

◆ **The desired MDC (minimum detectable change).** The MDC specifies the smallest change that you are hoping to detect with your sampling effort. The MDC should represent a biologically meaningful quantity given the likely degree of natural variation in the attribute being measured.

Following is an example of a change/trend type of management objective with a corresponding sampling objective:

◆ **Management objective:** I want to see a 20% increase in the density of *Lomatium cookii* at the Agate Desert Preserve between 1998 and 2000

◆ **Sampling objective:** I want to be 90% certain of detecting a 20% increase in density between 1998 and 2000 and I am willing to accept a 10% chance that I will make a false-change error.

This sampling objective specifies a power of 90%, a false-change error rate of 10%, and an MDC of 20%. The MDC is specified in relative terms, so the targeted MDC in absolute units will depend upon the estimated density in 1994. For example, if the mean density in 1994 is 10 plants/quadrat, the desired MDC is an increase of 2 plants/quadrat.

Why bother specifying false-change error rates, power, and some desired MDC when you are writing a sampling objective designed to detect change over time? The main advantage is that it helps you avoid designing low power monitoring studies. The sample size determination procedures discussed in Chapter 7 require the specification of false-change error rate, power, and the size of the change you are interested in detecting before you can determine how many sampling units to sample. If your pilot data indicate that you have low power to detect a biologically important change (high probability of a missed-change error), you can then correct your sampling design before you have gathered many years of monitoring data. See Appendix 3 for additional examples of sampling objectives paired with change/trend management objectives.

D. Setting Realistic Sampling Objectives

Sampling objectives should be written during the planning phase of a monitoring study. Targeted levels of precision, power, false–change error, and MDC should be based on existing knowledge of the species being monitored or information from similar species. It is a good idea to confer with managers or other stakeholders interested in the monitoring results to ensure they are comfortable with the targeted levels of precision, power, etc., specified in the sampling objectives. Chapter 5 describes the interplay between false-change error rates, power, and MDC (e.g., Figure 5.14)

The sampling objectives serve as a critical aid during the preliminary or pilot field sampling phase. Once pilot sampling data are available, information on the variability of the data can be plugged into sample size equations (Chapter 7) along with the information specified in the sampling objectives to determine how many sampling units should be sampled. If you are faced with a monitoring situation where there is a lot of variability between sampling units (despite all of your sampling design efforts to lower this variability) and the components of your sampling objective lead to a recommended sample size of more sampling units than you can afford to sample, then you need to reassess the monitoring study. Is it reasonable to make changes to some components of the sampling objective? For target/threshold types of management objectives, this may mean lowering the level of confidence or decreasing the precision of the estimate (i.e., increasing the confidence interval width) or both. For objectives directed towards tracking change over time, this may mean increasing the acceptable false–change error rate, decreasing the targeted power level, or settling on a larger specified MDC. Will these changes be acceptable to managers and other stakeholders? If you feel that making these modifications to the sampling objective is unreasonable, then you should take an alternative monitoring approach rather than proceed knowing that your monitoring project is unlikely to meet the stated objectives.

CHAPTER 7
Sampling Design

Phleum pratense
Timothy
by Jennifer Shoemaker

CHAPTER 7. Sampling Design

A. Introduction

Design is critical to any sample-based monitoring study. The consequences of poor study design are many: lost time and money, reduced credibility, incorrect (or no) management decisions, and unnecessary resource deterioration, to name just a few. Take your time during this stage to design a study that will meet your management and sampling objectives in the most efficient manner. Based on the pilot study you perform (the need for pilot sampling is discussed below), you may find that you cannot meet your objectives within the constraints of the time and money available. One solution to this dilemma is to change from sample-based monitoring to monitoring based on a qualitative technique or a complete census. Others could include choosing a different attribute to measure or changing your management and sampling objectives to reflect a less precise estimate (in the case of a target/threshold objective) or detection of a larger change (in the case of a change/trend objective).

Six basic decisions, which are discussed in detail in this chapter, must be made in designing a monitoring study based on sampling:

1. What is the population of interest?
2. What is an appropriate sampling unit?
3. What is an appropriate sampling unit size and shape?
4. How should sampling units be positioned?
5. Should sampling units be permanent or temporary?
6. How many sampling units should be sampled?

These decisions must be made based on site-specific information and objectives. There is no "right" quadrat size and shape, just as there is no "right" number of sampling units. In most situations these decisions can be made only through on-site assessment by pilot sampling.

B. What is the Population of Interest?

As we learned in Chapter 5, the population consists of the complete set of units about which we want to make inferences. We are using "population" in the statistical, rather than the biological, sense, although these populations may sometimes be the same, such as all of the plants of a particular species found on a certain mountain. The two populations, however, are often different. For example, the statistical population often consists of the complete set of quadrats we could place in a particular geographic area.

1. Target vs. sampled population

In sampling, the differences between the population you would like to make inferences about *(the target population)* and the population you actually sample *(the sampled population)* need to be understood. When monitoring a rare plant species, there is usually some target population with real physical boundaries that we would like to track. For example, if something like plant height or the number of flowers per plant is the subject of interest, our target population might be all the plants of a particular plant species on a preserve that has been set aside for that species' protection. Or, our target population might be the complete set of quadrats we could place in a particular wet meadow if the number of plants per quadrat is the subject of interest.

When biological populations are small and distributed in some uniform area, such as all plants within a fenced pasture, then we may be able to position sampling units throughout the entire target population of interest. However, two factors usually lead to defining a new sampled population: (1) irregular target population boundaries, and (2) target populations that cover a very large geographic area.

a. Small populations with irregular boundaries

When the target population is small, but has irregular boundaries, then we might fit some regular-shaped polygon, such as a square or rectangle, over the bulk of the population (as illustrated in Figure 7.1A). This newly defined area, often referred to as a macroplot, becomes our sampled population. The macroplot is usually permanently marked. The use of a macroplot facilitates the positioning of sampling units (see Section E) and ensures the same area is sampled each year.

We can make statistical inferences only to the boundaries of the sampled populations (i.e., to the area within the macroplot), not to the entire target population. This approach works well for small populations; a large population, however, would necessitate a very large macroplot, resulting in long distances between sampling units. The time necessary to travel to each sampling unit would make the design inefficient.

b. Very large populations

If the target population covers a very large geographic area, constraints of time and money, coupled with the tremendous variability usually encountered when sampling a very large population, often lead us to define some smaller geographic area(s) to sample. There are several ways this can be accomplished:

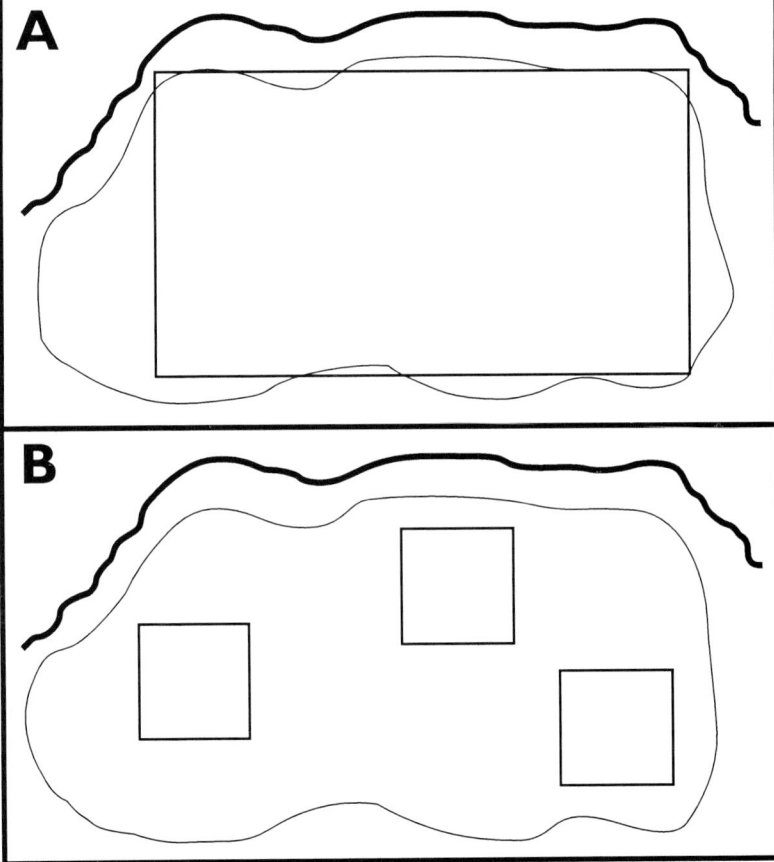

FIGURE 7.1. Positioning of macroplots (rectangles and squares) within irregularly shaped target populations (thin lines). The thick, irregular line denotes a river. A. A single macroplot is placed over the bulk of the target population. Inferences can be made only to the area within the macroplot (i.e., the macroplot is the sampled population). B. Three macroplots are randomly placed within the target population. Inferences can be made to the entire target population (i.e., the sampled population is the same as the target population).

◆ **One or more macroplots can be randomly positioned within the target population.** If several macroplots are randomly positioned in the target population, and sampling takes place within each macroplot, then we have something called a two-stage sampling design (Section E.6). Statistical inferences can be made to the entire population, and the sampled population and the target population are the same (see Figure 7.1B). If only a single macroplot is randomly positioned within the target population (Figure 7.2A), no inferences to the target population are possible because there is no way of determining how "representative" this macroplot is of the target population.

◆ **One or more macroplots can be subjectively positioned within the target population.** If macroplots are subjectively positioned in the target population, inferences can be made only to the area encompassed by the macroplots (Figure 7.2B). In other words, the sampled population is the area within the macroplots. This method may be preferable to randomly positioning macroplots if resource limitations prohibit the use of more than one or two macroplots.

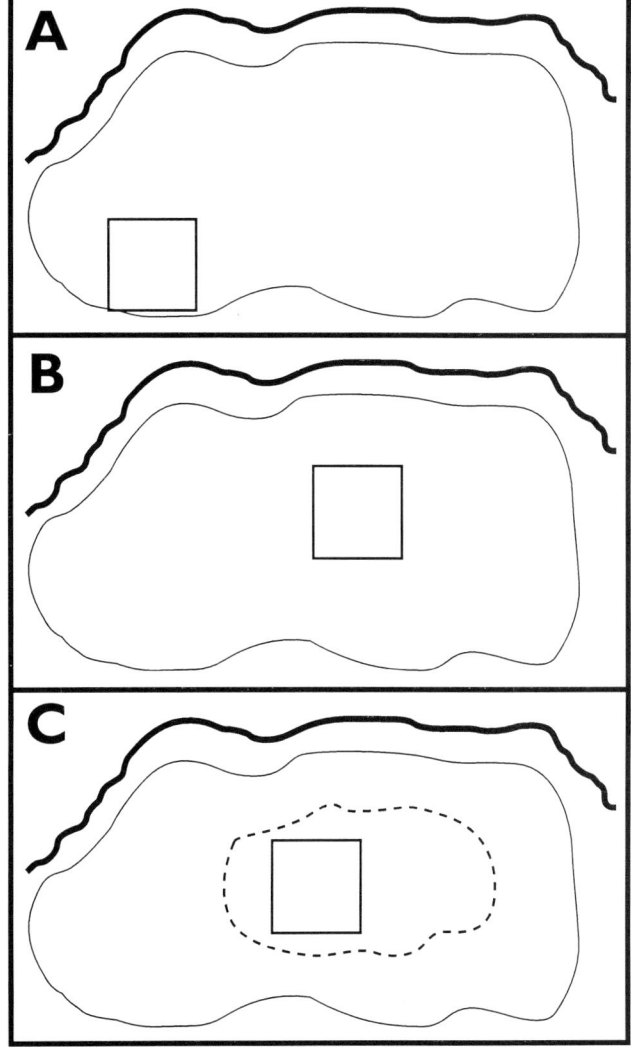

FIGURE 7.2. A single square macroplot placed in the target population. In all cases, inferences can be made only to the area within the macroplot (i.e., the macroplot is the sampled population). A. Random placement. B. Subjective placement. C. Random placement within a "representative" area (dotted line).

The suitability of this approach depends on the level of knowledge you have for the area of interest and the information needs for a particular project. Answers to the following questions will help guide you in this decision:

How comfortable will you be in making management decisions for the entire target population based on the information gathered within the subjectively positioned macroplots?

If the macroplot is positioned towards the middle of the population, will you miss changes that occur near the edge of the target population?

◆ **One or more macroplots can be randomly positioned within a selected region deemed to be "representative" of the target population.** This approach involves selecting an area considered to be representative of the target population (see section on key areas, below). Once this representative area is selected, we determine the macroplot position through a random compass bearing and random distance. This process, illustrated in Figure 7.2C, reduces the observer bias toward exact positioning of the macroplot. If a single macroplot is positioned in this way, statistical inferences can be made only about the area within the macroplot.

c. Macroplots vs. quadrats

Macroplots are relatively large areas, with sampling units, such as quadrats, lines, or points, randomly located within them. Quadrats are a type of sampling unit, within which the actual data are gathered.

d. Key areas

The key area concept is widely used, particularly in rangeland monitoring. Using this approach, key areas are selected (subjectively) that we hope reflect what is happening on a larger area. These may be areas we feel are representative of a larger area (such as a pasture) or critical areas (such as sites where endangered species occur). Monitoring studies are then located in these key areas.

Although we would like to make inferences from our sampling of key areas to the larger areas they are chosen to represent, this cannot be done statistically because the key areas were chosen subjectively. We could, of course, choose to sample the larger areas, but the constraints of time and money coupled with the tremendous variability usually encountered when sampling very large areas often makes this impractical. The key area concept represents a compromise.

The key area can be the target population, the sampled population, or both. The key area may be larger than the area that can be sampled given constraints of time and money. We would then end up with a situation like that depicted in Figure 7.2C, where we randomly locate a macroplot within a larger key area. In this case our target population is the key area, while our sampled population is the area within the macroplot. Since we will be making decisions based on what happens within this macroplot, we may wish to redefine our key area to be only that area within the macroplot. If we do this, then the target population (the new key area) and the sampled population will be the same, which is perfectly acceptable, as long as we clearly state this both in the study design and in the management objective.

Because statistical inferences can be made only to the key areas that are actually sampled, it is important to develop objectives that are specific to these key areas. It is equally important to make it clear that actions will be taken based on what happens in the key area, even when it can't be demonstrated statistically that what is happening in the key area is happening in the area it was chosen to represent. It is also important to base objectives and management actions on each key area separately. *Values from different key areas should never be averaged*, because this gives the impression that we have sampled a much larger area than is really the case and because this practice results in a "mean" value for which we can have no measure of precision.

Whether you choose to use the key area concept when monitoring plant populations depends on your objectives and the distribution of the target plant species. If a plant population you wish to monitor occupies a relatively small area, then you may be able to sample the entire area. You could then make inferences about that entire population without resorting to the key area concept. If, on the other hand, the plant population is diffusely spread throughout a large area, you may have to use the key area concept, depending on the amount of time and effort you wish to put into your sampling. Another possibility is to use a two-stage sampling design, shown in Figure 7.1B, and discussed in detail in Section E.6.

2. Finite vs. infinite populations

Before we leave the topic of populations, we need to address one more important feature of populations: whether they are finite or infinite (see Chapter 5, Section G). The distinction between finite and infinite populations can be important when determining how many units to sample, as discussed in Section G of this chapter.

C. What Is an Appropriate Sampling Unit?

The type of sampling unit you select depends on the vegetation attribute you are measuring, which should be detailed in a specific management objective. Density, cover, frequency, and biomass are the vegetation attributes most commonly monitored. Attributes related to individual plants (e.g., height, number of flowers per plant) are also often of interest.

1. Types of sampling units

Following are the types of sampling units that are relevant to the monitoring of plant populations.

Individual plants. Plants are the sampling units for attributes such as plant height, number of flowers per plant, or cover if the cover measurements are made on individual plants (e.g., tree stem diameters, bunchgrass basal area measurements).

Plant parts. Fruits might be the sampling units if the attribute is the number of seeds per fruit or the percentage of fruits containing some seed herbivore. Or, you may be interested in estimating the number of flowers per inflorescence, in which case the inflorescence is the sampling unit.

Quadrats (plots). Most estimates of density, frequency, or biomass require the use of quadrats, which represent the sampling units. Quadrats can also be the sampling units for cover measurements if visual estimates of cover are made within quadrats.

Lines (transects). When cover is measured using the line-intercept method, the line is the sampling unit. Lines can also serve as sampling units when points (for cover) or quadrats (for cover, density, or frequency) are positioned along lines and the points or quadrats are not far enough apart to be themselves considered the sampling units (because they are not independent of one another).

Points. When cover is measured with the point-intercept method and the points are randomly positioned, then the points are the sampling units.

Point frames or point quadrats. When cover is measured using point frames or point quadrats and these frames or quadrats are randomly positioned, then the point frames or point quadrats are the sampling units. Point frames are not recommended because they are inefficient for measuring cover in most vegetation types (see Chapter 8).

Distance (plotless) methods. There is a class of techniques to estimate density that are referred to as distance or plotless techniques. The sampling unit with these techniques is usually the individual distance between a randomly selected point and the nearest plant or between a randomly selected plant and its nearest neighbor. Distance measures are inaccurate for most plant populations (see Chapter 8).

2. Choosing the sampling unit

In many cases, simply determining the attribute you're going to measure determines the sampling unit. If you're going to measure density, frequency, or biomass, the sampling unit will be a quadrat. For cover, however, you have several choices. The sampling unit can be a line, a point, or a quadrat. (Chapter 8 gives information to help you decide which of these to choose.) If you are measuring something on individual plants, the sampling unit is the individual plant (although, as we will see later, you will often incorporate quadrats into your sampling design for this purpose as well).

3. Sampling units in multistage sampling designs

Certain sampling designs incorporate sampling at more than one level. These are called multistage sampling designs (Krebs 1989). The two-stage sampling design, discussed in Section E below, is one example. A random sample of *primary* sampling units is selected. Then, a subsample is taken from each of the primary sampling units. This subsample is made up of *secondary* sampling units (these are often called *elements* to differentiate between the two types of units).

D. What Is an Appropriate Sampling Unit Size and Shape?

The most efficient sampling unit size and shape depends on the type of vegetation attribute being measured and the growth form and spatial distribution of the sampling target. The most efficient design is usually the one that yields the highest statistical precision (smallest standard error and narrowest confidence interval around the mean) for either a given area sampled or a given total amount of time or money.

Sampling unit size and shape considerations are discussed separately for the following categories: (1) quadrats for density estimation; (2) quadrats for frequency estimation; (3) quadrats for cover estimation; (4) quadrats for biomass estimation; and (5) lines and points for cover estimation.

1. Quadrats for measuring density

Density is measured by counting some entity (e.g., individuals, ramets, stems) within quadrats. The efficiency of various quadrat sizes depends upon the following factors:

a. Quadrat size and shape

Objective of study: parameter estimation vs. pattern detection. For monitoring plant populations we are concerned with estimating true population parameters such as the true mean density or the true total population size. Differences in these true population parameters are what we are trying to track with this type of monitoring.

Detecting the intensity and scale of spatial pattern is a completely different objective that can lead to dramatically different sampling designs than those you use for tracking changes in population parameters. (Consult Greig-Smith (1983) for more information on detecting spatial pattern; this type of sampling will not be discussed further in this technical reference.)

Travel and setup time vs. searching and counting time. Which is more important: minimizing the number of quadrats or the total area (or proportion) of the population sampled? This depends on how large an area you are sampling (is it 1/2 mile between each quadrat location?) or how difficult it is to get from one quadrat position to another (are you sampling on a cliff face?). It also depends on how hard it is to search and count individuals. For large or conspicuous plants, such as shrubs, trees, or large herbaceous plants that occur at low densities, having a large sample area is not much of a problem because you can see all of the individuals, even from a distance. For small, obscure plants that may be hidden under the canopy of the other vegetation, you might have to search every square centimeter of habitat; in this case minimizing the total sample area may be critical.

Spatial distribution of individuals in the population. Very few plant populations are randomly distributed in the area they occupy. If they were, different shapes of the same quadrat size would perform similarly. Most plant populations, however, are aggregated or clumped in their distribution. In this situation rectangular quadrats will yield more precise estimates than square or circular quadrats of the same size. This is because rectangular quadrats are more apt to include some of the clumps of plants inside of them, thereby reducing the number of zero counts and reducing the number of very high counts. This decreases the variation among the quadrats and increases the precision of estimates. It is best if the quadrat length (i.e., the length of the long side of the quadrat) is longer than the mean distance between clumps.

As an example, consider the species *Primula wilcoxii*, which grows on the shaded side of terraces on a terraced slope in the foothills near Boise, Idaho. The terraces are approximately 1.5 meters apart. In this case, 1m x 1m quadrats would be a very poor choice, because many of these would fall between terraces, resulting in many zero values. Some of the 1m x 1m quadrats, however, would fall right on the terraces, and very high counts would be obtained for these quadrats. For this species at this terraced site, quadrats of 0.5m x 2.5m perform well.

Depending on the nature of your population, orientation of quadrats can be very important. You want to orient rectangular quadrats to capture the variability *within* the quadrats rather than *between* the quadrats. Thus, if there is some gradient such as elevation or moisture to which the plant population responds differently, you want to make sure your rectangular quadrats follow that gradient in order to incorporate the variability within the quadrats. Let's assume the plant population occurs on a north-facing slope. There are clumps of plants up and down the slope, but more clumps are near the bottom of the slope than near the top. You want to orient your quadrats with the long side going up the slope, rather than placing them along contour lines. This results in lower among-quadrat variance and higher precision. Similarly, you want to orient quadrats perpendicular to a stream rather than parallel to it.

Edge effects. The edge of a quadrat is its outer boundary. The more edge a quadrat has, the greater the difficulty in determining whether plants near the edge are in or out of the quadrat. Rectangular quadrats have more edge per unit area than squares or circles. Although this is an important issue, you can choose between two simple conventions that help to minimize the nonsampling error associated with plants landing right on the edge. You can either: (1) count every other plant that lands right on the edge as "in" and every other one as "out," or (2) specify that any plant that lands on the edge of two adjoining sides of the quadrat is considered "in" while any plant landing on the other two adjoining sides is "out." Of course, you need to be consistent in applying whichever convention you choose and to make sure, through training and documentation, that others involved in the monitoring during the first and all subsequent years use the same convention (especially with permanent quadrats).

Density of target population. If the density of whatever is being sampled (e.g., individuals, stems, or ramets) is relatively high, you will want to use smaller quadrats because you don't want to be counting hundreds to thousands of plants in many quadrats. On the other hand, if the density is relatively low, you will want to use larger quadrats to avoid sampling many quadrats with no plants in them.

Ease in sampling. The considerations here are the difficulties in searching the entire quadrat area and keeping track of what portions of a quadrat have already been searched. With larger quadrats, long, narrow rectangles are easier to search because you can start at one end and keep track of counts at intervals along the quadrat. With large, square quadrats, you will probably have to subdivide the quadrat area to ensure you don't double count.

Disturbance effects. If the quadrat size/shape is so large that you have to stand in the quadrat to search through it, you risk impacting the population through your sampling. This is particularly important when sampling permanent quadrats, because the changes you observe over time may simply be the result of your impacts to the quadrats and not reflect the true situation in the sampled population as a whole. It is also a problem when using temporary quadrats, however, especially if you impact areas of the quadrat before you have searched them.

Mathematical equations for determining optimal quadrat size. Krebs (1989) gives mathematical equations for determining optimal quadrat size using either the Wiegert's or Hendricks' methods. These equations incorporate the following components: (1) the variation among quadrats, (2) the cost of measuring one quadrat, and (3) the cost of locating one additional quadrat.

b. Computer-simulated sampling investigation

As stated, rectangular quadrats perform better than square or circular quadrats when sampling clumped populations. But there are still two unanswered questions: (1) What are the actual tradeoffs of changing quadrat size and shape on the number of quadrats to sample or on the total area sampled? (2) As you make quadrats larger, you will presumably have to sample fewer of them—but how many fewer? You can investigate this in the field, but you're somewhat limited in the number of different sizes and shapes you can try, and there are potential negative impacts from repeated sampling across the entire area.

One of the authors, Dan Salzer (in prep.), has investigated some of these sampling design decisions using computer-simulated sampling. On the computer he generated two populations, each with 4,000 plants. Plants in both populations exhibit a clumped distribution pattern, though they differ in the degree of clumping. One of the populations has plants that are distributed along a gradient. Salzer then used the computer to draw random samples from each population, using quadrats of different sizes and shapes. The results are summarized below.

The clumped-gradient population. We'll consider first the population of 4,000 plants depicted in Figure 7.3. This population is termed the "clumped-gradient population" because the plants are both clumped and distributed along a gradient (note that this gradient follows the x-axis: there are more clumps near the left side of the macroplot than there are near the right side). This population was subjected to 30 different sampling designs that differed in the width and length of the quadrats. The following quadrat widths were used: 0.25m, 0.5m, 1.0m, 2.0m, and 4.0m. The following quadrat lengths were used: 1m, 2m, 5m, 25m, and

50m. Every combination of quadrat width and quadrat length was used to sample the population (e.g., 0.25m x 1m, 0.25m x 2m...4m x 25m, 4m x 50m). Sampling was conducted so that the long side of the quadrat was oriented along the gradient (i.e., the long side was oriented parallel to the x-axis of the population as depicted in Figure 7.3).

For each of the 30 sampling designs, the entire population was sampled (i.e., all the quadrats that fit in the population, without overlapping) so that true, parametric values for the mean density and standard deviation could be calculated for every design. This is desirable for comparing various sampling designs, but is nearly impossible to achieve in a field setting. The true parametric values were plugged into a sample size formula to determine how many quadrats would need to be sampled to attain the desired precision. The precision level selected was an estimated mean density with a 95% confidence interval that was no longer than ±30% of the mean value. This brought performance of each sampling design into a common currency, the number of quadrats to sample, so that they could be compared with one another.

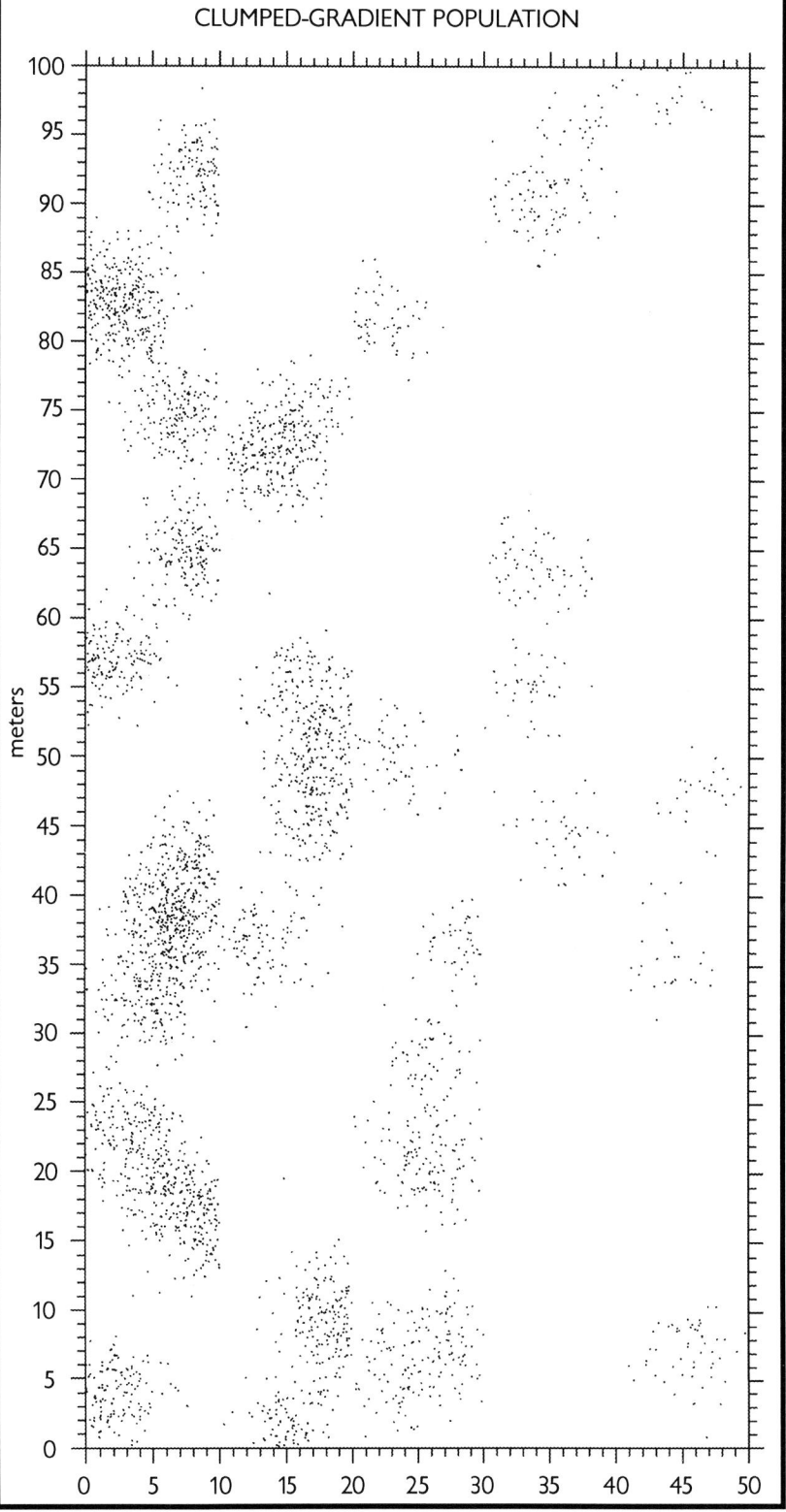

FIGURE 7.3. The "clumped-gradient population." A population of 4,000 plants aggregated into clumps and responding to a gradient that runs from left to right (along the x-axis). Note the much greater number of clumps near the left side of the population.

By knowing the size and number of quadrats being used, you can also calculate the proportion of the population sampled.

Table 7.1 shows the results for three designs that differ dramatically in the number of quadrats that need to be sampled and the proportion of the population sampled. The design that minimized sample size (ten 4m x 50m

quadrat width	quadrat length	# of quadrats	proportion of the population sampled
0.25m	1m	416	2.1%
1m	10m	99	19.8%
4m	50m	10	40%

TABLE 7.1. Results of three designs used to sample the clumped-gradient population shown in Figure 7.3. The long sides of the gradients were oriented along the gradient (the x-axis). All designs achieved the same level of precision but differed greatly in the number of quadrats required and the proportion of the population sampled.

quadrats) required sampling 40% of the population. The design that minimized the proportion of the population sampled (only 2.1% of the population using 0.25m x 1m quadrats) required sampling 416 separate quadrats.

Figure 7.4 displays the same four categories of information as Table 7.1: quadrat width, quadrat length, number of quadrats, and proportion of the population sampled. The results are summarized by width category below:

1m width. Six different sampling designs used a 1m wide quadrat. On the x-axis shown on Figure 7.4 is quadrat length, starting with 1m on the left and going all the way to 50m on the right. Recall that six different quadrat lengths were tested. The numbers you see next to the data points are the number of quadrats that need to be sampled to meet the desired level of precision. To get this precise an estimate of the mean density with a 1m x 1m

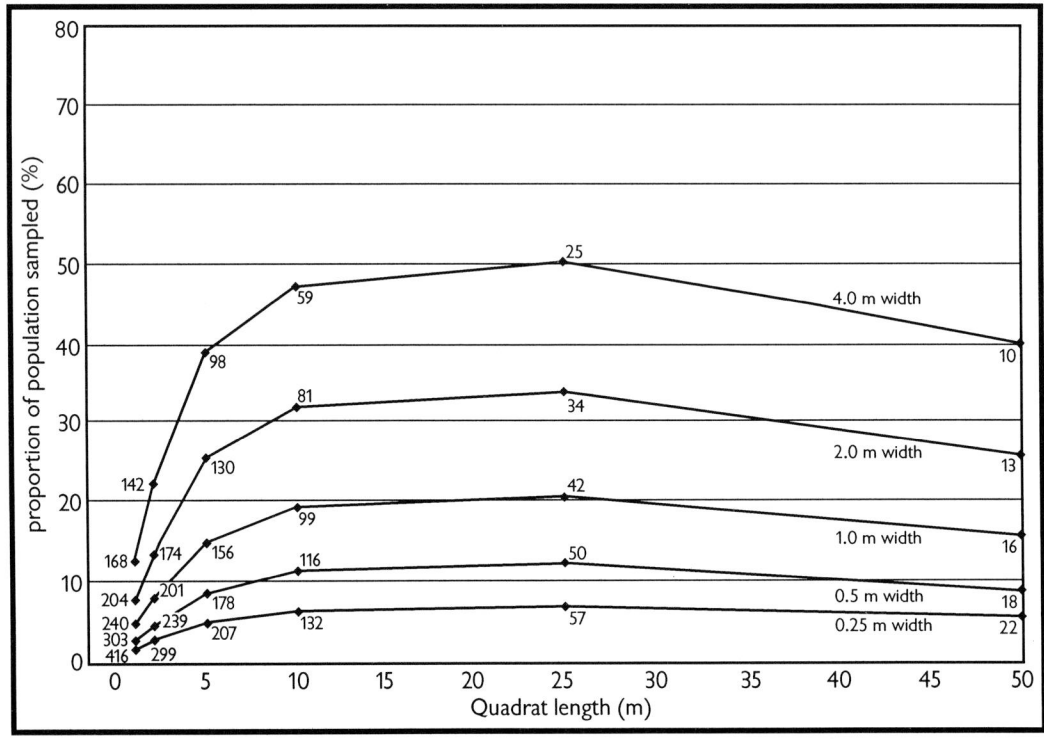

FIGURE 7.4. Comparison of 30 sampling designs for the clumped-gradient population in Figure 7.3. All designs achieve the same level of precision. Quadrats are oriented along the gradient (i.e., along the x-axis of the population). Numbers next to data points are the number of quadrats that must be sampled using that particular quadrat size and shape.

quadrat requires sampling 240 quadrats of this size. This same level of precision could be attained with only sixteen 1m x 50m quadrats.

2m width. By doubling the quadrat width to 2m, at each of the quadrat lengths the sample size numbers go down, but the proportion of the population sampled goes up. For example, instead of sampling 16% of the population with a 1m x 50m quadrat, you must sample 26% of the population with a 2m x 50m quadrat size.

4m width. By doubling the width again, to 4m, the proportion of the population sampled keeps going up. If we choose the 4m x 50m quadrat size, you now have to sample 40% of the population to achieve the desired level of precision.

0.5m width. Now look what happens if you reduce the width of the original 1m wide quadrat by one half. Even though sample sizes are a little bit higher with the 0.5m wide quadrat size than with the 1m wide quadrat, the proportion of the population that must be sampled is considerably less.

0.25m width. The same trend continues, with sample sizes a little bit higher than for the 0.5m width, but with a reduction in the proportion of the population sampled.

Choosing the best design for your situation. A comparison of these different designs shows that they vary considerably in both the number of quadrats to sample and the proportion of the population that is sampled. Some of these designs offer smaller sample sizes and smaller proportions of the population. For example, compare these two designs: sample twenty-five 4m x 25m quadrats (50% of the entire population), in which case you must count about 2,000 plants, or sample twenty-two 0.25m x 50m quadrats (5.5% of the population), in which case you must count only about 220 plants. Which of the 30 designs is best? It depends on the growth form of the individual plants and the habitat. How conspicuous are individual plants? Can they be spotted at eye-level or does it take careful searching of every square centimeter of sample area? How big a problem is edge effect? Are plants single-stemmed with small diameter stems clearly arising from a rooted point (edge effect not a problem)? Or are the target plants bunch grasses with a wide basal area and amorphous shapes (edge effect a problem)?

If plants are small and inconspicuous, with distinct single rooted stems, look for a design that has both a small sample size number and samples a small proportion of the population. The 22 0.25m x 50m quadrats would be a good choice in this case. Realize, however, that even if minimizing the sample area is critical, you will not want to sample 416 0.25m x 1m quadrats.

If plants are large and easily visible from eye level, you might choose a wider quadrat size, leading to a smaller sample size. The larger proportion of the population sampled might not carry much of a penalty (cost) if the portions of the quadrats between clumps can be searched rapidly.

Results for the same clumped-gradient population with quadrat orientation reversed. Figure 7.5 shows the results of sampling the same clumped-gradient population, but this time with the quadrats oriented in the opposite direction (i.e., with the long side parallel to the y-axis). Rather than looking at the individual sample sizes, concentrate on just the relative proportion of the population that must be sampled. Because we've positioned the long sides of our quadrat perpendicular to the gradient, quadrats located near the left of the macroplot will have high numbers of plants, while quadrats located near the right of the macroplot will have low numbers. This pattern of high and low quadrat counts is undesirable, producing a

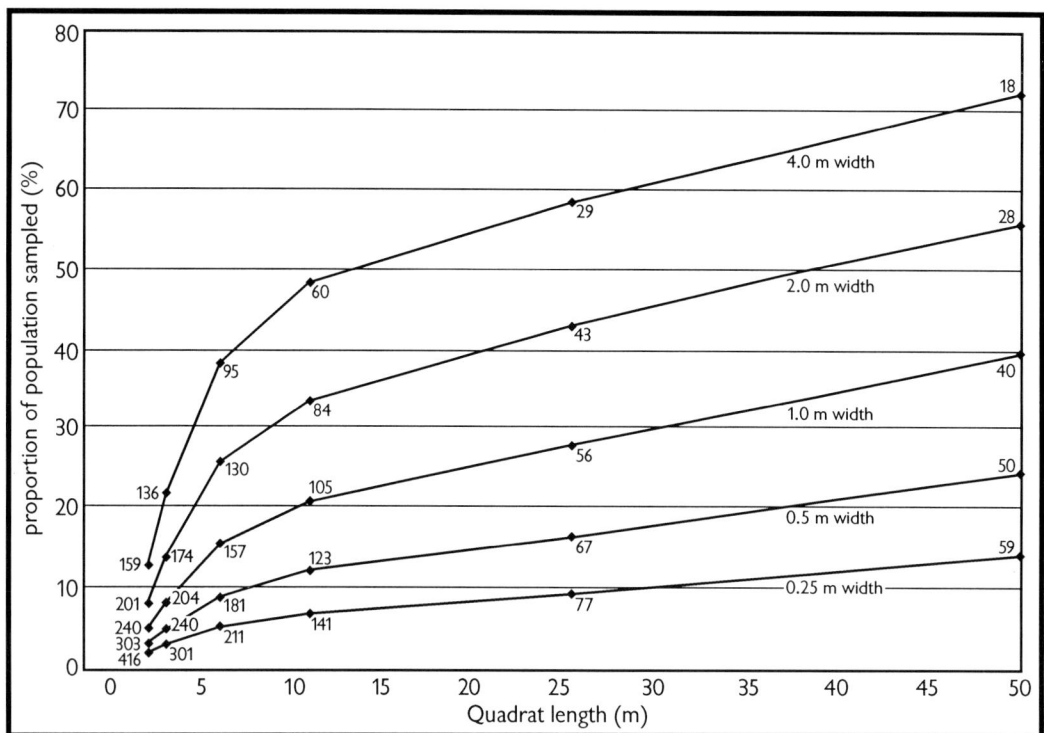

FIGURE 7.5. Comparison of 30 sampling designs for the clumped-gradient population of Figure 7.3. All designs achieve the same level of precision. Quadrats are oriented against the gradient (i.e., along the y-axis of the population). Numbers next to data points are the number of quadrats that must be sampled using that particular quadrat size and shape.

high standard deviation and wide confidence intervals. With the 4m x 50m quadrat, you need to sample over 70% of the population. You would be better off counting all of the plants in the macroplot (conducting a complete census) than using this quadrat size. Clearly it is better to use a narrower quadrat that is oriented in the opposite direction.

The dense-clumped population. Figure 7.6 shows a different population of 4,000 plants, where the clump centers themselves are randomly distributed. We call this the "dense-clumped population." Figure 7.7 shows the same comparison of 30 different sampling designs that was performed on the clumped-gradient population. Because of the tighter clumping of plants in the dense-clumped population, sample sizes are even greater for small square or short and wide quadrats than they were for the clumped-gradient population. This is because quadrats *with* plants tend to have higher counts and there are more quadrats with zero plants, a situation that drives up the standard deviation. It would take, for example, 578 1m x 0.25m quadrats to achieve the desired level of precision in the dense-clumped population as compared to 416 in the clumped-gradient population.

The pattern of narrower quadrats reducing the proportion of the population to sample continues as you make your quadrats narrower and narrower. Figure 7.8 shows a comparison of different quadrats from the dense-clumped population that are all the same *size* (all 1m² in area), but of different shapes. The graph on the left shows comparisons for the dense-clumped population. As you go from a square to a long, skinny quadrat, the number of quadrats that must be sampled declines from nearly 400 to less than 100. The graph on the right of Figure 7.8 shows the same quadrat comparisons, but for a 4,000 plant population where all the plants are randomly located (i.e., there is no clumping). If plants are randomly distributed, quadrat shape has no influence on the number of quadrats to sample. This, however, is seldom the case in nature.

Even though the narrower quadrat sizes perform better statistically, there are practical limitations that must be considered. For example, the 2cm wide quadrat performs best (see Figure 7.8), but this would be a ridiculous size to try to use in the field, because of the tremendous amount of "noise" introduced by edge effect.

c. Determining quadrat size and shape in real populations

The best way to determine the appropriate quadrat size and shape is to approach every new sampling situation without a preconceived idea of the quadrat size and shape you will use. Quadrat size and shape should be determined during pilot sampling. Wander around the population area and study the spatial distribution of the plant you will be sampling (use pin flags or flagging to improve the visibility of clumps). Attempt to answer the following questions: (1) At what scale(s) can you detect clumping? (2) How large are the clumps and what are the distances between clumps? (3) How long will quadrats need to be to avoid having many quadrats with zero plants in them? (4) How narrow will quadrats need to be to avoid counting hundreds or thousands of plants whenever the quadrat intersects a dense clump? (5) How wide an area can be efficiently searched from one edge of a quadrat?

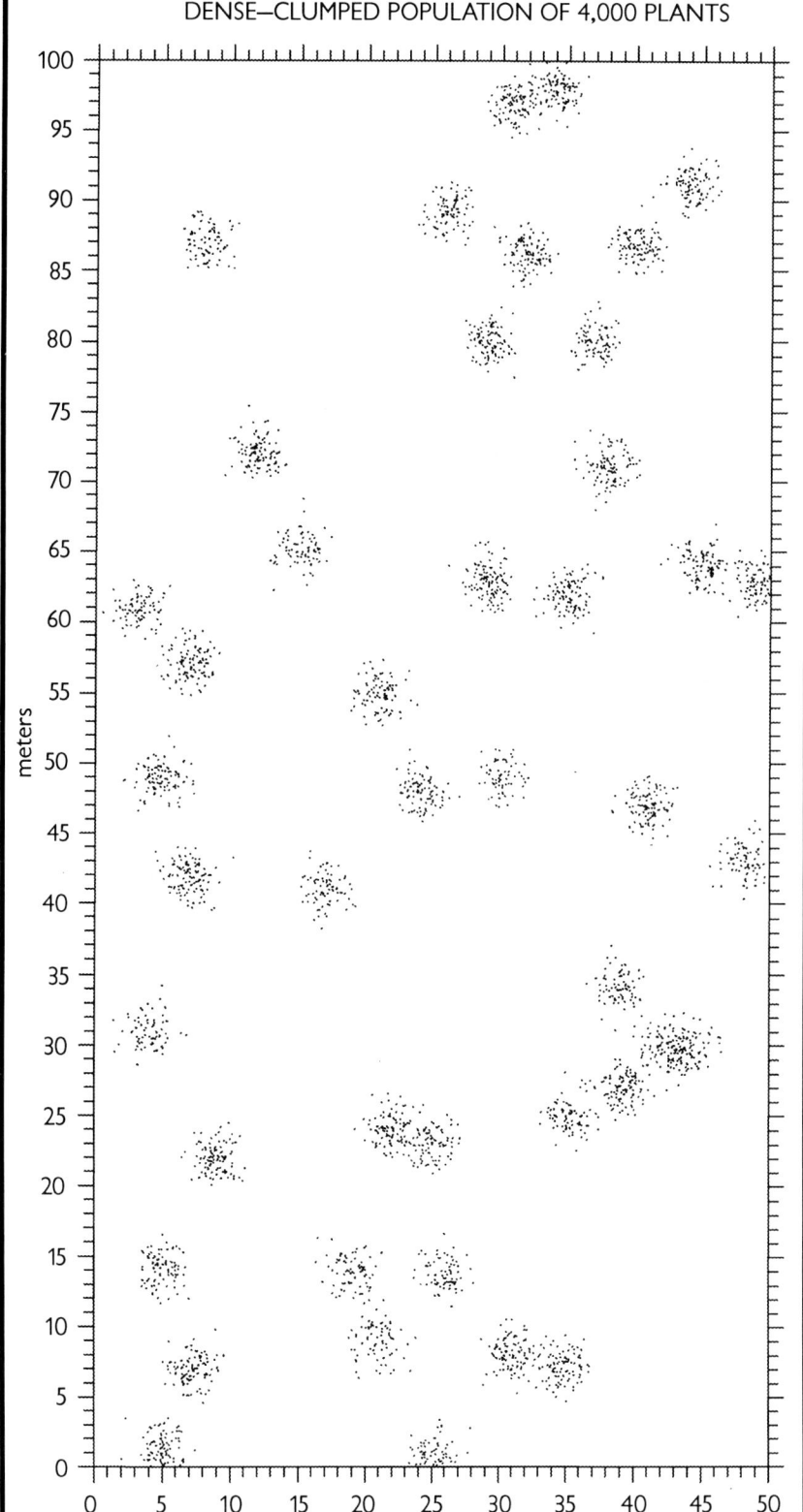

FIGURE 7.6. The "dense-clumped" population. A population of 4,000 plants aggregated into dense clumps.

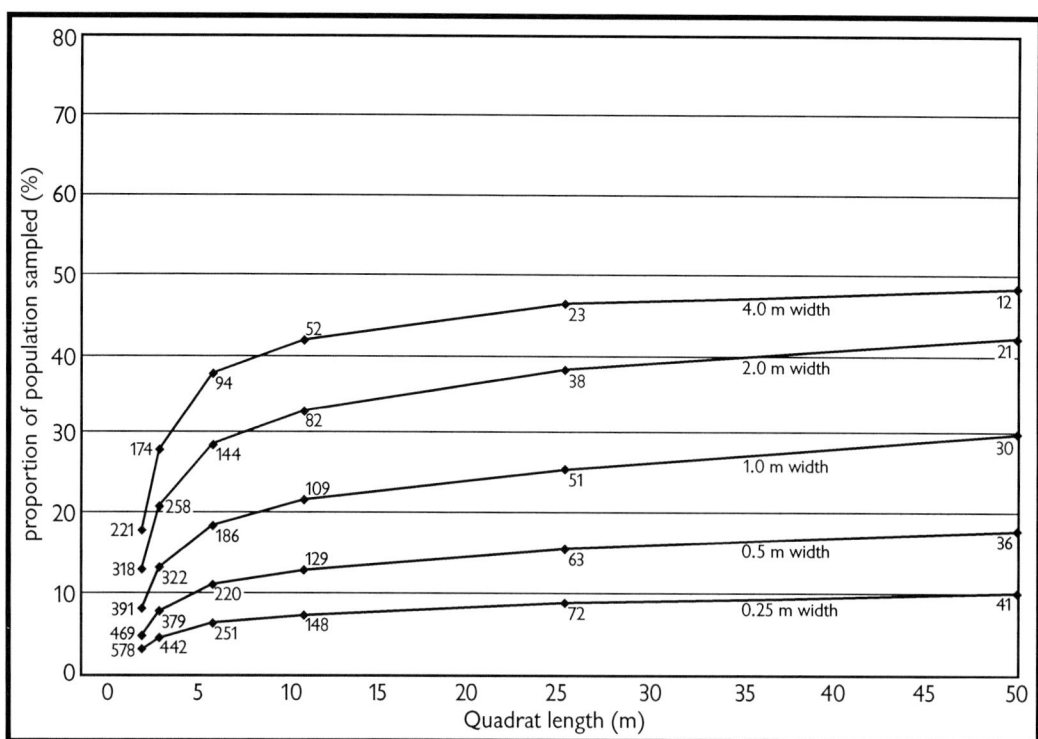

FIGURE 7.7. Comparison of 30 sampling designs for the dense-clumped population of Figure 7.6. All designs achieve the same level of precision. Quadrats are oriented along the x-axis of the population (although, since there is no gradient in this population, results would have been similar had the quadrats been oriented along the y-axis). Numbers next to data points are the number of quadrats that must be sampled using that particular quadrat size and shape.

(6) How big a problem will edge effect be? Appendix 17 gives a procedure for comparing the efficiency of different quadrat sizes and shapes through pilot sampling.

In many rare plant monitoring situations, a 0.25m or 0.5m quadrat width works well. Either is a convenient width to search in. A 1m or 2m wide quadrat is difficult to search because it is hard to see plants near the far edge (unless all the plants are fairly large and there is minimal associated vegetation to obscure your line of sight). The quadrat length should be determined by the size of the area that you are working in and the spatial distribution of the plants you are counting. You want to avoid getting many zeros so you want your quadrats to be long enough to hit several clumps. You also don't want your quadrats so long that you have to count thousands of plants—the time involved and the potential measurement error associated with counting that many plants would be too great.

2. Quadrats for estimating frequency

Frequency is most typically measured in square quadrats. With frequency sampling you are only concerned with whether the species of interest is present or absent within each quadrat–you make no counts. Because only presence or absence is measured, square quadrats are fine for this purpose (unlike the situation with density, you *want* at least 30% of your quadrats to have no plants in them; the reason for this is discussed below and in Chapter 8). With frequency sampling you are estimating the proportion of all possible quadrats in the population that have the species (or other attribute of interest) in them. Figure 7.9 shows the clumped-gradient population overlaid by a 2m x 2m grid. Of the total number of 1250 grid cells in this population, 540 cells have one or more plants in them. Thus, the true frequency is 43.2%. When you sample this population you will randomly select some subset of these 1250 quadrats. Let's say you sample 100 quadrats. If 45 of the 100 quadrats contain the plant, then your estimate of the true percent frequency would be 45%.

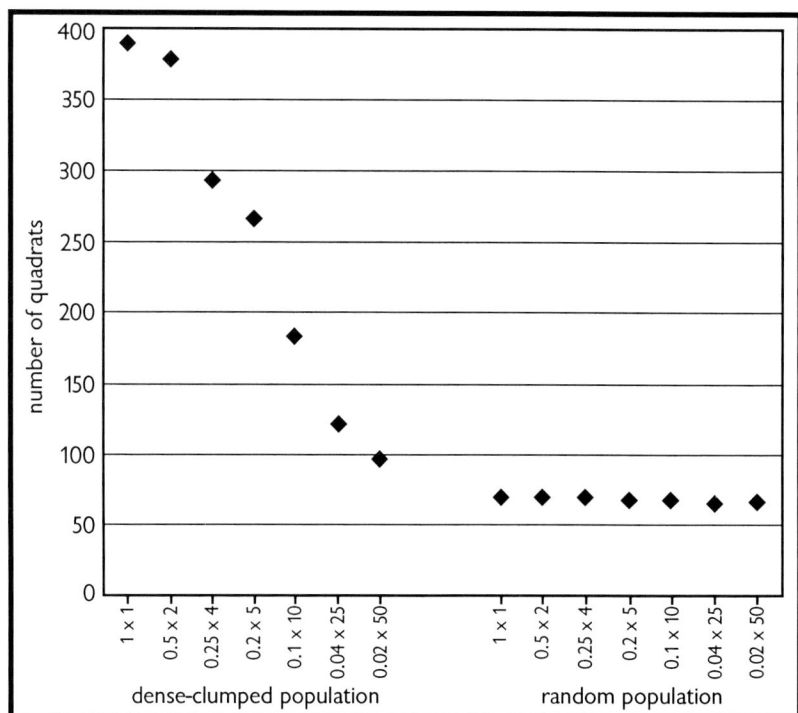

FIGURE 7.8. Comparison of sample sizes needed to achieve 95% confidence intervals within 30% of the mean using quadrats of the same size (area) but different shapes. Quadrat area is 1m². The graph on the left shows the necessary sample sizes for the dense-clumped population shown in Figure 7.6. The graph on the right shows necessary sample sizes for a population with individual plants distributed randomly.

Quadrat size has a strong influence on the resulting percent frequency values. If you make the quadrat large enough, you will have some individuals in every quadrat, giving you a frequency of 100%. This would not enable you to track any upward changes in frequency. On the other hand, if your quadrat is very small, you will end up with very low frequency values that will not be sensitive to declines in frequency.

Good sensitivity to change is obtained for frequency values between 30% and 70%. Because frequency values are measured separately for each species, what is an optimum size quadrat for one species may be less than optimum or even inappropriate for another. If you are measuring the frequency of more than one species, this problem is partially resolved by the use of a quadrat frame that includes nested quadrats of different sizes. For further discussion see Chapter 8.

3. Quadrats for estimating cover

Cover is sometimes measured by visually estimating canopy cover in quadrats. From the perspective of statistical precision, the same types of considerations as those given for density apply: long, thin quadrats will likely be better than circular, square, or shorter, wider rectangular quadrats. From a practical perspective, however, estimating cover accurately in long, thin quadrats is difficult. The amount of area in the quadrat is also a concern: the larger the area, the more difficult it is to accurately estimate cover.

For clumped populations the best approach is usually to randomly position transects in the population to be sampled, and to systematically (with a random start) place square or small rectangular quadrats of a size that facilitates accurate cover estimation along each transect. The transects, not the quadrats, are treated as the sampling units. Because the transects will intersect several clumps of the population, this ensures much of the variation will be incorporated within each sampling unit. If individual quadrats are treated as the sampling units, most of the variation will be between sampling units. This design is really a two-stage sampling design, with the transects serving as the primary sampling units, and the quadrats serving as the secondary sampling units. We treat this in more detail below.

4. Quadrats for estimating biomass

For the same reason as given for density, long, thin quadrats are likely to be better than circular, square, or shorter, wider rectangular quadrats (Krebs 1989). Edge effect can result in significant measurement bias if the quadrats are too small (Wiegert 1962). Since aboveground vegetation must be clipped in some quadrats, circular quadrats should be avoided because of the difficulty in cutting around the perimeter of the circle with handshears, and the nonsampling errors that will likely result. It is impractical in the field to estimate and clip biomass in long, narrow quadrats. For this reason, we recommend you use square quadrats. Like the case with estimating cover in quadrats, the quadrats for estimating and clipping biomass can be arranged along transects, with the transects treated as the sampling units.

5. Lines and points for estimating cover

Line interception and point interception are two techniques often used to estimate cover. From a theoretical standpoint both lines and points are plots: the

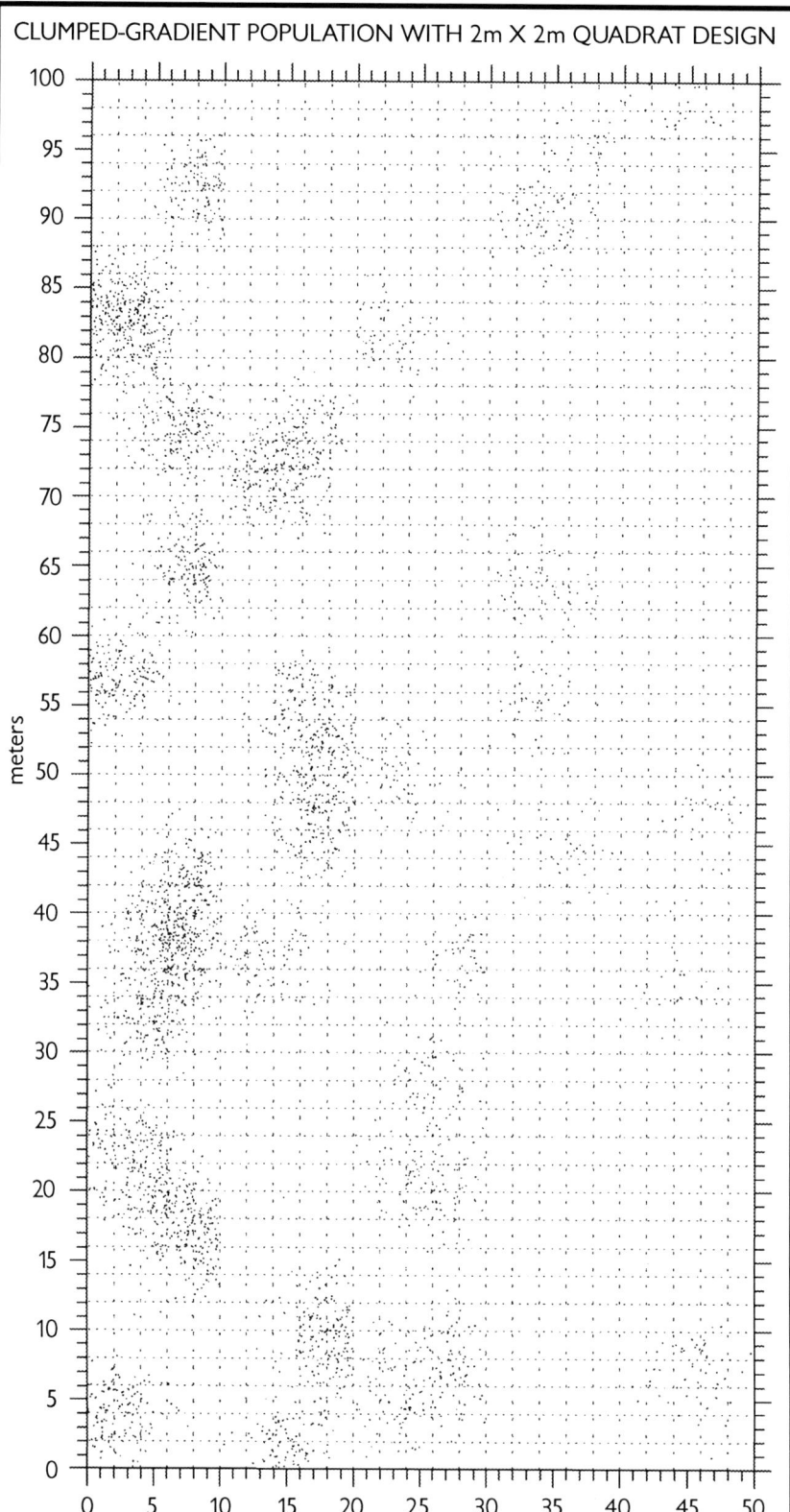

FIGURE 7.9. The clumped-gradient population with a grid of 2m x 2m quadrats overlaid on it. There are 1,250 possible quadrat locations for this size and shape of quadrat. Note that the quadrats do not overlap, yet cover the entire sampled population (the macroplot).

line is a quadrat with one dimension reduced to a line, and the point is a quadrat with both dimensions reduced to a point (Bonham 1989). These techniques are well-established in vegetation sampling. The major statistical considerations of these methods have to do with the width of the lines and the size of the points and with the placement of the lines and points in the area to be sampled.

a. Width of lines and size of points

The theoretical basis of line interception depends on reducing the width of the lines to zero (Lucas and Seber 1977; DeVries 1979; Floyd and Anderson 1987). Similarly, the size of points must also be as close as possible to being dimensionless (Goodall 1952). Making points as close to dimensionless as possible is important for obtaining good estimates of the true cover; it is less important if you are only interested in tracking change over time and you use the same size points each time. For line interception, read only along one edge of a measuring tape and ensure the tape is not inadvertently moved to include or exclude certain plants.

For points, the investigator should attempt to make the point as small as possible and to avoid selection bias. Because they effectively reduce the point to zero, crosshair sighting devices are preferable to metal rods to obtain reliable estimates of the true cover. These devices also eliminate the possibility of bias that can result in the placement of metal rods when frames are not used. A disadvantage of crosshair sighting devices is that only the vegetation stratum nearest to the sighting device can be sampled. If the plant species in which you are interested is under the canopy of shrubs or taller herbaceous plants, you will not be able to use a sighting device for this purpose (unless a second person is employed to move the upper story out of the way; this is acceptable as long as the movement doesn't change the probability of the understory plant being intercepted). Pins, even though they are not dimensionless, might be the better choice in this case, because you can move them down through all layers of vegetation and record "hits" in as many strata as you desire. You should, however, make the ends of these pins as sharp as possible. You must also ensure that pins are in some type of frame that eliminates the bias that results from attempting to manually place a pin vertically through vegetation (a tripod frame can be constructed that holds only a single pin).

b. Length of lines

Because each line is a single sampling unit, the precision of cover estimates will depend on the variation among lines. Lines should be long enough to cross most of the variability in the vegetation being sampled (for the same reasons discussed relative to quadrat size and shape for density sampling). Just as for quadrats, the optimum line length should be determined from pilot sampling.

E. How Should Sampling Units Be Positioned in the Population?

There are two requirements that must be met by a monitoring study with respect to positioning sampling units in the population to be sampled: (1) some type of random sampling method must be employed, and (2) the sampling units must be positioned to achieve good interspersion of sampling units throughout the population. Before discussing different methods of random sampling, let's discuss these two characteristics in more detail.

Random sampling. Critical to a valid monitoring study design is that the sample has been drawn randomly from the population of interest. Several methods of random sampling can be used, many of which are discussed below. The important point is this: all the statistical analysis techniques available to us are based on knowing the probability of selecting a particular sampling unit. If some type of random selection of sampling units is not incorporated into your study design, you cannot determine the probability of selection, and you cannot make statistical inferences about your population. Preferential sampling, the practice of subjectively selecting sampling units, should be avoided at all costs.

Interspersion. One of the most important considerations in sampling is good interspersion of sampling units throughout the area to be sampled (the target population). Although Hurlbert (1984) uses the term "interspersion" to apply to the distribution of experimental units in manipulative experiments, the term can also be applied to sampling units in observational studies. The basic goal is to have sampling units well interspersed throughout the area of the target population. The practice of placing all the sampling units, whether they be quadrats or points, along a single or even a few transects should be avoided. Arranging sampling units in this manner results in poor interspersion of sampling units and the sample will not adequately represent the target population. This is true even if the single transect or few transects are randomly located.

Eight types of random sampling are discussed below and summarized in a table at the end of this section.

1. Simple random sampling

A simple random sample is one that meets the following two criteria: (1) each combination of a specified number of sampling units has the same probability of being selected; and (2) the selection of any one sampling unit is in no way tied to the selection of any other (McCall 1982). Two methods of simple random sampling are described below.

a. Simple random coordinate method

As shown in Figure 7.10, random coordinates are selected for each of two axes. The point at which these intersect specifies the location of a sampling unit. Coordinates that fall out of the target population boundaries are rejected. This method will work for square sampling units, such as those used to measure

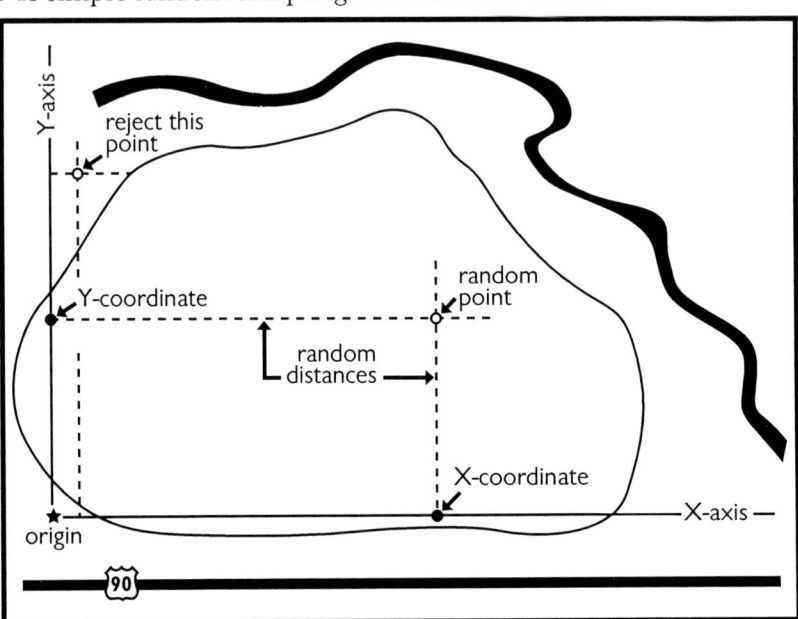

FIGURE 7.10. Locating points using the simple random coordinate method (adapted from Chambers and Brown 1983). Although this method will work to position points or square quadrats, the grid-cell method is much better for locating long, narrow quadrats or lines.

frequency,[1] but it will not perform well when the sampling units are lines or long rectangles. In the latter case, we have to decide in what direction from the randomly selected point we will orient the lines or rectangles. One way of doing this is to select a random compass bearing. There are three problems with this approach:

1. There is no unbiased way to deal with compass bearings that send a portion of a line or rectangular quadrat outside the target population. You can either reject such bearings, with the result that your sample will be biased toward the center of the population (i.e., you will be less likely to sample the edges of the population), or you can "reflect" the line or quadrat from the population edge back into the population, in which case you bias your sampling toward the edges of the population.

2. This technique introduces the probability of overlapping sampling units. If the sampling units are lines, this is not a great problem, since lines represent an infinite population regardless of their orientation. For quadrats, however, such overlap is highly undesirable, because we will not be able to use the finite population correction factor discussed later in this chapter.

3. You want transects and long, narrow quadrats to be oriented along any gradient such as elevation or moisture to which the target plant population responds differently. This incorporates most of the variability within the sampling units and minimizes the variability between them. If we orient these sampling units using random compass bearings we will end up with some sampling units oriented along the gradient, some oriented perpendicular to the gradient, and some oriented in intermediate positions. This does not make for an efficient design.

b. The grid-cell method

The grid-cell method eliminates the problems associated with the random coordinate method and is one of the most efficient and convenient methods of randomly positioning quadrats. The population area is overlaid with a conceptual grid (there is no need to actually lay out tapes and strings to achieve this), where the grid cell size is equivalent to the size of each sampling unit.

Consider the clumped-gradient population example introduced earlier. We've overlaid a grid of 4m x 10m quadrats on this population (Figure 7.11).[2] If we want to sample ten 4m x 10m quadrats from this population, we would first divide the population into 125 different 4m x 10m cells, as shown on Figure 7.11. Since we are *sampling without replacement*, 125 possible quadrat positions (5 along the x-axis times 25 along the y-axis) are possible, none of which overlap. Once one is sampled it will not be sampled again (at least not during the same sampling period).

---

[1] Although such a random selection procedure is justified for frequency sampling, the time required to position 100 to 200 or more frequency quadrats makes this procedure impractical. Instead, some type of systematic approach is usually used.

[2] The 4m quadrat width was chosen because it shows up well in diagrams. For most real-life sampling situations, 4m is too wide (because it is too difficult to search for plants without disturbing the inside of the quadrat, and in other respects this is often an inefficient width).

One way to draw a random sample (*n*) of 10 quadrats from the population (N) of 125 possible quadrats is to number each one of the quadrats from 1 to 125, put numbers from 1 to 125 on small slips of paper into a box, shake thoroughly, and select 10 slips from the box. Although valid, this is a time-consuming method. A much more efficient method of quadrat selection would be to select random points along both the x and y axes to serve as beginning points for each quadrat. Here is how to accomplish this.

Along the x axis there are five possible starting points for each 4m x 10m quadrat (at points 0, 10, 20, 30, and 40). Number each of these points 0 to 4 according-ly (in whole numbers). Along the y axis there are 25 possible starting points for each quadrat (at points 0, 4, 8...96). Number each of these points 0 to 24 (again in whole numbers) accordingly.

Now, using a random number table or a random number genera-tor on a computer or handheld calculator, choose at random 10 numbers from 0 to 4 for the x axis and 10 numbers from 0 to 24 for the y axis. (Directions on the use of random numbers tables and random number generators are given in Appendix 4). At the end of this process we will have

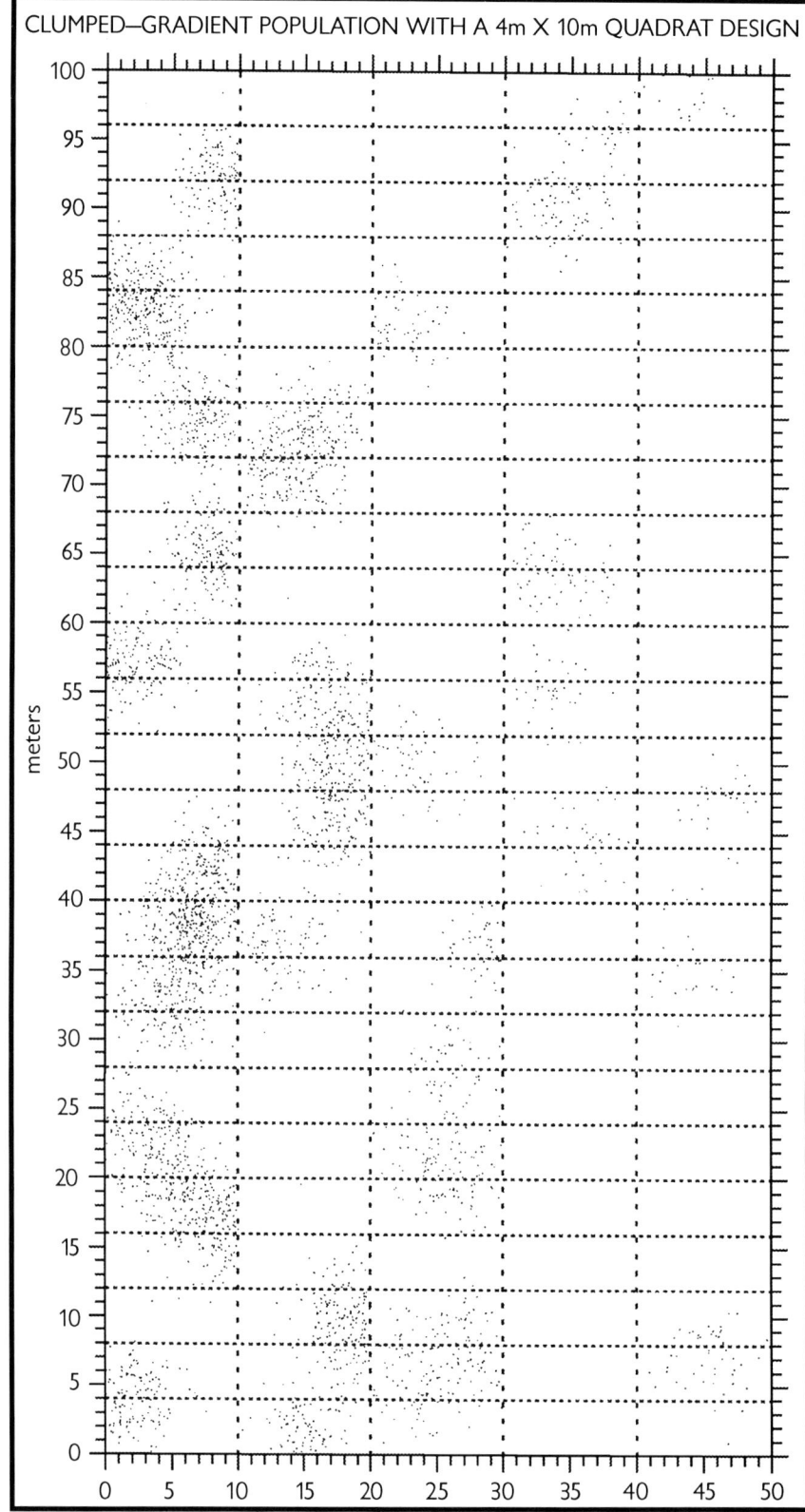

CLUMPED—GRADIENT POPULATION WITH A 4m X 10m QUADRAT DESIGN

FIGURE 7.11. The clumped-gradient population with a grid of 4m x 10m quadrats overlaid on it. There are 125 possible quadrat locations for this size and shape of quadrat.

10 pairs of coordinates. If any pair of coordinates is repeated, we reject the second pair and pick another pair at random to replace it. We continue until we have 10 unique pairs of coordinates.

Assuming that the x axis is on the bottom and the y axis is at the left, each pair of coordinates would represent the lower left corner of each quadrat. Thus, if we came up with the coordinates 0, 0 the quadrat would be placed with its lower left corner at the origin.

Now consider the population shown in Figure 7.12. This is the same population depicted in Figure 7.11, but instead of sampling with 4m x 10m quadrats we are now sampling with 4m x 25m quadrats. Now there are only two possible starting points along the x-axis (0 and 25). One way to select random positions along the x-axis with a random numbers table would be to consider every even number as the 0 position and every odd number as the 25 position. Alternatively, numbers 0-4 could represent the 0 position and 5-9 the 25 position. Or, you could flip a coin, with heads representing the 0 position and tails representing the 25 position.

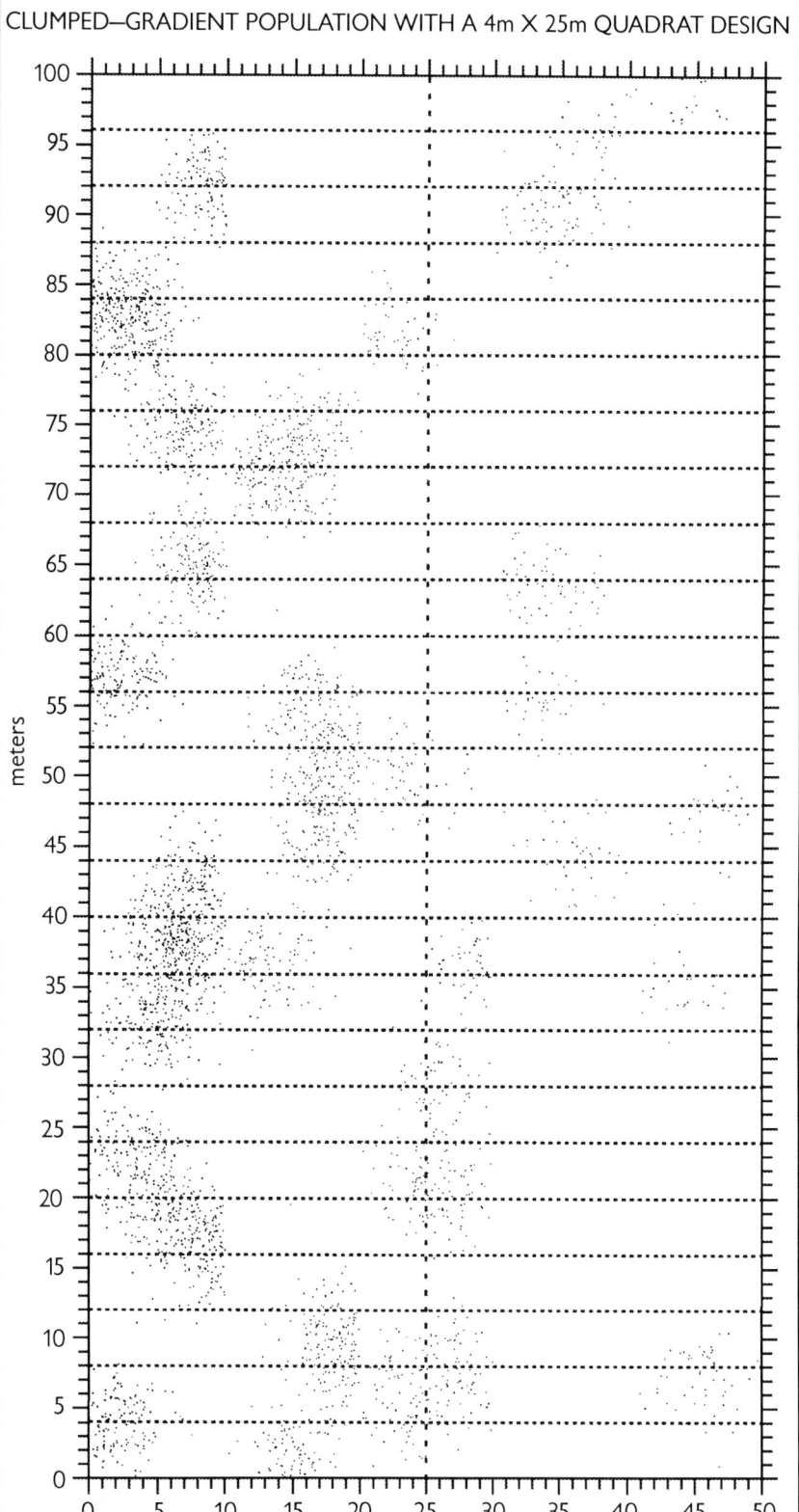

FIGURE 7.12. The clumped-gradient population with a grid of 4m x 25m quadrats overlaid on it. There are 50 possible quadrat locations for this size and shape of quadrat.

Now consider one last example (Figure 7.13). This time we decide to sample using 2m x 50m quadrats. In this case drawing a random sample is simplified because we only have to choose random locations along one axis (the y-axis).

c. Advantages and disadvantages of simple random sampling

As its name suggests, simple random sampling is the simplest kind of random sampling, and the formulas used to calculate means and standard errors are easier than with many of the more complex types of designs discussed below. Unless you are planning to use permanent quadrats to detect change, simple random sampling should only be used in relatively small geographical areas where a degree of homogeneity is known to exist. If the sampling area is large and/or the sample size is relatively large, as it often is for frequency or point-intercept simple random sampling, the time spent in locating quadrats or points and traveling between locations can be considerable. Also, simply by chance, some areas may be left unsampled. Figure 7.14 shows a simple random sample of 100 1m x 1m quadrats positioned within a 50m x 100m macroplot. By chance, some large portions of the macroplot did not receive any sampling units.

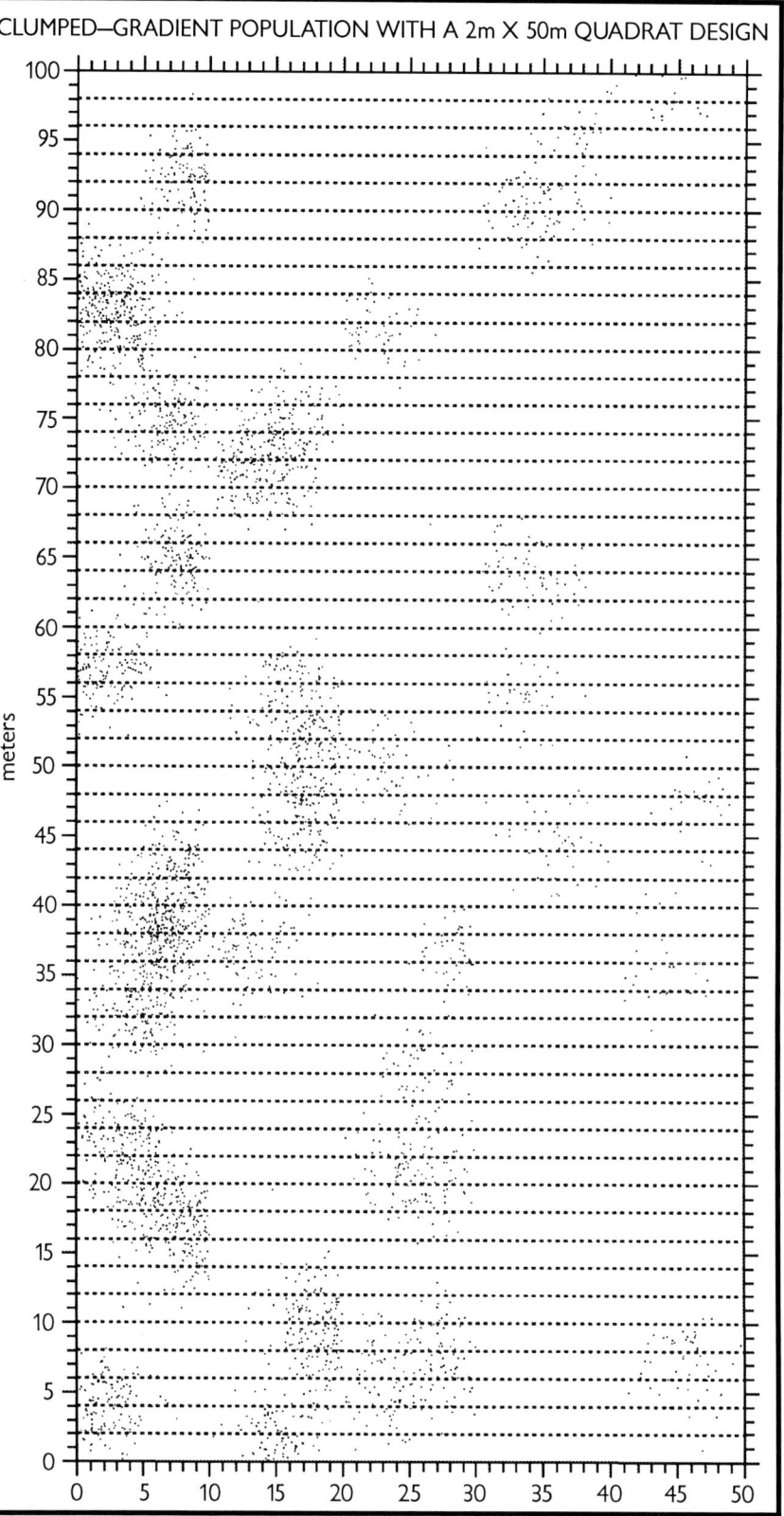

FIGURE 7.13. The clumped-gradient population with a grid of 2m x 50m quadrats overlaid on it. There are 50 possible quadrat locations for this size and shape of quadrat.

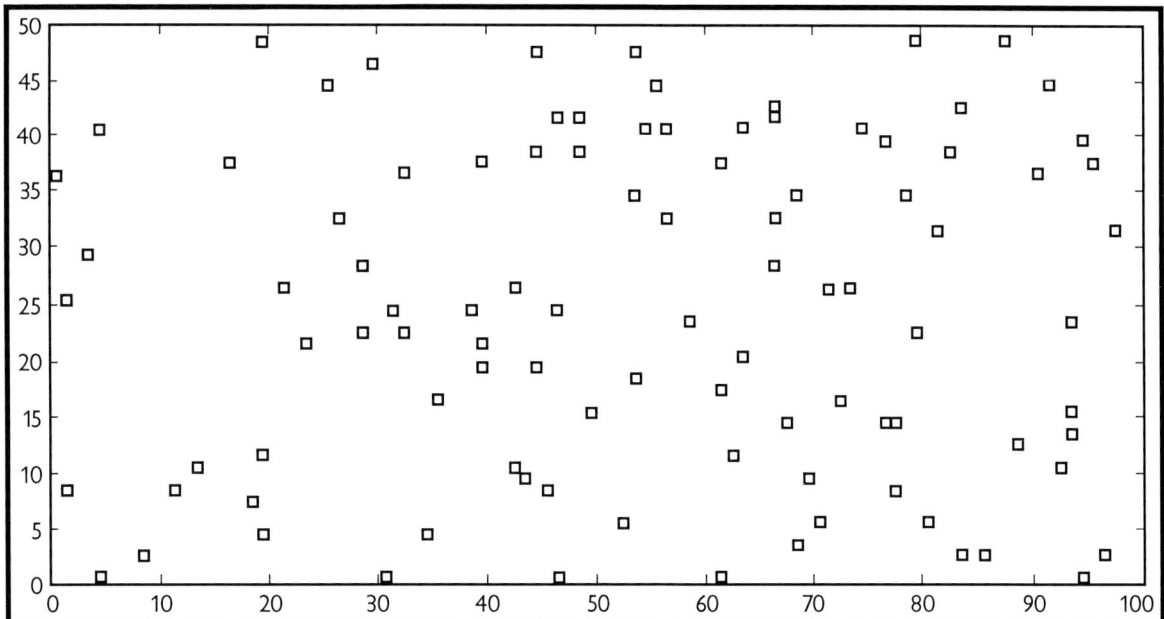

FIGURE 7.14. A simple random sample of 100 1m x 1m quadrats positioned within a 50m x 100m macroplot. Simply by chance, some large portions of the macroplot did not receive any sampling units.

A recent study using computer-simulated sampling (Salzer, in prep.) found that both restricted random sampling and systematic sampling designs (these are described below) result in more precise estimates than simple random sampling when sampling clumped distributions (the most common situation in plant and vegetation sampling).

2. Stratified random sampling

Stratified random sampling involves dividing the population into two or more subgroups (strata) prior to sampling. Strata are generally delineated in such a manner that the sampling units within the same stratum are very similar, while the units between strata are very different. Simple random samples are taken within each stratum.

a. Defining strata

Strata should be defined based on the response of the attribute you are estimating to habitat characteristics that are unlikely to change over time. Examples of characteristics that might be used to delineate strata are soil type, aspect, major vegetation type (e.g., forest or grassland), and soil moisture. You should avoid defining strata based on characteristics related to the attribute you are estimating, since this is likely to change with time, leaving you stuck with strata that are no longer meaningful. For example, if you are interested in estimating the density of species X and you note that the east half of the target population is much more densely populated than the west half, avoid basing your strata on this fact alone. If there is an obvious habitat feature responsible for this difference, such as aspect, then base your strata on this habitat feature. If there is no obvious reason for the difference you're probably better off using a simple random sampling procedure, because you might find that your management will result in more recruitment of species X into the west half of the target population, leaving you with a stratified random sampling procedure that is less efficient than simple random sampling.

Figure 7.15 shows a nature reserve with a valley running through it. A certain species of rare plant will be counted in 15 quadrats. The top figure (A) shows a simple random sample, where, by chance, 10 of the quadrats landed on the right plateau, three landed in the valley, and only two on the left plateau. The bottom figure (B) shows a stratified random design where an equal number of sampling units is allocated to each of the three strata.

b. Sampling within strata

Sampling units do not have to be allocated in equal numbers to each stratum. In fact, one of the benefits of stratified random sampling is that—when the attribute of interest responds differently to different habitat features—you can increase the efficiency of sampling over simple random sampling by allocating different numbers of sampling units to each stratum. Sampling units can be allocated: (1) equally to each stratum; (2) in proportion to the size of each stratum; (3) in proportion to the number of target plants in each stratum; or (4) in proportion to the amount of variability in each stratum.

Figure 7.16 shows a stratified random sampling scheme used in the National Wetlands Inventory. A sample of many plots, each 4 mi², was allocated to three strata in the state of North Carolina. Notice how the coastal stratum, because it has more habitat variability and greater suspected wetland density, is sampled more intensively. This differential sampling intensity, with greater effort allocated to strata with higher density and/or greater variability, is a common feature of stratified random sampling.

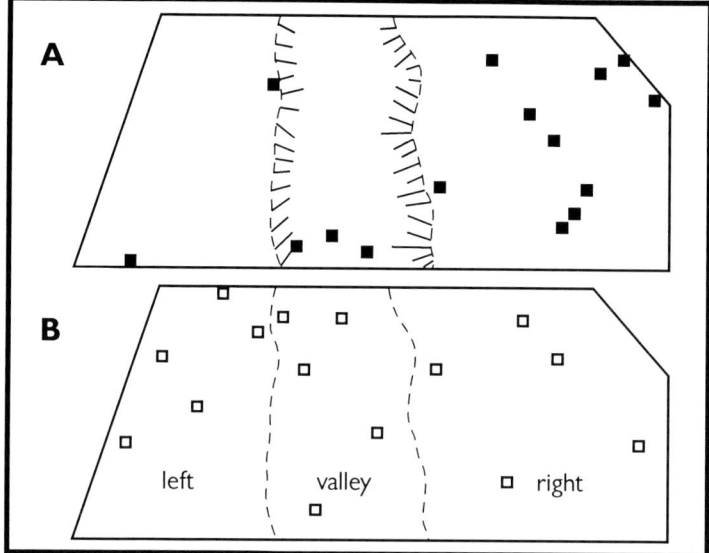

FIGURE 7.15. A diagram of a nature preserve with a valley running through it, on which a plant species is to be monitored. Fifteen quadrats are placed in the preserve. A. A simple random sampling design is used. Simply by chance the right plateau receives 10 quadrats, while the left plateau and valley receive only 2 and 3 quadrats, respectively. B. The preserve is divided into three strata, and 5 quadrats are randomly located within each stratum. This is a stratified random sample. Adapted from Usher (1991) with permission of Chapman and Hall, 115 Fifth Avenue, New York, NY.

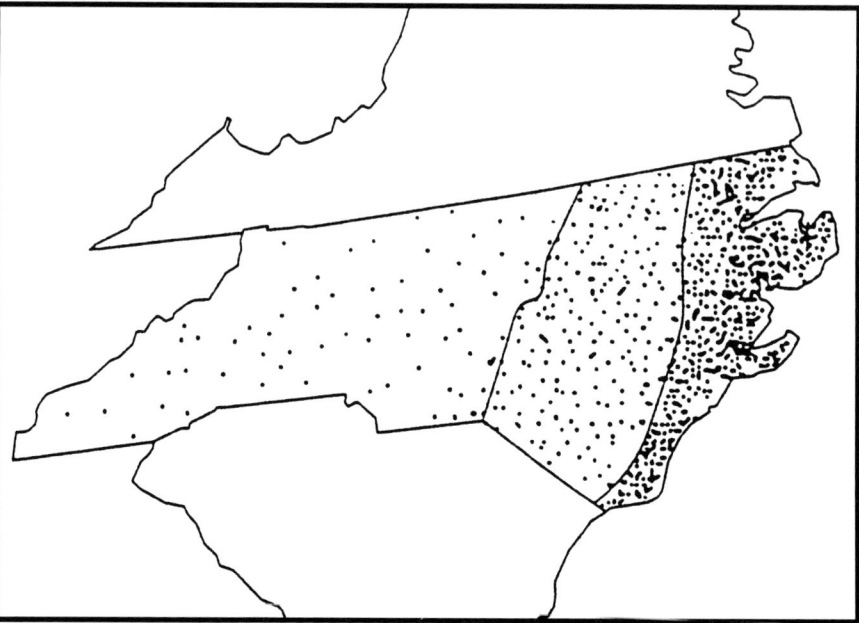

FIGURE 7.16. A stratified random sampling scheme. This example, from the National Wetlands Inventory (Dahl and Johnson 1991), shows how a sample of many plots, each 4 mi², was allocated to three strata in the State of North Carolina.

Refer to Appendix 9, reprinted from Platts et al. (1987), for the formulas necessary to calculate statistics and sample sizes when using stratified random sampling. Other good references include Cochran (1977), Krebs (1989), and Thompson (1992).

c. Advantages and disadvantages of stratified random sampling

The major advantage of stratified random sampling is an increase in the efficiency of population estimation over simple random sampling when the attribute of interest responds very differently to some clearly defined habitat features that can be treated as strata. The principal disadvantage is the more complicated formulas that must be used both to determine sample size allocation to each stratum and to estimate means and standard errors. Since we are taking a simple random sample within each stratum, the possibility exists that, simply by chance, areas within one or more strata may be left unsampled. Additionally, each stratum should be somewhat homogeneous and cover a relatively small geographical area; otherwise the method will be less efficient than systematic and restricted random sampling.

3. Systematic sampling

Systematic sampling is commonly used in sampling vegetation. The regular placement of quadrats along a transect is an example of systematic sampling. The starting point for the regular placement is selected randomly. To illustrate, let's say we decide to place ten 1m² quadrats at 5m intervals along a 50m transect. We randomly select a number between 0 and 4 to represent the starting point for the first quadrat along the transect and place the remaining nine quadrats at 5m intervals from this starting point. Thus, if we randomly select the 2m mark for the first quadrat, the remaining quadrats will be placed at the 7, 12, 17, 22, 27, 32, 37, 42, and 47m points along the transect. This is illustrated in Figure 7.17. The selection of the starting point for systematic sampling must be random.

Systematic sampling is commonly used to facilitate the positioning of quadrats for frequency sampling and of points for cover estimation. Using this approach, a baseline is laid across the population to be sampled, either through its center or along one side of it. Transects are run perpendicular to the baseline beginning at randomly selected points along the baseline (if the baseline runs through the middle of the population, transects

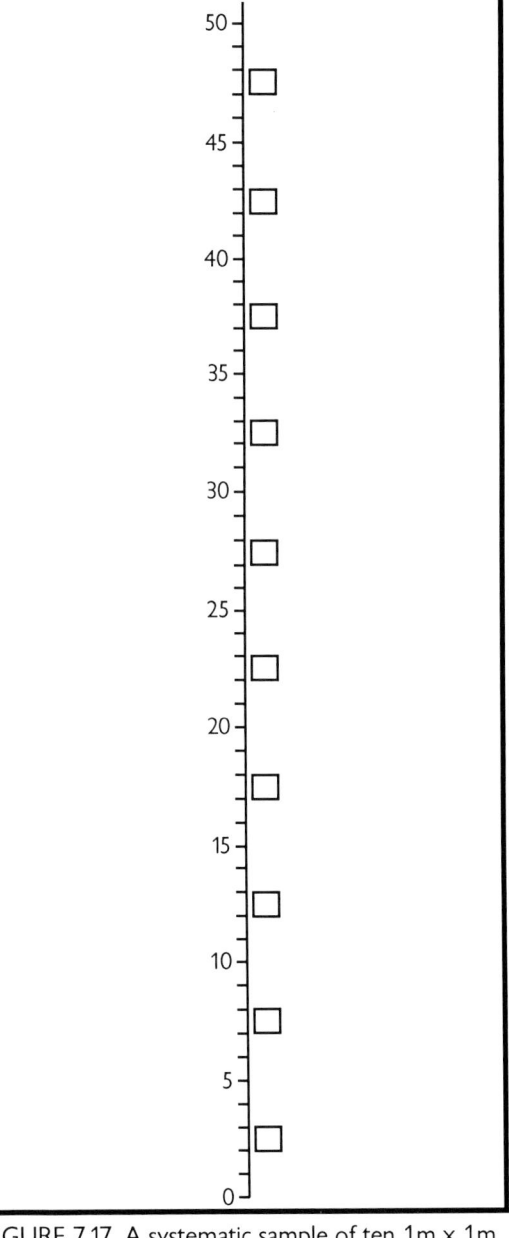

FIGURE 7.17. A systematic sample of ten 1m x 1m quadrats along a 50m transect. The 2m mark is randomly selected to be the beginning point within the first 5m segment. The remaining quadrats are then placed at 5m intervals after that (at 7m, 12m...47m).

are run in either of two directions; the direction for each one can be randomly determined by tossing a coin). Quadrats or points are then systematically positioned along each transect. The starting point for the first quadrat or point along each transect is selected randomly.

Figure 7.18 shows a 50m x 100m macroplot sampled by 100 1m² frequency quadrats, with a 100m baseline along the southern edge. The quadrats are aligned along transects. In this example both the transects and the quadrats were systematically positioned with a random start. In the case of the transects, a random number between 0 and 9 was selected. That number was 1. The first transect therefore began at the 1m mark along the baseline, with subsequent transects beginning at 11m, 21m, up to 91m. In the case of the quadrats, a random number between 0 and 4 was chosen for *each* transect, the first quadrat positioned at that point, and subsequent quadrats placed at increments of 5m from the first quadrat. Thus, for transect number 1 the first quadrat was located at the 3m mark, with subsequent quadrats located at the 8m, 13m...48m marks. This design ensures good interspersion of sampling units throughout the sampled population.

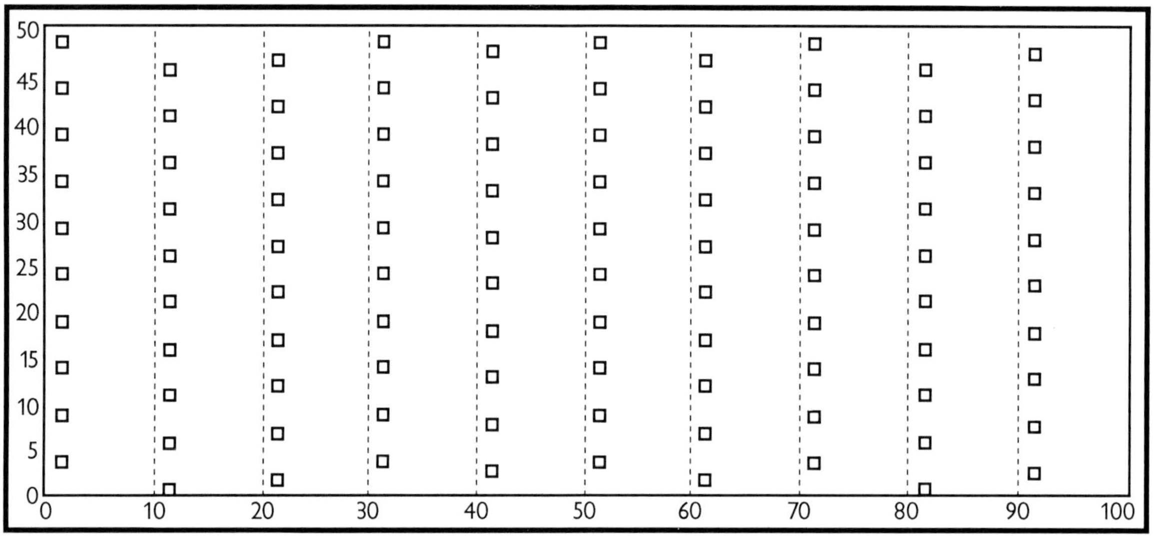

FIGURE 7.18. A 50m x 100m macroplot, sampled by 100 1m x 1m frequency quadrats. The quadrats are aligned along transects. Both the transects and the quadrats are systematically positioned with a random start. A random starting point is selected for the transects along the baseline, while separate random starting points are selected for the quadrats along each transect.

a. Analysis considerations

There are two different ways you can analyze the data from a design like that shown in Figure 7.18. You can treat the sample as if the quadrats had been selected as a simple random sample or you can calculate separate percent frequency values for each transect and then treat the transect as the sampling unit. Let's consider the implications of each of these types of analyses.

(1) Quadrats or points as the sampling units

Strictly speaking, systematic sampling is analogous to simple random sampling only when the population being sampled is in random order (see, for example, Williams 1978). Many natural populations of both plants and animals exhibit a clumped spatial distribution pattern. This means that nearby units tend to be similar to (correlated with) each other. If, in a systematic sample, the sampling units are spaced far enough apart to reduce this correlation, the systematic sample will tend to furnish a better mean and smaller standard error than is the case with a random sample, because with a random sample one is more likely to end up with at least some sampling units close together (see Milne 1959; discussion of sampling an ordered population in Schaeffer et al. 1979).

Milne (1959) analyzed data taken from random and systematic samples of 50 totally enumerated biological populations and found there was no error introduced by assuming that a centric systematic sample is a simple random sample and using all the appropriate formulas from random sampling theory (Krebs 1989:228). Milne's (1959) conclusion was that "with proper caution, one will not go very far wrong, if wrong at all, in treating the centric systematic area-sample as if it were random." Note, however, that Milne compared random samples to *centric* systematic samples, illustrated in Figure 7.19. The units of a centric systematic sample lie on equidistant parallel lines (these can be thought of as transects) arranged in a manner such that, in effect, the area is divided into equal squares (see dotted lines) and a sampling unit taken from each square. Thus, the sampling units are spaced a considerable distance apart.

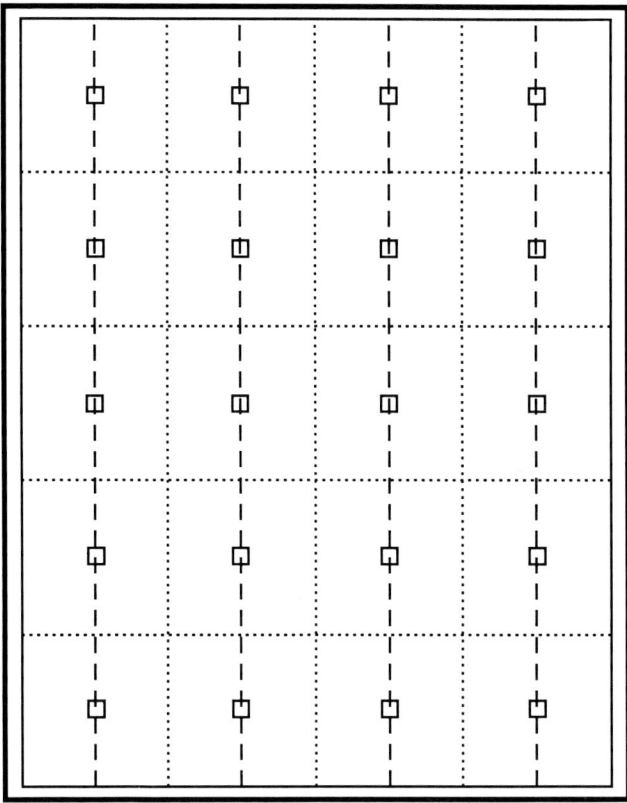

FIGURE 7.19. A centric systematic sample (adapted from Milne 1959). Small squares are sampling units, dashed lines are transects, and dotted lines show how the sampling units fall in the center of each subunit of area.

This spacing of sampling units (e.g., quadrats) is needed if one is to treat a systematic sample as if it were random. Indeed, the contiguous placement of quadrats along a transect or the separation of such quadrats by small distances (e.g., one "pace"), practically ensures that adjacent sampling units will be correlated. This will result in an underestimation of the standard error and questionable results. Certainly Milne's conclusion cannot be applied in this instance. The issue of how far apart to systematically place sampling units is discussed in detail in Section b, below.

(2) Transects of points or quadrats as the sampling units

For frequency or point cover data, you often want to treat the quadrats or points as the sampling units rather than the transects along which these quadrats or points are located. Estimates will be more precise and significance tests more powerful because of the larger sample sizes realized by using quadrats or points rather than transects as the sampling units. There are at least two situations, however, in which you might want to treat the transects as the sampling units. The first of these is when the quadrats or points are not far enough apart to be considered independent. This is more likely to be a problem in already established studies, where quadrats were placed contiguously or a very short distance apart. Hopefully, you will design *new* studies in such a manner that the quadrats are spaced far enough apart to achieve independence (how far is "far enough" is discussed in Section b, below).

The second situation in which you might want to treat the transects as the sampling units when systematically sampling frequency quadrats or cover points is when the transects are permanent. If you take care to permanently mark, not just the ends of the transect, but intermediate points as well, and to stretch the tape to approximately the same tension at each time of measurement (steel tapes with tensioners can be used to ensure this), you can still treat the quadrats or points as the sampling units (Chapter 8 discusses this in more detail). If, however, you have reason to believe that the average values per transect are more correlated between years then are the quadrat or point values, you may choose to analyze the transects rather than the quadrats or points as the sampling units.

b. Spacing of sampling units along a transect

When arranging sampling units (quadrats or points) along transects, a key question is how far apart they need to be in order to be considered independent. If the quadrats or points are far enough apart that they can be considered independent, we have the benefit of increasing our sample size dramatically (because the point or plot is the sampling unit instead of the transect) while keeping the field efficiency of locating sampling units rapidly along a transect.

Independence means that the sampling units are not correlated. For example, if quadrats are not correlated, high mortality in Quadrat A does not necessarily mean there will be high mortality in Quadrat B, at least not because of its proximity to Quadrat A. However, whenever quadrats are located fairly close together they will often respond similarly. For example, if Quadrats A and B are close enough together that they are both in a canopy gap caused by a fallen tree, they will likely change similarly. If your design has quadrats located closely along a transect, each plot is in close proximity to two others, and changes in each plot will probably be correlated with two others (or more). In simple random sampling, there will always be some quadrats located close together simply by chance. The difference is that this correlation only affects some of the quadrats, and the degree of correlation fluctuates randomly with the spatial location of the randomly placed plots.

Determining how far apart to place sampling units along a transect in order for them to be considered independent can be difficult. It is easier to define what is *not* far enough apart. Clearly, quadrats that are positioned contiguous to one another along a transect are not far enough apart to be considered independent. The same can be said of quadrats or points that are spaced so close together they may fall on the same individual plant. But what should the minimum spacing be? Some factors to consider are the average size of gaps (especially in forests), the average size of individual plants, the size of areas of clones, and the size and distribution of microsites. In general, sampling units should be far enough apart that they do not fall into the same microsite, gap, or clone. This, however, is scale dependent. If you are sampling an area that only covers a typical gap, your plots by necessity will all fall within that gap.

Probably the best way to determine spacing of sampling units along transects is to consider the degree of interspersion of your design. The concept of interspersion was introduced at the beginning of Section E of this chapter. The goal is to have sampling units as well interspersed throughout the area of the target population as possible. Once you have delineated the area you intend to sample, strive for a design in which the spacing between transects is about the same as the spacing between sampling units. If you do this, it is likely that the issue of independence will take care of itself.

c. Systematic sampling of biomass and cover when ocularly estimated in quadrats

An important use of systematic sampling is to estimate biomass and cover when cover is visually estimated in quadrats. We've already noted that from the standpoint of statistical precision we are better off with long, narrow quadrats when estimating density, biomass, or cover. We've also noted the impracticality of using long, narrow quadrats for anything but density. A good compromise for biomass and cover is to place square (or small rectangular) quadrats systematically (with a random start) along transects and to treat the transects as the sampling units. Thus, we are able to clip plots or estimate cover efficiently in the quadrats, while at the same time crossing the variability in the population, making for more precise estimates of means. For analysis, we would take the mean of the quadrat values for each transect and use this set of transect means as our sample.

d. Relationship of systematic sampling to two-stage sampling

When we use transects as the sampling units, whether for frequency quadrats, cover points, biomass quadrats, or cover quadrats, we are really conducting two-stage sampling. The transects are the primary sampling units, and the quadrats or points are the secondary sampling units. There are standard deviations associated with both the primary sample of transects and the secondary sample of quadrats or points. Two-stage designs take into account both sets of standard deviations. The result is a much more complex set of equations that standard statistical programs will not calculate. Although we *could* subject these data to the more complex formulas of two-stage sampling, there is no need to do so. Cochran (1977:279) points out that we can ignore the standard deviation of the secondary sample as long as we do not use the finite population correction factor in our analysis. We can simply use the mean of each transect's collection of quadrats as our unbiased estimate of the transect value. We then treat the collection of transect values as a simple random sample. This allows us to use standard statistical computer programs to perform our analysis.

e. Advantages and disadvantages of systematic sampling

One of the principal advantages of systematic sampling is the fact that it enables the investigator to sample evenly across a whole area. This results in good interspersion of sampling units throughout the area containing the target population. Systematic sampling is more efficient than simple random sampling, particularly if the area being sampled is large, because of decreased setup and travel time.

Systematic sampling is undesirable if the pattern of the sampling units intersects some pattern in the environment (e.g., dune ridges and slacks; Goldsmith et al. 1986). If some periodic pattern does exist, the data analysis will not reveal this, and your estimates, particularly of standard errors, will be wrong. Although this type of periodic pattern is rare in nature, it is a possibility you should be alert to.

Another advantage of systematic sampling is that it enables us to use square quadrats to accurately estimate cover and biomass, while taking advantage of the benefits of lines in crossing the variability inherent in the population. By treating the transects as the sampling units we get the best of both worlds.

For density estimation, Salzer (in prep.) has shown through Monte Carlo simulations that systematic designs outperform simple random sampling in terms of precision when sampling clumped populations. Be aware, however, that systematic sampling for density estimation can lead to questionable results if the sampling design creates a situation where there are only a small number of potential samples. For example, consider the macroplot shown in Figure 7.20. Ten 1m x 50m quadrats are systematically positioned in the macroplot with a random starting point at the 2m position on the x-axis, and the quadrats spaced at 10m intervals after that. In this case, since the position of all quadrats is fixed once the first quadrat is positioned, there are only 10 possible samples to draw from, depending on which of the 10 possible starting points is randomly selected in the first 10m segment of the population (0, 1, 2, 3, 4, 5, 6, 7, 8, or 9). The sampling distribution (distribution of all possible sample mean values) for this sampling design might resemble a uniform (flat) distribution instead of the smooth, bell-shaped curve of the normal distribution, because there are only 10 different sample means possible. This can lead to inaccurate measurements. The next type of sampling design, restricted random sampling, solves this problem. Except for this somewhat uncommon situation, however, systematic sampling is preferred over restricted random sampling.[3]

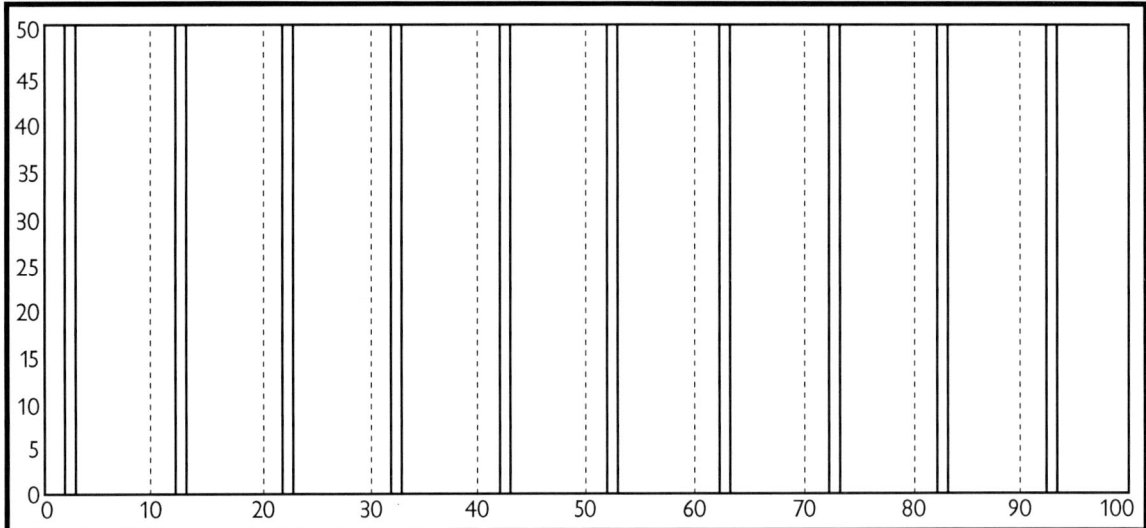

FIGURE 7.20. A systematic sample of 10 1m x 50m quadrats in a 50m x 100m macroplot. Note that there are only 10 possible samples, corresponding to which of the 10 possible starting points in the first 10m segment of the baseline (x-axis). In this case, the sample started at the 2m mark.

4. Restricted random sampling

a. Description

In restricted random sampling, you determine the number of sampling units, n, you will need to meet your monitoring objective (sample size determination is discussed in Section G, below), then divide your population into n equal-sized segments. Within each of these segments, a single sampling unit is randomly positioned. The sample of n sampling units is then analyzed as if it were a simple random sample.

Figure 7.21 is an example of a restricted random sampling procedure. This is the same 50m x 100m macroplot as we used in our discussion of systematic sampling. In this case, however, we divide the x-axis into ten 10m segments. Within each of these segments we randomly select a single quadrat location. This gives us 10 possible random locations within every 10m

---

[3] If there are 25-30 or more possible samples, there is no problem in using systematic sampling.

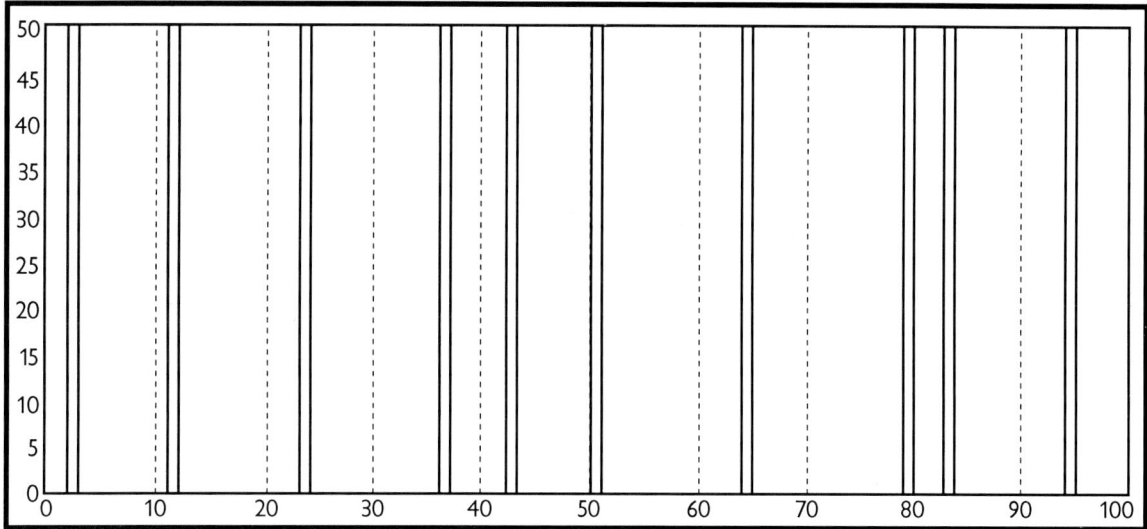

FIGURE 7.21. A restricted random sample of 10 1m x 50m quadrats in a 50m x 100m macroplot. One quadrat is randomly positioned within each 10m segment of the baseline (x-axis).

segment of the x-axis. Every quadrat location in the macroplot still has an equal probability of selection. The same technique can also be applied to the y-axis if there is more than one possible quadrat position along that axis.

The restricted random sampling procedure can also be used when the sampling unit is a line instead of a quadrat. Divide the population into equal-sized segments and allocate a single line to each segment.

If percent cover is being measured using the point-intercept sampling method in which points are arranged along lines, then you may want to use a combination of the restricted and systematic designs. If, for example, you decide to run 10 transects, each with 50 points, perpendicular in one direction from a baseline, you could divide the baseline into 10 equal segments, randomly locate beginning points for each transect within each of these 10 segments, and then systematically space the points along each transect (like Figure 7.21, except with points systematically positioned along one edge of each quadrat).

Restricted random sampling is similar to both stratified random and systematic sampling. It is similar to stratified random sampling in that we have effectively stratified our macroplot into 10 strata. But, unlike stratified random sampling, the strata are arbitrary and we take only one sampling unit in each stratum. Like systematic sampling, we divide our population into equal sized segments. With systematic sampling, however, only the first sampling unit is randomly determined; all the others are spaced at equal intervals from the first.

b. Advantages and disadvantages of restricted random sampling

Like systematic sampling, restricted random sampling results in very good interspersion of sampling units throughout the target population. Furthermore, Salzer (in prep.) has shown through simulation studies that restricted random sampling results in more precise estimates of density than simple random sampling. He has also demonstrated the procedure to be more robust than systematic sampling when the number of possible systematic samples are few, because with restricted random sampling designs you don't constrain the number of potential samples you can draw from. The principal disadvantage of restricted random sampling is that you can, purely by chance, end up with sampling units positioned side-by-side. This can leave larger portions of the sample area unsampled than is the case with a systematic design. When the number of potential systematic samples is large enough (more than 25-30), you are probably better off choosing a systematic sample. Otherwise, use the restricted random design.

5. Cluster sampling

a. Description

Cluster sampling should not be confused with cluster analysis, a technique used in classification and taxonomy. Cluster sampling is a method of selecting a sample when it is difficult or impossible to take a random sample of the individual elements of interest. With cluster sampling, we identify groups or clusters of elements and take a random sample of these clusters. We then measure every element within each of the randomly selected clusters.

In rare plant monitoring cluster sampling is most often used when the objective is to estimate something about individual plants, such as the mean height of each plant or the mean number of flowers per plant. For example, you may want to track the average height of plant X in population Y. There are too many plants in the population to make measuring all of them feasible. Five quadrats are randomly placed in the population and the heights of all plants within these quadrats are measured (Figure 7.22).

b. Advantages and disadvantages of cluster sampling

The advantage of cluster sampling is that it is often less costly to sample a collection of elements in a cluster than to sample an equal number of elements selected at random from the population (Thompson 1992). It is most efficient when different clusters are similar to each other and incorporate much variability within. Because plants near each other tend to be similar, this condition will not be realized with square clusters (Thompson 1992). Therefore, just as with simple random sampling for density estimation, cluster sampling using long, narrow quadrats to delineate clusters will be more efficient than using square quadrats. Cluster sampling and two-stage sampling are the only two efficient designs that can be used to sample individual plant characteristics. A disadvantage is that all elements within each cluster must be measured. If the clusters contain large numbers of the element of interest, two-stage sampling, described below, will be more efficient. Another disadvantage is that it is difficult to figure out how many clusters should be sampled versus how large each cluster should be. Additional disadvantages are the more complex calculations required and the fact that statistical software packages do not include these calculations. See Platts et al. (1987), reprinted in Appendix 9, for the formulas needed to analyze cluster sampling data.

6. Two-stage sampling

Two-stage sampling is similar to cluster sampling in that we identify groups of elements about which we wish to make inferences. We then take a random sample of these groups. However, instead of measuring every element in each group as we would if doing cluster sampling, we take a second sample of elements within each group. The groups sampled are called *primary sampling units* while the elements sampled are called *secondary sampling units*. The secondary sampling units can be either a simple random sample of elements or a

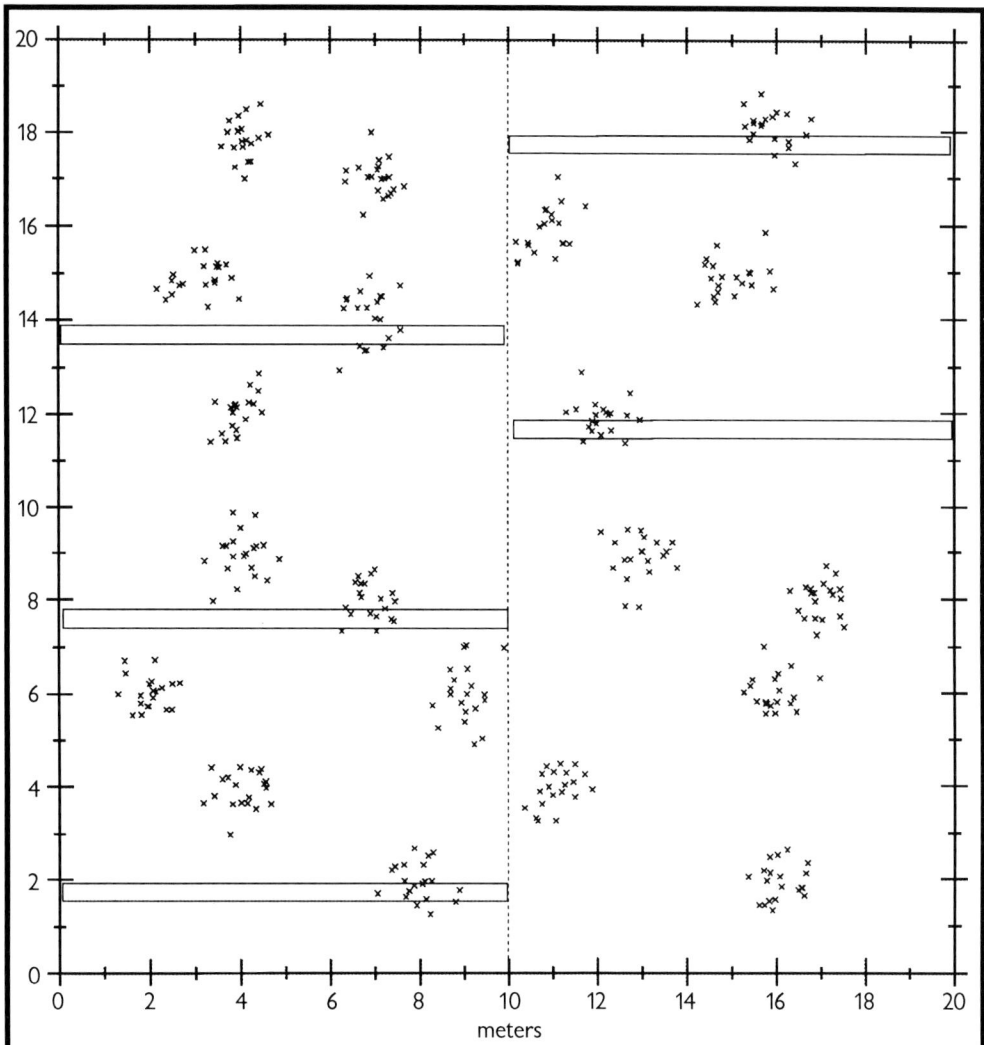

FIGURE 7.22. An example of cluster sampling to estimate the mean height of plants in a population. Five quadrats are randomly placed in the population and the heights of all plants within these quadrats are measured.

systematic sample of elements. Figure 7.23 shows a two-stage sampling design. Like cluster sampling, the main use of two-stage sampling is to estimate some value associated with individual plants.

a. Examples

For example, to estimate the number of flowers per plant produced by species X, we might randomly locate a sample of quadrats in the target population. Within each quadrat we then take a random sample of plants and count the number of flowers on each plant selected. The quadrats are the primary sampling units and the plants are the secondary sampling units.

Another example of two-stage sampling involves macroplots and quadrats. You are interested in the mean density of plants/quadrat, and you want to be able to make statistical inferences to a large area. The area is relatively homogeneous, with no logical basis of stratification. Seven 50m x 100m macroplots (primary sampling units) are randomly distributed throughout the population, and fifteen 0.20m x 25m quadrats (secondary sampling units) are randomly sampled within each macroplot.

Both these examples involve simple random sampling at both stages. Either or both of the stages may involve different types of sampling. A common type of two-stage sampling involves simple random sampling at the primary stage and systematic sampling at the second stage. We have already seen examples of this: when quadrats or points (secondary sampling units) are systematically located (with a random start) along transects, and the transects (primary sampling units) are run from randomly selected points along a baseline. Of course, the transects could be positioned using another type of design such as restricted random sampling or systematic sampling. The point is that the two stages can involve different sampling designs.

The use of quadrats or points along transects is covered in more detail under systematic sampling (discussed above). In this situation we usually treat the data as if they came from a one-stage sampling design. We simply use the mean of each transect's collection of quadrats as our unbiased estimate of the transect value. We then treat the collection of transect values as a simple random sample. This allows us to use standard statistical computer programs to perform our analysis.

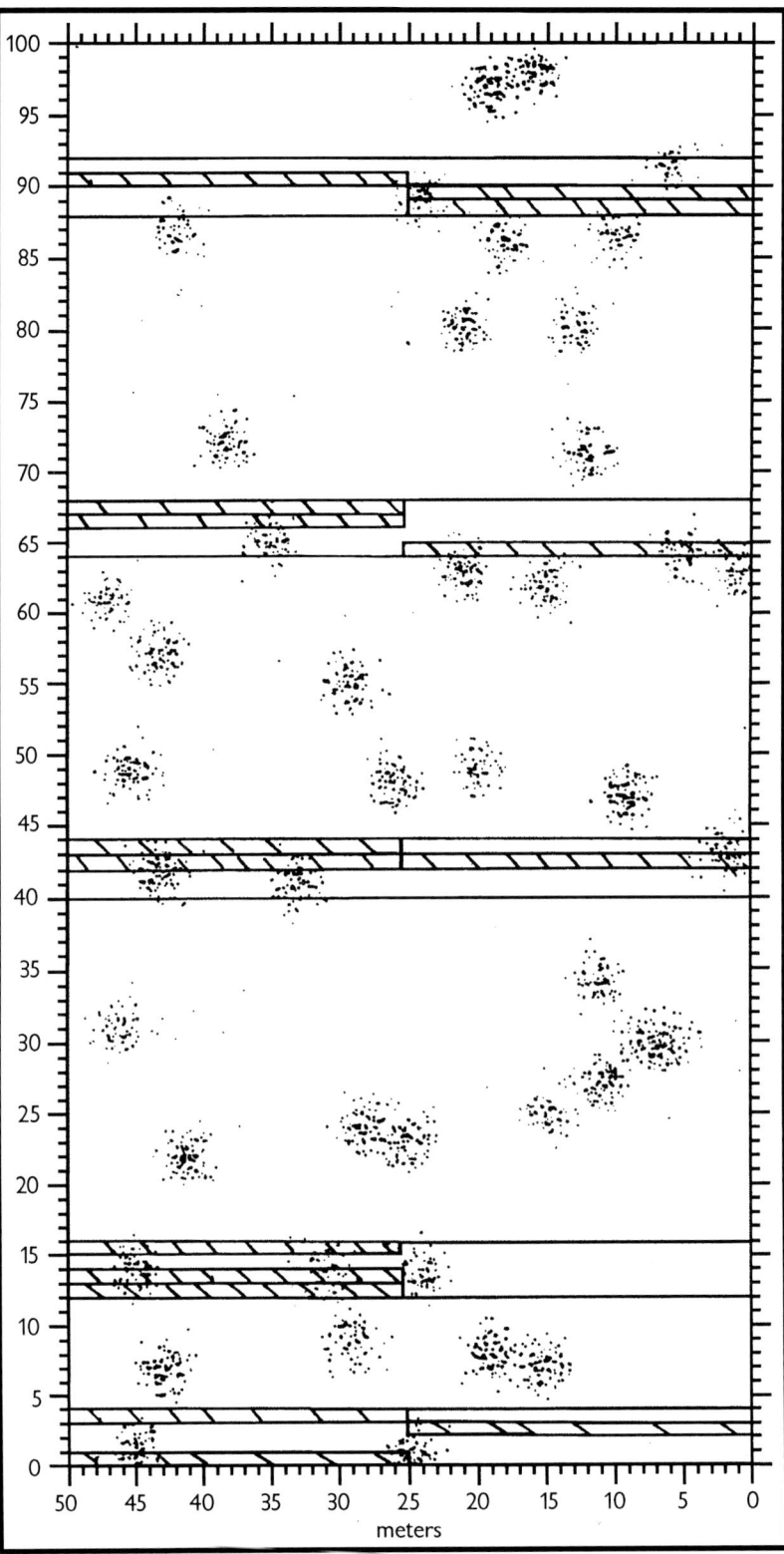

FIGURE 7.23. Two-stage sampling to estimate the number of flowers per plant on a particular species of plant. Five 4m x 50m quadrats (primary sampling units) are randomly located in the sampled population and three 1m x 25m quadrats (secondary sampling units) are randomly located within each of the five larger quadrats. The number of flowers per plant is counted within all of the selected 1m x 25m quadrats.

b. Advantages and disadvantages of two-stage sampling

The practical advantage of two-stage sampling, compared to a simple random sample of the same number of secondary units, is that is often easier or less expensive to observe many secondary units in a group than to observe the same number of secondary units randomly spread over the population (Thompson 1992). Travel costs are therefore reduced with two-stage sampling. Two-stage and cluster sampling designs are the only two efficient designs that can be used to sample individual plant characteristics.

Because sampling occurs at both stages, there are standard deviations associated with estimates of the values at both stages (unlike cluster sampling which has no standard deviation associated with the values measured at the second stage). This results in more complicated formulas in arriving at estimates of values and standard errors (although the standard deviation of the secondary sample can be ignored as long as the finite population correction is not applied to the standard error of the primary sample—see Cochran 1977). Refer to Platts et al. (1987), reprinted in Appendix 9, for formulas for calculating means and standard errors from two-stage sampling. More detailed discussions can be found in Cochran (1977:279), Krebs (1989), and Thompson (1992).

7. Double sampling

Double sampling, sometimes called two-phase sampling, involves the estimation of two variables. Because one of these variables, the variable of interest, is difficult and expensive to measure, it is measured in only a relatively small number of sampling units. In order to improve the rather poor precision of the estimate that normally results from a small sample, an auxiliary variable that is much easier to measure is estimated in a much larger number of sampling units. Often, but not always, the variable of interest is measured in a subsample of the sample of units in which the auxiliary variable is measured.

a. Examples

The idea of double sampling will become clearer with examples. The technique is often used in estimating aboveground biomass in rangelands. Because it is slow and expensive to clip, dry, and weigh biomass in many sampling units, observers train themselves to visually estimate biomass. Once trained, the observers randomly locate quadrats within a target population and visually estimate the biomass in all the quadrats. For example, 100 quadrats are so estimated. Then, in a subsample of these quadrats, say 10, the visual estimates are made as in the other quadrats, but after these estimates are recorded, the aboveground biomass is clipped, dried, and weighed. Thus, for these 10 quadrats we have two estimates of biomass, one from the visual estimate, the other from the actual weighing of the clipped biomass.

In forest surveys to estimate the volume of trees in a stand, visual estimates of volume by trained observers can rather easily be obtained from a large sample of standing trees, while accurate volume measurements that require felling are limited to a small subsample of trees (Thompson 1992).

In both these cases, the subsample on which the variable of interest is actually measured is more *accurate*. But the *precision* of the estimate can be greatly improved by considering the measurements on the auxiliary variable. The improvement in precision depends upon how well the auxiliary variable correlates with the variable of interest. In the two examples given above, this relates to how well the trained observers actually estimate biomass or tree volume.

b. Advantages and disadvantages of double sampling

If the auxiliary variable is relatively quick to be measured and is highly correlated with the variable of interest, double sampling is much more efficient in estimating variables that are difficult to measure than directly measuring the variable. A disadvantage is that the formulas for data analysis and sample size determination are much more complicated than formulas for simple random sampling. Refer to Cochran (1977) or Thompson (1992) for the formulas needed to analyze double-sampling data.

8. Taking a simple random sample of individual plants

Let's say we want to estimate something about a population of individual plants, such as their mean height or the mean number of flowers per plant, and the population is too large to measure this variable on every single plant in the population. Easy, you say; let's just take a simple random sample of plants, measure the variable on the sample, and calculate the mean and standard error for the sample. We can then construct a confidence interval around the estimate at whatever confidence level we choose (e.g., a 95% confidence interval). Although it might seem logical to take a simple random sample of plants, for most plant populations this is not feasible.

a. Sampling the nearest plant to a random point—an incorrect method

One way that is often—and incorrectly—used is to select a random sample of points in the population and to take the nearest plant to each of these points. Unfortunately, this works only if the population of plants is randomly distributed, a condition rarely met by plant populations. If, as is typically the case, the population of plants occurs in clumps, this technique most decidedly will not result in a simple random sample of plants. Consider Figure 7.24, which shows the distribution of individuals of a hypothetical plant species along a 10m transect. Note that 9 of the 10 individuals are clumped in the last 3 meters of the transect, while a single individual occurs at the 3m mark. A randomly positioned point along this transect would have about a 50% probability of being closest to this isolated individual and about a 20% chance of being closest to the individual at the 7m mark.

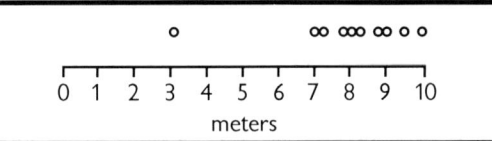

FIGURE 7.24. Distribution of individuals of plant species X along a 10-meter transect. A randomly positioned point on the transect will be far more likely to be closest to the individual at the 3m mark than to any of the other plants.

The probability of the point lying closest to any of the other eight individuals is much less. Thus, in a clumped population of plants, a "random" sample of individuals chosen by taking the individuals closest to randomly located points will be biased toward those individuals that are isolated from the majority of the population. These individuals may either be much larger than the majority of plants in the population because of reduced intraspecific competition or much smaller because they occupy suboptimal habitat. Let's say we're interested in estimating the mean height of such a population. By biasing our estimate toward the isolated plants in the population we may greatly under- or overestimate the mean height of the population. The same is true for any other attribute associated with individual plants that we may wish to estimate. Number of fruits per plant is one of many examples. Obviously, for populations of plants that follow a clumped distribution—which is by far the majority of populations—such a sample of plants cannot be used to adequately characterize the population.

b. Complete enumeration method

How, then, can you take a random sample of individual plants? One way is to completely enumerate every plant in the population by, for example, mapping every plant and numbering each one from 1 to *n*. A simple random sample could then be taken by drawing random numbers between 1 and *n*. This, of course, would be extremely time-consuming except for small populations, in which case you might be able to measure the attribute on every plant in less time and not have to sample at all. For example, if you are interested in mean height of plants, you could simply measure the height of every plant in the population. If, however, you need to estimate the mean number of flowers per plant and each plant has several hundred flowers, this method might make some sense, although for most practical purposes it is far too time-consuming.

c. Systematic random sample method

Another possibility is to take a systematic random sample of plants. With this method you gather information from every *n*th plant in the population. This method will work if you are planning to conduct a complete census of the population, but you are also interested in estimating some attribute from a subset of the plants (e.g., number of flowers/plant). Before you start you need an estimate of the following two types of information: (1) the approximate size of the population, and (2) the approximate number of individual plants you will need to sample (calculated as a proportion of the total population size). Based on your estimate of total population size and your sample size calculations from pilot sampling, you decide to count the number of flowers on every 10th plant encountered. You choose a random number between 1 and 10. Say the number is 4. Then, starting at one edge of your population you systematically count the plants. You place a pin flag next to plant number 4, another next to plant number 14, and so on until you've counted all the plants. You can then come back and count the flowers on the flagged plants. This sample can properly be analyzed as a simple random sample.

d. Probability proportional to area method

Neither complete enumeration or systematic random sampling is practical in most plant population sampling situations if the objective is to take a random sample of individual plants. One additional possibility will be mentioned for the sake of completeness; it, too, is largely impractical in most situations. This method involves taking a sample of plants by including in the sample those plants closest to randomly positioned points. Instead of (incorrectly) treating this as a simple random sample, however, we take into account the fact that the probability of selecting these plants is unequal. Figure 7.25, from Stehman and Overton (1994), shows the spatial distribution of a population (triangles). Around the triangles are polygons of different size and shape. The polygons are called Thiessen polygons. A random point placed in one of these polygons is closer to the triangle within the polygon than to any other triangle. A procedure that selects the closest object to a random point will select objects with probability proportional to the area of the object's Thiessen polygon (Stehman and Overton 1994). Formulas are available to calculate means and standard errors when conducting this kind of sampling, which is called probability proportional to size sampling (see, for example, Thompson 1992). Unfortunately, this requires more information regarding the distribution of the plant population than we are likely to have without completely enumerating the population. Because of this, and because of the rather complicated formulas necessary for calculating statistics, the method is not discussed further here. It may, however, be suitable for sampling large, mature plants such as trees and shrubs, where aerial photography is available to construct Thiessen polygons around individual plants.

e. The most practical approach to estimating attributes of individual plants

Because of the difficulties involved in selecting a random sample of plants, cluster sampling or two-stage sampling designs are usually employed instead. As we have seen, these designs involve using quadrats as primary sampling units, with individual plants as elements (in cluster sampling) or secondary sampling units (in two stage sampling).

9. Summary of the different types of random sampling

Table 7.2 reviews the uses of the random sampling designs discussed above and summarizes the advantages and disadvantages of each design.

FIGURE 7.25. Thiessen polygon for selecting the nearest object to a random point (from Stehman and Overton 1994). The Thiessen polygon encloses the region surrounding an object (triangle symbol) in which any point is closer to that object than to any other object. A sampling procedure selecting the closest object to a random point will select objects with probability proportional to the area of the object's Thiessen polygon. Reprinted from G.P. Patil and C.R. Rao, Environmental Statistics, page 269, 1994, with kind permission from Elsevier Science - NL, Sara Burgerhartstraat 25, 1055 KV Amsterdam, The Netherlands.

Table 7.2. Summary of random sampling types.

Sampling Type	Recommended Uses	Advantages	Disadvantages
Simple Random Sampling	Useful in relatively small geographic areas with homogeneous habitat, when the number of sampling units is not likely to be large.	The formulas necessary to analyze data are the simplest of all sampling types.	By chance, some areas within the target population may be left unsampled. The travel time is considerable when the sampling area and/or sample size is large. Restricted random sampling and systematic random sampling outperform simple random sampling when populations have a clumped distribution.
Stratified Random Sampling	Useful when the attribute of interest responds very differently to some clearly defined habitat features. Since it involves taking a simple random sample within each stratum, each stratum should consist of a relatively small geographic area with homogenous habitat, and the number of sampling units in each stratum should not be too large.	Results in more efficient population estimates than simple random sampling when the attribute measured varies with clearly defined habitat features.	The mathematical formulas required for analysis are more complex than those used for simple random sampling. When the geographic area within any stratum is large and/or the number of sampling units is likely to be large, then one of the other types of sampling listed below will be more efficient. By chance, some areas within each stratum may be left unsampled.
Systematic Sampling	Useful for any sampling situation, as long as the first sampling unit is selected randomly and the sampling units are far enough apart to be considered independent. Can also be used as part of cluster and two-stage sampling designs.	When the conditions given in the cell to the left are met, this is the best type of sampling design to use. There is better interspersion of sampling units than with simple random sampling. The data can be gathered much more efficiently than with simple random sampling and still be analyzed using the formulas for simple random sampling.	In the uncommon event that the number of possible samples is limited to fewer than about 25-30 (see text), systematic sampling may lead to questionable results; in this situation you should use restricted random sampling.

Sampling Type	Recommended Uses	Advantages	Disadvantages
Restricted Random Sampling	Although more useful than simple random sampling in most situations, restricted random sampling should be used only when the number of potential samples is fewer than 25-30. Otherwise, systematic sampling is the better choice.	Like systematic sampling, restricted random sampling results in better interspersion of sampling units than with simple random sampling. If the number of potential samples is less than 25-30, restricted random sampling is better than systematic sampling. The data can be analyzed using the formulas for simple random sampling.	This design is not as efficient as systematic sampling when the number of potential samples is greater than 25-30.
Cluster Sampling	Cluster sampling is used to select a sample when it is difficult or impossible to take a random sample of the individual elements of interest. A cluster of elements is identified and a random sample (usually using systematic sampling) is taken of the clusters. Every element within each cluster is then measured. In plant population monitoring, cluster sampling is most often used to estimate something about individual plants (e.g., mean height, number of flowers/plant). In this situation, quadrats are the clusters.	It is often less costly to sample a collection of elements in a cluster than to sample an equal number of elements selected at random from the population. Except in rare situations, it is not practical to take a random sample of individual plants. Instead, the attribute of interest is measured on every plant in a sample of quadrats (which function as the clusters).	All the elements within each cluster must be measured. If the clusters contain large numbers of the element of interest, two-stage sampling is more efficient. Other disadvantages include the difficulty in determining how many clusters should be sampled versus how large each cluster should be, the more complex calculations required for analysis, and the fact that most statistical software packages do not include these calculations.
Two-stage Sampling	Similar to cluster sampling in identifying groups of elements (such as plants) and taking a random sample (usually using systematic sampling) of these groups. In two-stage sampling, however, a second sample of elements is taken within each group. Like cluster sampling, the main use of two-stage sampling is to estimate some value associated with individual plants.	Same advantages as cluster sampling. The two types are the only efficient means of estimating some attribute associated with individual plants. When the number of plants in each group (quadrat) is large, two-stage sampling is more efficient than cluster sampling.	There are standard deviations associated with both stages of sampling (unlike cluster sampling which has no standard deviation associated with the values measured at the second stage). This results in more complicated formulas in arriving at estimates of values and standard errors (although the standard deviation of the secondary sample can be ignored as long as the finite population correction factor is not applied to the standard error of the primary sample).
Double sampling	Useful when the variable of interest (e.g., actual measurements of biomass) is difficult to measure, but is correlated with an auxiliary variable (e.g., ocular estimates of biomass) which is more easily measurable. The second variable is measured in a large number of sampling units, while the first variable is measured in only a subset of the sampling units. The samples are often taken using systematic sampling.	If the auxiliary variable is relatively quick to be measured and is highly correlated with the variable of interest, double sampling is much more efficient in estimating a variable that is difficult to measure than directly measuring the variable.	The formulas for data analysis and sample size determination are much more complicated than for simple random sampling, and most statistical software programs do not include the necessary calculations.
Taking a random sample of individual plants	This can only be accomplished in rare situations. When the objective is to measure something on individual plants, it is best to use either cluster or two-stage sampling. See text for further information.	In those few situations where it is possible to take a random sample of individual plants, the calculations necessary for analysis are simpler than those for either cluster or two-stage sampling.	It is not practical to take a simple random sample of individual plants in most monitoring situations.

F. Should Sampling Units Be Permanent or Temporary?

A critical decision in sampling is whether to make your sampling units temporary or permanent. When sampling units are temporary, the random sampling procedure is carried out independently at each sampling period. For example, your sampling objective involves detecting change in density over time of a plant species in a 50m x 100m macroplot. In the first year of sampling you place twenty-five 0.5m x 25m quadrats within the macroplot by randomly selecting 25 unique sets of coordinates and counting the number of plants in each quadrat. In the second year of sampling, you place another twenty-five 0.5m x 25m quadrats by randomly selecting a *new* set of coordinates and counting the number of plants in each quadrat. The sampling units (quadrats) in this example are *temporary* and the two samples are *independent* of each other.

Using the same sampling objective, you could decide to use permanent quadrats. In the first year of sampling you randomly place the 25 quadrats as described above and count the number of plants in each quadrat. This time, however, you permanently mark the locations of the 25 quadrats. In the second year of sampling, you count the number of plants in the same quadrats. In this example the sampling units are *permanent* and the two samples are *dependent*.

1. For many sampling situations permanent sampling units far outperform temporary sampling units

The principal advantage of using permanent instead of temporary sampling units is that for many plant species the statistical tests for detecting change from one time period to the next in permanent sampling units are much more powerful than the tests used on temporary sampling units. This advantage translates into a reduction in the number of sampling units that need to be sampled to detect a certain magnitude of change.

a. Examples

To see why this is so, let's consider the process used in comparing the samples between two time periods when using permanent quadrats. If we were using temporary quadrats, we would calculate separate means and standard errors for the two samples and compare these using a statistical test (such as a *t* test) for independent samples. *With permanent quadrats, however, we calculate only one mean and one standard error.* This requires some explanation. Each quadrat at time 1 is paired with the same quadrat at time 2. The data from which we calculate the mean and standard error consists of the set of differences between each of the quadrats at time 1 and its corresponding quadrat at time 2. For example, we randomly position five permanent quadrats in a population and count the number of plants in each quadrat in 1993 and again in 1994 (Figure 7.26). Data from these permanent quadrats yield the values in Table 7.3.

Note that the permanent quadrats are extremely effective at detecting the lack of change from year to year (because the difference between 1993 and 1994 was zero in every quadrat, there is no variation between sampling units, and the standard error is

quadrat number	number of plants in 1993	number of plants in 1994	difference between 1993 and 1994
1	5	5	0
2	5	5	0
3	5	5	0
4	6	6	0
5	3	3	0
			mean difference 0
			standard error 0

TABLE 7.3 Density data taken from the permanent quadrats in Figure 7.26.

actually 0). Had temporary quadrats been used in both years it is quite likely that the estimates for each year would have been different just because of chance. For this reason more temporary sampling units (perhaps many more) would have been required to reach the same conclusion.

Because we are interested only in the change that takes place *within* each permanent sampling unit between two time periods, the difference *between* sampling units at either time period is not nearly as important as it is when using temporary quadrats. Consider the following example. In order to detect change in cover of species X between

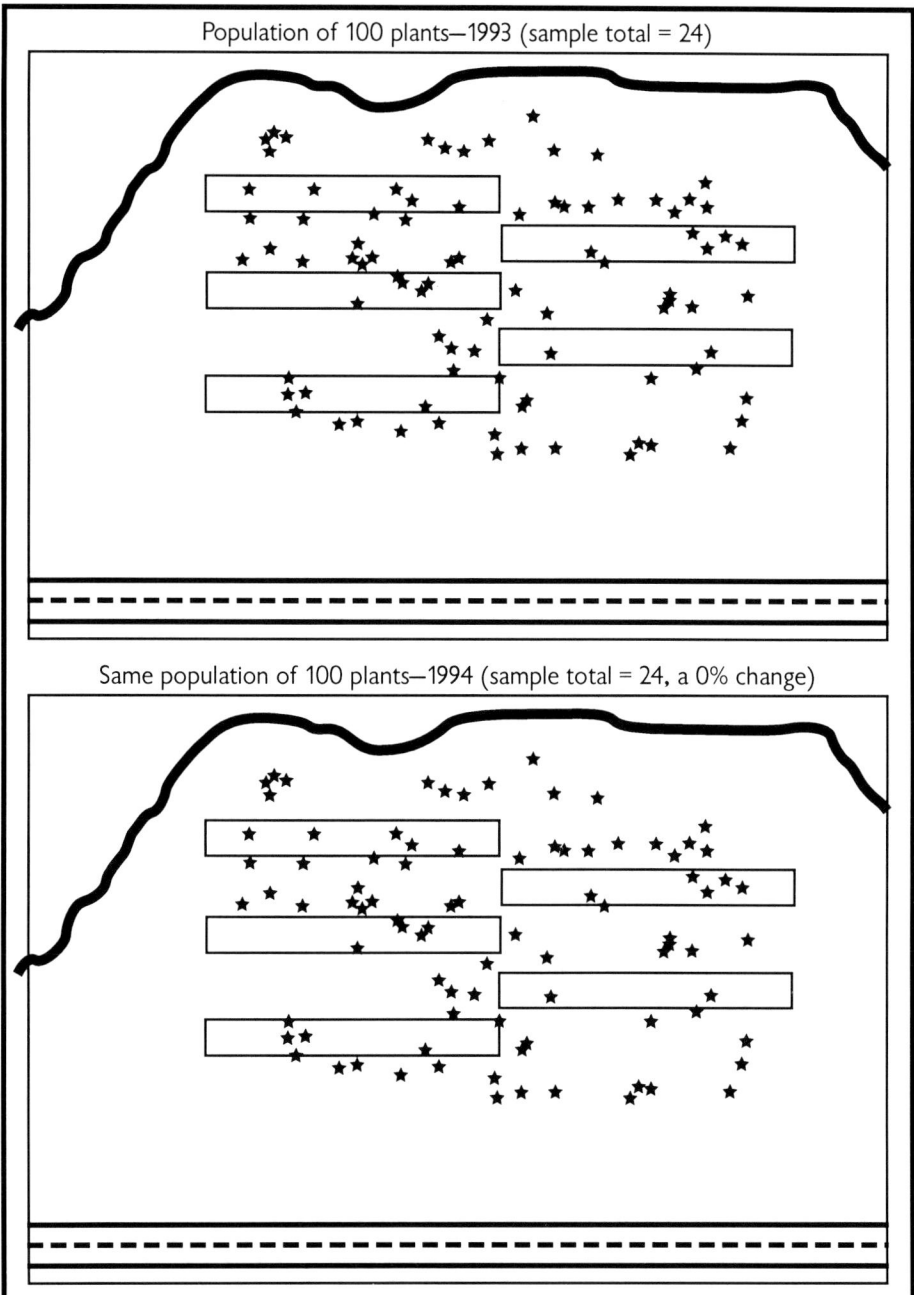

FIGURE 7.26. A population of 100 plants sampled at two times using five permanent quadrats.

two time periods, 10 transects were randomly positioned in the target population in 1990. The beginning, middle, and end points of each transect were permanently marked. Fifty points were systematically positioned (with a random start) along each transect and "hits" recorded on canopy cover of species X. The estimate of cover along each transect is then this number of hits divided by the total number of possible hits, 50. Thus, a transect with 34 hits would have a cover estimate of 68 percent or 0.68. The data from these two years are shown in Table 7.4.

Even though the cover estimates are highly variable between transects for both 1990 and 1994 (for example the mean cover for 1990 is 0.44 with a 95% confidence interval of 0.27 to 0.62), the standard error of the mean difference is relatively small. A 95% confidence interval around this mean difference is -0.02 to -0.12. In fact, in lieu of doing a paired

statistical test (such as a paired
t test), you could simply look at
the 95% confidence interval
around the mean difference to
see if it includes 0. If not, then
you can declare the change
significant at (at least) a *P*
value of 0.05.

If you had collected these data
using temporary transects (i.e.,
independent samples at both
sampling periods), you would
have concluded that no change
took place. In fact, with the
large degree of variability
between transects, you would
have needed unreasonably large
numbers of transects to detect
the change that only 10

transect number	cover in 1990	cover in 1994	difference between 1990 and 1994
1	0.22	0.20	-0.02
2	0.32	0.26	-0.06
3	0.06	0.06	0.00
4	0.86	0.80	-0.06
5	0.62	0.58	-0.04
6	0.54	0.50	-0.04
7	0.50	0.32	-0.18
8	0.28	0.24	-0.04
9	0.36	0.18	-0.18
10	0.68	0.64	-0.04
			mean difference -0.07
			standard error 0.02

TABLE 7.4 Cover values taken along 10 permanent transects of 50 points each in 1990 and 1994.

permanent transects were able to detect. This is displayed graphically in Figures 11.10 and 11.11 (in Chapter 11).

b. When to use permanent sampling units

Permanent sampling units will be the most advantageous when there is a high degree of correlation between sampling unit values between two time periods. This condition often occurs with long-lived plants (e.g., trees, shrubs, large cacti, or other long-lived perennial plants). If, however, there is low correlation between sampling units between two time periods, then the advantage of permanent quadrats is diminished. This could occur, for example, with annual plants, if their occurrence in quadrats one year is not greatly dependent on their occurrence in the previous year. Even for these plants, however, permanent quadrats may still outperform temporary quadrats if seedling recruitment most often takes place near parent plants.

Let's take a look at two very different situations involving permanent quadrats. Figure 7.27 compares sample sizes needed to detect different levels of change in a clumped population of 4000 plants using permanent and temporary quadrats. All sampling was done with 0.25m x 50m quadrats. In this example, there was no recruitment of new plants; all change between year 1 and year 2 was due to plant mortality. This created a strong correlation between quadrat counts between the two time periods for the low mortality changes. The x-axis shows the percent change in mean plant density (equivalent to percent mortality in this example). The y-axis shows the number of quadrats that needed to be sampled to detect the true population change with false-change and missed-change error rates both set at 0.10. When the change in mean plant density between the first and second sampling periods was less than 50%, permanent quadrats were much more effective than temporary quadrats at tracking the change. For example, for detecting a 5% change, 22 permanent quadrats performed as well as 338 temporary ones!

The advantage of permanent quadrats occurs when plant counts between two time periods correlate with one another. This is true in the situation depicted in Figure 7.27 because no new plants show up in new locations. The opposite extreme, illustrated by Figure 7.28, shows population changes due to 100% mortality of the original population combined with various levels of recruitment from plants in completely new positions. Permanent quadrats no longer provide any advantage over temporary ones, and the disadvantages of permanent quadrats would lead you to a temporary quadrat design.

Most populations will show a combination of mortality and recruitment, as opposed to the extreme situations shown in Figures 7.27 and 7.28. For most plant species, permanent quadrats will provide greater precision with the same number of quadrats or equivalent precision at smaller sample sizes, because the locations of new plants will likely be correlated with the location of old plants given typical patterns of sexual and asexual reproduction. You must balance the magnitude of this increase in precision (or reduction in sample size) against the disadvantages of using permanent sampling units, discussed below.

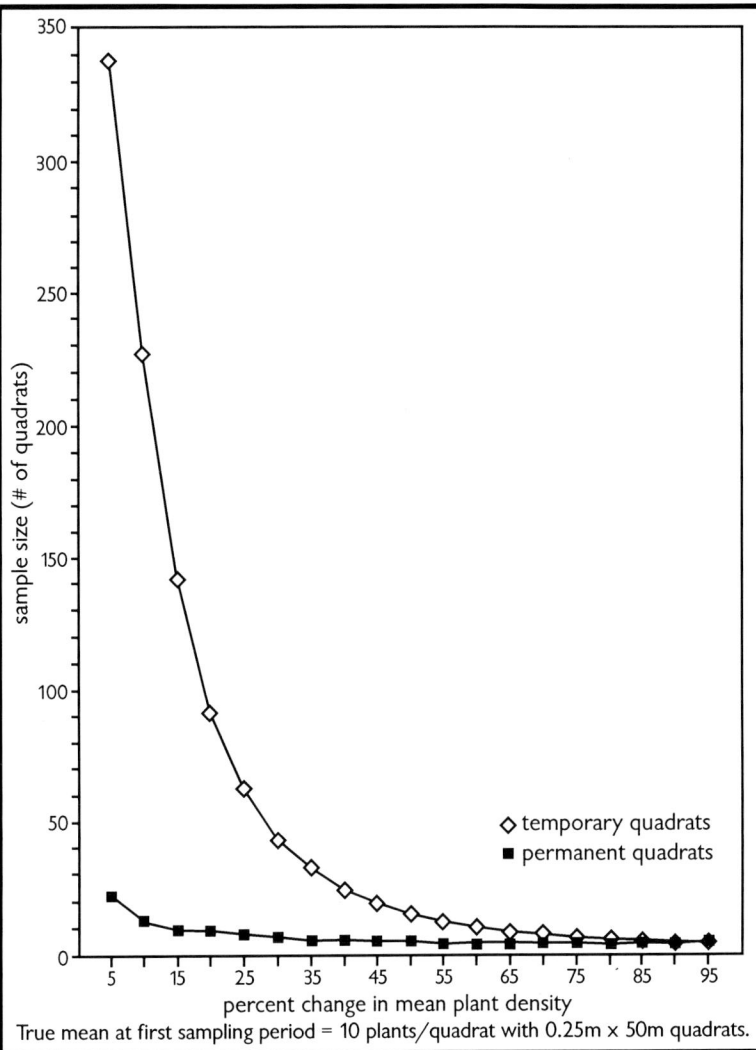

FIGURE 7.27. Sample sizes needed to detect different degrees of population decline from an artificial clumped population of 4,000 plants using temporary vs. permanent quadrats. All changes are due to mortality of the original population without any recruitment of new plants. Note the much better performance of permanent quadrats in detecting changes below 50%.

c. Permanent frequency quadrats and points

The discussion so far has centered on the use of paired quadrats for estimating density. This type of sampling is analyzed by means of a paired *t* test (this will be covered in Chapter 11—Statistical Analysis). The paired *t* test would also be used to analyze changes in paired quadrats used to estimate cover and to analyze changes in permanent transects, such as those used for line intercept sampling or for point or quadrat sampling in systematic sampling designs (when the transects, as opposed to the quadrats or points, are treated as the sampling units).

When frequency quadrats or points are treated as the sampling units, a different set of tests is used to determine if a statistically significant change has taken place. The chi square test is used when these types of sampling units are temporary (i.e., randomly located in each year of measurement), while McNemar's test is used when the quadrats or points are permanently located in the first year of measurement. These tests are discussed in Chapter 11, but it's important here to point out that—just as for permanent designs that use transects or quadrats for estimating density or cover—it is sometimes much more efficient to make use of permanent frequency quadrats or points.

Salzer (in prep.) concludes that under certain population change scenarios, permanent frequency quadrats offer large reductions in sample size over those required for temporary quadrats. In the most extreme example, 87 permanent quadrats perform as well as 652 temporary quadrats in detecting a 5.5% decline in frequency (with the false- and missed-change error rates both set at 0.10). In other situations there is little or no difference between permanent quadrat designs and temporary quadrat designs.

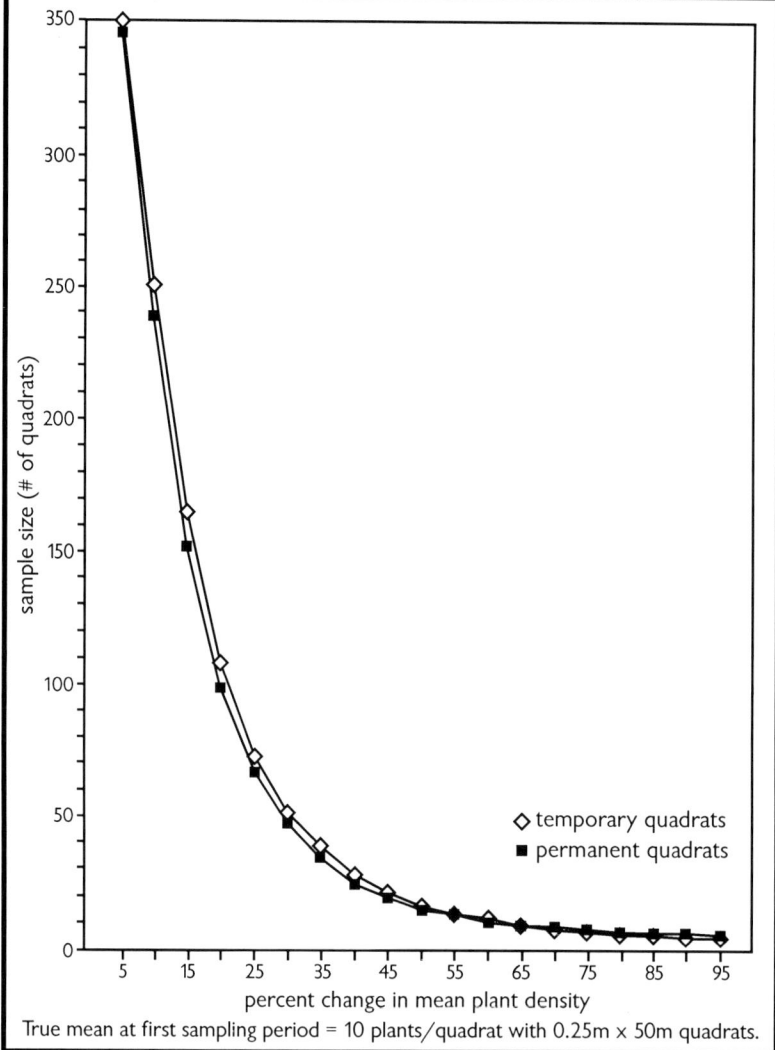

FIGURE 7.28. Sample sizes needed to detect different degrees of population decline from an artificial clumped population of 4,000 plants using temporary vs. permanent quadrats. All changes result from 100% mortality of the original population with various levels of random recruitment. Temporary and permanent quadrats perform about the same in this situation.

The sample size differences between temporary and permanent frequency designs depend on the particular nature of population changes. For this reason, the determination of whether to use permanent or temporary frequency quadrats must be evaluated on a case-by-case basis, taking into account the life history of the target plant species, the sample size advantages of using the permanent design, and the disadvantages associated with designs using permanent quadrats (discussed in the following section). All of the above discussion applies equally to the estimation of cover using the point intercept method, when the points are treated as the sampling units.

Appendix 18 contains more information on the use of permanent frequency designs and should help you decide when to use one.

2. Disadvantages of permanent sampling units

There are several disadvantages associated with using permanent sampling units. One is the time and cost required to permanently mark the units. Permanent markers (such as t-posts and rebar) can be expensive and awkward to pack long distances. The markers used to permanently mark sampling units are susceptible to loss or damage from such things as vandalism, animal impacts, and frost-heaving. Duplication of markers and back-up methods of relocating permanent sampling unit locations can help with this problem but can be very time-consuming.

Even when the markers are still in place, they may be difficult to find. Metal detectors and global positioning units can help you find the markers, but these add costs and field time. If frequency quadrats and points are the permanent sampling units, and you are positioning these at systematic intervals along transects, you must ensure the quadrats or points are repositioned as close as possible to the positions in the year in which the study was set up. This is especially critical for small frequency quadrats when the rooted portion of the target species is small and for points when the cover of the target species is likely to be sparse. Permanently monumenting not only the transect ends but also intermediate points in between and carefully stretching the tape at each measurement period can help to ensure the transect is in the same location. For frequency quadrats you can also monument two corners of each quadrat using large nails. This adds additional insurance but also results in more labor.

Impacts either from investigators or from animals may bias your results. By going back to the same sampling unit locations each year, you might negatively impact the habitat in or near the permanent sampling units. In addition, permanent markers may also attract wildlife, domestic livestock, wild horses, or burros. This might lead to differential impacts to the vegetation in or near the sampling units. If markers are too high (for example, t-posts or other fence posts), livestock may use the markers for scratching posts and differently impact the sampling units. Wildlife impacts may also occur. Raptors, for example, might use the markers as perches; this could result in fewer herbivores in the sampling units than elsewhere in the target population, with resulting differences in the plant attribute being measured. Using shorter markers, such as rebar no greater than 0.5m high, will at least partially resolve this problem (but see below for safety concerns).

Permanent markers are not feasible in some situations because of the nature of the habitat or for safety reasons. For example, sand dune systems do not lend themselves to the use of permanent markers because drifting sand can quickly bury the markers. You wouldn't want to use permanent steel posts or rebar in areas frequented by off-road vehicles because of the risk to human life.

Another disadvantage of a design using permanent sampling units is that you usually need 2 years of data to determine adequate sample size. The only exception to this is when you have some basis to estimate the degree of correlation (the correlation coefficient) of sampling units between years when estimating means (e.g., density sampling) or a model of how the population is likely to change when estimating proportions (e.g., frequency sampling). We'll discuss this at more length in the next section.

G. How Many Sampling Units Should Be Sampled?

An adequate sample is vital to the success of any successful monitoring effort. Adequacy relates to the ability of the observer to evaluate whether the management objective has been achieved. It makes little sense, for example, to set a management objective of increasing the density of a rare plant species by 20% when the monitoring design and sample size will not likely detect changes in density of less than 50%.

1. General comments on calculating sample size

Deciding on the number of sampling units to sample (which we refer to as "sample size") should be based on the following considerations:

a. Sample size should be driven by specific objectives

If you are targeting point-in-time estimates (parameter estimation), you need to specify how precise you want your estimates to be. If you are trying to detect changes in some average value, you need to specify the magnitude of the change you wish to detect and the acceptable false-change and missed-change error rates (refer to Chapter 6 for further guidance).

b. Sample size should be based on the amount of variability in actual measurements

You should assess this variability during pilot sampling. Once you have tried various sampling unit sizes and shapes and have decided upon a particular one, start randomly positioning the sampling units in the population. After you have sampled some initial bunch of sampling units, stop and do some simple number-crunching with a hand calculator to see what the variation in the data looks like. You can plug standard deviations into sample size equations or computer programs, and the output will inform you as to whether you have sampled enough. If you haven't, sample size equations (or a computer program) will calculate the number of sampling units you need to sample in order to meet your objective. We discuss the process of sequential sampling in detail below.

c. Assumptions of formulas and computer programs

The sample size formulas and computer programs assume that the sampling units are positioned in some random manner and that a distribution of sample means (a sampling distribution) from your population fits approximately a normal distribution. If your population is highly skewed, this latter assumption will not be true for small sample sizes. We discuss this issue in more detail in Chapter 11.

d. Infinite vs. finite populations

We introduced this concept in Chapter 5. Most computer programs and standard sample size equations assume that the population you are sampling from is infinite. This will always be the case if you are estimating cover using either points or lines, because these are considered dimensionless. If, however, you are sampling a relatively small area, and you are making density, frequency, cover, or biomass assessments in quadrats, then you should account for the fact that you are sampling from a finite population. This means there is some finite number of quadrats that can be placed in the area to be sampled.

The sample size formulas provided in Appendix 7 include a correction factor called the Finite Population Correction (FPC). If you are sampling more than 5% of a population, applying the FPC "rewards" you by reducing the necessary sample size. In addition to describing how to apply the FPC to sample size determination, Appendix 7 also describes how to apply it to the results of two-sample significance tests. Appendix 16 shows how to use the finite population correction factor when sampling to detect a difference in proportions using permanent sampling units.

e. Relationship of sample size to precision level

Precision increases with sample size, but not proportionately. This is illustrated in Figure 7.29. For this example, the statistical benefits of increasing sample size diminish once you reach about $n=30$; any benefits to using more than 30 sampling units relate to adequately capturing the variability in the population being sampled. This also serves to highlight the most important aspect of good sampling design: you should seek to increase statistical precision and power not by simply increasing sample size, but by reducing the standard deviation to as small a value as possible.

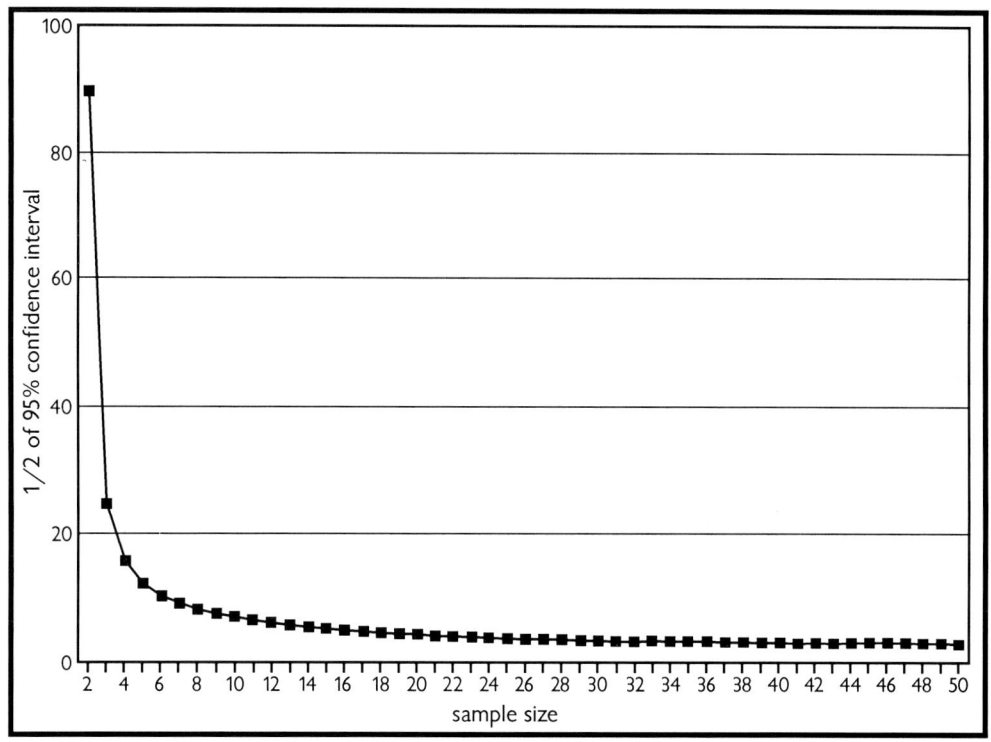

FIGURE 7.29. Influence of sample size on level of precision. Sample sizes necessary to achieve different levels of precision at a constant standard deviation of 10. Note that there is no effective improvement in precision after about n = 30.

f. Problems with some sample size formulas

Most formulas that are designed to determine sample sizes for "point-in-time" estimates (parameter estimation) with specified levels of precision do not account for the random nature of sample variances. They do not include a "level of assurance" (also known as a tolerance probability) that you will actually achieve the conditions specified in the sampling size equations and obtain a confidence interval of a specified width. Blackwood (1991) discusses this topic in lay person's terms and reports the results of a simulation that illustrates the concept. Kupper and Hafner (1989) provide a correction table to use with standard sample size equations for estimates of single population means or population totals. A modified version of this table and instructions on how to use it are included in Appendix 7.

2. Information required for calculating sample size

Appendix 7 gives equations for calculating sample sizes for the following sampling objectives: (1) estimating means and totals; (2) detecting change between two time periods in a mean value; (3) detecting differences between two means when using permanent sampling units;

(4) estimating a proportion; and (5) detecting change between two time periods in a proportion using temporary sampling units. Appendix 16 gives directions for using the computer programs STPLAN and PC SIZE: CONSULTANT to calculate sample sizes to meet all of these objectives (except estimating a proportion) and, in addition, gives instructions on calculating the sample size required to detect change between two time periods in a proportion when using permanent sampling units (these programs are discussed further in Section H). Both appendices include completely worked-out examples. The following discussion briefly summarizes the information required to use either the equations or computer programs to calculate sample size.

Estimating means and totals. You must specify the precision desired (confidence interval width), the confidence level, and an estimate of the standard deviation.

Detecting change between two time periods in a mean value. You must specify the false-change error rate, the power of the test, the magnitude of the smallest change you wish to detect, and an estimate of the standard deviation (the population standard deviation is usually assumed to be the same for both time periods).

Detecting change between two means using permanent sampling units. You must specify the false-change error rate, the power of the test, the magnitude of the smallest change you wish to detect, and an estimate of the standard deviation (this is the standard deviation of the differences between the paired sampling units, *not* the standard deviation of the population being sampled in the first year).

Estimating a proportion. You must specify the precision desired (confidence interval width), the confidence level, and a preliminary estimate of the proportion to be estimated (if you don't have any idea of what proportion is to be expected you can conservatively estimate the sample size by assuming the proportion to be 0.50).

Detecting change between two time periods in a proportion using temporary sampling units. You must specify the false-change error rate, the power of the test, the magnitude of the smallest change you wish to detect, and a preliminary estimate of the proportion in the first year of measurement (using a value of 0.50 will conservatively estimate the sample size).

Detecting change between two time periods in a proportion using permanent sampling units. You must specify the false-change error rate, the power of the test, the magnitude of the smallest change you wish to detect, and an estimate of the sampling unit transitions that took place between the two years.

Your management and sampling objectives already include most of the information required to calculate sample size using either the equations of Appendix 7 or the computer programs STPLAN and PC SIZE: CONSULTANT, following the instructions of Appendix 16. What is missing is an estimate of the standard deviation for those situations where you wish to estimate a mean value or detect change between two mean values and a preliminary estimate of the population proportion when estimating a proportion or detecting change between two proportions using temporary sampling units. For proportions you have the flexibility of simply entering 0.50 as your preliminary estimate of the population proportion and calculating your sample size based on this. Alternatively, you can use an estimate derived from pilot sampling. When dealing with mean values, however, you must have an estimate of the standard deviation. This is the subject of the next section. (Detecting change between two time periods in a proportion using permanent sampling units is a special case that will be discussed separately below and in Appendix 18.)

3. Sequential sampling to obtain a stable estimate of the mean and standard deviation

In several places in this chapter we have stressed the need for pilot sampling. The principal purposes of pilot sampling are to assess the efficiency of a particular sampling design and, once a particular design has been settled upon, to assist in determining the sample size required to meet the sampling objective. Pilot sampling enables us to obtain stable estimates of the population mean and the population standard deviation. By dividing the sample standard deviation by the sample mean we get the coefficient of variation. Comparing coefficients of variation enables us to determine which of two or more sampling designs is most efficient (the lower the coefficient of variation, the greater the efficiency of the sampling design). The estimate of the standard deviation derived through pilot sampling is one of the values we use to calculate sample size, whether we use the formulas of Appendix 7 or a computer program.

Sequential sampling is the process we use to determine whether we have taken a large enough pilot sample to properly evaluate different sampling designs and/or to use the standard deviation from the pilot sample to calculate sample size. The process is accomplished as follows.

Gather pilot sampling data using some arbitrarily selected sample size. The selection of this initial sample size will depend upon the relative amount of variation in the data—if many of the sampling units yield numbers similar to one another, then you may want to perform the first sequential sampling procedure after $n = 8$ or 10. If there is a lot of variation among the sampling units, then you may want to start with a larger number (e.g., $n \geq 15$), or consider altering the size and/or shape of your sampling unit prior to doing the first iteration of the sequential sampling procedure.

Calculate the mean and standard deviation for the first two quadrats, calculate it again after putting in the next quadrat value, and then repeat this procedure for all of the quadrats sampled so far. This will generate a running mean and standard deviation. Look at the four columns of numbers on the right of Figure 7.30 for an example of how to carry out this procedure. Most hand calculators enable you to add additional values after you've calculated the mean and standard deviation, so you don't have to re-key in the previous values.

Plot on graph paper (or use a computer program as discussed later) the sample size vs. the mean and standard deviation. Look for curves smoothing out. In the example shown in Figure 7.30, the curves smooth out after $n=35$.

In graphing your results beware of y-axis scaling problems. If your first few quadrats are very deviant from each other, you may scale your y-axis with too broad a range, which will give a false impression of the lines smoothing out. The top and bottom graphs of Figure 7.31 both graph the same data set (only the order of the data was changed). Because the first few quadrats in the upper graph contained large values, the scale of the y-axis was set from 0 to 7. The result is that there appears to be a smoothing out of the curves at around 15 quadrats. In the bottom graph, the first few quadrats contained smaller values, so the scale of the y-axis was set from 0 to 2.5. This graph gives a much clearer view of the true situation: the

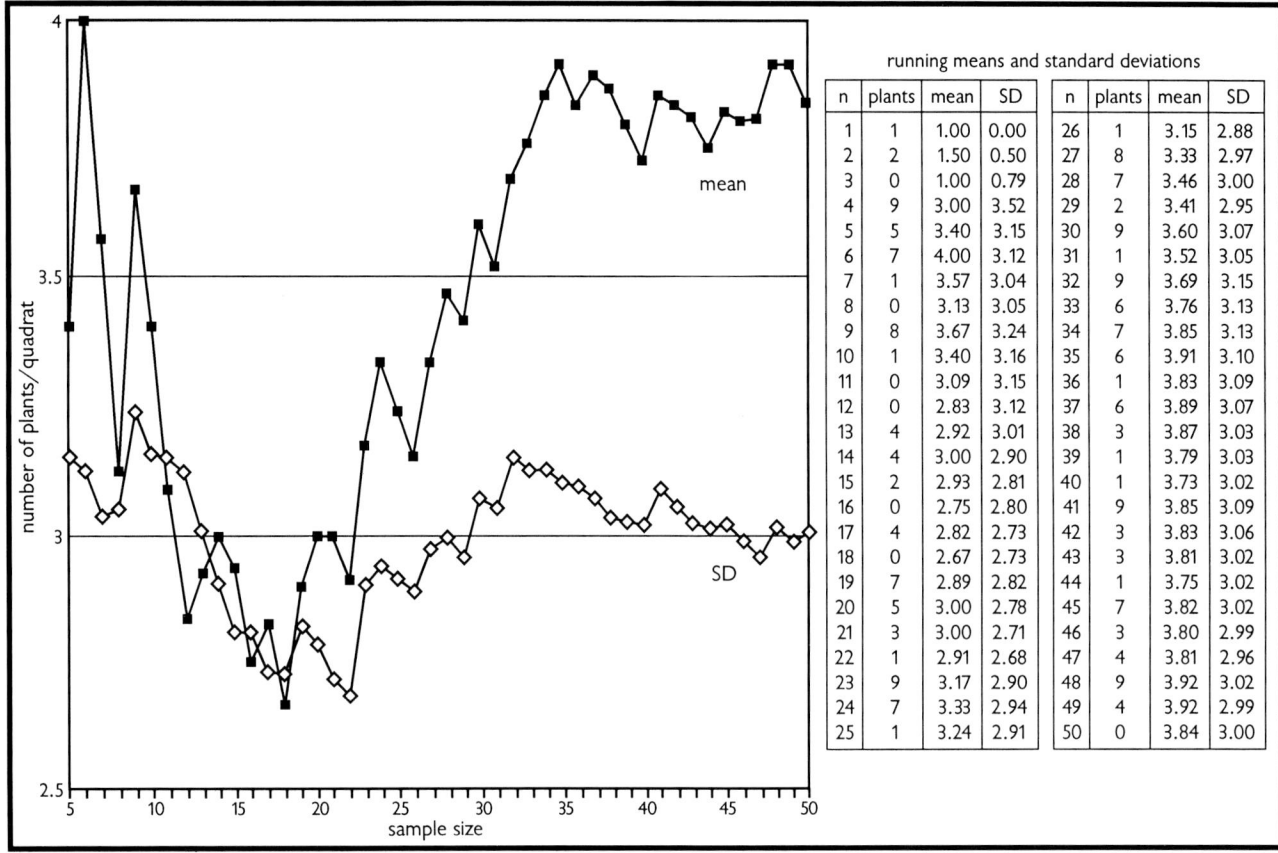

FIGURE 7.30. A sequential sampling graph. Running means and standard deviations are plotted for increasing sample sizes. Note how the curves smooth out after n = 35.

curves have not really smoothed out even after 100 quadrats.[4] If early quadrat values are too extreme you may want to start plotting with *n*=5 rather than *n*=2 to avoid too great a y-axis range. The decision to stop sampling is a subjective one. There are no hard and fast rules.

A computer is valuable for creating sequential sampling graphs. Spreadsheet programs such as Lotus 1-2-3 and Excel enable you to enter your data in a form that can later be analyzed and at the same time create a sequential sampling graph of the running mean and standard deviation. This further allows you to look at several random sequences of the data you have collected before making a decision on the number of sampling units to measure. Figures 7.32 and 7.33 both show the results of sampling the entire "400-plant population" (introduced in Chapter 5) using a 0.4m x 10m quadrat size (the population is contained in a 20m x 20m macroplot; there are 100 possible quadrat positions in the population with this size quadrat). The only difference between these two graphs is the ordering of the data: the data were randomly reordered prior to creating each graph.

Figure 7.34 shows sequential sampling graphs where the number of sampling units gathered far exceeded the number where the curves flattened out.

---

[4] The sequential sampling graph at the bottom of Figure 7.31 illustrates a poor sampling design. Because 1m x 1m quadrats were used, most of the quadrats had 0 plants in them. Sampling several consecutive quadrats with 0 plants brings the running mean and standard deviation down until a quadrat is located with several plants in it. This brings the running mean and standard deviation up sharply (see the spikes on the graph). This phenomenon by itself should alert you to the fact that the sampling design is inadequate.

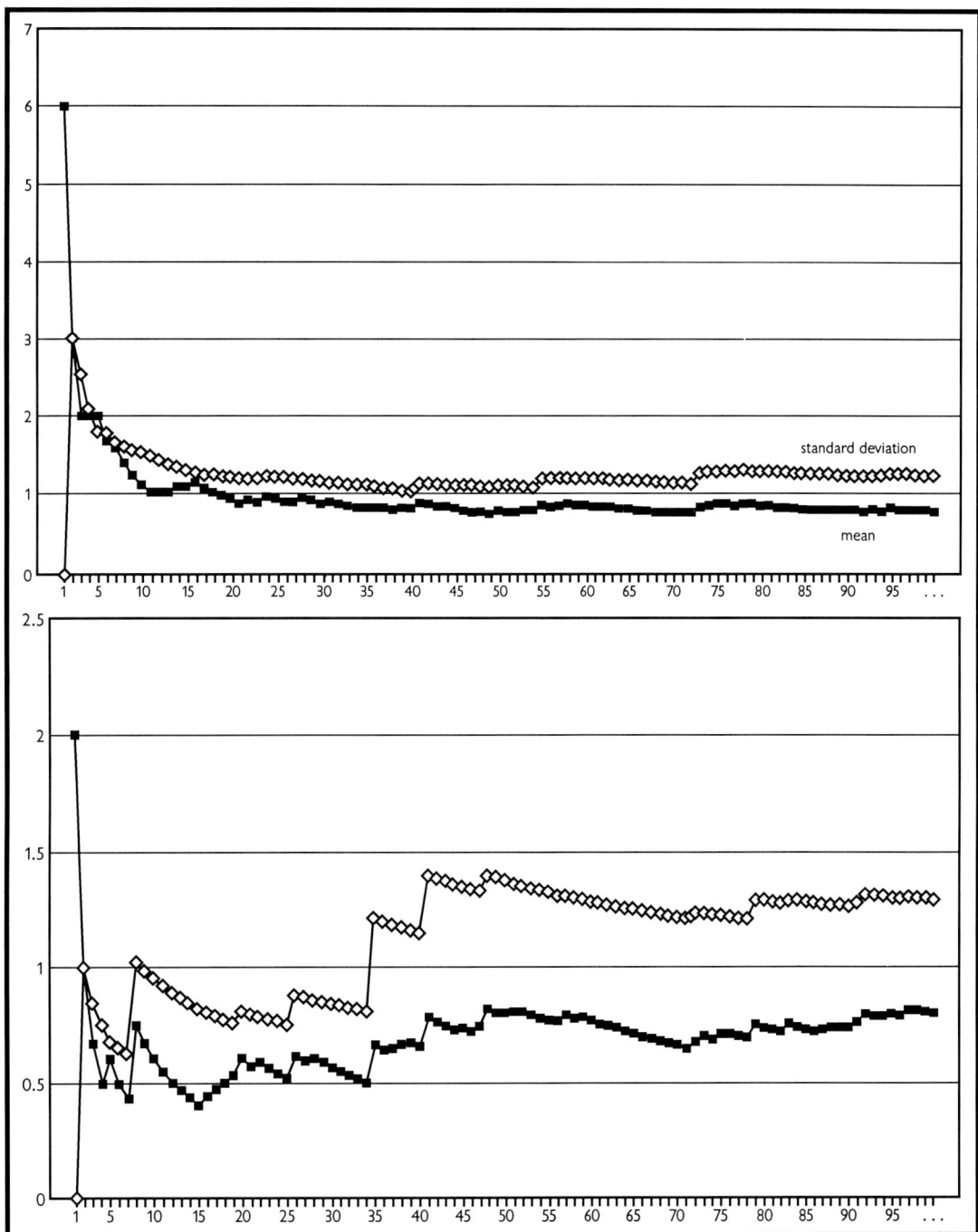

FIGURE 7.31. Sequential sampling graphs for <u>Astragalus</u> <u>applegatei</u> at the Euwana Flat Preserve. The upper graph shows what can happen when the y-axis is set at too large a range, because of initial large values. This can make it appear that the running mean and standard deviation has smoothed out when in fact they haven't. The bottom graph illustrates the real situation: neither statistic has smoothed out even by n = 100. This is a poor sampling design. See text for further elaboration.

If there are too many zeros in your data set, then sequential sampling graphs will not make sense. We saw this to some extent in Figure 7.31. A more extreme example is shown in Figure 7.35. Graphs like this should alert you to major problems with the sampling design.

Use the sequential sampling method to determine what sample size *not* to use (you don't use the sample size below the point where the running mean and standard deviation have not

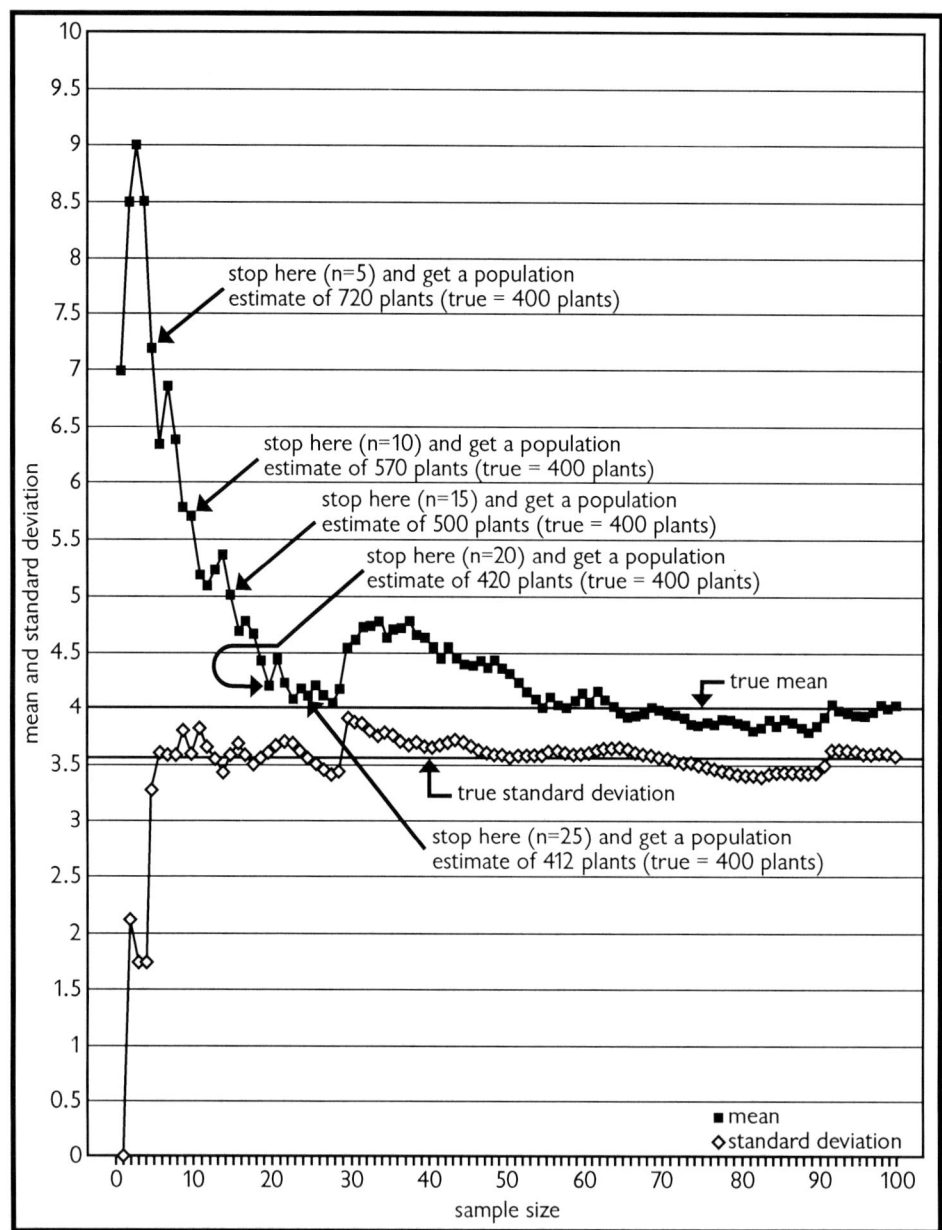

FIGURE 7.32. Sequential sampling graph of the 20m x 20m "400-plant population" introduced
in Chapter 5. The population was sampled using a 0.4m x 10m quadrat.
The entire population consists of 100 quadrats. Notice how far estimates are
from the true mean value if they are made prior to the curves smoothing out.

stabilized). Plug the final mean and standard deviation information into the appropriate
sample size equation or computer program to actually determine the necessary sample size.

4. Alternatives to sequential sampling to obtain an estimate of the standard deviation

Pilot sampling, using the sequential sampling procedure described above, is by far the best
means of deriving an estimate of the standard deviation to plug into a sample size equation
or computer program. There are, however, two other methods that will be briefly discussed.

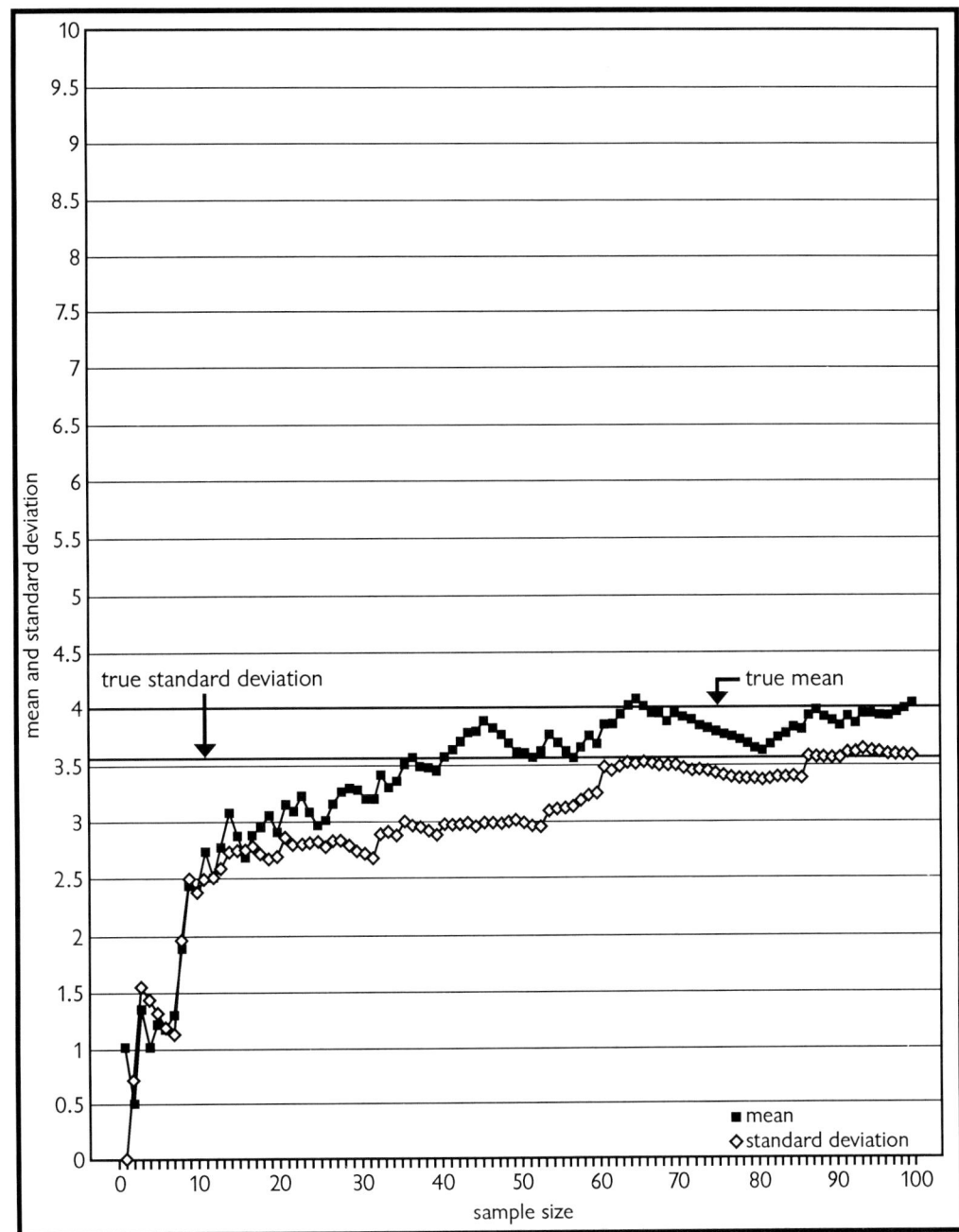

FIGURE 7.33. Sequential sampling graph for the 20m x 20m "400-plant population." Sampling unit size is the same as in Figure 7.32. The only difference between this graph and Figure 7.32 is that the data were randomly reordered. If we'd used the initial values shown in this graph (prior to the curves leveling off), we would have seriously underestimated the true mean value, as opposed to overestimating it as was the case in Figure 7.32.

a. Use data from similar studies to estimate the standard deviation

Although not as reliable as a pilot study, you may have conducted a study using the same study design, measuring the same vegetation attribute, and in the same vegetation type. The standard deviation of the sample from this study can be used as an estimate of the standard deviation of the population that is the focus of the current study.

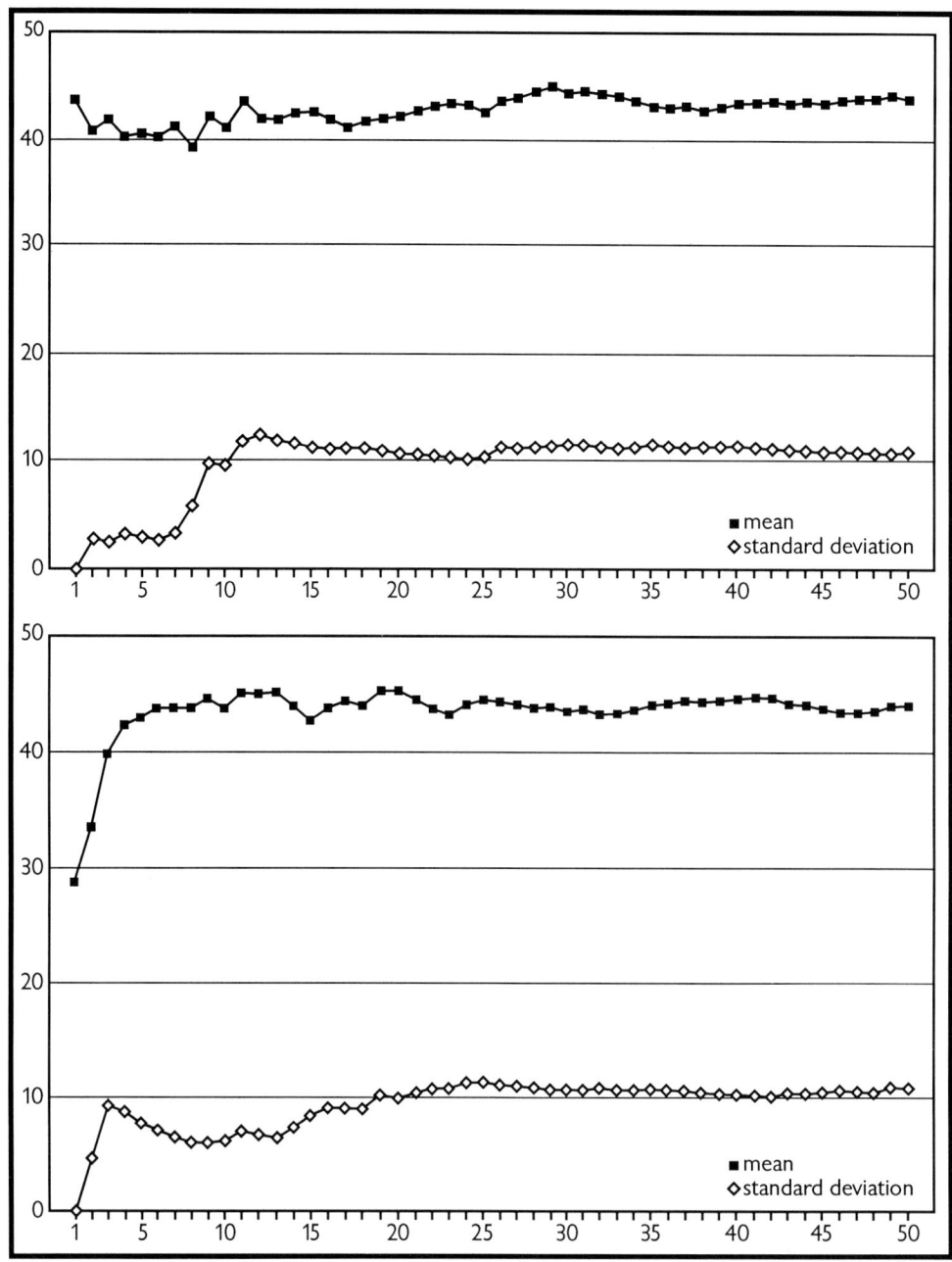

FIGURE 7.34. Sequential sampling graphs of vegetation height measurements at Mt. Hebo. These are graphs of the same data but in different orders. Note how the graphs have flattened out long before the sampling ended.

b. By professional judgment

As pointed out by Krebs (1989) an experienced person may have some knowledge of the amount of variability in a particular attribute. Using this information you can determine a range of measurements to be expected (maximum value - minimum value) and can use this to estimate the standard deviation of a measure. Table 7.5, adapted from the table in Dixon and Massey (1983), and reproduced in Krebs (1989), gives the appropriate conversion factor to be multiplied by the range value to come up with an estimate of the population standard deviation.

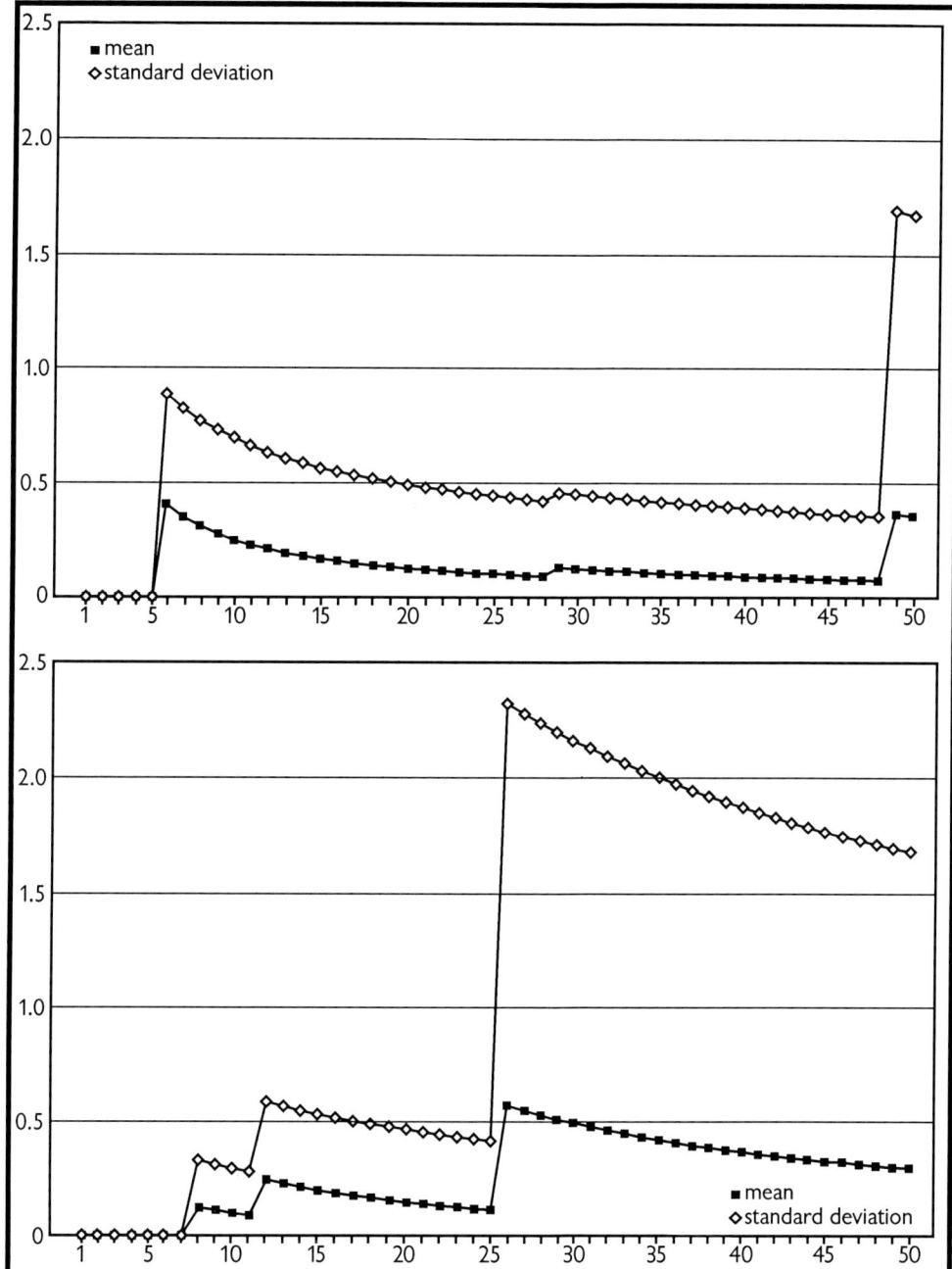

FIGURE 7.35. Sequential sampling graphs of bracken fern stem density at Mt. Hebo. Both graphs plot the same data set in different order. This is an example of a poor sampling design. Because 1m x 1m quadrats were used, most of the quadrats had 0 plants in them. Sampling several consecutive quadrats with 0 plants brings the running mean and standard deviation down until a quadrat is located with several stems in it. This brings the running mean and standard deviation up sharply and results in the spikes shown on the graphs. This pattern should alert you to the need to change your sampling design.

To illustrate how to use this table, let's assume we know from experience with the plant species we're working with that we expect, in a sample of size 30, a range of 0 plants per quadrat to 100 plants per quadrat (this process assumes a normal distribution so we'd better not have too many quadrats with 0's in them). The range in this case is 100 plants - 0 plants = 100 plants. The conversion factor for a sample of size 30 is 0.245. Our estimate of the population standard deviation is, therefore, 100 plants x 0.245 or 24.5 plants per quadrat.

Although this method *can* be used, it should be emphasized again that data from a pilot study are more reliable and are preferable to this method.

5. Estimating the standard deviation when using permanent sampling units

Estimating the standard deviation for a design that uses permanent sampling units is difficult because it is the standard deviation of the difference between the sampling units between the two years that must be plugged into the sample size equation or computer program, and this is a value that you will not have until you have collected data in two years. Thus, your pilot study must span two years before you can accurately estimate the sample size required to meet your sampling objective. You would like, however, to make a reasonable estimate from the first year's data of the standard deviation of the difference. This will give you a good chance of having used a large enough sample size the first year, with the result that you will not have to add more sampling units the second year and will be able to use the first year's data in your analysis. Following are some methods you can use for this purpose.

sample size	conversion factor	sample size	conversion factor
2	0.886	19	0.271
3	0.591	20	0.268
4	0.486	25	0.254
5	0.430	30	0.245
6	0.395	40	0.231
7	0.370	50	0.222
8	0.351	60	0.216
9	0.337	70	0.210
10	0.325	80	0.206
11	0.315	90	0.202
12	0.307	100	0.199
13	0.300	150	0.189
14	0.294	200	0.182
15	0.288	300	0.174
16	0.283	500	0.165
17	0.279	1000	0.154
18	0.275		

TABLE 7.5. Estimating the standard deviation of a variable from knowledge of the range for samples of various sizes. Multiply the observed range (maximum - minimum value) by the table values to obtain an unbiased estimate of the standard deviation. This procedure assumes a normal distribution. From Dixon and Massey (1983) and reproduced in Krebs (1989).

You can estimate the standard deviation using the alternative methods discussed under the section above. Remember, however, that it is the standard deviation of the difference that must be estimated, so if you use data from previous studies they must be studies that used permanent sampling units. If you use the expected range to estimate the standard deviation, it must be the range of the differences, not the range of the data for any one year.

There is another way you can calculate the necessary sample size by having only the first year's pilot data. This method requires that you have some knowledge of the degree of correlation (correlation coefficient) expected between the permanent sampling units between years. Sample Size Equation #3 in Appendix 7 gives a formula by which you can estimate the standard deviation of the difference between years by using the standard deviation of the first year's sample and the correlation coefficient. This is something you might have from similar studies on the same plant species (although in that case you'd probably already have an estimate of the standard deviation of the difference between years that you could use). Based on your knowledge of the life history of the species you are dealing with, you might make an initial estimate of correlation. For example, if you're monitoring a long-lived perennial and you don't anticipate a lot of seedling recruitment (or if you expect seedling recruitment to be very close to parent plants), you might estimate that the correlation coefficient between years is relatively high, say about 0.80 or 0.90. You then plug this coefficient into the formula, along with your estimate of the standard deviation of the first year's data.

Whichever method you use to estimate the standard deviation of the difference, once you've collected the second year's data, you will still need to plug the actual observed standard deviation of the difference into Equation #3 of Appendix 7 or STPLAN. You can then modify your initial estimate of sample size accordingly.

6. Calculating the sample size necessary to detect changes between two time periods in a proportion when using permanent sampling units

Appendix 7 gives no formulas to calculate the sample size necessary to detect changes between two time periods in a proportion when using permanent sampling units. Appendix 16 does, however, describe how to use the program STPLAN to calculate sample size when you have 2 years of data from these types of permanent sampling units. Appendix 18 describes a method that you can use to derive an estimate of the sampling unit transitions that might be expected based on a single year's data and an ecological model of the plant species you are monitoring. You are strongly encouraged to read Appendix 18 if you are considering using permanent frequency quadrats.

H. Computer Programs for Calculating Sample Size

Believe it or not, most of the general statistical programs do not include routines for calculating sample size, despite their expense. Thomas and Krebs (1997) reviewed 29 computer programs for calculating sample size. They also maintain a World Wide Web site with information on how to order these programs. Refer to Chapter 11, Section L, for the address.

For beginner to intermediate level use, Thomas and Krebs recommend one of the following three commercial programs: PASS, NQUERY ADVISOR, or STAT POWER. The first one on this list, PASS, was the one most preferred by a graduate student class. Refer to their website for information on the cost of these programs and how to order them. Thomas and Krebs also give relatively high marks to the program GPOWER, primarily because it is free.

The documentation for GPOWER is extremely limited, and the user must have familiarity with Cohen's (1988) treatment of power analysis (Thomas and Krebs 1997). For these reasons we do not recommend the program for the sample size determination and power analysis needed for the types of monitoring treated in this technical reference. Instead, we suggest you consider the following two programs (unless you have the money to purchase the commercial program PASS): STPLAN and PC SIZE: CONSULTANT. STPLAN, currently in version 4.1, is free. PC SIZE: CONSULTANT costs $15 as shareware. Both can be downloaded from the World Wide Web. See Chapter 11, Section L, for the addresses.

STPLAN will calculate sample sizes needed for all the types of significance testing discussed in this chapter, but will not calculate those required for estimating a single population mean, total, or proportion. It will also calculate sample sizes for permanent frequency quadrat designs. PC SIZE: CONSULTANT will calculate sample sizes for all of the significance tests discussed in this chapter, as well as sample sizes required to estimate a single population mean or total. It will not, however, calculate sample size for estimating a single population proportion. Both programs are DOS-based and not, therefore, particularly "user friendly." They are not difficult to learn, however, and documentation files are included when you download the programs. Appendix 16 gives instructions on the use of these two programs for calculating sample sizes.

Literature Cited

Blackwood, L. G. 1991. Assurance levels of standard sample size formulas. Environmental Science and Technology 25(8):1366-1367.

Bonham, C. D. 1989. Measurements for terrestrial vegetation. New York, NY: John Wiley & Sons.

Chambers, J. C.; Brown, R. W. 1983. Methods for vegetation sampling and analysis on revegetated mined lands. Ogden, UT: U.S. Forest Service Intermountain Research Station, General Technical Report INT-151.

Cochran, W. G. 1977. Sampling techniques, 3rd ed. New York, NY: John Wiley & Sons.

Cohen, J. 1988. Statistical power analysis for the behavioral sciences. Hillsdale, NJ: Lawrence Erlbaum Associates.

Dahl, T. E.; Johnson, C. E. 1991. Status and trends of wetlands in the conterminous United States, mid-1970's to mid-1980's. Washington, D.C.: U.S. Department of the Interior, Fish and Wildlife Service.

DeVries, P. G. 1979. Line intersect sampling—statistical theory, applications, and suggestions for extended use in ecological inventory. In: Cormack, R. M.; Patil, G. P.; Robson, D. S., eds., Sampling Biological Populations, Vol. 5: Statistical Ecology, pp. 1-70. Fairland, MD: International Cooperative Publishing House.

Dixon, W. J., Massey, F. J., Jr. 1983. Introduction to statistical analysis, 4th ed. New York, NY: McGraw-Hill.

Floyd, D. A.; Anderson, J. E. 1987. A comparison of three methods for estimating plant cover. Journal of Ecology 75: 229-245.

Goldsmith, F. B.; Harrison, C. M.; Morton, A. J. 1986. Description and analysis of vegetation. In: Moore, P.D.; Chapman, S. B., eds. Methods in plant ecology, 2nd edition, pp. 437-524. Palo Alto, CA: Blackwell Scientific Publications.

Goodall, D. W. 1952. Some considerations in the use of point quadrats for the analysis of vegetation. Australian Journal of Scientific Research 5: 1-41.

Greig-Smith, P. 1983. Quantitative plant ecology, 3rd ed. Berkeley, CA: University of California Press.

Hurlbert, S. H. 1984. Pseudoreplication and the design of ecological field experiments. Ecological Monographs 54: 187-211.

Krebs, C. J. 1989. Ecological methodology. New York, NY: Harper & Row.

Kupper, L. L.; Hafner, K. B. 1989. How appropriate are popular sample size formulas? The American Statistician 43:101-105.

Lucas, H. A.; Seber, G. A. F. 1977. Estimating coverage and particle density using the line intercept method. Biometricka 64:618-622.

McCall, C. H., Jr. 1982. Sampling and statistics handbook for research. Ames, IA: Iowa State University Press.

Milne, A. 1959. The centric systematic area-sample treated as a random sample. Biometrics 15:270-297.

Platts, W. S.; Armour, C.; Booth, G. D.; Bryant, M.; Bufford, J. L.; Cuplin, P.; Jensen, S.; Lienkaemper, G. W.; Minshall, G. W.; Monsen, S. B.; Nelson, R. L.; Sedell, J. R.; Tuhy, J. S. 1987. Methods for evaluating riparian habitats with applications to management. Ogden, UT: U.S. Forest Service Intermountain Research Station, General Technical Report INT-221.

Schaeffer, R. L.; Mendenhall, W; Ott, L. 1979. Elementary survey sampling. North Scituate, MA: Duxbury Press.

Stehman, S. V.; Overton, W. S. 1994. Environmental sampling and monitoring. In: Patil, G. P.; Rao, C. R., eds. Environmental Statistics, pp. 263-306. Handbook of Statistics 12. New York, NY: North-Holland.

Thomas, L.; Krebs, C. J. 1997. A review of statistical power analysis software. Bulletin of the Ecological Society of America 78(2): 128-139.

Thompson, S. K. 1992. Sampling. New York, NY: John Wiley & Sons.

Usher, M. B. 1991. Scientific requirements of a monitoring programme. In: Goldsmith, F. B., ed. Monitoring for Conservation and Ecology, pp. 15-32. New York, NY: Chapman and Hall.

Wiegert, R. G. 1962. The selection of an optimum quadrat size for sampling the standing crop of grasses and forbs. Ecology 43:125-129.

Williams, B. 1978. A sampler on sampling. New York, NY: John Wiley and Sons.

CHAPTER 8
Field Techniques for Measuring Vegetation

Atriplex nuttalli
Nuttall saltbush
by Jennifer Shoemaker

CHAPTER 8. Field Techniques for Measuring Vegetation

The techniques described in this chapter are primarily used for herbs, graminoids, and shrubs. Specific methods for tree species were deliberately excluded because many references are available on methods for trees of commercial importance. Several texts we recommend describing density and basal area estimation for trees include Dilworth and Bell (1973), Husch et al. (1982), Dilworth (1989), Schreuder et al. (1993), Avery and Burkhart (1994), and Shivers and Borders (1996).

A. Qualitative Techniques

1. Presence/absence

Presence/absence techniques note whether the species still occurs at a site. The key advantages are that no special skills are required (anyone who can recognize the species can do the monitoring) and that the monitoring requires very little time. The main disadvantage is that presence/absence observations provide no information on trend, except when the population disappears.

A presence/absence approach may be useful for large or showy plants that grow along roads and are visible during a drive-by visit. You can enlist specialists from other disciplines to monitor the presence or absence of the species while they are performing other work. The technique can effectively monitor occurrences across the landscape and is especially appropriate for species with many small populations.

You can improve the consistency and usefulness of observations with a short form to report population visits. Fields to include are observer, date, and time spent at site. You might also add a field for noting whether the survey was a drive-by or walk-through, a comment field for specific threats or problems, and a field for listing photographs. You can make it easier for other specialists to do this work (and make it more likely that they will do it) by putting together a packet of maps and data sheets for them to carry in their vehicles. Recommended is a map of the entire resource area or district showing population areas marked in red and the outlines and names of all overlying topographic quadrangles. This should be accompanied by a packet of photocopies of portions of topographic maps, each clearly labeled (e.g., "lower right of Cobalt Quad") and with the population locations shown. Make it easy to flip through (use 8.5" x 11" sheets in a binder) and easy to locate things (e.g., alphabetical tabs for the photocopied topographic maps).

2. Estimates of population size

Estimates of population size require only a small amount of additional time and effort over that needed for presence/absence. The advantage of estimates is that they provide a gross index of population trend. The key disadvantage is that because of variability among observer estimates, only large changes can be monitored with confidence.

Establishing some guidelines will improve the repeatability of estimates. You will need to decide, for example, if all individuals will be included or only large or reproductive ones. Estimates that include small, cryptic individuals can be especially variable among observers. Conversely, estimates that include only reproductive individuals may vary year to year because of the variability of reproduction in response to annual weather patterns. The best choice of which types of individuals to include in a visual estimate of population size will depend on the ecology of the species and the situation, but you must ensure the counting units are specifically identified.

If the population is very large or spread over a large area, consider using several marked macroplots in which the number of plants is estimated. These should be small enough that an observer can view the entire macroplot from a single vantage point.

Another option in estimating population size is to use classes rather than require the observer to provide a number. In most situations where this approach is used, class boundaries are closer at the low end (e.g., 1-3, 4-10, 11-30, 31-60, 61-100, 101-200, 201-500, 501-1000, 1001-5000, and so on). A logarithmic series (1, 2, 4, 8, 16, 32, etc.) has also been used (Muir and Moseley 1994). An alternative logarithmic series sometimes used is 10, 100, 1000, etc. Note that at low numbers, you could simply count plants rather than estimate.

3. Estimation of population condition

You can develop standard field observation sheets to aid observers in making consistent notes about population condition. The types of data fields included will vary by species, habitat, and situation. Examples of potential fields include the following:

◆ Estimated number of individuals
◆ Percentage of individuals in stage class: seedling, immature, mature, senescent
◆ Percentage of individuals in flower, fruit, vegetative state
◆ Association of stage classes with habitat features (e.g., location of seedlings)
◆ Evidence and level of herbivory
◆ Signs of disease
◆ Pollinators observed
◆ Dispersal agents

4. Site condition assessment

This technique evaluates the condition of the habitat through repeated subjective observations. Assessments can focus on a single activity, potential disturbances, or site characteristics.

Existing conditions may have to change dramatically before it is clear from verbal descriptions that a change has occurred. Training of observers and the use of photographs illustrating condition categories may reduce between-observer differences. Because of variability of visual estimates among observers, site condition assessments are often more effective at capturing the appearance of a new disturbance than estimating changes in an existing disturbance. Observers may, however, miss new conditions for several visits until they become obvious. A careful observer may note an exotic invasion when there are only a few plants, but many observers will miss an infestation until it becomes quite large.

Site condition assessments are most effective when observers articulate their qualitative assessment quantitatively. For example, requiring an observer to estimate the size or areal extent of a weed population, even using broad size classes, provides a better measure of the situation than general descriptive terms such as "common."

Site condition assessments should be done with a standard field sheet used every time the study area is visited. Standard fields and questions should prompt the observer to look for certain conditions and to assess conditions in as quantitative a manner as possible.

The types of observations are specific to the habitat, species, and issues; thus, a specifically tailored field sheet must be developed for each situation. Two examples of site condition assessment forms are shown in Appendix 10. Examples of data fields include the following:

Associated vegetation (successional changes)
Exotics
Disturbances:
 ◆ Fire
 ◆ Flooding
 ◆ Slope movement
 ◆ Animal disturbances (burrowing, trampling)
 ◆ Mining (exploration, material removal, other)
 ◆ Logging
 ◆ Domestic livestock grazing
 ◆ Off highway vehicles
 ◆ Recreation
 ◆ Road construction or maintenance
 ◆ Weed control
Condition of facilities:
 ◆ Fences
 ◆ Signing
 ◆ Road barriers

5. Boundary mapping

Boundary mapping involves measuring or monumenting the boundaries of the population and tracking changes in spatial location or size. Highly accurate maps illustrating boundaries and features of populations can be generated by computer-aided drawing and design programs (CADD) and standard survey equipment, such as a theodolite or transit with an electronic distance measurer (EDM) (see Sections N and O, this chapter). Global Positioning Systems (GPS) may also be used, although their accuracy is variable (Section O, this chapter). A fairly accurate, quick, and inexpensive hand-drawn map can be made with a plane table and alidade or Reinhardt Redy-Mapper (Section N, this chapter).

For some species, mapping the locations of population areas on a low-level aerial photograph may be adequate. For example, *Primula alcalina*, an eastern Idaho endemic, is found on low terraces associated with spring-fed streams. These habitat areas are fairly small (ranging from 10-200m²), but can be easily distinguished and located on a 1:4000 scale aerial photograph. All population areas within a 250ha meadow were mapped; the longevity of each cluster can now be monitored by periodic remapping (Elzinga 1997).

B. Photoplots and Photopoints

Photographs should be a routine part of all monitoring projects and can be the primary method for some. Two general photographic approaches are common. *Photoplots* are photographs of a defined small area (a plot), usually the size of the photograph frame or slightly smaller, taken from above at a specified height. *Photopoints* are landscape or feature photographs retaken each time from the same spot and filling the same frame so that differences between years can be compared.

1. Photoplots

Photoplots can be qualitative records of condition within a limited area from year to year. Their key value is to provide a visual permanent record of the past, allowing factors and changes to be evaluated that might not have been considered when the monitoring was initiated. Photoplots can be used to evaluate invasion by exotic or weedy species, successional changes, soil disturbance, and trampling.

Photoplots are usually defined on the ground with a standard-sized frame. Typical ones are shown in Figure 8.1. A permanent monument in two corners of the frame ensures that the same area is rephotographed every year.

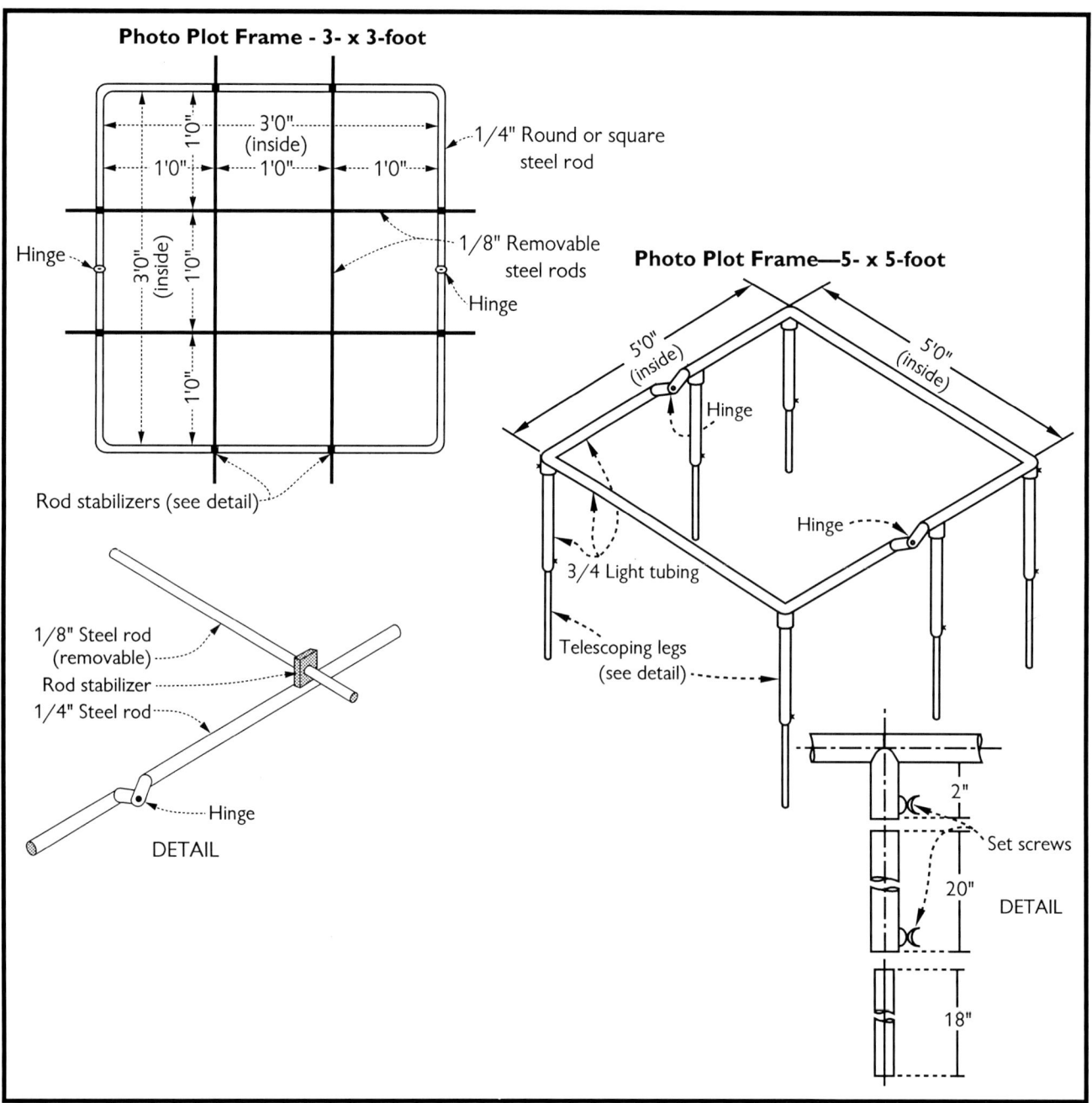

FIGURE 8.1. Examples of photoplots that have been used in rangeland studies by the Bureau of Land Management. Frame size and shape will depend on vegetation characteristics, objectives, and camera lens size.

If you can identify individuals within the plot on the photo, photoplots can function as a density sampling unit. Counting individuals can be deferred to the "slow" time of year. You can use this approach if field time is very limited, but recognize that total time (including office) will be much longer for this approach compared to completing counts while in the field (Bonham 1989). You should test the method on the target species before using it extensively because serious problems often appear unexpectedly. Individuals are usually less obvious on a photo than they are in real life (enlarging the photo or projecting it as a slide onto a screen can sometimes help). Counts will likely also be underestimates of total density because individuals hidden under taller plants will not be counted. Finally, if you are estimating density through a sample of photoplots, counts in square photoplots are likely to be less efficient for estimating density than counts in rectangular quadrats (Chapter 7).

You can also use photoplots as permanent sampling units for cover. Cover can be measured on a photo in two ways. One is to lay a grid over the photo with a known number of intersections, and note the number of "hits" on the target species. The drawback of this approach is that species with low cover may be missed completely (Foster et al. 1991; Meese and Tomich 1992), and it may be difficult to identify small individuals (Leonard and Clark 1993).

Another method is to define canopy polygons on the photo and planimeter the area encompassed by the polygons. The drawback of this approach is that plants with lacy canopies are usually overestimated. Boundaries may be difficult to delineate for some species (Winkworth et al. 1962). If the overestimation is consistent from year to year, it will not affect the monitoring value of the method (because trend is what is of interest), but observers will probably draw polygons around lacy or open canopies differently.

The scale of the photograph will affect the estimate of cover. If, for example, the photograph scale was 1:100, the ground area covered on the photograph by even a small diameter pin or crosshair would be very large, thus dramatically overestimating cover. In general, the smaller the relative surface covered by a pin or crosshairs, the closer the measure will be to the true cover of the vegetation (see more on this in Section H.2.c). If cover is being measured on the projected image of a slide, the pin or crosshair bias will vary depending on the projected scale.

Your methodology must account for these biases. In general, the ratio of pin area to ground surface area should be as small as possible, and the scale used in the photographs or projection kept constant throughout the monitoring project.

Several photoplot methods have been published. Schwegman (1986) describes a frame made of PVC pipe. A camera with a 28mm lens is suspended on the frame 1.4m above the ground. The camera frame is attached to a 1m² gridded frame that rests on the ground surface. Frames can also be constructed to suspend the camera over an offset plot, so that the observer can remain a few meters away and not trample the area near the plot (Windas 1986).

Stereo pairs can be made of photoplots with a stereo adapter for the lens or by taking two frames. Wimbush et al. (1967) used two cameras with 28mm lenses, placed 76mm apart for a stereo pair of a 125cm x 80cm plot from a height of 120cm. Ratliff and Westfall (1973) placed a camera with a stereo adapter about 130cm above the ground surface to photograph a stereo pair of a square foot frame. This gave about a 1:7 scale on a standard 3.5in x 5in photograph. Wells (1971) used two cameras, each with a 25mm wide angle lens, mounted 15cm apart, to make stereo pairs. The frame supporting the cameras was 132cm above the ground, resulting in a stereo frame of a quadrat 1m x 1.5m. Pierce and Eddleman (1970) created stereo pairs of a 1m² plot by taking two frames, 18cm apart, with a camera with a 55mm lens, suspended 152cm above the ground.

Photoplots taken with a telephoto lens may be especially effective for large plants growing on steep erosive slopes that cannot be physically accessed by an observer. The Salmon District BLM in Idaho, for example, used a series of photographs taken with a 400mm telephoto lens to make a long linear photographed plot from the bottom to the top of the slope. Several of these permanent photographed plots were established. The target species, *Physaria didymocarpa* var. *lyrata*, was clearly visible on the sparsely vegetated slope, especially when in bloom. Individuals could be relocated from year to year and the total number of individuals counted within each plot. Although a good idea in theory, and one that worked well most years in practice, in some years poor retakes (either because of poor quality photographs or failure to retake the exact same frames) resulted in complete loss of monitoring data for an entire year.

2. Photopoints

Use photopoints abundantly as a standard part of monitoring for documenting the following:

1. **Location of study site.** Consider taking photos at the parking spot and along the walking path to the study site. At the monitoring site, photographs taken from the boundary of the population or study site facing both toward and away from the site can help relocate boundaries if other monuments are lost.

2. **Transects and macroplots.** Photographs taken at each end of a transect or at the four corners of a macroplot can help to relocate the transect or plot and provide a visual record of general conditions.

3. **Habitat conditions.** Photographs of general habitat can help you monitor changes in plant cover, weed invasion, and disturbances.

4. **Population conditions.** Plant height, flowering effort, plant size, and levels of herbivory are some of the conditions that can be illustrated with photopoints.

Todd (1982), Rogers et al. (1984), and Brewer and Berrier (1984) provide overviews and suggestions for establishing and using photopoints. Two examples of the use of photopoints for monitoring long-term change are Sharp et al. (1990) and Turner (1990). Hart and Laycock (1996) provide an annotated bibliography of 175 publications that use repeat photography, giving the number of repeat photographs, the dates, and the habitat type and State in which the photographs were taken.

3. Hints for monitoring with photopoints and photoplots

1. A good 35mm camera is essential for quality monitoring photographs. A camera that allows control of both shutter speed and aperture is best. Disposable cameras are convenient, but should only be used for recording images that will not be retaken, such as photographs of the parking area and the route to a monitoring site.

2. Lenses should be chosen with care. Generally lens sizes of 28-75mm are appropriate for photoplots, and lenses from 50-200mm for photopoints. For photoplots, a wide lens is best. These open up to f1.6 or f1.8 allowing you to take quality photographs in low light conditions. The wide diameter of the glass allows maximum light to pass through the lens and can dramatically improve the quality of the photograph and the depth of field (see #4, below). These lenses are more expensive but may be worth the investment if photoplot monitoring will be extensive or if quality is critical. Generally avoid fish-eye

lenses because of the distortion. Also avoid telephoto lenses unless they are specifically required for a given project; these generally do not give as sharp an image as smaller lens sizes and do not function well under low light conditions.

3. Some cameras come with lenses that zoom in and out with the touch of a button, but the actual focal length is unknown. It is difficult to retake the exact same frame with this kind of camera. Even with a manually operated zoom lens, it can be difficult to get the exact focal length unless you are at an end of the zoom scale. Standard lenses, rather than zoom ones, allow for better repeatability.

4. Use the smallest aperture (the largest f-stop) possible, given the light conditions and restrictions on shutter speed. Small apertures give the photograph increased depth of field, meaning that a larger range of distances from the camera is in focus. This can be especially important for monitoring photoplots.

5. Use the slowest shutter speed possible to maximize the depth of field. Shutter speed should probably be no slower than 1/60th of a second unless the camera is supported by a tripod and the air is very still (no moving vegetation). If even a slight breeze is blowing, increase the shutter speed to reduce blurring caused by moving vegetation.

6. A tripod improves the quality of photographs in nearly all situations, and is especially critical for low-light conditions (such as dark woods). A tripod can also help to maintain a standard camera height, if this is recorded. This reduces the different camera angles caused by varying heights of different photographers.

7. Take three or four frames of the same picture, each at a slightly different exposure. Multiple frames are cheaper than return trips to retake photographs because the first ones are all overexposed or underexposed.

8. Most professional photographers prefer slide film to print since both high quality prints and slides can be made from slide film. Slower films (ASA 25-100) give better clarity and less graininess, but faster film (ASA 200-400) may be needed for shady areas.

9. Use the first frame of a series as a record frame (a picture of a clipboard with date, time of day, location, and subject). This will save many hours of trying to match boxes of slides with field notes. Use chalkboard or beige paper with the information written in heavy black marker. Avoid reflective white dry-erase board or bright white paper. These are often unreadable in a photograph because of glare.

10. Use record frames whenever changing subjects, locations, or film. The first frame of a film should always be a record frame.

11. If photographs won't be curated immediately, include a record board in a bottom corner of each frame.

12. When taking general landscape photographs, include enough horizon in the picture to aid relocation.

13. If the photopoint frame does not include any horizon (e.g., pointed to the ground) use pairs of photographs—the first from the photopoint containing something recognizable or the horizon, and the second of the desired frame.

14. Map photopoint locations on an aerial photograph or a topographic map. A symbol such as ♂ illustrates both the photopoint and the direction of the photograph.

15. Some studies have used permanent monuments such as rebar or T-posts to mark photo-point locations. These are recommended for situations that lack visual indicators to use in relocating photopoints. A riparian area, for example, probably contains diverse enough habitat features to allow you to relocate photopoints by using the previous photographs. In other types of sites such as a large meadow, a dense forest, or a sagebrush grassland site with little topography, photopoints may be difficult to relocate from the photograph. In these situations, a permanent monument can save much time.

16. Keep a photo log in your field notes. An example is given in Appendix 15.

17. Invest in a camera that records the date on every photograph.

18. Curate photographs immediately after developing. Write identifying information in pen-cil on a label placed on the back of each photo. You can write directly on the slide frame. You can also purchase special pens from photo supply companies designed for writing on the back of photographs. Do not use pen or marker; these may bleed through the photo-graph, or smear onto another photograph if photos are stacked. Invest in archival-quality plastic sleeves for photographs, slides, and negatives and store them in labeled three-ring binders. Photographs kept in boxes or envelopes are seldom looked at again.

C. Video Photography

The most common use of video photography is as a visual record of the site, similar to the use of still cameras and photopoints. Video can provide a good visual overview of the site (and verbal, as well, if commentary accompanies the film), providing a better sense of features and conditions than photographs. The disadvantage is that video footage is difficult to retake, and a video cannot easily be used in the field to compare to current conditions. This drawback can be overcome, however, with the creation of stills from the portions of the video that best represent the features that are being monitored. These stills can then be retaken with a regular 35mm camera equipped with a zoom lens to match focal length of the video image.

Plant cover in quadrats or transects can be recorded by video photography. This application has been most widely used in marine studies (Whorff and Griffing 1992; Leonard and Clark 1993), primarily because diving costs associated with underwater sampling are expensive. Video requires extensive laboratory time for processing and analyzing the images, but only a fraction of the field time that most other sampling techniques require (Leonard and Clark 1993). The drawback of using video for sampling vegetation is that the resolution of the image may make species identification difficult, and limit the detection of small species (Leonard and Clark 1993).

D. Remote Sensing Techniques

Remote sensing encompasses a range of techniques which involve the collection of spectral data from a platform that does not touch the object of interest. This definition is somewhat vague because of the range of remote sensing techniques, from taking a photoplot with a camera suspended from a hot air balloon to satellite-based imagery.

Satellite imagery includes several types of spectral data and several platforms. In general, the resolution of satellite imagery does not lend itself to the site-specific plant population monitoring situations addressed in this technical reference. Although there may be some use for satellite imagery to identify community types known to contain the target species, gap analysis tests have shown that the level of resolution is often inadequate to identify small habitat islands (Stine et al. 1996). Thus the use of satellite imagery to identify or stratify habitat should proceed with caution. For some examples of the use of satellite imagery in landscape-level monitoring, as well as excellent overviews of applications of satellite imagery in natural resource management, see Luque et al. (1994), Sample (1994), Lyon and McCarthy (1995), Verbyla (1995), and Wilke and Finn (1996).

Aerial videography can be used for mapping and monitoring landscape features such as plant communities. Most of the systems described in the literature involve the use of video cameras associated with low-level fixed-wing flights (Bartz et al. 1993; Nowling and Tueller 1993; Redd et al. 1993). The cameras record spectral data (not necessarily from the visual range), which are immediately processed into digital information associated with a certain pixel size. Pixel size is determined by flight altitude. These pixels are then classed based on their spectral signature. Sizes range from 3m x 3m to 50cm x 50cm, with the cost increasing as the pixel size decreases.

Aerial photography captures visual spectral data (sometimes infrared), generally from a fixed-wing aircraft or helicopter. Most agency offices have access to recent air photo coverage of their entire administrative unit at 1:12,000 to 1:24,000 scales. Most photo series are in stereo-pairs. With some practice, these can be viewed in 3-D through a small tool called a stereoscope (see Section N, this chapter). Most foresters routinely use stereo-pairs in their work, and can usually assist in locating photograph pairs in your office and lending stereoscopes.

These photographs can be extremely valuable for identifying community and population boundaries, for stratifying sites, and for documenting study locations. Aerial photographs can also help identify features and disturbances that are not apparent from the ground. In some offices, older photo series may be available to compare with newer ones. This comparison can provide a historical perspective on changes in disturbances and human use, ground cover, and even species composition.

Low-level aerial photography (scales of 1:500 to 1:6000) are usually commissioned for a specific project. Although expensive, if low-level aerial photographs can be used for monitoring in place of ground measurements, the savings in personnel time may make aerial photography competitive with more conventional monitoring techniques. Low-level photography is especially applicable to woody species, very large herbaceous perennials, and overall community cover and habitat condition assessments. It obviously will not work well for small species, or for a species that is hidden in a photograph by a taller canopy. For examples where low-level photography was successfully used to monitor a community, see Knapp et al. (1990) and Jensen et al. (1993).

We also recommend Avery and Berlin's (1992) *Fundamentals of Remote Sensing and Airphoto Interpretation* as a guide to photo interpretation. The book has over 440 black and white photographs as examples, including 160 stereo-pairs and 50 color photographs. The layout and fascinating photographs make this book attractive and extremely readable. Of primary interest for vegetation monitoring is a chapter entitled "Forestry Applications", which also addresses community cover mapping from aerial photographs in non-forested types. The authors describe how individual range plants and grassland types can be identified at scales of 1:500 - 1:2500, and individual trees and large shrubs at 1:2500 - 1:10,000. Typical diagnostic features used to identify different species are plant height, shadow, crown margin, crown shape, foliage pattern, texture, and color. For some forested areas, diagnostic keys to species identification have been developed; the authors include references to these guides.

E. Complete Population Counts

Some populations can be completely counted or censused and, where possible, this is the preferred monitoring method. No statistics are required to analyze the results or the precision of the estimate. The change or number observed is real (provided the count is accurate and plants are not missed); there is no sampling error. The only question remaining is whether the change is biologically significant.

To use a census approach, a counting unit has to be consistently recognizable (ramet, genet, or some consistent arbitrary unit). For a non-clonal species, individual plants (genets) may be relatively easy to delineate and recognize, but for a clonal species, such as a grass, it is much more difficult to define a counting unit. For a clonal species like aspen, the count may focus on obvious units like trunks (which, since aspen is clonal, are actually ramets), but you must still decide whether the count excludes any size classes (such as small ramets or seedlings). If a consistent counting unit is a problem, an alternative sampling approach (such as cover or frequency) is a better option.

In theory, any population can be censused. In practice, however, accuracy of counts can be very poor because of missed individuals. This can occur even when the plant is large and obvious. For example, a census of a large (up to 60cm tall) *Penstemon* species done with four individuals walking a grid pattern resulted in a count of 63 plants. When approximately 20% of the area was sampled, the sampled area alone contained 93 plants (Elzinga, unpublished data). The discrepancy was probably due to misses of non-reproductive and small individuals.

Factors that make counts difficult include a large population area, a large population, dense associated vegetation, the presence of similar species, small stature of the target species, and many of the target species in cryptic stage classes (such as seedlings). Before using a census approach, ensure that counts are accurate by using two or more observers and comparing the results.

You can improve census counts by using some type of systematic search of the population area (e.g., 0.1 hectare macroplots or parallel lines marked by pin-flags) and by setting standards (parallel swathes of a certain width, macroplots searched for a given amount of time each year). Boundaries of the population or macroplot should be marked so future counts cover the same area.

F. Density

Density is the number of counting units per unit area. To define density as the number of individuals per unit area is suitable for animals, for which an individual is a readily recognized entity. For plants, however, the above-ground expression can be of individuals (genets) or it can be an intermixing of a few individuals with many above-ground members (ramets). An aspen clone, for example, is an individual, but most people would consider each stem an individual, in spite of the fact that stems are interconnected underground and are technically ramets.

Thus, a critical question in the measure of density, just as it is in censusing, is to define the counting unit. A counting unit has to be consistently recognized by all observers for density to be used as a monitoring method (Appendix 11).

1. Advantages and disadvantages

Density is most effective when the change expected is recruitment or loss of individuals (or of the counting unit). Estimated density (in terms of number per unit area) is theoretically the same for all quadrat shapes and sizes, although the precision of the estimate will vary (sometimes dramatically) among sampling units of different shapes and sizes. The fact that density is reported as a per area measure allows comparison between sites even if the quadrat shape used for sampling differs. This is in contrast to another measure described below, frequency, which is dependent on plot size and shape.

In practice, the density estimate may vary with plot size because of the effects of boundary decisions, which are most pronounced in small quadrats or long narrow ones (see more on this in Section 2, below). Because most observers will consistently include boundary plants, estimates of density in small quadrats or in long, narrow ones (high perimeter to area ratio) are usually higher than estimates from larger or square quadrats. A key monitoring design decision when using density is to select a quadrat size and shape that will efficiently estimate density with acceptable precision (see Chapter 7), while controlling these boundary errors (see Section 2, below, for ways of reducing boundary errors).

Density is most sensitive to changes caused by mortality or recruitment. It is less sensitive to changes that are vigor-related, especially those that are sub-lethal (e.g., a reduction in production that is not accompanied by an increase in mortality or a decrease in recruitment). Figure 8.2 shows that a population can change dramatically without a large change in density. In this example, cover and the ratio of reproductive individuals compared to non-reproductive have declined dramatically, but simple counts would have detected a decline of only two individuals. Density may be an especially poor monitoring measure when

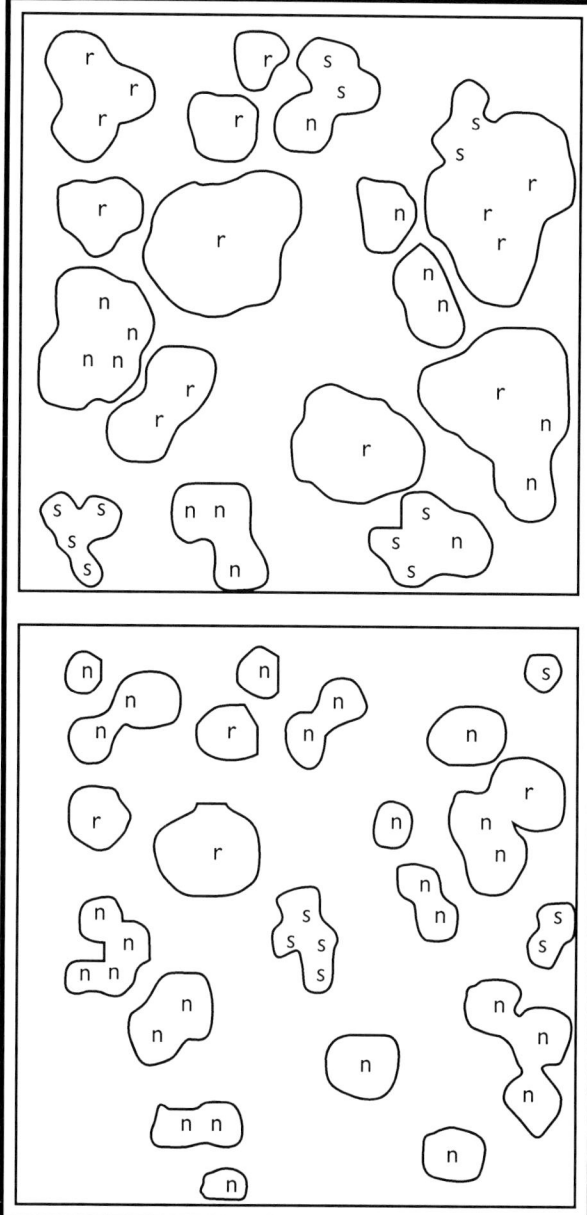

FIGURE 8.2. Two views of the same quadrat, the top measured in 1995 and the bottom in 1996. Outlined polygons denote canopy cover; letters represent individuals. Note that density declined from 39 individuals to 37 individuals. In 1995, there were 14 reproducing individuals (r), 14 non-reproducing individuals (n) and 11 seedlings (s) in the plot. In 1996, there were 4 reproductive individuals, 26 non-reproducing individuals and 7 seedlings. Note also the dramatic decline in cover from 1995 to 1996. The changes illustrated in this plot are not well captured by density measures of total individuals. Even a count of seedlings versus adults would not have captured the dramatic change in reproductive fraction.

individuals are long-lived, and respond to stress with reduced biomass or cover, rather than mortality. Density may also be a poor measure for plants that fluctuate dramatically in numbers from year to year, such as annuals.

Observer bias is generally low if the counting units are few and easily recognized, but errors are common when quadrats contain cryptic individuals or numerous plants. The most common non-sampling errors originate in "high speed" counts that overlook small individuals. Establishing a minimum search time per quadrat can reduce the temptation to hurry the measurements, although the actual time required per quadrat will vary depending on the number of counting units occurring within it.

2. Design and field considerations: quadrats

a. Quadrat design

The density of herbaceous plants is usually counted within the boundaries of a quadrat, each of which is a sampling unit. Quadrat design is discussed at length in Chapter 7. A few of the points are reiterated here:

1. The size of the quadrat should not be impractical, i.e., the quadrat should not be too large either in terms of number of individuals to be counted or search time required.

2. Size and shape of the quadrat must be tailored to the specific plant distribution observed in the field. For most situations, the most efficient quadrat shape will be a rectangle.

3. You should attempt to include at least some "clumps" of the target species in your initial trial quadrat sizes and shapes. The most efficient plot shape and size in terms of number of quadrats needed will be one in which the density in each quadrat is very similar (little variability between quadrats). Good guesses on size and shape can be made by first observing the distribution of the plants in the field. Pin flags can be placed throughout the population in areas of concentration to get a better picture of the distribution of the species at the site. You want to design a plot size and shape that intersects those areas of concentration. Appendix 17 describes a procedure to compare the efficiency of different quadrat sizes and shapes based on pilot sampling.

b. Counting unit

Density is usually based on a count of plants rooted within a quadrat. It works best with plants with distinct and fairly small diameter stems. As the size of the area of the plant intersecting the ground increases (such as trees with larger diameter trunks or bunchgrasses with large basal areas), deciding whether a plant along the boundary is in or out of the quadrat becomes more complicated. For some species, using a "rooted density only" rule is also problematic. For example, if the counting unit is a shoot of grass, individual tillers are sometimes not clearly rooted, but are clearly individual shoots. For many matted plants, trying to determine the rooted zone requires lifting and pulling at the top mat, possibly causing injury to the plant; thus, for these matted plants using the canopy outline for boundary decisions may be better than the rooted area. (A more appropriate technique for matted plants may be cover.) For most species, however, avoid using the outline of the canopy as the boundary to determine whether a plant is in or out of the quadrat (Figure 8.3), because changes in canopy (vigor) will affect the density measure and increase the complexity of the interpretation. For

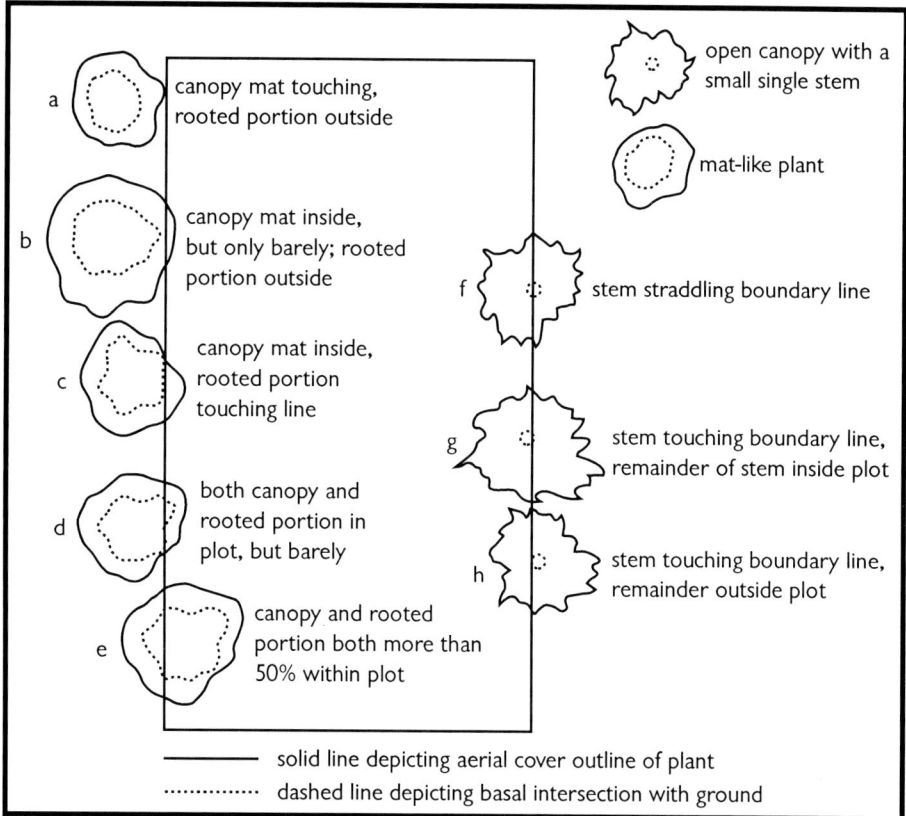

FIGURE 8.3. Boundary decisions. Which plants should be considered within the plot? Most investigators use one of two rules: (1) all boundary plants are counted in on two contiguous sides and out on the other two sides; or (2) every other boundary plant is counted. Plants c-h would be considered by most observers to be boundary plants. Plants a and b would generally not be considered boundary plants (because only the canopy intersects the plot), although occasionally a specific situation may require that the canopy boundary be used rather than the basal boundary (see text for additional discussion).

most species the best counting unit is a rooted individual, but for some species other rules may have to be developed (and documented).

In addition to identifying the counting unit, you should consider the value of using stage classes such as seedling, non-reproductive, and reproductive. Doing counts by stage class requires more time, but in many situations the additional information warrants the extra effort. Figure 8.2 clearly shows that measuring density in stage classes can rectify the insensitivity of density as a measure of some kinds of change. In this example, the number of plants in the quadrat only declines by two—from 39 plants the first year to 37 the second—but demographic structure displays a dramatic change, declining from 14 reproducing plants the first year to 4 the second year. Dividing the population into seedlings and non-seedlings can provide additional information for interpreting changes in density, although in this example adults increased by 2 individuals and seedlings declined by 4.

c. Boundary decisions

Boundary decisions are important in density measures, since a plant must be counted in or out (Figure 8.3). You must establish boundary rules and apply them consistently each time the monitoring is done. Plants with a single thin stem are fairly easy to determine if they are in or out of a quadrat, but plants with a large basal diameter (e.g., bunchgrasses and tree), may be half in and half out. How will these be addressed?

Some viable alternatives are as follows:

1. Plants are considered in if any part of the plant boundary is touching the plot boundary along two adjacent sides of a rectangular plot, and considered out if any portion of the plant boundary is touching the other two sides of the plot. This provides an accurate estimate of density and is the recommended approach for reducing boundary bias. For monitoring in permanent plots, you must specify which sides are interpreted in which way (compass direction works well), and measure along those sides the same way consistently. The sides must be split so an equal portion of the perimeter is treated as the "in" sides compared with those considered the "out" sides. In other words, if the plot is rectangular, you would consider straddler plants "in" along one long side and one short side of the rectangle, and "out" along one long side and one short side (adjacent sides).

2. Plants are counted as in or out alternately along the boundary. This provides an unbiased estimate of density, but in a very large or long quadrat, you may have trouble keeping track of whether you last counted an "in" or an "out" plant.

3. Plants are considered in if more than 50% of the plant boundary (canopy or basal) is within the plot. This is illustrated by plants e and f in Figure 8.3. While this method will give an accurate measure of density, we do not recommend it because additional subjective observer decisions are required. Observers may have consistent bias in their estimates of 50% (over-inclusion is the most common), introducing an unknown observer error. Plants with irregular basal outlines are especially difficult to consistently determine if they are to be counted "in" or "out".

Some non-viable alternatives:

1. Count all plants that touch the line, even if most of the plant boundary is outside of the plot. This is illustrated in Figure 8.3, plants c and h. If you use this approach, you will overestimate density (number of individuals per unit area) because the length and width of the plot are essentially increased by the average diameter of the boundary of the plant. This is easiest to visualize with the matted plant in Figure 8.3.

2. Include only plants that are completely within the plot, including those that just touch the line. This is illustrated by plant g in Figure 8.3. This gives the opposite result as approach (1), above—an underestimation of true density.

Both approaches have been used in monitoring studies, and if you have a current study using one of these designs, it is not a fatal error. If the purpose of the study is to measure change over time, and the boundary rule resulting in over or under estimation has been consistently applied, you may still be able to interpret changes in terms of trend in the population. One problem in interpretation that may arise however, is that as plant boundaries change due to changes in vigor, the impact of that change on density estimated using either of these boundary rules will be much larger than if boundary decisions were made using an unbiased approach. Thus, interpreting changes in density measured in a monitoring project using one of these boundary rules will be partially obscured by changes in vigor. Another problem is that both methods create difficulties in comparing density estimates at different sites since the estimate of density is partially a function of plant diameter, which can vary from site to site.

3. Design and field considerations: distance measures

An alternative to estimating density in quadrats is a suite of techniques called distance measures. Several variations on the theme have been developed, but they all involve the measure of the distance of an individual from a point or from another individual, and estimating density from the average distance measure. Figure 8.4 shows the four most commonly used distance measures in vegetation sampling. These measures are most often used for large or scattered individuals such as trees, for which the use of quadrats is not practical. They have, however, occasionally been used in grasslands on common herbaceous plants (Becker and Crockett 1973). Distance measures are based on the concept of a mean area per plant. Once this is known, the value can be used to calculate a density per unit area.

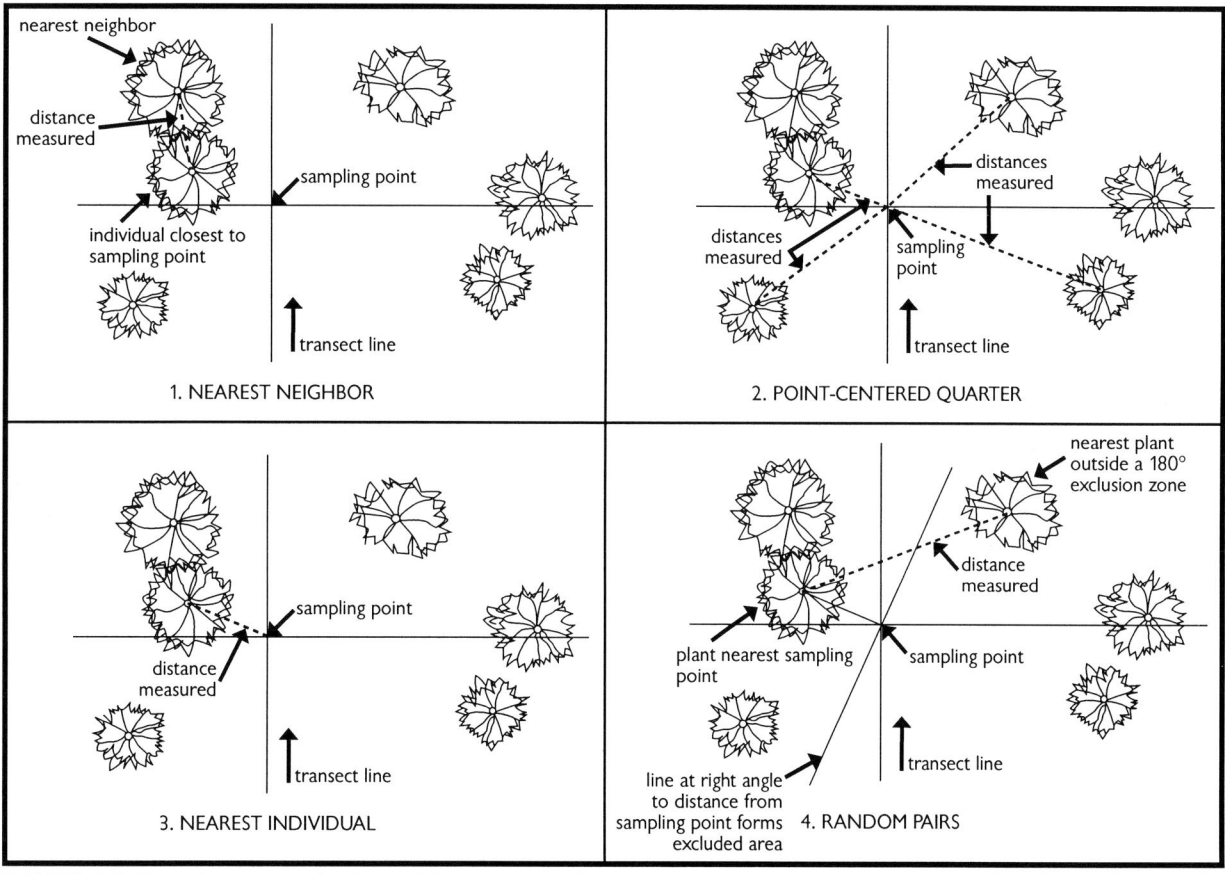

FIGURE 8.4. Four distance methods used for measuring density in plant populations with randomly distributed individuals: (1) nearest neighbor; (2) point center quarter; (3) nearest individual; and (4) random pairs. None of these methods are appropriate for species that have contagious (clumped) distributions.

These techniques, however, are only suitable for use on plants with random distributions. Most plants do not grow randomly in space, but occur in clumps, the result of short-distance dispersal of propagules or micro-variation in habitat. One technique, the wandering quarter method (Figure 8.5), was designed for individuals with non-random aggregated distributions (Catana 1963). A similar approach, the T-square method, was proposed by Diggle (1975) and by Blyth (1982).

Field tests of the latter two methods give mixed results. Lyon (1968) found that the wandering quarter method gave an accurate estimate of density in a shrub community in which all individuals had been enumerated. To achieve a reasonably precise estimate of density, however, actual counts were quicker than sampling with points and distance measures.

McNeill et al. (1977) sampled an area in which all individuals had been marked and mapped. They found that quadrats were superior in terms of the accuracy of the estimate and field efficiency compared to several distance measures. Becker and Crockett (1973) concluded that the wandering quarter method underestimated a clumped species and overestimated a single-stalked, well-dispersed species.

In a simulation study of 24 distance-based density estimators, Engeman et al. (1994) determined that the approach proposed by Diggle (1975) did not provide unbiased estimates of the mean when sampling clumped distributions. They also argued that the method is relatively inefficient in the field because of the difficulty in defining the area of exclusion (see Figure 8.5). They concluded that the best estimators were those that measured three distances per point (point to nearest individual, nearest individual to nearest neighbor, and nearest neighbor to its nearest neighbor), and estimators that measured from the point to the third nearest individual.

The value and performance of distance measures depends on the field situation. The best estimators for clumped distributions are complex, either requiring three measures per sampling point, or determining which individual is the third farthest from the sampling point. This complexity dramatically reduces the field efficiency of these methods. Distance measures may be appropriate when the individuals are so widely spaced that using quadrats is not practical (as for some trees), but for most monitoring situations involving rare plants, quadrat-based density estimates are more efficient and free from the potential biases of distance methods.

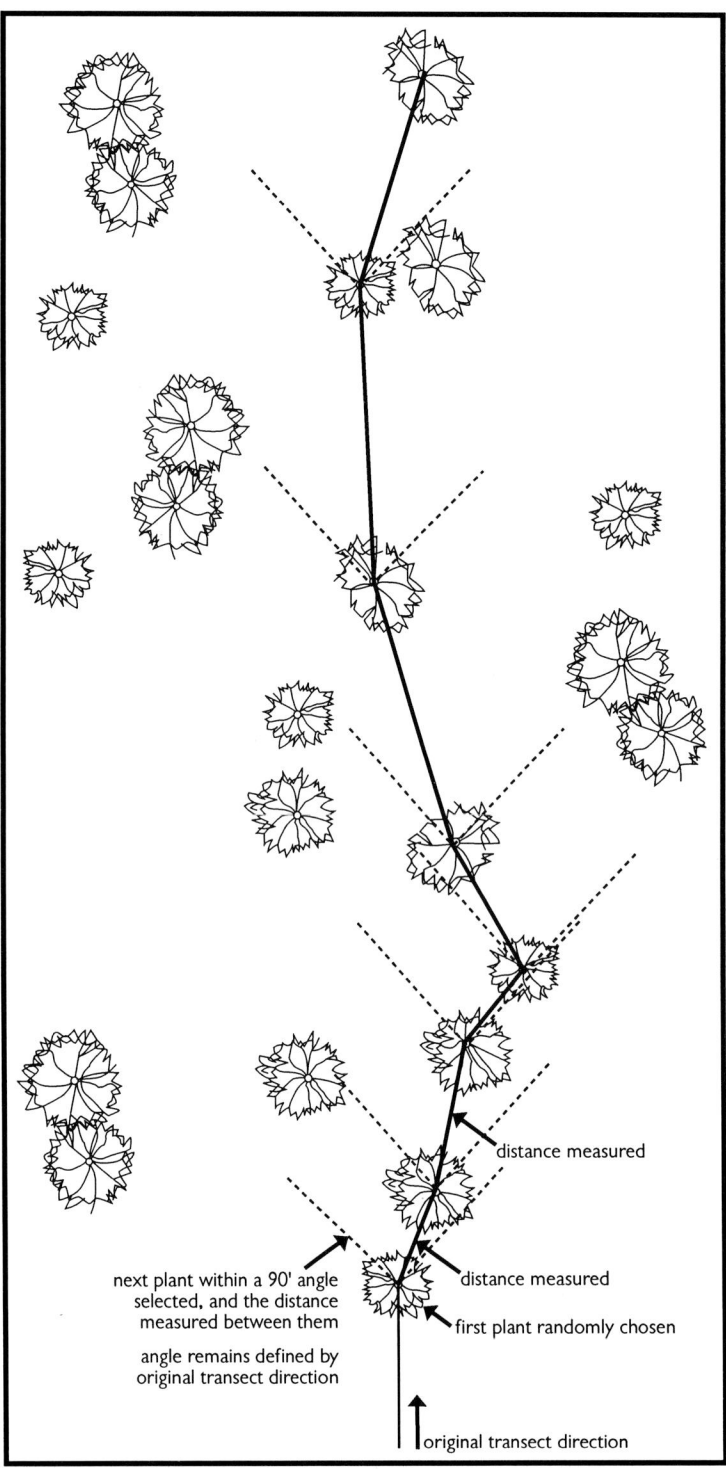

distance measured

distance measured

next plant within a 90' angle selected, and the distance measured between them

first plant randomly chosen

angle remains defined by original transect direction

original transect direction

FIGURE 8.5. Wandering quarter distance measure, which can be used in plant populations with individuals contagiously distributed. This is the only distance measure recommended for general use since few plant populations are randomly distributed.

G. Frequency

Frequency is usually measured in plots, and can be defined as the percentage of possible plots within a sampled area occupied by the target species. You can visualize frequency by imagining the sampling area overlaid with a grid of cells the same size as the frequency plot. The percentage of cells occupied by the species is the frequency. Occupation is defined by occurrence; the abundance of the species within the plot does not matter, only whether it is present. Because the target species will more likely occur in very large plots compared to small ones, frequency is a measure dependent on plot size and shape. Frequency values from different studies are not comparable unless the plots used were identical.

1. Uses, advantages, and disadvantages

Frequency is appropriate for any species growth form (Appendix 11). It is especially sensitive to changes in spatial arrangement. It may be appropriate for monitoring some annuals, whose density may vary dramatically from year to year, but whose spatial arrangement of germination remains fairly stable. Rhizomatous species, especially graminoid species growing with similar vegetation, are often measured by frequency because there is no need to define a counting unit as with density. Frequency is also a good measure for monitoring invasions of undesirable species.

Another advantage of frequency methods over methods for measuring cover (Section H., below) is the longer time window for sampling. Once plants have germinated, frequency measures are fairly stable throughout the growing season, compared to cover measures, which can change dramatically from week to week as the plants grow.

The key advantage of frequency methods is that the only decision required by the observer is whether or not the species occurs within the plot. Technicians can usually measure frequency with minimal training on methodology and species identification. If the species is easy to spot, frequency plots can be evaluated very quickly.

The disadvantage is that frequency is a measure affected by both the spatial distribution and the density of a population (Grieg-Smith 1983). Because of this, changes can be difficult to interpret biologically since we will not know if a change is due to changes in density, distribution, or both (Figure 8.6). Unlike other vegetation attributes, like density or cover, frequency is difficult to visually estimate for a whole site. Thus, the biological significance of changes may also be difficult to express to managers and user groups because they cannot easily visualize the change.

2. Design and field considerations

a. Arrangement of plots

You can locate frequency plots randomly or along transects. Plots arranged systematically along randomly located transects are more efficiently located in the field than plots that must be located individually at random coordinates; thus, using transects is the most common approach. Locating plots along transects also allows for permanent monumentation of plots. Monumenting individual frequency plots is prohibitively time-consuming, but by monumenting the ends of a transect, and providing periodic monuments along the transect to ensure later accurate relocation, quadrats placed along a transect can be relocated fairly accurately and can be considered permanent. This results in two major benefits. The first is that the design is usually much more powerful for detecting change (see Chapter 11). The other

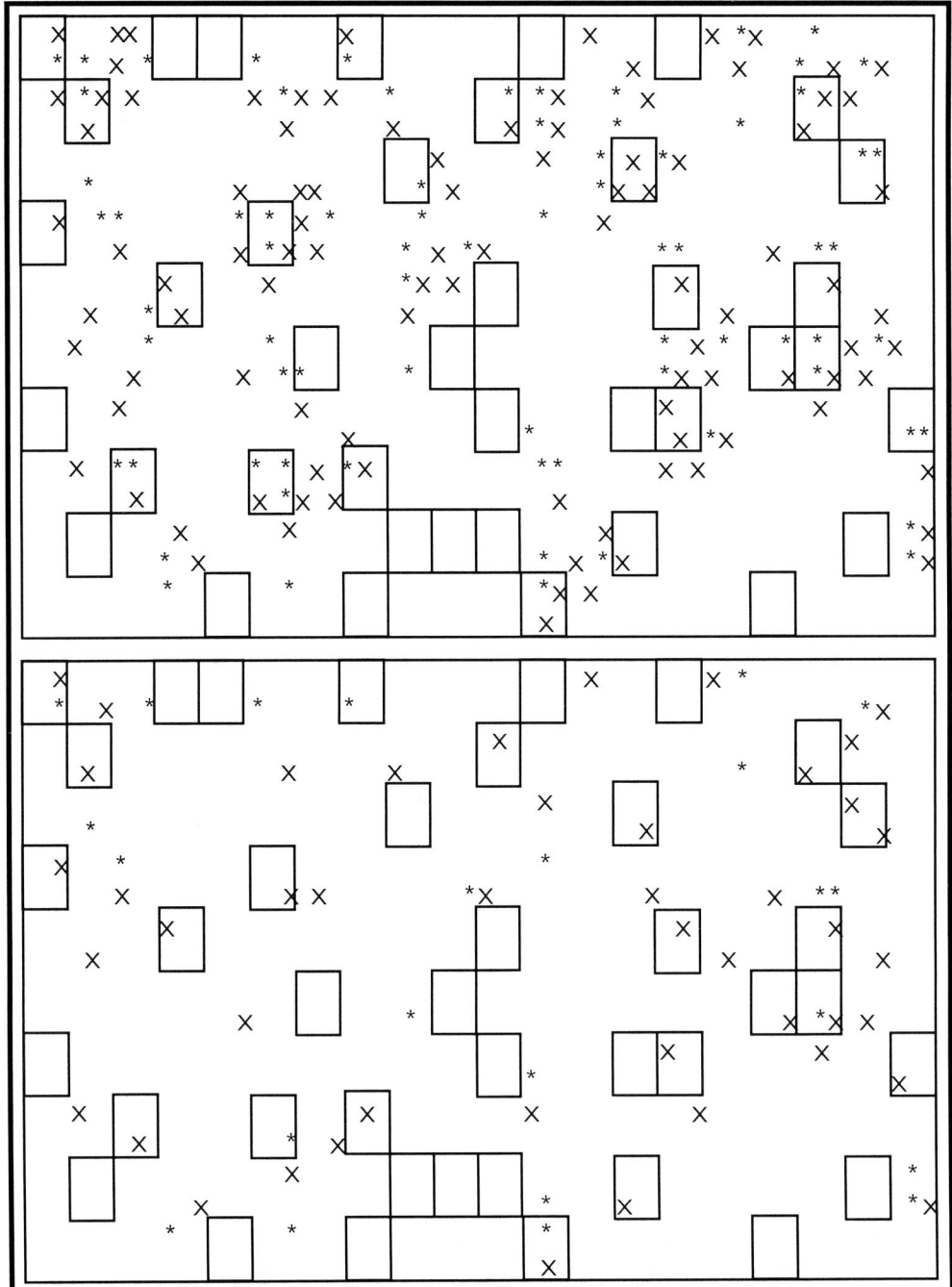

FIGURE 8.6. This macroplot was sampled with 40 permanent frequency plots. The first year, density in the macroplot was 198 individuals—72 seedlings (*) and 126 adults (X). The second year, density declined to 71 individuals—23 seedlings, 48 adults. Frequency between the two years declined from 57.6% to 50.0%.

benefit is an increase in biological understanding when you can relate changes to spatial data. Because you know the location of the changed quadrats, you may be able to determine causes of change (opening of the canopy, wet microsite, invasion of weeds, etc.).

The approach that has been commonly used in rangeland monitoring is to locate plots systematically along transects, usually at 1m or 2m meter increments. This design is appropriate if the sampling unit is considered the transect, but inappropriate if the plot is intended as the sampling unit. Spacing at 1m or 2m intervals is usually inadequate to ensure that each quadrat is an independent observation, but the actual spacing required to provide independence will vary by site (see Chapter 7).

b. Boundary rules

The key decision in frequency measures is whether the species occurs in the plot. While this is relatively straightforward for single small-stemmed plants, it is more difficult for larger plants and matted ones. You must establish counting rules and apply them consistently. Some researchers have used the rule that if a perennating bud occurs within the plot, the plant is counted (Bonham 1989). Under this rule, shrubs and trees with live buds that fall within the volume of the plot (as projected upward in space) would be considered in the plot. Most researchers, however, use rooted occurrence. Developing boundary rules similar to those described for density is important for all plants, but especially so for plants with wide bases such as bunchgrasses.

c. Stage classes

As with density, you must decide whether to evaluate occurrences by size or stage class (such as seedling, non-reproductive, reproductive). Using stage classes increases the amount of time required to evaluate each plot, but it can dramatically increase the understanding of frequency change in many cases. At a minimum, consider separating classes by seedling and non-seedling.

d. Plot size and shape

Plot size determines the frequency value. The larger the plot, the greater the likelihood that an individual will occur within the plot, resulting in a larger overall frequency value. If plots are large enough, all of them will contain the target species (100% frequency). This leaves no sensitivity to upward change. If too small, there will be little sensitivity to downward change. Frequency values, at least the first year, should be between 30% and 70%. If you are concerned about change in only one direction, or that the change may be dramatic, you may wish to change these target percentages. For example, if you are only concerned about declines, you may want to target your initial measure to between 50% and 80% to provide a wide margin of sensitivity to declines.

Nested plots are often used by federal agencies in range sampling; a common frame size is 50cm x 50cm, with four smaller plot sizes nested within the 50cm x 50cm frame (5cm x 5cm, 25cm x 25cm, and 25cm x 50cm). The Nature Conservancy uses a nested frequency frame with square frames measuring $0.01m^2$, $0.1m^2$, and $1.0m^2$ for plant community monitoring studies. While this is useful in sampling communities in which many species are measured, when measuring a single species, it is more efficient to sample a single quadrat size designed for the particular density and distribution of that species, rather than a standard-sized nested quadrat. Nested quadrats designed specifically for the target species, however, may be useful if frequencies change dramatically from year to year, such as for annuals or short-lived perennials (Appendix 11). A nested design that gives about 20% for one plot size and 80% for another provides a greater range for measuring large upward or downward changes the following year compared to a single plot size that gives approximately 50% frequency the first year. Nested plots may also be advantageous for measuring populations by stage classes. If, for example, seedlings are more abundant than adult plants, a smaller plot for seedlings nested inside a larger plot for adult plants may be very efficient. Finally, using nested plots in a pilot study is the best approach to determine the best plot size to use.

The advantages of long narrow plots for estimating density, cover, and biomass are clearly shown in Chapter 7. For these types of estimates, square plots are inefficient because a few plots will contain large values of the target species, while most plots will have none of the target species. With frequency data, however, only two values are possible—present or absent—and you want at least 30% of your plots to contain no plants. For this reason, use square plots when sampling frequency, and adjust the size of the plot to reach the desired frequency range. An exception to this rule may be sparsely distributed rare plants for which frequency plots would have to be very large to contain plants 30% of the time. For these plants, a more rectangular frequency plot may be advantageous. If you use a rectangular quadrat, you must be sure to orient the long sides of the quadrat in the same direction in each year of measurement.

A special case is a plot size reduced to a point. These data can be considered a frequency measure but are most often interpreted as a measure of percent cover.

H. Cover

Cover is the vertical projection of vegetation from the ground as viewed from above. Two types are recognized. Basal cover is the area where the plant intersects the ground; aerial cover is the vegetation covering the ground surface above the ground surface (Figure 8.7). You can visualize aerial cover by considering a bird's-eye view of the vegetation.

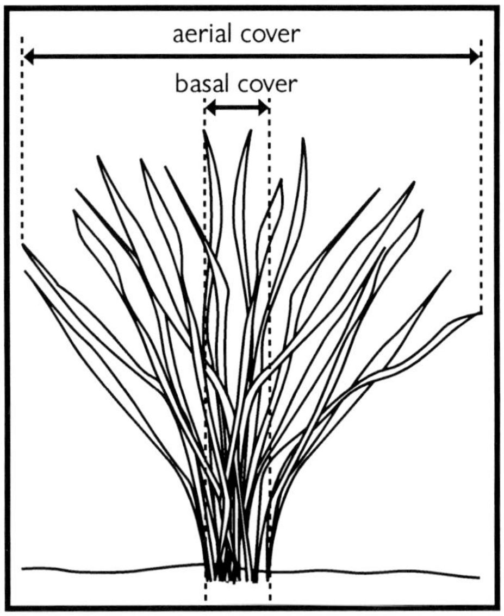

FIGURE 8.7. Basal cover compared to aerial cover.

1. Uses, advantages, and disadvantages

You can most easily measure the cover of matted plants and shrub species with a well-defined canopy, but cover measurements are applicable for nearly all types of plants (Appendix 11). Cover measurements often used for grasses because of the difficulty in counting grass plants or tillers. Cover is one of the most common measures of community composition because it equalizes the contribution of species that are very small, but abundant, and species that are very large, but few. Of the three measures—density, frequency, and cover—cover is the most directly related to biomass. A key advantage of cover as a vegetation measure is that it does not require the identification of the individual (as density does), yet it is an easily visualized and intuitive measure (unlike frequency).

A disadvantage of cover measures (especially canopy cover) is that they can change dramatically over the course of a growing season, while both frequency and density measures are fairly stable after germination is complete. The change in cover over the course of the growing season may make it hard to compare results from different portions of large areas where sampling takes several weeks or a few months.

Another disadvantage is that cover measures are sensitive to both changes in number (mortality and recruitment) and in vigor (annual biomass production). Because you may be unable to determine whether measured cover changes are due to density or production changes, cover trends can be difficult to interpret. Real trends in density may be obscured in species with highly variable annual production. For example, increases in cover can obscure significant mortality (Figure 8.8). For plants with less annual variability, such as shrubs and matted perennials, cover changes will be due primarily to mortality or recruitment. Because basal cover is generally less responsive to annual weather events than canopy cover, annual variability for all species will be highest with measures of canopy cover.

2. Design and field considerations

Because cover can change dramatically over the course of a growing season, sampling must be done at the same stage of the growing season during each measurement event. Comparable stages will probably not occur on similar calendar dates given variation in annual weather. Several techniques have been developed for measuring cover. The most common techniques use lines, points, or plots. All three approaches have been used in plant ecology for over 50 years, and many studies have compared their relative strengths (Bonham 1989). Growth form and the objectives of the study are the key determining factors.

a. Visual estimates in plots

Cover data collected in plots is usually based on a visual estimate of cover class. Many cover class systems have been developed (Table 8.1); all are fairly similar, but the Daubenmire (1959) and the Braun-Blanquet (1965) systems are probably the most commonly used. Many later systems (e.g., Bailey and Poulton 1968; Jensen et al. 1994) split the lowest classes into even finer units. This is because in community studies, the most common application of plot cover methods, many species fall into these low cover classes. For rare plant monitoring studies, a cover class system that is specific to the target species may be more appropriate than any presented here.

The key problem with visual estimation of cover in plots is the introduction of an unknown level of observer bias. Kennedy and Addison (1987) determined that more than 20%

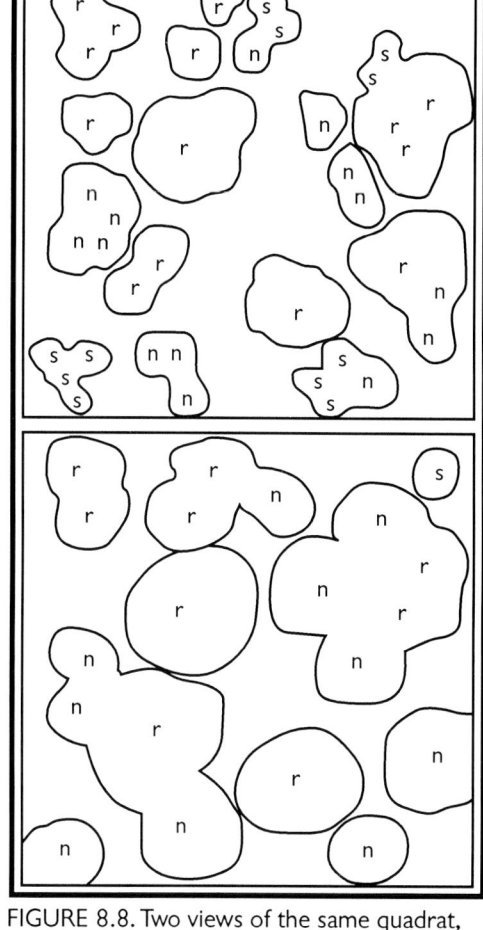

FIGURE 8.8. Two views of the same quadrat, the top measured in 1995 and the bottom in 1996. Note that the density declined (from 39 individuals to 21), while cover actually increased. Also note the scarcity of seedlings in 1996, which would not have been detected by cover methods unless cover was measured separately for adults and seedlings.

Class	Braun-Blanquet (1965)	Daubenmire (1959)	Domin-Krajina (Shimwell 1972)	EcoData (Jensen et al. 1994)	Bailey and Poulton (1968)
				0 - 1% (+)	
	very small		solitary	1 - 5% (c)	0 - 1%
1	small 1-5%	0 - 5%	seldom	6 - 15%	2 - 5%
2	6 - 25%	6 - 25%	very scattered	16 - 25%	6 - 25%
3	26 - 50%	26 - 50%	1 - 4%	26 - 35%	26 - 50%
4	51 - 75%	51 - 75%	6 -10%	36 - 45%	51 - 75%
5	> 75%	76 - 95%	11 - 25%	46 - 55%	76 - 95%
6		96 - 100%	26 - 33%	56 - 65%	96 - 100%
7			34 - 50%	66 - 75%	
8			51 - 75%	76 - 85%	
9			76 - 90%	86 - 95%	
10			91-100%	100%	

TABLE 8.1. Cover estimation classes recommended by various authors.

change in cover must be observed before the change can be attributed to factors other than observer bias and annual variation. Greig-Smith (1983) states that observer bias can be as high as 25% of the mean. Hope-Simpson (1940) concluded that a cover change of up to 23% could be attributed to observer disagreements. In a comparison of estimates by two trained observers measuring 5m x 5m plots, it was found that for 39.5% of the species there was a difference of one class assigned by each observer, and for 3% of the species the observers differed by two classes (Leps and Hadicova 1992). Clymo (1980) found that estimates of cover of wetland vegetation in 25cm x 25cm plots could vary 10-fold among observers. Fine and lacy-leaved species are more variably estimated compared to broad-leaved ones (Goebel et al. 1958; Clymo 1980; Sykes et al. 1983). Accurate estimates are especially difficult when the target species is intermingled with similar species, such as a rare sedge that occurs in a meadow with dense cover of several similar grasses and sedges. Estimates are most variable among observers at moderate levels of cover (40-60%), but are least accurate at the lowest cover values (Hatton et al. 1986).

Using cover estimation in quadrats remains popular, however, because of the ease and speed at which data can be collected. Cover estimation is also more effective for locating and recording rare species (with cover values of less than 3%) than are point and line intercept methods (Meese and Tomich 1992; Dethier et al. 1993).

Some techniques have been used to improve the reliability and repeatability of visual estimates. Cover is more similarly evaluated in small quadrats than in larger ones (Sykes et al. 1983). Use of frames that include a known number of grid squares can also increase the similarity of estimates among observers. In a study of sessile marine species, Dethier et al. (1993) used a 50cm x 50cm frame divided into 25 10cm x 10cm squares, each of which was considered 4% cover. Incompletely filled squares were grouped. This method resulted in visual estimates that were more similar among observers than estimates made with 50 point intercepts in each frame, and required only half the field time. Another approach that has been successful in reducing the variability between observers is training with pieces of cardboard of known cover values.

Because the level of variability among observers differs for different species, you may want to assess the impact of observer variability during a pilot study by conducting trials using several observers. If variability is extremely high, take steps to reduce variability, or use another method of estimating cover.

Sampling design considerations for estimating cover in quadrats are theoretically similar to that of density: using long, narrow, and fairly large plots will reduce the between-quadrat variability and increase the efficiency of the sampling. This design is impractical for most field situations because you cannot accurately and consistently estimate cover in plots with large area, especially long narrow ones that cannot be completely viewed from one observation point. A better approach is to use smaller quadrats that facilitate accurate cover estimates. Place these along a transect line and treat the transect line as the primary sampling unit in a two-stage sampling design (see Chapter 7 for more information).

In rare cases, the small plot may be a good plot design (size and shape) to use, and may be appropriate as the primary sampling unit. If so, you may still want to arrange plots along transects for ease in locating them in the field. If they are far enough apart to be considered independent, and if distributing them along transects results in adequate interspersion, you can consider each quadrat the sampling unit (see Chapter 7).

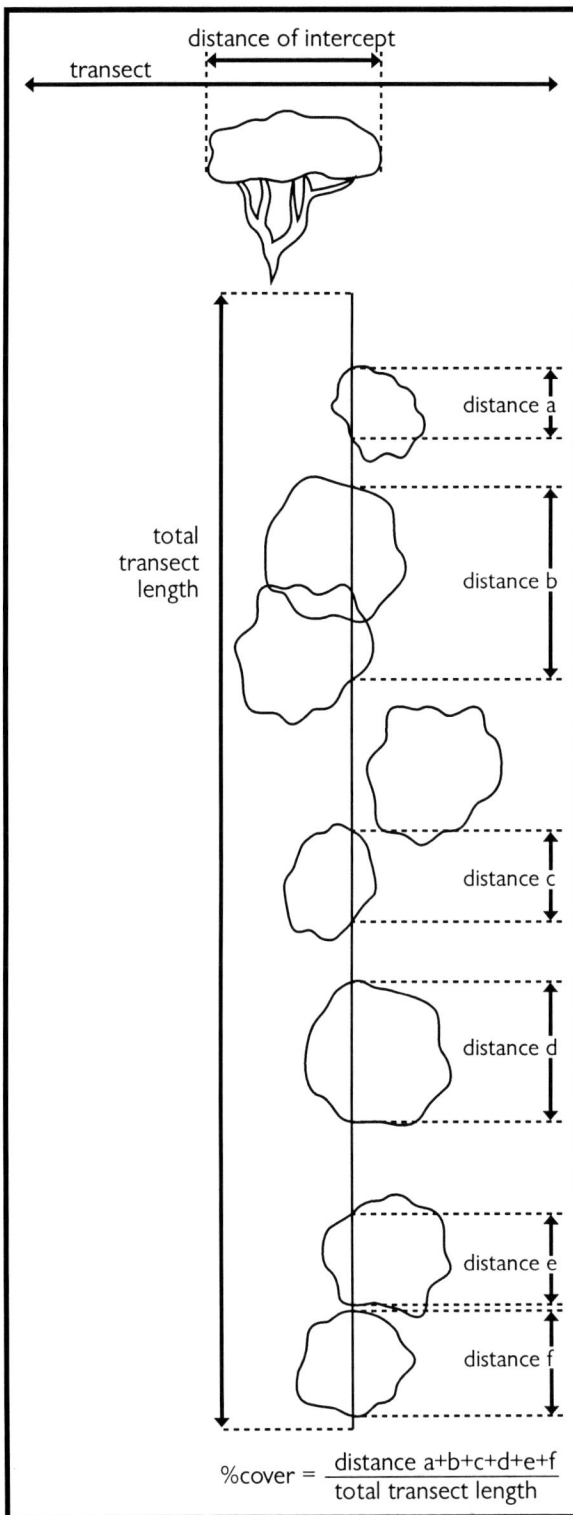

FIGURE 8.9. Line intercept method of measuring cover for a single shrub species.

b. Line intercepts

Canopy cover is measured along a line intercept transect by noting the point along the tape where the canopy begins and the point at which it ends (Figure 8.9). When these intercepts are added, and then divided by the total line length, the result is a percent cover for that species along the transect. Line intercept techniques are effective for species with dense canopies, such as some shrubs and matted plants. Line intercept is more difficult to use for plants with lacy or narrow canopies, such as grasses and some forbs and shrubs, because of the large number of small interceptions requiring evaluation.

Few plants form complete canopies, lacking any gaps. Typical gaps are formed by dead centers in bunchgrasses, fractured canopies in matted plants, gaps between blades of grass, and gaps between branches of shrubs. One approach for dealing with gaps along line intercepts is to measure small increments one at a time (such as a 1cm distance along the tape). This approach forces the observer to evaluate each centimeter and reduces errors caused by sloppiness. It is also very time-consuming. Alternatively, the observer can assume a closed canopy until the gap exceeds a predetermined width; Bonham (1989) suggests 2cm. In practice, observers often treat gaps differently when sampling line intercepts; thus, gap rules must be clearly documented in the description of the sampling methodology to ensure consistency among observers.

Another problem with line intercept is the potential for observer bias because the sighting line is not perpendicular to the tape or plumb. One option is to suspend the tape over the vegetation and use a plumb bob to locate canopy starts and stops. For overhead vegetation, a pole with a level can be used. The most accurate method for locating canopy boundaries of both low and overhead vegetation is to use some type of optical sighting device (described under points, below).

A final problem with line intercept is that repeatable measures are difficult to achieve if the wind is blowing. Not only is there the problem of trying to locate the intersection of the tape with a moving target, there is also the problem of the tape bowing in the wind, and of the vegetation laying at an angle and presenting a larger surface area than would be available under still conditions.

The sampling unit for line intercept is always the transect. During sampling design, the length of the transect should be considered. Longer transects will cross more small-scale variability, reducing the number of transects needed for a given precision of the cover estimate. Longer transects, however, require more time to measure, and may also be problematic in dense vegetation.

c. Point intercepts

Cover is measured by point intercept based on the number of "hits" on the target species out of the total number of points measured (Figure 8.10). Measuring cover by points is considered the least biased and most objective of the three basic cover measures (Bonham 1989). Observers need only decide whether the point intercepts the target species. Canopy gaps and cover estimations do not need to be evaluated.

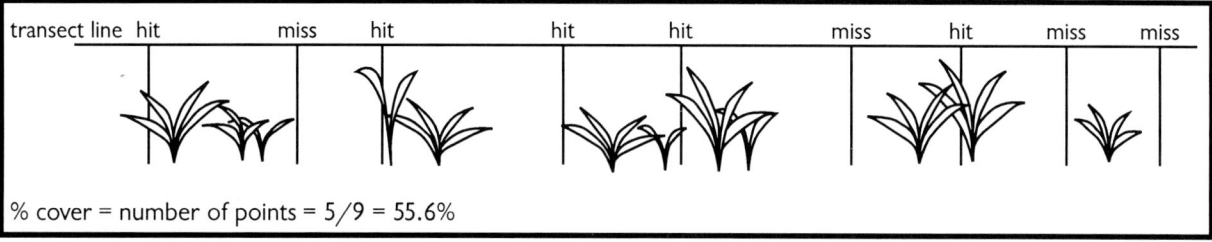

FIGURE 8.10. Point intercept method of measuring cover.

Points are measured either with pins that touch the vegetation or with a cross hair optical sighting method. Pins have the advantage of being inexpensive and easy to use. Their key disadvantage is the error associated with the diameter of the pin, resulting in overestimation of actual cover, especially for narrow or small-leaved species (Warren-Wilson 1963). This is generally not a problem in most monitoring situations where change is of interest rather than the actual cover value. It is important, however, to use the same pin diameter for successive measurements. Pins need to be used with some type of pin frame. Few observers can lower a pin freehand without bias.

Optical sighting devices have been developed that use a mirror system to enable the observer to remain standing while looking at the ground. Crosshairs in the field of view identify a point. These devices are usually mounted on a tripod, and can be set for a specific angle of intercept. They are quick, accurate, and fairly easy to use, but are costly ($500-1000). Another disadvantage is that a second observer needs to move canopy vegetation from the line of sight if the target species is an understory plant.

Cheap sighting tubes with crosshairs made of fine wire, fishing line, or dental floss can be constructed, but often require the observer to bend awkwardly to look downward and are also difficult to maintain at a constant angle. Buell and Cantlon (1950) and Winkworth and Goodall (1962) give complete directions for their versions. For braver do-it-yourselfers, Morrison and Yarranton (1970) describe the construction of a high-quality optical device from a rifle telescope, a right angle prism, and a homemade frame.

The angle of the point intercept has a dramatic effect on the cover measure. Most cover measures are perpendicular to the ground, but species with narrow upright leaves are rarely encountered with this angle. Other angles have been used to increase the number of "hits" on these types of plants (Bonham 1989). The monitoring methodology should always specify the angle used. Note that angled pins eliminate the advantage of intuitive visualization of canopy cover as a bird's eye view.

The fact that the angle of the pin affects the probability of intersection with vegetation suggests a problem if wind changes the angle of the vegetation by making plants "lay over." This can result in a dramatic increase in the percentage of points intercepting a target species, especially fine-leaved or grass-like ones, and has clear implications for using point intercept to measure change over time.

The cover most often measured by points is canopy cover. Cover can also be measured within defined layers (i.e., the cover of individuals over 50cm tall and those less than 50cm tall), or by different species. For both approaches, recording more than one interception at each point is likely, depending on how many layers have been defined or how many species occur at the point. You can also use points to measure multiple layers of a species by recording all the interceptions with the species as a pin is lowered to the ground (Goodall 1952). Note that this measure is no longer a measure of canopy cover since the pin may intercept the same individual or same species more than once at each point. Most researchers have interpreted multiple interception measures as an index of biomass, volume, or composition (Goodall 1952; Poissonet et al. 1973).

A key disadvantage of point intercept is that species with low cover values are often not sampled efficiently because points so rarely intersect the species of interest. This is intuitively obvious: a species with 1% cover would likely only be intercepted once or twice (or not at all) in a sample of 100 points. To estimate a species with 8% cover (within ±10% of the mean with a 95% confidence), Walker (1970) calculated that about 2000 point intercepts would be required. Dethier et al. (1993) found that when 50cm x 50cm plots were visually sampled and with 50 point intercepts, the latter failed to detect 19% of the species found by visual estimates; all of these species had less than 2% cover. Leonard and Clark (1993) and Meese and Tomich (1992) also noted the difficulty of using point intercept to sample species with low cover.

Detecting small changes, especially at low cover values, is also problematic, as it is with any method. Brady et al. (1995) found in simulations of sampling a shortgrass prairie community with 100 points distributed along 100m transects, that 10 transects (1000 points) would detect a change from 12% to 10% cover only 40% of the time. A larger decline from 12% to 6% cover, however, was detected by only three transects 90% of the time in their simulation trial. Using 20cm x 25cm plots with known cover, Dethier et al. (1993) calculated that 1118 points would be needed in a permanent plot to distinguish a change from 3% cover to 4% cover 95% of the time, but only 92 points would be needed to distinguish a change from 40% cover to 50% cover.

These problems are most apparent in community sampling, where each point requires recording the species intercepted. In a study that monitors the cover of a single species, such as a rare species with low cover, many points would require no evaluation beyond the fact that the species is not there. You can imagine that a 100m transect of 100 point intercepts could be evaluated very quickly if it only crosses a clump of the target species once. Most points could simply be checked visually; only those points that are close to intercepting the target species would require a setup of the point frame or optical sighting device.

The sampling unit depends on the arrangement of points. Points can be sampled in frames (which then form the sampling unit), as single randomly located points (each point a sampling unit), or as points located along a transect (either points or the transect forming the sampling unit). The original method of point interception used a linear point frame of 10 pins as the sampling unit (Levy and Madden 1933), with suspended pins that could be gradually lowered until contacting the vegetation. Point frames can also be rectangular, with a grid of points (Floyd and Anderson 1982). The literature is replete with variations on point frame size and shape, but, in general, frames are usually not the most efficient approach. Goodall (1952) demonstrated that sampling point intercepts using frames was much less efficient than sampling random points, and later studies have supported this conclusion (Evans and Love 1957). Depending on the vegetation, time required for measuring single points as the sampling unit can be 1/3 to 1/8 that required for point frames, since many points must be measured in the latter to achieve the same precision as independent random points.

Rarely are point sampling units each located randomly. Sample size is usually several hundred sampling units, a prohibitive number to locate randomly throughout an area. The better approach is to arrange points along transects. If points are far enough apart, they can be considered independent sampling units (Chapter 7). If fairly close together, the transects can be considered the sampling unit.

How many points should be placed along each transect? Fisser and VanDyne (1966) found that it was best to sample with fewer points and more lines when using the transect as the sampling unit. This design maximizes the number and interspersion of sampling units throughout the sampled area. The number of points you place along the transect, however, controls the resolution of the cover value if your sampling unit is the transect. For 10 points, for example, only cover values of 0%, 10%, 20%, 30%, etc. are possible. With 50 points, cover values can be measured in increments of two: 2%, 4%, 6%, etc. At a minimum, you want enough points so that you will intersect at least some individuals of the species of interest along each transect line. This may require many points (50 to 100 or more) for some species with very low cover.

If the point is the sampling unit, using more points and fewer transects may be advantageous. The largest field expense when points are located along transects is the location and establishment of each transect. Maximizing the number of points per transect would minimize setup time. Transects may, however, have to be quite long in this design to have points located far enough apart to be considered independent sampling units (Chapter 7). If the structure of the vegetation presents challenges to establishing long transects, several short transects may actually require less time.

d. Permanent sampling units for measuring cover

All three methods of monitoring changes in cover—plots, line intercept, and point intercept—can be used with either permanent or temporary sampling units. Remember that the value in permanent units is that because of the high correlation between sampling units, the difference observed over time at each sampling unit is of interest. A key consideration with measuring cover with permanent sampling units is whether you can actually make them permanent. Measuring the exact same line intercept each time is much more difficult than measuring within a permanent density plot in which all four corners of the plot are permanently monumented. Changes in tape tension (sag), bowing in the wind, and slightly different placement due to brush are examples of factors that may reduce the correlation between each measurement. If you intend to use a permanent design, consider the following factors:

Plant morphology. Failure to intersect a particular plant on the second measurement that was recorded during the first measurement results from (1) tape and point movement and a miss of the exact location; (2) decline in cover of the plant so that your sampling unit no longer intersects it; or (3) the plant died (or is dormant). Only the first is a problem; the second two scenarios are true changes. You can perhaps guess which scenario is most likely based on the size and morphology of the target species. If the plant you are sampling has a fairly large area, you will likely intersect the same individual at the second measure because you have the area of the plant as room for "error." Thin-leaved species are more problematic than matted species because only minor movements of a point or a line will result in missing a plant that has not changed.

Field conditions. Exact relocation of a permanent transect in places that are difficult to travel through, such as dense brush, is unlikely. You can be more confident of accurate tape relocation in a short-grass prairie.

The sampling unit. If points are the sampling unit, your individual points must be correlated from year to year. If transects or plots are the sampling unit (either as a line intercept, a line of points, or a collection of plots), the transects or plots must be correlated. Points are much more difficult to relocate than transects or plots.

Several field techniques can reduce placement error. You should monument transects with permanent markers at each end and at points along the transect. The number of intermediate markers depends on the field circumstances. In dense brush, a transect may require a marker every few meters to ensure accurate relocation, while at a meadow site every 10 or 20 meters may be sufficient. Shorter transects are less affected than long transects by tape stretch, bowing and sagging, or using alternative pathways around large vegetation. For cover estimation in plots, marking one or two individual plot corners as well as the transect ends will ensure that plots are relocated accurately. This monumentation adds to the time required to establish a study, and these costs must be weighed against the benefits gained from a permanent design compared to a temporary design (Chapter 7).

In general, permanent sampling units for cover, especially using points as the sampling unit, may be difficult to achieve in field settings, although they usually do increase the efficiency of the design for measuring change (Goodall 1952). If you intend to use a design with permanent sampling units, test the degree of physical correlation by conducting a measure, picking up the tape, then having a second observer re-establish the tape line and complete the measurements. If the correlation between the two measures is not good, you should use a sampling design with temporary sampling units.

e. Comparison of plots, points, and lines

You must choose transects, plots, or points as the sampling unit for measuring cover. The best sampling unit depends on the total cover of your species, its distribution in the field, and its morphology.

Transects vs. Plots. Daubenmire (1959) found that the cover estimates from 40-50 quadrats was nearly identical to that measured by 350m of line intercept. Standard error of the quadrat samples, however, was high (likely because many did not contain the target species). Bonham (1989) states that line intercept is more accurate than quadrats when working with different sized plants. Hanley (1978) found that at low cover (8%), line intercepts required about half the time to achieve the same precision as randomly placed quadrats, but at 26% cover, the two methods became more comparable (34 minutes for quadrats compared to 29 minutes for lines).

Points vs. Plots. Dethier et al. (1993) created simulated plots containing a known cover of 13 species and compared cover measured by point intercept to cover visually estimated to the nearest percent in the plot. Cover estimations done with the aid of subdividing the plots into 4x5cm rectangles were close between observers, and closer to the true value of cover than measured points. In the field, point intercept failed to detect 19% of the species that were detected by cover estimation. Differences among observers were less for cover estimation than for point measurements.

Transects vs. Points. Floyd and Anderson (1987) found that point interception achieved the same precision as line interception in one-third the time. Line transect and points gave similar results in a study by Heady et al. (1959), but points required only about half the field time and less office time compared to line intercept. At low cover (3% or less) line intercept gave better results. Brun and Box (1963) found that point intercepts required less than two-thirds the time of line intercepts to achieve the same precision. In contrast, Whitman and Siggeirsson (1954) found that points and line were similar in time requirements.

I. Production and Other Vigor Indicators

Production is the annual output of vegetative biomass. It is most commonly measured as a harvest of aboveground standing crop, usually at peak (before plants start senescing and losing leaves). This approach underestimates total annual production, missing biomass consumed by herbivores, loss that occurs throughout the growing season, below-ground production, and regrowth after harvest.

Vigor indicators are many and include height, basal diameter, number of flowers, number of inflorescences, number of leaves, number of stems, number of leaf whorls, diameter of rosette, and volume of plant (height x cover).

1. Uses, advantages, and disadvantages

Production varies each year depending on the favorability of growing conditions and therefore may not be sensitive to the type of trend that is of interest in most rare plant monitoring projects. Production is also usually sampled destructively by harvesting, drying, and weighing and, for most rare plants, this type of monitoring is not appropriate. Because of these constraints, production is not discussed at length here. If you are monitoring a more common species and wish to use a production measure, information on the subject is abundant (Malone 1968; Sandland et al. 1982; Ahmed et al. 1983; Bonham 1989; Ruyle 1991).

Vigor indicators are also strongly influenced by annual weather patterns, but they may be appropriate for some monitoring questions. As nondestructive measures of vigor, they are appropriate for use on rare plants. Most are easy to measure, with little observer bias.

2. Design and field considerations

The key consideration often ignored in production and vigor studies is explicit definition of the sampling unit. Sampling units can be plots (e.g., grams produced/m^2, number of flowers/m^2), or an individual (e.g., grams/individual, number of flowers/individual), or a part of the individual (seeds/fruit). The different sampling units have very different design considerations. See Chapter 7 for an extensive discussion of the difficulty of selecting a random sample of individuals and the use of cluster sampling or two-stage sampling to address that difficulty.

J. Choosing an Attribute and Technique

1. Limitations of morphology and life history

Some species cannot be measured by some techniques. For example, you would find it difficult to count individuals of a clonal mat-like species to estimate density. Appendix 11 summarizes the appropriateness of various techniques for the common growth forms and life histories.

2. Logistical constraints

Some species may be measured by several methods, but logistical and practical constraints limit the options. Some considerations include the following:

Investigator impact. Species occurring on steep erosive slopes are susceptible to investigator impacts from monitoring, but even stable flat sites can be impacted. Trails can be worn along permanent transects. Transects that require close examination, with investigators perhaps spending time on their knees, will be especially affected. In a macroplot sampled by randomly placed plots, plants can be easily crushed by travel between plots. Frequency is often the least destructive method, simply because noting whether the species occurs within the plot takes less time than counting or measuring plants within a plot.

Available equipment. Most of the methods described in this technical reference can be done with inexpensive equipment, but the availability of more sophisticated technology such as accurate Global Positioning System (GPS) units or survey equipment may improve the speed and accuracy of field sampling.

Available expertise. Frequency and point cover methods require the fewest decisions on the part of the observer and can be implemented by technicians with minimal training.

K. Locating Sampling Units in the Field

Chapter 7 describes how to establish random coordinates for sampling units using several random number generators. Once these have been identified, how do you locate the points in the field, and how accurately do these points need to be located?

Within a rectangular sampling area or macroplot, plot or transect locations can be identified by their coordinates. For example, your plots are 40m x 1m, and you have identified a target sampling plot that is found at the second position along the baseline and at the tenth position along the perpendicular (see Appendix 4 for methods to randomly select plots). Figure 8.11 shows this plot at 40m along the x-axis (the baseline) and 9m along the y-axis. Usually, it is most efficient if you have tapes laid along both your baseline and the y-axis. Even better is a tape along each side of your rectangular sampling area so that you can measure to plots from any side. You may even want to place pin flags regularly along your boundaries. You could place them every 10m if you are still in the pilot stage and might be trying plots of different configurations, or at the increments determined by your selected plot size. These flags form the grid in which you will locate your plot corner at x = 40m and y = 9m. There are several ways to physically locate a plot corner:

◆ **Pacing.** One is to simply pace from the 40m point along the baseline up approximately 9m using a compass. If you did this, you would only need the baseline tape from which to pace all of your plot corners. In practice, the additional tapes may save you steps by allowing you to pace from any of the four sides. Pacing is an acceptable way to find plot corners; plots do not have to be located *exactly*. Pacing will not work, however, if the "slop" inherent in pacing allows you to place a sampling unit with bias. If you are working in an area with scattered pockets of prickly brush, for example, you might (inadvertently, of course!) shorten your pace somewhat to avoid a brush pocket. Another example of bias is to adjust your pace to try to intersect the target species. Placement bias is less of a problem with very long or large quadrats because you probably cannot judge whether features will be included in the plot once it is established. Bias can also be lessened by consciously avoiding looking at the ground as you near the destination. If, in spite of your best efforts, you think you have biased placement with pacing, consider taping your distances.

◆ **Pin Flags**. If you have placed pin flags along your four boundaries and can see all four boundaries, you can approximate your coordinate location using those pin flags. If, for example, you placed pin flags every 10m along all four boundaries (alternating colors would be especially helpful), you could then move until you were between the two flags at the 40m mark along the top and the bottom x-axis, and about a meter short of the flags that marked the 10m points along the left and right y-axis. Again, if you think you might bias plot locations using this method, opt for taping.

◆ **Taping**. The last method is to measure from two boundaries and place your plot at the intersection. You can measure fairly accurately (within 10-30cm) with a tape measure or a pocket electronic distance measurer (Section O). Measuring is more time-consuming than either of the other methods, but will help eliminate bias.

An important benefit in using a taping method is that it can function as a backup monumenting system for permanent plots and transects. In most cases, you would monument each permanent plot or transect. If these markers are lost, but the baseline markers remain, you could relocate the plot fairly accurately by retaping the distance.

An alternative approach is translating x and y coordinates to distance and azimuth from a single base point (Figure 8.11). This method is especially useful in brushy or wooded areas where establishing a baseline is difficult. In meadow systems, the distance could be paced from the base point to the plot corner using a compass to determine azimuth. In brushy areas, or where increased accuracy is desired, you could have one observer at the base point with a Sonin (for distance measures) and a compass (for azimuth) who could direct the other observer to the correct location. Alternatively, you could use a survey instrument with an electronic distance measurer set up at the base point. Plot locations found with these instruments, especially survey instruments, are very accurate, and much faster than taping through brush. Surveyed plots that are permanent sampling units can be relocated accurately if individual plot monumentation is lost. The time-consuming part is the conversion of x-y coordinates to azimuth and distance, and instrument set-up time (especially for the survey instrument). Awbrey (1977) provides a complete discussion of this approach. A similar concept is to dispense with x-y coordinates and simply use a random distance and azimuth as the plot location. One problem with this approach is that the distances along azimuths radiating from a central point are clustered near the center (like the spokes of a wheel near the hub) and farther apart toward the outside edge. Laferriere (1987) provides a complete discussion of this approach and some solutions to the clustering problem. Another problem with this approach is that starting points located in this way may result in projecting transects and long quadrats beyond the boundaries of the sampling area (Chapter 7).

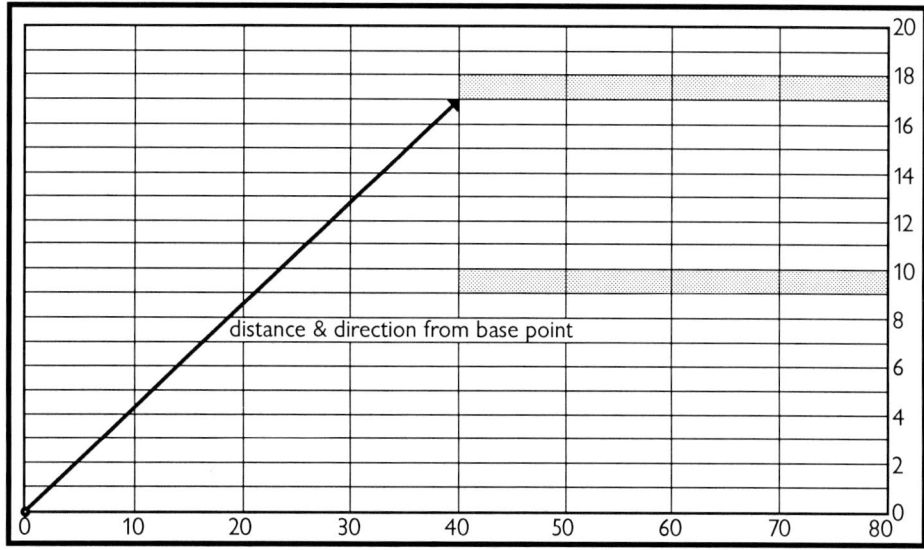

FIGURE 8.11. Sampled area gridded into sampling unit locations (potential plot placements). Plots can be located by pacing or measuring from base lines at the edges of the area to be sampled or by converting x-y coordinates into distance and azimuth from a single base point.

L. Relocating Study Areas

Critical to the success of a monitoring project is relocating study areas. Often study areas are not documented in project notes because the initial investigator assumes s/he will be returning the following year to do the measurements. Many studies have not been continued because study areas could not be relocated after the originator leaves.

Description of the location of study areas should include the following:

1. Driving instructions from a well-known landmark, including direction and mileage to the nearest 0.1 mile. A hand-drawn map is helpful if there are a number of roads in the area, especially if topographic maps are outdated.

2. Walking directions including compass direction and distance (paced) to study area. Again a map is helpful.

3. Study site location marked on a USGS 7.5-minute quadrangle and on a recent aerial photograph.

4. Compass direction from the study site toward at least three prominent, permanent landmarks such as mountain tops. If dense forest vegetation requires that trees be used, at least six trees should be included, and these trees monumented.

5. Photos should be taken as needed, such as at the parking spot, along the path to the site, or in several directions at the study site. Each photo should include compass readings to describe direction of the photo and should be marked for location on the topo map or aerial photograph.

6. If available, a GPS unit may be useful for recording the study location.

M. Monumentation

Securing monumentation of plots, macroplots, transects, or population boundaries is a critical part of the success of a monitoring project. This section contains some hints to secure monuments.

1. Assess potential for loss

No monument is completely safe, but some are more at risk than others. Visible markers such as brightly painted stakes will always be removed in areas frequented by people. Some markers such as pin flags are attractive to animals and may be pulled up from the ground by deer nibbling on the flagging. Flagging, especially the biodegradable type, is attractive to animals, and rarely lasts more than a few weeks in areas with grazers and browsers. Wooden lathe is easily broken, and rarely lasts more than a few weeks with grazers. Plastic pipe degrades in sunlight, and is especially subject to breakage in the winter when the cold makes the material brittle.

Natural catastrophes should be considered. Fire is possible in almost any habitat. Use only metal monuments for studies that are needed for more than one year. Don't depend on trees for monuments, although they can be used for a backup system.

2. Stakes and T-posts

Monuments such as T-posts or fence posts are often stolen. Cutting the top 12 inches from a T-post reduces its value and lessens the chance that it will be stolen. T-posts should be sunk as deeply as possible in the ground to make it difficult to pull-up by a casual vandal.

Inexpensive stakes can be made from angle iron or rebar. Lengths of 60-70cm are good for deep soils. Lengths should be shorter if you plan to use them on more shallow soils. At least 1/2 - 2/3 of the total length should be below ground. If there is the chance of injury to live-stock or horses, the upper third should be bent or looped over. This can easily be done with a large box end wrench once the stake has been pounded into the ground. If the ground is soft, bend the stakes before pushing them into the ground.

Stakes can be easily pulled out, and if not sunk deeply enough into the ground can be knocked out by livestock. Where vandalism is a problem, special stakes are available that can be used for monuments or anchoring lines. Duck-bill tree anchors have hinged winged plates that are closed as the stake is driven downward, but open up as the stake is pulled up. The lower half of screw or auger stakes look like ice augers. The effort of turning these stakes out of the ground deters all but the most determined vandal. Both stakes are commonly available from forestry supply companies.

3. Marking trees

Use only marking paint designed for outdoor use. Use bright unnatural colors if this does not conflict with aesthetics or attract vandals. More subdued colors can be used, but because these can be difficult to see, additional travel and monumenting information will be needed. Paint a concentrated spot on both sides of the tree and a ring all the way around the tree below the spots. Paint can be used for marking trees if the study is to last 1 year or if study sites will be visited annually. For longer-term marking, supplement paint with another marker (such as a blaze or a tag), since paint can fade or be sloughed off with bark. Blazes are pre-ferred where damage to the tree and visible impacts are acceptable, since they are easily spotted from a distance and will not fall off the tree or be pulled out by vandals. Blazes can last decades on some trees.

You can use tags to supplement blazes and paint with information or to mark trees. Tags can occasionally fall from the tree and are difficult to spot from any distance. Numbered metal tags are commercially available from forestry supply companies. Tags should be affixed at eye level or breast height (4.5ft). Aluminum nails should always be used. These pose minimal hazard to sawyers or mill operators. Even in areas where logging is unlikely, it is a safety issue and a courtesy to use aluminum nails. Trees last far beyond the life of many studies and protective designations, and there is no justification for using a nail that might potentially endanger a person in the future. Aluminum nails are readily available from forestry supply companies.

The heads of tag nails should be slanted downward with about an inch protruding to allow for tree growth. This allows the tag to slide to the head of the nail, and reduces the chance that it will be enveloped by bark.

You should identify and map marked trees in the field notes and the methodology section of the Monitoring Plan (Chapter 10). Including information on species and diameter for each marked tree makes it easier to relocate them later. Tree diameters should be measured at the forestry standard of 4.5ft from the ground (diameter breast height).

4. Landmark references

All monuments should be supplemented with references to visible permanent landmarks. Obvious landmarks on the site, such as a rock outcrop, can be used to identify the location of monuments. Directions from the landmark to the monument should include both measured distance and compass direction (note whether declination or magnetic). A photograph of the landmark from the monument that includes the monument in the foreground helps in relocation.

On sites lacking nearby landmarks, triangulation can be used to identify the location of a monument in relation to distant landmarks. This involves measuring the compass direction toward landmarks such as mountain tops and permanent (you hope) man-made objects such as water towers or microwave towers. By measuring the direction to two objects, your location on the ground is fixed by the angle formed by those objects and your location. The site can then be relocated in the future. Triangulation is most accurate when the angle formed by the two triangulation points is approximately 90°.

5. Adding "insurance"

No monuments that are required for continuation of the study (e.g., permanent quadrat corners or transect ends) should be without insurance in case the primary monuments are lost. One option is to bury physical markers such as large nails or stakes. A single buried nail next to a monument may be disturbed or dug up when the monument is disturbed. Better insurance is to use four buried nails, each exactly 1m from the primary monument on the four compass directions. A metal detector can then be used to locate the nails if the primary monument is removed.

A second option is to survey the primary monument using survey or forestry grade survey instruments. You can survey the monument from a permanent known point or from two or more inconspicuous secondary monuments.

N. Field Equipment

Standard field equipment is listed in Appendix 12. We recommend storing all or most of the equipment in a box that can be taken to the field. This precludes forgetting an item and allows trial of other methods if the planned method fails during the pilot study.

1. Tapes

Tapes come in a variety of lengths, increments, materials, and cases. Which type of tape you wish to purchase is largely a matter of personal preference. Some considerations are:

Length. Short tapes are less expensive, lighter, and easier to use than long tapes, but you will usually need some long tapes for monitoring. At least one tape that is a minimum of 100m long is recommended.

Increments. Tapes can be purchased in both English and metric units. We recommend all rare plant monitoring be done in metric units, to allow for potential publication and exchange of information. Metric is the unit of choice for scientific studies. Government agencies, especially in the forestry and range shops, have long used English units, but those conventions are changing. It may still be difficult, however, to find a metric tape in your office. Tapes can be purchased with English units on one side and metric units on the other.

Increments can be in millimeters, centimeters, decimeters, and with marks at the meter and half meter points. The most versatile tape for field use has centimeters marked and numbered, decimeters identified with heavier marks and numbered, and meters numbered and marked with heaviest marks and/or alternative colors. Tapes that are not numbered at every increment take longer to read and increase misreadings.

Materials. Tapes come in steel, fiberglass, and cloth. Steel is most accurate over the life of the tape. Stretching is virtually non-existent, although the tape length will change depending on temperature. This amount of error is probably insignificant for most rare plant monitoring work. Steel tapes are expensive, heavy, and difficult to use because of kinking. Some steel tapes come with a nylon coating that reduces the tendency to kink. Steel tapes are the tape of choice for work requiring extreme accuracy. They may be appropriate for permanent transects, where repositioning the transect in exactly the same place is important. Devices called tensioners ensure the same amount of tension is placed on the tape each time a transect is measured (available from survey suppliers).

A universally useful metal tape is the steel loggers tape, which has a hook or a ring at the end and a retractable case. These tapes are useful if a number of measurements will be taken and time spent unwinding and winding a tape becomes burdensome. Hooks can be set into a tree or a stake, and released with a flick of the wrist, retracting automatically. These features may make the relatively expensive loggers tape worth its initial cost in the long run. We recommend that the first 10cm of the tape be wrapped with electrical tape or some other protecting tape to prevent the end of the tape from wearing as it snaps back into the case. Another useful tip is to replace the standard hook at the end of the tape with a bent horseshoe nail. These press into a tree more easily and provide better controlled release than a standard hook.

Fiberglass tapes will stretch over the life of the tape and when under tension. The amount of stretch is related to material, age, use, and tension. Some manufacturers offer fiberglass tapes with as little as 0.01% stretch per pound of tension over 4.5 lbs., a standard similar to steel tapes. These tapes are advertised to retain accuracy over the life of the tape. Fiberglass tapes are light, durable, and easy to handle. Cloth tapes are also light and easy to use, but are less durable than fiberglass and they stretch.

Cases. Most tapes come in open reel cases, allowing for rapid pick-up. Some tapes come in enclosed metal cases, which provide better protection but are difficult to repair if the tape twists inside the case and binds up. Surveyor's rope, or rope chain, is a type of tape that is designed to be pulled from site to site rather than rolled up. These usually come in 50-100m lengths. Their advantages are that they are designed to withstand dragging and can save roll-up time when sites are close together. They can be coiled rather than reeled up, and they are generally lighter than reel tapes. The disadvantage is that surveyor's rope is usually marked in increments of 5-10cm, which is not useful for making measurements to the nearest centimeter.

2. Paint and flagging

Paint should be specially designed for outdoor marking use, and is available from forestry supply companies. You may wish to choose an unusual color (compared to the standard orange or red). Florescent colors are recommended; they are easily spotted and can even be seen by persons who are color blind. Yellow paint can provide an intermediate color choice for stakes that need to be relocated, but shouldn't be glaringly attractive to vandals. Avoid blue paint, especially in forested areas, since it is a standard color for marking cut trees on logging units.

Flagging comes in a wide variety of colors and patterns; a stock of an unusual type can be useful for unique markings in a project area full of orange and pink flagging. Biodegradable flagging is available, but it is more likely to be eaten by animals and often becomes brittle, breaking in cold weather.

3. Compass

A compass that will be used primarily for route finding should have the following characteristics: (1) mechanism for adjusting declination; (2) a housing with vertical lines to aid in map work; (3) azimuths by degrees, 0-360; and (4) folding mirror to increase accuracy of sightings. Most compasses used by resource specialists are of this type.

A compass that will be used primarily to make sightings on objects should be an optical bearing type. These are similar to clinometers, with a viewing hole in an otherwise enclosed housing. To use these, you sight on an object while reading the azimuth through the viewing hole. While this type of compass provides very accurate azimuth sighting, it is difficult to use for map work because it lacks the built-in protractor and movable housing of the folding mirror type of compass, and declination is not adjustable (0° [360°] always reads magnetic north).

4. Field notebooks and data sheets

While data from most monitoring studies will be collected on pre-printed data sheets (see Chapter 9), field notebooks will still be needed to keep a record of general observations and notes. It is strongly recommended that you keep a field notebook as a log of daily field activities. Field notebooks are also necessary for recording information on plant collections (see Section P).

Field notebook systems vary from biologist to biologist. Standard bound field notebooks and binder system notebooks are available from forestry supply companies. While bound notebooks have the advantage of keeping all data in a single volume and eliminating potential problems of "fixed" field notes (inserting pages to "doctor" data), they have a key disadvantage of potential for loss of an entire field season's worth of notes. At a minimum, photocopy field notebooks daily if possible, and store photocopies in a safe place.

Binder systems have three key advantages. The first is that sheets can be removed after each day's field work, and stored in a safe place. This can be especially advantageous if field work is done in a remote place where daily photocopying is not possible. The second advantage is that sheets that fit the binder system can be used in a laser printer to prepare pre-printed sheets. You may wish, for example, to have a section in your field notebook just for tracking collections or photographs. A preprinted entry sheet can save time for these standard types of field notes. The final advantage is cost; binder systems can be used over several field seasons by purchasing more filler paper.

Both bound and binder system notebooks are available with waterproof paper. This paper is recommended even in arid climates where a field day in the rain is rare, since the paper will not be destroyed by being dropped in a creek or soaked in a backpack by a leaky water bottle. Waterproof paper can also be used in a laser printer or photocopy machine. The laser and photocopy ink will not smear, fade, or run on this kind of paper.

Waterproof paper is best matched with special pens, available from forestry supply companies for $5-10. These pens will not smudge, rub, or wash out. Standard ink pens should never be used; they can bleed and will wash out if the field notes are soaked. Pens are recommended as standard field and scientific practice, since they make doctoring of notes difficult. Incorrect entries and notes should be struck with a single slanted line (for one character) or a single horizontal line for several letters. Also suitable for field notes are hard pencils (number 4 or 5 leads), which will actually make an impression on the paper. Soft pencils (standard number 2 leads) are also not suitable for any type of field notebooks or data sheets because the lead can fade and smudge to the point of illegibility, and will become unreadable if wetted.

5. Handy tools

Clipboards. Look for a clipboard that contains an area for storage of additional data sheets, and a metal cover that can be quickly flipped over the data sheet in the event of inclement weather.

Pocket stereoscope. This tool enables you to look at stereo pairs of aerial photographs in stereo (3 dimensions). It can be handy in the field for locating study plots on aerial photographs when landmarks are scarce.

Rock picks. This geologist's tool can be very handy for digging up plants to collect as herbarium specimens.

Hip chain. A hip chain is used primarily by foresters and surveyors. It is a box, worn on your belt, that measures the amount of fine string that is fed out as you walk, thus enabling you to measure long distances without a using a tape or counting paces. Measured distances are not accurate enough for fine measurements, but the hip chain can be an excellent tool to measure distance from a known landmark to the study site, or to provide rough measurements of population boundaries. The main advantage is that measurement does not require using your hands, or keeping track of paces.

Clinometers. These look similar to an optical bearing type compass. They measure slope, heights, and vertical angles.

Plane table and alidade. A plane table is a flat mapping surface, about a meter square, attached to a tripod. The alidade is basically a telescope mounted on a straight edge. The plane table is set up and leveled at a central location. Direction to a point is measured using the alidade. Distance can be measured with the alidade and a stadia rod based on the principle of similar triangles. Straight edges on the mapping table are used to generate a hand-drawn map to scale.

Reinhardt Redy-Mapper™. This tool allows you to quickly and easily map population boundaries in the field to scale. It is basically a pocket plane table that hangs around your neck. It consists of a 25cm x 25cm sheet of hard plastic with a translucent disk attached at the center. The disk accepts pencil lead, and is the drawing surface for the project (it can be cleaned after the map is transferred to paper). Angles and distances are determined by compass and tape or by pacing; the mapping tool facilitates translating those angles and distances to scale. The Redy-Mapper can be used while traversing the boundaries of a population, or can be used to map from a central location. Although this tool has been partially replaced by electronic tools (Electronic Distance Measurers, GPS units, and computer-generated maps), it is still useful. It is much less expensive, quicker, and more accurate than many of the available GPS units. With this tool, you can map a population boundary almost as fast as you can walk it.

Electronic Distance Measurer (EDM). An EDM is a survey tool that reads distance and direction between stations, and records the values electronically. The information from an EDM can be downloaded into standard survey software or drawing programs to generate maps. The instrument is fast and precise, but requires user training both in field techniques and software applications. An EDM can be used to map population boundaries, permanent plot locations, individual plants, and site features.

O. New Tools

1. Global Positioning System (GPS)

These electronic systems interface with satellites to enable the user to locate or relocate a spot on the earth's surface. Their accuracy depends on the system and access to satellites. Expensive survey grade instruments can be accurate to within millimeters, but most units owned by resource management agencies and available to botanists range in accuracy from 0.5m to 50m. Factors such as availability of satellites, terrain, and controls can diminish their accuracy. At the time of this writing, real-time use is impacted by Department of Defense scrambling of satellite signals; two units or a fixed base station are needed to overcome this for precise locations. Units containing a crypto code to decipher scrambled satellite signals are now available on a limited basis to government agencies. These handpacks are small, lightweight, and fairly precise (within 4m). Post-processing of GIS data using information from base stations can increase this precision.

While they are likely not accurate enough to map individuals of many types of plant populations, the more accurate units may be useful for mapping large, widely spaced, long-lived individuals, such as trees or cacti. GPS can also be used for mapping population boundaries and locations. Many GPS systems can download information electronically into a mapping program.

The technology in this field is changing rapidly, and accuracy of relatively inexpensive instruments is improving quickly. GPS units will likely become more widely available and commonly used in the next few years.

2. Electronic data recorders

Three types of electronic data recorders are discussed in Chapter 9: tape recorders, portable computers and field data-loggers.

3. Pocket electronic distance measures (EDMs)

These are manufactured primarily for construction use, but have been used for outdoor resource work as well, and are now available from forestry supply companies. Most units rely on sound waves combined with an invisible light beam to measure distance. They can measure distances up to about 80m, less in heavy brush or timber. For outdoor use, best results are achieved with both a transmitter and a reflector (indoors, the transmitter can measure the reflected sound waves from a solid object, like a wall). Accuracy of these units can be as good as 2mm. Accuracy, however, can be negatively affected by other sounds; they work poorly in the rain, along water courses, and in noisy urban areas. The units are also difficult to aim; sighting along the edge of the device can help. To ensure accuracy, three consistent measures should be taken for each distance.

Applications for these instruments are many. Distances from a baseline can be measured without a tape. Pocket EDMs could be used for line intercepts, using the electronic distance from a reflector set up at the end of the transect line. They can also be used to determine limiting distance (plant in or out of plot). Almost any distance currently measured with a tape could be measured with a pocket EDM. Time savings are potentially tremendous, especially for long distances or in dense vegetation where pulling a straight tape is difficult.

Price for these units is under $200. Your office forestry shop may have them, since they are becoming increasingly popular with foresters.

P. Collecting and Pressing Plants

1. Plastic mounts

Plants that are not succulent or wet from precipitation can be preserved as field mounts (Burleson 1975). Plants are cleaned of dirt and dead leaves and arranged on the adhesive side of a sheet of plastic acetate. A second sheet of acetate is carefully and firmly rolled onto the first. With some practice, good clean mounts can be made using this method. The advantage of this method is that you can mount plants permanently in the field, providing a quick and inexpensive record of species encountered. Leaves and flowers also retain much of their original color. Plants rarely mold (Burleson 1975). A disadvantage is that collections cannot be manipulated (flowers teased open, etc.) for later identification. Only the features that you expose at the time you bind the sheets of acetate together will be visible.

Collections preserved in acetate sheets are not suitable for herbarium deposition because these collections tend to degrade over time (10-20 years). Rare species should be preserved following standard collection and curation methods (as described in the next section).

2. Pressed and dried collections

a. Ethics of collection

Any collection impacts a population, although impacts of collections made in large populations are insignificant. Rare plants may be especially prone to collection damage because of their small populations and the propensity of botanists to collect rare species. This destructiveness must be weighed against the information gained. Plants should not be collected in populations of less than 50 individuals. In smaller populations, small portions of plants may be collected if absolutely necessary.

It is especially important when collecting rare species to make each collection of herbarium quality. It is senseless to destroy an individual of a rare species if the collection is so poor that it is not worth storing for future use.

b. Collecting plants

Within any population, plants will vary significantly in size and reproductive status. In general, choose individuals that are of moderate size. If you chose an exceptionally large or small individual, note it. Try to choose an individual with both fruit and flowers.

If the plant is small (can reasonably be fitted onto a 11 1/2" x 16 1/2" sheet of herbarium paper), the rule is to collect the entire plant, root and all. An incomplete plant is of little use as a herbarium specimen since any feature may be evaluated in a taxonomic study.

If the plant is larger, is should still all be collected and pressed in portions; e.g., lower third, mid-third, and upper third. If the plant size is completely unmanageable, for example a woody species, collect branches that contain leaves, fruit, or flowering structures, and first and second year bark. The height of the individual from which the collection was made should be noted.

c. Pressing plants

Press plants immediately after they have been collected; plants are much easier to arrange and press if they are not wilted. Collections carried around for a day tend to become pretty bedraggled. If immediate pressing isn't an option, place the plants in a vasculum (an airtight metal container designed to hold collected plants). A large tin can or pickle bucket will also work, and, in a pinch, plastic bags. The hard-sided container is preferable because it keeps plants from getting smashed, but most botanists use plastic bags simply because they take so little room. Small plastic bags are useful for keeping plants separate by collection site, habitat, or species. Information written in pencil (not pen) will survive several days in a plastic bag with moist plants. Plants placed in the refrigerator or a field ice chest will keep much better than those that get warm. Plants that have wilted in the field can be partially revived by placing them in a plastic bag with some soaked paper towels and storing in a refrigerator overnight.

Standard plant presses are a sandwich made of two pieces of wood lattice (to allow moisture to escape), with blotters and pieces of cardboard layered inside. The blotters help to absorb moisture and the cardboard helps to flatten the plants as well as let moisture vent. Two compression straps hold the sandwich together very tightly. You can purchase plant presses for $30-40 or make them. The standard size is 18in x 12in.

The plants are usually pressed within newspapers. The keys to a good pressing job are patience, practice, and a penchant for neatness. Roots should be carefully but completely cleaned. Leaves have to be individually smoothed flat. Plants with small compound leaves or many leaves require special care. Leaves should be pressed so as to represent both upper and lower leaf surfaces. Arrange flowers in various positions showing all of the features. Some flowers should be opened and pressed in a "mutilated" form; once the plant is dried, it will be very difficult to open flowers and peer at stamens or other small parts. Think about all of the characteristics you may want to see later for identification and descriptive purposes, and remember the collection you are pressing may later be glued to a sheet of herbarium paper.

Another difficulty is that you are trying to flatten a 3-dimensional plant onto a 2-dimensional sheet of paper. Thick roots may need to be longitudinally sliced in order to press. Bushy plants are also problematic. You may need to do some judicious pruning of your collection, but you must be careful not to change the aspect of the plant. Leave clues that this plant once had more branches or flowers than it does now. Or prune severely one inflorescence, or one portion of the plant to show features clearly, and leave the rest unpruned to show the general form of the plant.

Wetland plants pose a unique problem because they are sometimes very succulent or watery. Very wet plants, if pressed, may mold before they dry or stick to the newspaper. Be sure to shake all excess water from the plant. If the plant seems to hold large amounts of internal water, allow it to dry slightly before pressing, although be cautious of wilting. Once in the press, accelerate drying by placing the press outside, in full sun, in a breeze. In high humidity, you may need to rig a plant dryer, which can be as simple as a light bulb placed beneath the press. Presses can also be placed overnight in gas ovens, using the heat of the pilot light.

Extremely succulent plants, such as cacti, are especially problematic. Tissue must first be killed; a recommended method is blanching in boiling water. They must then be dried rapidly. Some plants can be hollowed out to allow rapid drying and flat pressing. Fosberg and Sachet (1965) give further ideas for dealing with these difficult species.

d. Collection notes and numbers

To keep track of collections, and the field notes which correspond to each collection, you should keep collection numbers. These are sequential numbers assigned to each collection you make. Write the number in the field notebook and on the newspaper sheet in the press. Collections without supporting information are nearly useless; thus, most collectors carry a field notebook and take careful notes. At minimum, each collection should have the following information: (1) collection number; (2) date collected; (3) location, including township, range, section, county and state and a general driving or walking description; (4) habitat information, including slope, aspect, substrate, elevation, shade, and moisture regime; (5) associated species and vegetation type; (6) abundance of the species and approximate size of population area; (7) notes on flower color, plant size, variability; and (8) ownership of site. Other information that may be useful include keying notes, if field keyed, threats noted, and observed ecological information such as herbivory, pollination, and insects. Some botanists prepare preprinted forms and carry them in a binder field notebook.

This information helps in identification and later study. It is also critical for the herbarium label that will accompany your collection. Labels will be needed if you intend to send your collection to specialist at a herbarium for verification, but even in an agency or personal herbarium they serve important functions. They summarize habitat and distributional information for the plant, which provides information about the plant and its ecology. The locational information on a label allows the site where the plant was collected to be relocated. Historic and, occasionally, extant populations of rare plants have been located by looking at the locational information on the labels of herbarium specimens.

An example of a label is shown in Figure 8.12. Labels are usually 3" x 5." Most biological supply companies sell pre-gummed labels and labels that can be fed through a computer printer.

Plants of Idaho

Carex nebrascensis Dewey
Beaked Sedge   **Cyperaceae**

Lemhi County, Agency Creek Road, approximately 0.5 mile above Chief Tendoy Monument. Collection is from a beaver dam complex along Agency Creek, T23N R17E S10. 5550'

Along shores and mucky areas of beaver ponds. Common to abundant, forming small to large patches. With Glyceria sp., Salix lutea, S. boothii. Substrate is organic silt and muck overlaying gravel and cobble. Soils are deep and organic, wet to flooded with up to 20cm standing water.

Collected by Caryl Elzinga, #4033, 23 May 1992.

FIGURE 8.12. Typical herbarium label.

e. Mounting collections

Mounted specimens are easier to handle and examine and are more likely to be used than plants residing in a newspaper. Identification, however, is usually easier with unmounted material, which is why most herbaria prefer to receive specimens unmounted. An unmounted specimen can be examined on both sides and is more easily maneuvered under a dissecting scope. Occasionally, small portions of the plant, such as a single flower, will be rehydrated to aid certain identification.

Standard paper size for mounting is 11 1/2" x 16 1/2." Mounting paper and herbarium paste are available from biological or herbarium supply houses. Herbarium stock is recommended because the special paper content will not deteriorate nor damage specimens over the long-term. Herbarium glue is designed to be unattractive to insects, which can do major damage to herbarium specimens. Normal household glue is an attractant, and also deteriorates over time. Plants can also be secured with narrow strips of cloth tape. Well-mounted specimens could last for a hundred years or more. Larger herbaria may have collections dating from the 1700s.

Loose material such as seeds or small flowers should be placed in an envelope attached to the herbarium sheet. Normal letter envelopes are not recommended, since the seams are not glued all the way to the edges and small seeds can escape from the corners. The paper will also deteriorate over time. Supply companies sell special envelopes that avoid these problems.

Literature Cited

Ahmed, J.; Bonham, C. D.; Laycock, W. A. 1983. Comparison of techniques used for adjusting biomass estimates by double sampling. Journal of Range Management 36:217-221.

Avery, T. E.; Berlin, G. L. 1992. Fundamentals of Remote Sensing and Airphoto Interpretation. New York, NY: MacMillan.

Avery, T. E.; Burkhart, H. E. 1994. Forest Measurements. 4th edition. New York, NY: John Wiley and Sons.

Awbrey, R. T. 1977. Locating random points in the field. Journal of Range Management 30: 157-158.

Bailey, A. W.; Poulton, C. E. 1968. Plant communities and environmental relationships in a portion of the Tillamook Burn, northwest Oregon. Ecology 49:1-14.

Bartz, K. L.; Kershner, J. L.; Ramsey, R. D.; Neale, C. M. U. 1993. Delineating riparian cover types using multispectral, airborne videography. Proceedings of the 14th biennial workshop on color aerial photography and videography for resource monitoring. Bethesda, MD: American Society for Photogrammetry and Remote Sensing: 58-67.

Becker, D. A.; Crockett, J. J. 1973. Evaluation of sampling techniques on tall grass prairie. Journal of Range Management 26 (1): 61-67.

Blyth, K. 1982. On robust distance-based density estimators. Biometrics 38: 127-135.

Bonham, C. D. 1989. Measurements for terrestrial vegetation. New York, NY: John Wiley and Sons.

Brady, W. W.; Mitchell, J. E.; Bonham, C. D.; Cook, J. W. 1995. Assessing the power of the point-line transect to monitor changes in basal plant cover. Journal of Range Management 48: 187-190.

Braun-Blanquet, J. 1965. Plant sociology: the study of plant communities. London: Hafner.

Brewer, L. and D. Berrier. 1984. Photographic techniques for monitoring resource change at backcountry sites. USDA Forest Service. Northeastern Forest Experiment Station. General Technical Report NE-86. 13p.

Brun, J. M.; Box, W. T. 1963. Comparison of line intercepts and random point frames for sampling desert shrub vegetation. Journal of Range Management 16: 21-25.

Buell, M. F.; Cantlon, J. E. 1950. A study of two communities of New Jersey Pine Barrens and a comparison of methods. Ecology 31: 567-586.

Burleson, W.H. 1975. A method of mounting plant specimens in the field. Journal of Range Management 28: 2410-241.

Catana, A. J. 1963. The wandering quarter method of estimating population density. Ecology 44: 349-360.

Clymo, R. S. 1980. Preliminary survey of the Peat-Bog Knowe Moss using various numerical methods. Vegetation 42: 129-148.

Cochran, W. G. 1977. Sampling techniques, 3rd ed. New York, NY: John Wiley & Sons.

Daubenmire, R. F. 1959. A canopy-coverage method. Northwest Science 33: 43-64.

Dethier, M. N.; Graham, E. S.; Cohen, S.; Tear, L. M. 1993. Visual versus random-point percent cover estimations: "objective" is not always better. Marine Ecology Progress Series 96: 93-100.

Diggle, P. J. 1975. Robust density estimation using distance methods. Biometrika 62: 39-48.

Dilworth, J. R. 1989. Log scaling and timber cruising. Corvallis, OR: Oregon State University.

Dilworth, J. R.; Bell, J. F. 1973. Variable probability sampling: Variable plot and 3P. Corvalis OR: Oregon State University.

Elzinga, C. 1997. Habitat Conservation Assessment and Conservation Strategy for *Primula alcalina*. Unpublished paper on file at U.S. Department of Interior, Bureau of Land Management, Salmon District, Salmon, ID.

Engeman, R. M.; Sugihara, R. T; Pank, L. F.; Dusenberry, W. E. 1994. A comparison of plotless density estimators using Monte Carlo simulation. Ecology 75 (6): 1769-1779.

Evans, R. A.; Love, R. M. 1957. The step-point method of sampling: a practical tool in range research. Journal of Range Management 10: 208-212.

Fisser, H. G.; VanDyne, G. M. 1966. Influence of number and spacing of points on accuracy and precision of basal cover estimates. Journal of Range Management 19: 205-211.

Floyd, D. A.; J. E. Anderson. 1982. A new point frame for estimating cover of vegetation. Vegetation 50: 185-186.

Floyd, D. A.; J. E. Anderson. 1987. A comparison of three methods for estimating plant cover. Journal of Ecology 75: 221-228.

Fosberg, F. R.; Satchet, M. 1965. Manual for tropical herbaria. Utrecht, Netherlands: International Bureau for Plant Taxonomy and Nomenclature.

Foster, M. S.; Harrold, C.; Hardin, D. D. 1991. Point versus photo quadrat estimates of the cover of sessile marine organisms. Journal of Experimental Marine Biology and Ecology 146: 193-203.

Goebel, C. J., DeBano, L. F.; Lloyd, R. D. 1958. A new method of determining forage cover and production on desert shrub vegetation. Journal of Range Management 11: 244-246.

Goodall, D. W. 1952. Some considerations in the use of point quadrats for the analysis of vegetation. Australian Journal of Scientific Research, Series B 5: 1-41.

Greig-Smith, P. 1983. Quantitative plant ecology. 3rd edition. Berkeley: University of California Press.

Hanley, T. A. 1978. A comparison of the line interception and quadrat estimation methods of determining shrub canopy coverage. Journal of Range Management 31: 60-62.

Hart, R. H.; Laycock, W. A. 1996. Repeat photography of range and forest lands in the western United States. Journal of Range Management 49: 60-67.

Hatton, T. J.; West, N. E.; Johnson, P. S. 1986. Relationship of the error associated with ocular estimation and actual total cover. Journal of Range Management 39: 91-92.

Heady, H. F.; Gibbens, R. P.; Powell, R. W. 1959. A comparison of the charting, line intercept, and line point methods of sampling shrub types of vegetation. Journal of Range Management 12: 180-188.

Hope-Simpson, J. F. 1940. On the errors in the ordinary use of subjective frequency estimations in grasslands. Journal of Ecology 28: 193-209.

Husch, B.; Miller, C. I.; Beers, T. W. 1982. Forest mensuration. New York, NY: John Wiley and Sons.

Jensen, J. R.; Burkhalter, S. G.; Althausen, J. D.; Narumalani, S.; Mackey, H. E. Jr. 1993. Integration of historical aerial photography and a geographic information system to evaluate the impact of human activities in a cypress-tupelo swamp. Proceedings of the 14th biennial workshop on color aerial photography and videography for resource monitoring. Bethesda, MD: American Society for Photogrammetry and Remote Sensing: 125-131.

Jensen, M. E.; Hann, W.; Keane, R. E.; Caratti, J.; Bourgeron, P. S. 1994. ECODATA--A multiresource database and analysis system for ecosystem description and analysis. In: Jensen, M. E.; Bourgeron, P. S., eds. Eastside forest ecosystem health assessment, volume II: Ecosystem management: principles and applications. General Technical Report GTR-PNW-318. Portland, OR: U.S. Department of Agriculture, Forest Service: 203-216.

Kennedy, K. A.; Addison, P. A. 1987. Some considerations for the use of visual estimates of plant cover in biomonitoring. Journal of Ecology 75: 151-157.

Knapp, P. A.; Warren, P. L.; Hutchinson, C. F. 1990. The use of large-scale aerial photography to inventory and monitor arid rangeland vegetation. Journal of Environmental Management 31: 29-38.

Krebs, C. J. 1989. Ecological methodology. New York, NY: Harper & Row.

Laferriere, J. E. 1987. A central location method for selecting random plots for vegetation surveys. Vegetation 71: 75-77.

Leonard, G. H.; Clark, R. P. 1993. Point quadrat versus video transects estimates of the cover of benthic red algae. Marine Ecology Progress Series 101: 203-208.

Leps, J.; Hadicova V. 1992. How reliable are our vegetation analyses? Journal of Vegetation Science 3: 119-124.

Levy, E. B.; Madden, E. A. 1933. The point method for pasture analysis. New Zealand Journal of Agriculture 46: 267-279.

Luque, S. S.; Lathrop, R. G.; Bognar, J. A. 1994. Temporal and spatial changes in an area of the new jersey pine barrens landscape. Landscape Ecology 9: 287-300.

Lyon, J. G.; McCarthy, J. (eds.). 1995. Wetland and environmental applications of GIS. Boca Raton, FL: Lewis Publishers.

Lyon, L. J. 1968. An evaluation of density sampling methods in a shrub community. Journal of Range Management 21: 16-20.

Malone, C.R. 1968. Determination of peak standing crop biomass of herbaceous shoots by the harvest method. American Midland Naturalist 79: 429-435.

McNeill, L.; Kelly, R. D.; Barnes, D. L. 1977. The use of quadrat and plotless methods in the analysis of the tree and shrub component of woodland vegetation. Proceedings of The Grassland Society of Southern Africa 12: 109-113.

Meese, R. J.; Tomich, P. A. 1992. Dots on the rocks: a comparison of percent cover estimation methods. Journal of Experimental Marine Biology and Ecology 165: 59-73.

Morrison, R. G.; Yarranton, G. A. 1970. An instrument for rapid and precise sampling of vegetation. Canadian Journal of Botany 48: 293-297.

Muir, P. S. and R. K. Moseley. 1994. Responses of *Primula alcalina*, a threatened species of alkaline seeps, to site and grazing. Natural Areas Journal 14: 269-279.

Nowling, S., and P. T. Tueller. 1993. A low-cost multispectral airborne video image system for vegetation monitoring on range and forest lands. Proceedings of the 14th biennial workshop on color aerial photography and videography for resource monitoring. Bethesda, MD: American Society for Photogrammetry and Remote Sensing: 1-8.

Pierce, W. R. and L. E. Eddleman. 1970. A field stereo-photographic technique for range vegetation analysis. Journal of Range Management 23: 218-220.

Poissonet, P. S.; Poissonet, J. A.; Bodron, M. P.; Long, G. A. 1973. A comparison of sampling methods in dense herbaceous pasture. Journal of Range Management 26: 65-67.

Ratliff, R. D. and S. E. Westfall. 1973. A simple stereophotographic technique for analyzing small plots. Journal of Range Management 26: 147-148.

Redd, T. H.; Neale, C. M. U.; Hardy, T. B. 1993. Use of airborne multispectral videography for the classification and delineation of riparian vegetation. Proceedings of the 14th biennial workshop on color aerial photography and videography for resource monitoring. Bethesda, MD: American Society for Photogrammetry and Remote Sensing: 202-211.

Rogers, G. F., H. E. Malde, and R. M. Turner. 1984. Bibliography of repeat photography for evaluating landscape change. Salt Lake City, UT: University of Utah Press.

Ruyle, G. B., ed. 1991. Some methods for monitoring rangelands and other natural area vegetation. Extension Report 9043. Tucson, AZ: University of Arizona.

Sample, V. E., ed. 1994. Remote sensing and GIS in ecosystem management. Washington, D.C.: Island Press.

Sandland, R. L.; Alexander, J. C.; Haydock, K. P. 1982. A statistical assessment of the dry-weight rank method of pasture sampling. Grass Forage Science 37: 263-272.

Schreuder, H. T.; Gregoire, T. G.; Wood, G. B. 1993. Sampling methods for multiresource forest inventory. New York, NY: John Wiley and Sons.

Schwegman, J. 1986. Two types of plots for monitoring individual herbaceous plants over time. Natural Areas Journal 6: 64-66.

Sharp, L. A., K. Sanders and N. Rimbey. 1990. Forty years of change in a shadscale stand in Idaho. Rangelands 12: 313-328.

Shivers, B. D.; Borders, B. E. 1996. Sampling techniques for forest resource inventory. New York, NY: John Wiley and Sons.

Stine, P. A.; Davis, F. W.; Csuti, B.; Scott, J. M. 1996. Comparative utility of vegetation maps of different resolutions for conservation planning. Biodiversity in managed landscapes. R. C. Szaro, and D. W. Johnson, eds. New York, NY: Oxford University Press: 210-220.

Sykes, J. M.; Horril, A. D.; Mountford, M. D. 1983. Use of visual cover assessments as quantitative estimators of some British woodland taxa. Journal of Ecology 71: 437-450.

Todd, J. E. 1982. Recording changes: a field guide to establishing and maintaining permanent camera points. R6-10-095-1982. Portland, OR: U.S. Department of Agriculture, Forest Service, Pacific Northwest Region.

Turner, R.M. 1990. Long-term vegetation change at a fully protected Sonoran desert site. Ecology 7:464-477.

Verbyla, D. L. 1995. Satellite remote sensing of natural resources. Boca Raton, FL: Lewis Publishers.

Walker, B. H. 1970. An evaluation of eight methods of botanical analysis on grasslands in Rhodesia. Journal of Applied Ecology 7: 403-416.

Warren-Wilson, J. 1963. Errors resulting from thickness of point quadrats. Australian Journal of Botany 11: 178-188.

Wells, K. F. 1971. Measuring vegetation changes on fixed quadrats by vertical ground stereophotography. Journal of Range Management 24: 233-236.

Whitman, W. C.; Siggeirsson, E. J. 1954. Comparison of line interception and point contact methods in the analysis of mixed grass range vegetation. Ecology 35: 431-435.

Whorff, J. S.; Griffing, L. 1992. A video recording and analysis system used to sample intertidal communities. Journal of Experimental Marine Biology and Ecology 160: 1-12.

Wilke, D. S.; Finn, J. T. 1996. Remote sensing imagery for natural resource monitoring. A guide for first time users. New York, NY: Columbia University Press.

Wimbush, D. J.; Barrow, M. D.; Costin, A. B. 1967. Color stereo-photography for the measurement of vegetation. Ecology 48:150-152.

Windas, J. L. 1986. Photo-quadrat and compass-mapping tools. Natural Areas Journal 6: 66-67.

Winkworth, R. E.; Goodall, D. W. 1962. A crosswire sighting tube for point quadrat analysis. Ecology 43: 342-343.

Winkworth, R. E.; Perry, R. W.; Rosetti, C. O. 1962. A comparison of methods of estimating plant cover in an arid grassland community. Journal of Range Management 15: 194-196.

Zar, J. H. 1996. Biostatistical analysis, 3rd3rd edition. Upper Saddle River, NUJ: Presntice Hall.

CHAPTER 9
Data Collection
and
Data Management

Hilaria jamesii
Galleta
by Jennifer Shoemaker

Chapter 9. Data Collection and Data Management

A. Introduction

This chapter covers the different methods of recording and managing actual field monitoring data. Some methods lead to the orderly and efficient processing of information, smoothing the way for data summary, data analysis, and report completion. Other ways lead to tortuous routes that cause frustration and headaches to the data managers stuck with processing messy and confusing data sets. Poorly gathered and poorly managed monitoring data usually stem from an unawareness that an enormous amount of time (days, weeks, or even months) can be saved by following some of the guidelines presented in this chapter.

Successful data collection and data management needs to start with the planning of a monitoring study and continue for as long as data sets are archived in computers or hard files. Good data collection methods lead to efficiency in the field and in the office. Detailed documentation of field methods and descriptions of codès or abbreviations helps to ensure the integrity of data from the field to the final interpretation of monitoring results.

This chapter is divided into the following two sections: (1) recording data in the field, and (2) the entry and storage of data in the office. This chapter is intended to be read by all persons involved in any phase of a monitoring project.

B. Recording Data in the Field

Three options exist for gathering ecological monitoring data in the field: (1) tape recorders, (2) portable computers or data loggers, or (3) field data forms or field notebooks. The use of field data forms is covered in more detail than other methods since field data forms are still the most common way that field data are gathered.

1. Tape recorders

a. Advantages

Portable tape recorders can reduce the amount of time spent in the field, especially when a person is working alone and needs to record a large amount of data. Voice-activated recorders reduce the amount of button-pushing and shorten transcription time by eliminating the quiet time between data points. Detailed site descriptions and other field observations can be verbally recorded in less time than it takes to write them in a field notebook.

b. Disadvantages

Prior to any data summary steps, the audio recording will need to be transcribed, either onto some kind of data form or directly into a computer. Most portable tape recorders will cost at least $75 and require either many batteries or a battery charger with rechargeable batteries. Like other electronic devices, tape recorders will occasionally fail and data could be lost if the tape is damaged. It is also difficult to scan the recorded data to look for any patterns or problems or to verify which sample areas have been sampled or which types of data have been gathered. Few tape recorders are designed to operate in poor weather conditions.

c. Tips to improve data collection

We strongly advise that transcription of the tapes occur within hours of recording or, at most, within a few days. Carry a blank field data sheet or some check list to serve as a guide to consistently gather all categories of information in the same sequence. Always carry plenty of spare batteries. Periodically reverse the tape and play back a section to be sure that it is recording properly. Use fresh tapes.

2. Portable computers or data loggers

This category includes any device that allows field data to be recorded in an electronic form. This includes laptop, notebook, and palmtop computers, and hand-held data loggers.

a. Advantages

Recording data directly into a portable computer or electronic data logger can be the most efficient means of collecting field data. This method eliminates the time-intensive data entry and data-proofing steps that go along with data recorded on field data forms. Field data can be entered in a pre-designed format that will facilitate data summary and analysis steps. Some portable computers support the use of DOS- or Windows-based software (e.g., spreadsheet programs), making data exchanges with desktop computers easy. Hewlett Packard palmtop computers are relatively inexpensive and can work well as field data recorders. These palmtop computers come with a built-in Lotus spreadsheet program and cost around $600. Husky Hunter and Corvallis MicroTech make data loggers that are DOS-compatible, show several lines of data, and are extremely durable under field conditions (water proof, dust proof, etc.). They are also quite expensive (about $3,000 - $4,000). Some data recorders allow the entry of bar codes so that a wand can be passed over a sheet with bar codes to input species identification or other labels.

b. Disadvantages

Most portable computers will cost at least $500 and require either many batteries or a battery charger with rechargeable batteries. Some electronic data loggers use non-standardized computer programs that can make the transfer of data to a DOS-based computer difficult. The viewing screen on most portable computers is quite small, which can make it difficult to scroll around a large data entry template. Most portable computers are heavy and awkward to use in the field although some of the palmtop computers are quite light. Few portable computers are designed to operate in poor weather conditions. Data can be lost due to hardware or software problems or if a computer's batteries run dead.

c. Tips to improve data collection

Palmtop-size computers can be used in the rain when placed in a gallon zip-locked plastic bag. Data entry and screen viewing works fine through the plastic. Carry plenty of spare bags and periodically inspect for leaks. Federal agency staff should check on the availability of data recorders from other departments. Some of the forestry and field survey personnel have heavy duty, weather proof data recorders that you might be able to borrow, though these will often require some difficult programming steps to adapt them to your specific uses. Be sure to make some kind of back-up of the data at least every day by transferring the field data to another computer, to a floppy diskette, to a flash-memory card, or by printing a hard copy of the data.

3. Field data forms

a. Advantages

Field data forms or field notebooks are inexpensive and lightweight. They can be made of waterproof paper.

b. Disadvantages

If data need to be summarized or analyzed with a computer then data will need to be transcribed from the field data forms. The data-entry and data-proofing steps can consume more time than the field data collection. Wet field data sheets lead to writing smears or streaks and pages may become stuck together.

c. Tips to improve data collection

Print field data forms on waterproof paper. Several paper suppliers sell waterproof paper that can be used in standard printers and photocopiers. Field data forms should be designed to promote efficiency in field collection and computer data entry. The time required to complete data-entry and data-proofing steps is profoundly influenced by the design of the field data form. Transcribing data from a poorly designed, sloppily written field data form can take more than 10 times longer than transcribing data from a well-designed, clearly legible data form.

Each set of data should have a cover sheet that stays with the field data at all times
The cover sheet should provide information on what, why, where, who, how, and when types of information. Detailed information should be provided on the location of study plots, the species or community being studied, the personnel involved, the types of management treatments that have occurred or are being planned, a description of any codes that are used, and a thorough description of the field methodology. See Figure 9.1 for a list of the types of information that should be included on the cover sheet, and Appendix 15 for a blank field monitoring cover sheet. In addition, each field data form should have a complete "header" section that links the form to the project described on the cover sheet. The header should be completely filled out on every page. The header should include at least the following items:

1. Date.
2. Location (general area and specific sampling location).
3. Title/project description.
4. Species or community name.
5. Treatment category (if applicable).
6. Observer (person(s) doing the sampling).
7. Transect or macroplot number (if this information applies to entire data sheet).
8. Page number _____ of _____ total pages.
9. Room for additional comments.

Field monitoring cover sheet

1. **Include header from the field data form**. This header should include the following categories of information.
   a. Title or project description name
   b. Location
   c. Species or community name
   d. Type of study (density, cover, frequency, etc.)
   e. Personnel
   f. Date(s)
   g. Treatment (if applicable)
   h. Macroplot or transect, or other location identifier if this information applies to the entire data sheet.

2. **Management objective**. (see Chapter 4).

3. **Sampling objective**. (see Chapter 6).

4. **Location and layout of the study area**. Sketch location, including access. Denote key area, macroplot, or transect locations with macroplot numbers, names, and treatments, as applicable, and the approximate bounds of the population being studied. If sampling units are placed along transect lines, show how they were placed. Provide approximate scale.

5. **Detailed description of data collection methods**. This should include sufficient detail that someone unfamiliar with the project can understand how the data were gathered. Consider the following issues:
   a. What are the bounds of the population study area?
   b. If you are sampling within macroplots, what is the size and shape of the macroplots and how were they positioned?
   c. What is the sampling unit (e.g., quadrats, lines, individual plants)?
   d. What is the size and shape of the individual sampling units (quadrats, lines)?
   e. How are sampling units positioned in the population of interest?
   f. Are sampling unit positions permanent or temporary? If permanent, describe markers and methods used to ensure that positions will be accurately relocated.
   g. Describe any boundary rules for plant counts or measurements that occur along the edge of sampling units.
   h. For density measurements—describe the counting unit (e.g., genet, ramet, stem, flowering stem) and any rules that are used to discriminate among adjacent counting units.
   i. For cover measurements—define whether basal or canopy cover is measured and define gap rules. If ocular estimates of cover are made in cover classes, define those classes. For point-intercept cover measurements, describe the point diameter and type of tool being used.
   j. Include a full description of any codes used on the field data sheets, including species acronyms.

FIGURE 9.1. Categories of information to include in a cover sheet that should accompany all field data forms.

Pre-print as much information as possible

Time can be saved in both the field collection and data entry phases by pre-printing as much reference information as possible on the field data form. This eliminates the need for a lot of repetitive writing and cuts down on mistakes. When plant communities are being sampled and a large number of species codes are being used, include the full genus and species name on the field data sheet in addition to the code. The code will be used during data entry, and having the full name listed with the code eliminates serious data summary problems such as two species being inappropriately grouped together or the data for a single species being split between two or more categories. If a list of species known to occur in a particular plant community is available, or if only a subset of the species are being tracked, then pre-print the species codes, and the genus and species names, on the field data form. See the nested frequency data sheet example in Appendix 15 for an example of this type of field data form. Pre-printing species codes and names saves a lot of writing time in the field, minimizes data transcription errors, and greatly speeds up data entry because the sequence of species stays the same from page to page. The sequence of species can either be alphabetical, by taxonomic or growth form groupings (e.g., all grasses together, all forbs together, etc.), by relative abundance, or through some combination of these methods (e.g., list the four most common species first with the remainder of the list sorted alphabetically).

Species codes frequently consist of four letters, the first two letters of the genus and the first two letters of the species (e.g., LIOC = *Lilium occidentale*). To avoid using duplicate codes for different species that share the same four-letter acronym consult a book such as "The National List of Plant Names" (USDA, SCS 1982) or consult the PLANTS National Database, maintained by the National Resources Conservation Service (the database can be accessed and downloaded via the Internet; the address is <http://plants.usda.gov/plants/>. If a plant or animal is only identified to genus, and some master list of codes is not available, avoid the use of 'SP' as an abbreviation for 'SPECIES' (e.g., *Bromus* species = BRSP). Instead, adopt some convention such as 'ZZ' or 'Z1' (e.g., BRZ1) to use whenever a plant is only identified to genus. This will reduce the number of duplicate codes (many species names actually start with the letters 'sp'), and it also more clearly indicates when species identity is unknown. Six-digit species codes (composed of the first three letters of the genus and species) reduce the number of duplicate codes.

It is important to define any numerical or character codes that are used on the field data sheet. These codes should always be defined in the field data cover sheet and, when possible, they should appear on the field data sheet itself. For example, if plant counts are being made in randomly positioned quadrats and the particular habitat type (e.g., mound, intermound, pool) that each quadrat lands in is being recorded, use a numerical code to define the habitat type rather than writing the full habitat type at each quadrat location. Placing the code descriptions near the top of the data sheet ensures that habitat type information will be recorded and summarized properly.

Recording unanticipated information

Not all data form needs can be anticipated. Unexpected observations can lead to the need to incorporate additional information onto a field data form. For example, a subset of plants being counted in quadrats may have some peculiar attribute such as yellowish, dried leaves or evidence of flower head herbivory. There is sometimes a tendency to incorporate many detailed comments onto field data forms, taking advantage of any available blank space. Sometimes the same characteristic is described in different ways (e.g., "some flower heads eaten," "inflorescence damaged," "three seed heads with signs of herbivory"). This could create confusion during the data-entry process. Which comments are important and should be entered into the computer with the regular monitoring data? Which comments are insignificant and should be ignored? Which different comments mean the same thing? How should the additional data be used during the data summary process?

The best way to incorporate additional information is to consider how the inclusion of this type of information will impact data summary and analysis steps. Will it be useful to have a tally of all plants showing some characteristic (such as evidence of flower head herbivory) separate from plants that do not show the characteristic? If so, create a "Notes" column along the margin of a field data form and create a numerical code to assign to any observation exhibiting the characteristic. The code should be described at the top of the field data sheet (e.g., 1 = flower head herbivory noted). These additional data are then easily incorporated into the data set during data entry, the observations can be sorted by this additional field, and separate summary statistics can be generated very easily.

Design forms for numeric data entry

Try to design field data forms so that nearly all data entry will be numeric. Data entry is most efficient when data can be entered from the 10-key numeric keypad portion of a computer keyboard. Using a combination of character and numeric data slows down data entry.

Maintain legibility

Take adequate time to make sure that all hand writing is clearly legible. You should not assume that you will be the only one who will be reading the completed field data sheets. Poor hand writing can significantly slow down the data entry process and can introduce errors into the data sets.

Examples of field data forms

Examples of well-designed field data forms are included in Appendix 15. Data forms are included for gathering the following types of vegetation monitoring data: (1) density, (2) frequency, (3) ocular estimates of cover, (4) point-intercept cover, and (5) line-intercept cover. Two versions of each data sheet are shown; one blank (labeled A), and one with some sample data entered (labeled B). We also include data forms for documenting studies and photopoints.

C. Entry and Storage of Data in the Office

If the quantity of data gathered is small, sometimes the data can be efficiently summarized straight off the field data form using a hand calculator. Calculations should be repeated, to ensure that no mistakes were made in entering and summarizing the data. Often, however, monitoring data will need to be input into a computer system for data summary and analysis. If the data were gathered on a portable computer, then the data are ready to go. If, however, the data were gathered on field data forms or with a tape recorder, then data entry is the next step. This topic is divided into the following five sections: (1) selecting a computer software program, (2) storing data files—filenames and directories, (3) adequately documenting data files, (4) proofing entered data sets, and (5) making backups of entered data.

1. Selecting a computer software program

There are primarily four categories of software applications where monitoring data can be entered: (a) word processors, (b) relational databases, (c) spreadsheets, and (d) statistical software programs.

a. Word processors

Word processors used to be the worst place to enter or store ecological monitoring data. Most word processors did not distinguish data files from regular text files containing memos or reports. Data summarization procedures were not available or they were extremely limited in most word processors. In recent years, however, data table formats have been added to many word processors and some of these support limited spreadsheet type operations. Check carefully to make sure that data can be easily exported to other software applications prior to entering field data into a word processor.

b. Relational databases

Programs such as Dbase, Paradox, Oracle, and Microsoft Access are examples of relational databases. Relational databases are designed to organize and manage large amounts of information. Custom data entry screens can be created where the user enters data into blank highlighted fields. Most relational databases include some basic data summary procedures (e.g., calculating totals or averages). Entering and storing monitoring data in a relational database may be a logical alternative if data from individual observations (i.e., height of an individual plant or the number of plants in a certain permanent quadrat) are frequently referenced or reported. Data gathered as part of a large-scale monitoring network (e.g., ECODATA, EMAP) should be stored in a relational database to facilitate data management and data processing (Stafford 1993). Relational databases usually have sophisticated tabular reporting features but limited graphical reporting features. Most relational databases can import and export data easily with other software programs.

c. Spreadsheets

Programs such as Lotus 1-2-3, Quattro Pro, and Excel are examples of spreadsheet programs. The data entry screen in a spreadsheet is a rectangular matrix of labeled columns and rows. There are many time-saving data entry procedures built into spreadsheet programs. For example, if data were gathered from plots numbered 1 to 100, a few key strokes can generate a list of plot numbers from 1 to 100 so that 100 individual plot numbers do not have to be entered. Large sections of data can easily be copied or moved within a spreadsheet. Most spreadsheets include at least some basic data summary procedures and some include more advanced summary and analysis routines. Descriptive reference information (species, location, dates, treatments, definition of codes, etc.) can be placed in the spreadsheet above the actual data matrix. Spreadsheet programs usually offer sophisticated tabular and graphical reporting features. They also can import and export data files in many different formats. Data entry onto spreadsheets may be the most efficient means of transferring data from field data forms into a computer file. Even if data are going to be stored in a relational database, it may be more efficient to enter the data in a spreadsheet and then transfer the data to the relational database.

d. Statistical programs

Programs such as SYSTAT, SAS, SPSS, and StatGraphics are examples of statistical software programs (Chapter 11) and they offer powerful data summary, data analysis, and graphing procedures. They all have some kind of data entry mode, usually a screen resembling a spreadsheet. However, the data entry screens in statistical packages may not include many of the time-saving routines found in real spreadsheets. The spreadsheet-like format does not usually allow you to add the type of descriptive reference information that you can enter onto spreadsheets. Compare the features in your statistical program with your spreadsheet program before entering a lot of data directly into the statistical program. Most statistical programs readily import data from spreadsheets and relational databases.

2. Storing data files—filenames and subdirectories

The naming and storing of files doesn't seem like a problem when there are only a few data files to input. At first, a data manager may decide to place all data files in a single computer directory called something like "DATA." He or she may name individual files with whatever seems like a logical name at the time the file is created, without adopting any standard conventions for naming files. Confusion starts to increase as more and more data files are created. Soon it becomes difficult to find a particular file and numerous files may need to be opened until the right one is located. Some files may be accidentally deleted because the data manager thought another file contained data superseding the deleted file.

Creating an efficient, standardized system of naming and storing computer data files early in the development of a monitoring program will save a data manager many hours, days, weeks, or months of frustrating data management. Figure 9.2 shows an example of a protocol for naming data files for DOS-based programs that limit filenames to a total of 11 characters (8 before the period and 3 after the period). Windows 95 and Unix programs do not share this 8 + 3 filename limitation.

One efficient method of storing monitoring data is to create separate subdirectories for different sites. This could be done either by establishing a DATA directory with different subdirectories for each site (e.g., all data files from the Middle Fork of the John Day Preserve are stored in C:\DATA\MFJD*.*) or by creating a DATA subdirectory under a site directory (e.g., C:\MFJD\DATA*.*).

Avoid creating many separate files for related monitoring data. Keep related information from different sampling areas or from the same sampling area over different years in the same file. The data will need to be brought together for data summary and analysis purposes, and

---

Sample protocol for naming data files

DOS file names use eight digits, a period, and a three digit extension.

Column 1 (Type of data):
  F = Frequency (incl. nested frequency)   example
  D = Density (#/unit area)   DCALE891.WK1
  P = Population demography | | | | \
  C = Cover (other than line intercept)   1 2-5 6 7-8 10-12
  L = Line intercept
  R = Reproductive information
  T = Tree data (dbh, height)
  S = Stem counts
  B = Basal area

Column 2-5 (Acronym of element):
  Acronyms are 4-letter codes that typically list the first two letters of the genus and the first two letters of the species. Use a standardized list of codes (see text).

Column 6 (last digit of first year of data collection):
  Example: "1989" would be coded "9", "1983" would be "3"
  Note: Only one column is used because of space limitations.

Column 7,8 (last year of data collection):
  Example: "1991" would be coded "91"

Column 9 (period)

Column 10-12 (extension code for software)
WK1 = Lotus 123

Example using the protocol shown above

DCALE891.WK1 "means"
     D = Density data
  CALE = Castilleja levisecta (species acronym)
     8 = 1988 (first year of data collection)
    91 = 1991 (most recent year of data collection
  WK1 = Lotus 123 file

FIGURE 9.2. Example of a file-naming protocol when working with computer files where the filename is limited to 8 primary digits plus a 3-digit extension.

having the data in a single file all along can reduce data management headaches. Figure 9.3 shows a sample format for recording data from multiple macroplots and multiple years in a single file.

3. Adequately documenting data files

Each data file should include reference information about the data in that file (Stafford 1993). This information should detail the how, when, what, where, and who information included in the field data cover sheet and in header sections of the field data forms. This kind of information should be included in a file header that appears in the computer file above the rows of actual monitoring data. Any codes contained in the data set should be listed and described in the file header. A detailed description of the methods used to gather the data should be included in the file header or a reference to another source for this information should be provided. See Figure 9.3 for an example of a completed data file header.

4. Proofing data sets

If data were entered into a computer file from field data forms, then the data need to be checked for any keystroke errors introduced during data entry. Having someone read off the data from the original data form while another person follows along either at the computer file or on a printed hard copy is one method that works well. Any corrections are noted on the computer printout.

FILENAME: C:\DATA\AGAT\FLOCO891.WK1
PRESERVE NAME: Agate Desert
ELEMENT NAME: Rogue Valley Mounded Prairie
DATE OF OBSERVATION: 25 May 1989, 14-16 May 1990, 15-21 May 1991
SITE DESCRIPTION: Research Plots LOCO Burn
MACROPLOT NUMBERS: 2, 3, 5-8
TYPE OF MEASUREMENT: Nested Frequency
NUMBER OF QUADRATS: 50
QUAD SIZE/CODED VALUE: $1 = 0.01 \ m^2$, $2 = 0.1 \ m^2$.
DATA CONTACT: Darren Borgias
COMMENTS:
    HABIT = Habitat codes; 1 = mound, 2 = flank, 3 = intermound, 4 = pool
    THAT = thatch measured in cm
    Grouped species codes:
        TRNA = Trifolium native (T. variegatum)
        TREX = Trifolium exotic (T. subterraneum, T. arvense, T. dubium)
        TRSP = Trifolium species (unidentified)
        UNK1 = unknown composite
    Thatch information can be found to the right of the spreadsheet

For a full list of species codes and a detailed description of field methodology see: Borgias, D. 1993. Fire effects on the Rogue Valley Mounded Prairie on the Agate Desert, Jackson CO.

YEAR	MPLOT	QUAD	HABIT	POSC	TACA	BRSP	VUSP	POBU	HOGE	DEDA	AICA
89	2	1	2	0	1	2	2	0	1	0	0
89	2	2	2	0	1	1	0	0	1	0	0
89	2	3	1	0	1	1	0	0	0	0	0
89	2	4	3	0	1	1	0	0	2	0	0
89	2	5	2	0	1	1	1	0	0	0	0
89	2	6	2	0	1	1	0	0	1	0	0
89	2	7	2	0	1	1	0	0	0	0	0
89	2	8	2	0	1	1	1	0	0	0	0
89	2	9	1	0	1	2	0	0	0	0	0
89	2	10	1	0	1	1	1	1	0	0	0

Continued for the rest of 90 and 91 | Continued for the rest of the macroplots | Continued for the rest of the quadrats

FIGURE 9.3. Example of a spreadsheet file showing the reference information provided in the file header.

Using a dual entry procedure is an alternative quality control option for catching keystroke errors (Stafford 1993). Data are entered twice by different key stroke operators. Any mismatches between the two entered copies are noted and the original field data sheets are checked to determine which copy is in error.

5. Making backups of entered data

It is essential that backup copies are made of all computer data files. Hardware, software, and user failures occur on an unpredictable schedule and large amounts of grief can be saved if a regular backup schedule is maintained. Daily backups can easily be made to a floppy disk. Weekly backups of all files made to a tape drive can make data recovery much easier following a hard disk failure. You should "leap frog" backup tapes so that you are not copying to the only backup copy of the data. It is a good idea to keep one copy of the backup at another location (in case that a catastrophic fire consumes the office copy).

Literature Cited

Stafford, S. G. 1993. Data, data everywhere but not a byte to read: managing monitoring information. Environmental Monitoring and Assessment 26: 125-141.

USDA, SCS. 1982. National List of Plant Names. Volume 2. Synonymy. Washington DC: SCS-TP-159.

CHAPTER 10
Communication and Monitoring Plans

Senecio layneae
Layne's butterweed
by Mary Ann Showers

CHAPTER 10. Communication and Monitoring Plans

Communication doesn't start when the monitoring results have been analyzed. Beginning with the planning stage, those who will be making decisions based on the monitoring and those who may be affected by those decisions must be included in the design of the monitoring project. You can increase the likelihood of seeing needed management actions implemented by involving all interested parties in developing the management objective and designing the monitoring, and reaching agreement that all parties will abide by the results (Hirst 1983; Johnson 1993). Objectives, written as Management Objectives-Management Response pairs (Chapter 4), should clearly identify the management changes that will be implemented based on monitoring results (Gray and Jensen 1993). This point cannot be stressed enough, especially when potential decisions may adversely affect other parties or interests. If you fail to include all who should be involved in the initial stages of objective setting and monitoring design, you can expect problems implementing new management once monitoring is completed.

A. Participants

Several classes of participants needed in the development of a monitoring project are described in Box 1. The number of people and groups to involve in a monitoring project depends on the potential impacts of the management changes that may occur based on monitoring results. Developing objectives for plant populations in areas that are not affected by commodity extraction or recreational use may require little interaction with interest groups or other agency specialists. Large populations, or populations in high use/high visibility areas, may require extensive communication efforts before monitoring is initiated.

Establishing communication and considering alternative points of view can be time-consuming and difficult. An apparently easier route is collecting "really good data" to prove your point and get management changed. In practice, however, monitoring that is specialist-driven usually fails to result in a management change for three reasons. The most common is that the specialist spearheading the monitoring leaves, and the monitoring project is suspended because it lacked the knowledge and support of managers. A second reason is that other priorities take precedence over the monitoring project. In order for monitoring to be completed, managers must support the time and resources it requires. Third, a lack of consensus on objectives and methodology almost ensures that monitoring data will not be used to make a decision. You need to involve people from the beginning to ensure a cooperative effort and the application of monitoring results to the decision-making process (Hirst 1983).

Communication about monitoring projects associated with non-controversial management actions can safely be limited to decision-makers and internal resource specialists. For example, often you will know too little about populations and their interactions with management activities to develop Management Objective-Management Response pairs that identify a specific management response. Many management responses in the examples in Appendix 3 specify a second stage of more intensive monitoring and perhaps research if the population is declining or failing to increase. Such two-stage monitoring requires only the involvement of the decision-maker and resource specialists within the administrative unit in the first stage because implementing increased monitoring or research is rarely controversial.

You may, however, enlist involvement and/or review by a broader spectrum of participants even in non-controversial projects. Review by user groups during the development of objectives will inject fresh perspectives. Review during the design phase by academic specialists, statisticians, experienced professional botanists, and peers may help you avoid potential technical problems.

B. Monitoring Plans

1. Importance

Communication with these participants is facilitated by a monitoring plan that explains the rationale for the monitoring project, documents objectives and the management response, and describes the monitoring methodology in enough detail to direct continued implementation. Monitoring plans serve five important functions:

- A plan provides a full description of the ecological model, the objectives, and the proposed methodology.

- Draft monitoring plans provide a means to solicit input from many participants.

- A final monitoring plan consolidates all information into a single document that can be easily accessed and referenced.

- A final monitoring plan documents the location and techniques of the monitoring in sufficient detail that a successor can continue the monitoring.

- A final monitoring plan documents the agency's commitment to implementing a monitoring project and the management that will occur based on monitoring results. A monitoring plan can also be signed by all participants to demonstrate their support for the project and acceptance of the proposed management changes that may result.

2. Elements of a monitoring plan

Monitoring plans must be complete, providing all the information needed to judge the quality of your proposed monitoring and to continue it in your absence. Box 2 summarizes the elements to include in an extensive monitoring plan for a complex project. Less complex projects may require less extensive explanations and fewer elements. A short (1-2 page) nontechnical summary at the beginning of the plan will be useful to decision-makers, non-specialists, and user groups.

3. When to write a monitoring plan

Do all monitoring projects require a monitoring plan? Does a qualitative monitoring project that simply involves taking a picture of the population each year require a full-scale document such as the one summarized in Box 2? Some form of documentation of the management objective, sampling objective (if sampling), management response, location, and methodology is necessary for all monitoring projects, no matter how small or simple. (The field monitoring cover sheet in Appendix 15 requires many of these elements, and may be adequate for some situations if an introduction that describes the objectives is included.)

BOX 1: Participants in a Monitoring Project

Decision-makers (managers, or management teams). This is the most important audience. They will decide the amount of resources to devote to the monitoring project and, once monitoring is completed, decide whether management should change or continue. Each manager's "comfort level" varies for making decisions based on monitoring data. Some managers feel confident making decisions based on photographs and their specialist's judgement. Others require much more information.

Agency specialists (in-house). Other resource specialists may have information critical to the design of the monitoring (e.g., the area containing the population is likely to be rested from grazing for the next three years; the timber stand is set aside from cutting because it is in a protected watershed). These other specialists also tend to be advocates for the resource they manage and may potentially disagree with the management changes resulting from monitoring. Including these specialists in the design creates ownership in the monitoring and reduces the potential for in-house disagreements later.

Regulatory decision-makers (U.S. Fish and Wildlife Service, state agencies). Participation by these agencies is required for species listed under the Endangered Species Act or state laws and may be helpful for other species of concern.

Non-regulatory agencies. State agencies that maintain statewide conservation databases, such as the Heritage Program or conservation programs, often have information about the same species on private lands, on other Federal lands, or on lands in other States. Many of these database agencies also maintain a monitoring database; participation in it can reduce redundancy in monitoring efforts. Local Natural Resource Conservation Service (formerly Soil Conservation Service) personnel and County Extension Agents may function as advocates for agricultural interests. Their participation and support of the monitoring project increase the credibility of the monitoring data with traditional Federal land users such as grazing permittees.

Traditional Federal land users. These are primarily commodity producers such as miners, loggers and timber companies, and livestock operators. If the monitoring potentially will affect these interests, you should include them throughout the process. Not only does their involvement from the beginning diffuse much of their disagreement when assessing results, it will also make the monitoring much better. Because their economic interests are potentially at stake, they will be interested more in false-change errors (e.g., concluding that a decline took place when it really did not), whereas you may be more concerned with missed-change errors (e.g., failing to detect undesirable changes that in fact did occur). The explicit balancing of the two errors is important. In addition, individuals involved in commodity production on Federal lands often know facts about a population area or an activity that you do not. A rancher, for example, may know that cows have not used an area for the last 10 fall seasons because of a non-functioning water source. A logger may know that his grandfather cut a patch of timber using horses in the 1930s. These bits of information may improve your ecological model.

Non-traditional Federal land users. Newer users of the Federal lands such as off-road recreationists, hikers, hot-spring visitors, and others whose use of the Federal land may be affected by changes in management resulting from monitoring should be included.

Environmental groups, Native plant societies. You should include groups that have an interest in native flora and biodiversity, especially if local representatives are available. Native plant societies not only have a special interest in the preservation of the diversity of native vegetation within a State, but may also have specialized skills or volunteer labor that will improve the quality of monitoring.

Professional and academic botanists. These people may have much to contribute to the development of ecological models, objectives, and monitoring designs. Their contribution to and review of the monitoring strategy will improve the quality and increase the credibility of the monitoring effort.

The flow chart in Chapter 2 suggests writing the monitoring plan before the pilot study. There is a valid concern, however, that if the pilot study demonstrates that the monitoring approach needs significant revisions, the monitoring plan will need to be rewritten. If the primary audience is in-house (other specialists, your successor), draft the plan as an informal communication tool, and finalize it after the methodology proves effective. If, however, the primary purpose of the monitoring plan is to communicate with outside groups and interests,

---

BOX 2: Elements of a monitoring plan

I. Introduction (general).

    Species, need for study, management conflicts.

II. Description of ecological model.

    Life history, phenology, reproductive biology, causes of distribution, habitat characteristics, management conflicts or needs, and effects of other resource uses on the species (e.g., herbivory of flower heads by cattle). The model should describe known biology (based on natural history observations) and conjectural relationships and functions. Sources of information and relationships that are hypothesized should be identified. The purpose of this section is to help identify the sensitive attribute to measure and to describe the relationships between species biology and management activities. This section is the biological basis for the development of objectives.

III. Management objective(s).

    Includes rationale for the choice of attribute to measure and the amount of change or target population size.

IV. Monitoring design.

    A. Sampling objective.

        Includes rationale for choice of precision and power levels (if sampling).

    B. Sampling design.

        Describe methods clearly. What size is the sampling unit? How are sampling units placed in the field? How many sampling units?

    C. Field measurements.

        What is the unit counted (for density)? How are irregular outlines and small gaps of vegetation treated (for line-intercepts)? How are plots monumented (if permanent)? Include all the information needed for someone else to implement or continue the monitoring in your absence.

    D. Timing of monitoring.

        What time of year, both calendar and phenologically? How often?

    E. Monitoring location.

        Include clear directions, maps and aerial photographs describing the study location, and the location of individual sampling units (if permanent).

    F. Intended data analysis approach.

V. Data sheet example.

VI. Responsible party.

VII. Funding.

VIII. Management implications of potential results.

---

and to gather peer and expert review, complete the plan before the pilot study. Portions of the plan such as the introduction and description of the ecological model will remain useful even if the monitoring project changes significantly.

Clearly, a significant investment of resources is required to complete all the elements of a monitoring plan, and most botanists prefer field work to writing plans. The temptation is great to skip this stage and get on with "more important" work, like counting plants in plots. Resist the temptation. A monitoring plan is worth the time commitment and is critical to successful long-term implementation of monitoring.

Literature Cited

Gray, J. S.; Jensen, K. 1993. Feedback monitoring: a new way of protecting the environment. Trends in Ecology and Evolution 8: 267-268.

Hirst, S. M. 1983. Ecological and institutional bases for long-term monitoring of fish and wildlife populations. In: Bell, J. F.; Atterbury, T., eds. Renewable resource inventories for monitoring changes and trends: Proceedings of an international conference; 1983 August 15; Corvallis, OR. Corvallis, OR: Oregon State University, College of Forestry: 175-178.

Johnson, R. 1993. What does it all mean? Environmental Management and Assessment 26: 307-312.

CHAPTER 11
Statistical Analysis

Achillea millefolium
Common yarrow
by Jennifer Shoemaker

CHAPTER 11. Statistical Analysis

With two exceptions, quantitative data collected through monitoring must be subjected to some type of statistical analysis. The two exceptions involve the following two types of data: (1) data gathered from a complete census, and (2) data gathered by sampling techniques that do not incorporate some type of random selection process. A census provides you with complete information about the population. The means, totals, or proportions resulting from a complete census are the actual population values (assuming no measurement error, such as errors in counting or in identifying plants). If there is no sampling error, no statistical analysis is necessary. Any changes in these population values between years are real. All that remains is to determine whether the changes have any biological significance.

At the opposite extreme from a census are data gathered without using some type of random sampling procedure (as discussed in Chapter 7). The fact that statistics cannot be applied to nonrandom sampling procedures makes proper analysis and interpretation of the data virtually impossible; this should reinforce the need to use a random sampling procedure in designing and implementing monitoring.

Statistics are extremely important to sample-based monitoring. They enable us to make management decisions even when we have access to only part of the information. For example, you might like to know the true number of plants in a given area. Because the area is large, however, and the plants far too numerous to count, the best you can do is take a random sample of quadrats within this area and estimate the total number of plants from this sample. The use of statistics enables you to derive an unbiased estimate of this total and, more important, assess how good this estimate is.

No doubt, you will use calculators with statistical functions or computer software programs to analyze your data. For that reason, this chapter emphasizes principles and concepts and contains a minimum of mathematical formulas. Formulas for many of the statistics and tests discussed below can be found in Appendix 8.

A. Using Graphs to Explore the Nature of Your Data

Several types of graphs can be used to examine your data prior to analysis: normal probability plots, density plots, box plots, and combinations of these. These are particularly important in the initial stages of designing your study. Graphs of pilot study data can, for example, help show you whether you are using the correct quadrat size, or whether your data meet the assumptions of parametric statistics. (Parametric statistics are those statistics used to estimate population parameters such as means and totals; we discuss the assumptions you must make when using them in Section G, below).

Graphing your monitoring data is likely to reveal patterns in your data that will not be apparent if all you do is calculate standard summary statistics like the mean and standard deviation. Figure 11.1 shows four samples, each of which has a mean of 100 and a standard deviation of 10. Without graphing the individual data points, we would probably assume these four samples had the same or very similar distributions. The graph in Figure 11.1 shows how wrong we would be. An excellent and concise discussion of using graphs for exploratory data analysis can be found in Ellison (1993). Following are some of the most valuable of these graphs, along with examples of each.

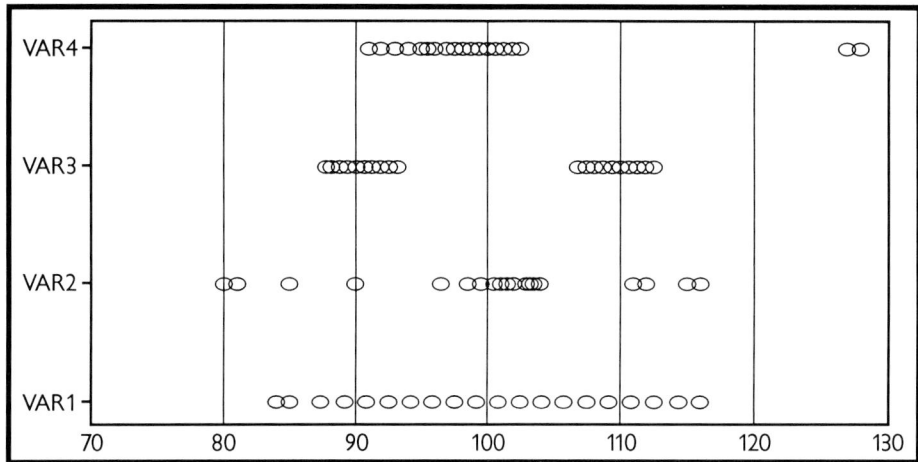

FIGURE 11.1. Four different samples of size 20, each of which has a mean of 100 and a standard deviation of 10. For these samples, the two summary statistics alone (mean and standard deviation) are insufficient to fully characterize the population. The differences in data distributions become apparent only after the individual data points are plotted.

1. Normal probability plot

A normal probability plot is a good way to inspect your data to determine if they approximate a normal distribution. Most statistical packages produce these plots. The observed values are plotted against the values that would be expected if the data came from a normal distribution. If the data come from a normal distribution, the plotted values fall along a straight line extending from the lower left corner towards the upper right corner.

Figure 11.2 shows a normal probability plot of cover values obtained from a sample of 40 randomly placed transects, each of which had 50 point cover estimates (the transects are the sampling units). These data approximate a normal distribution.

Figure 11.3 is a normal probability plot of plant height data. Because the plotted values do not fall along a straight diagonal line, we know they do not conform to a normal distribution. The reason for this is that there are a lot of very small values and a few large values, a distribution that is common in biology. If you were to take the logarithms of the data, the resulting values would more closely approximate a normal

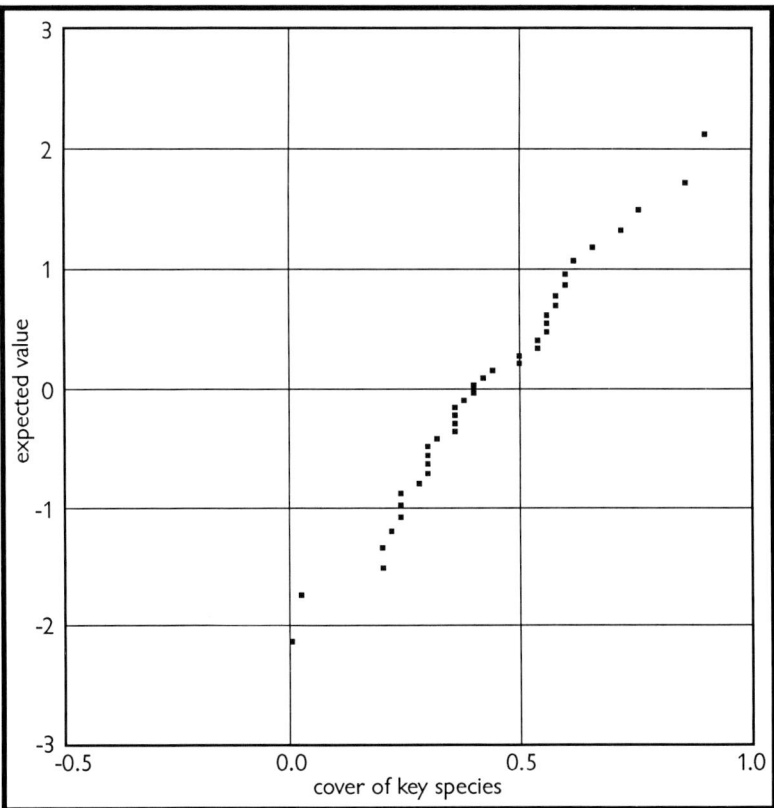

FIGURE 11.2. Normal probability plot of cover data. If data are from a normal distribution, the plotted values fall along a straight line extending from the lower left corner toward the upper right corner. These data approximate a normal distribution.

distribution. For this reason, the distribution is called a lognormal distribution. This plot has alerted us that we need to be careful when applying parametric statistics to this data set.

2. Density plots

Density plots show the distribution of a continuous variable. Histograms and dit plots are two types of density plots.

a. Histogram

A histogram is a type of density plot. Each bar in a histogram illustrates the density of data values found between the lower and upper bounds of the bar. Figure 11.4 is an example of a histogram. Although commonly used, histograms have three disadvantages (Ellison 1993):

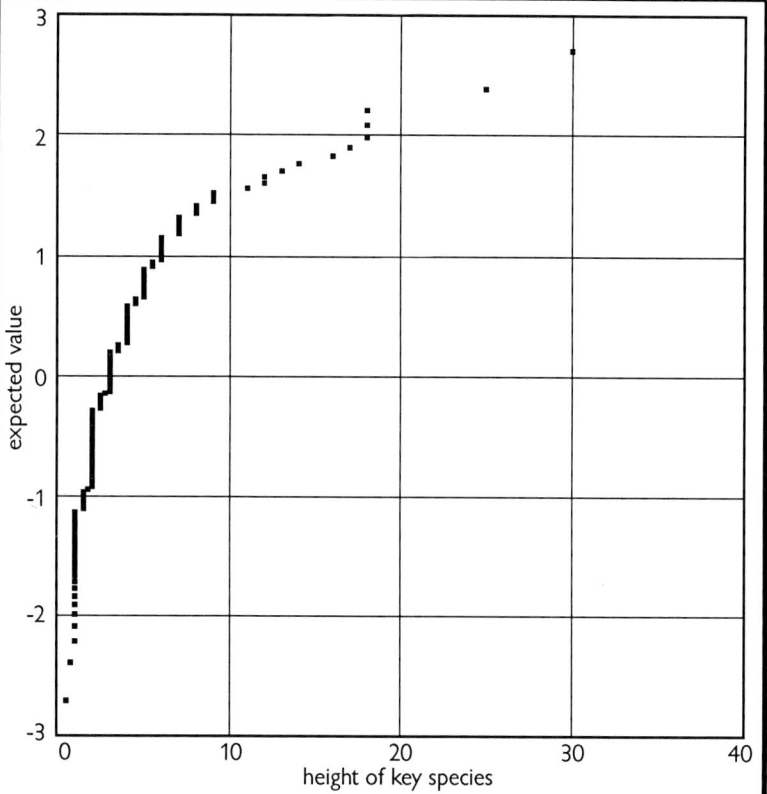

FIGURE 11.3. Normal probability plot of plant heights. If data are from a normal distribution, the plotted values fall along a straight line extending from the lower left corner toward the upper right corner. These data are not from a normal distribution.

1. The raw data are hidden within each bar. Consider the histogram of cover data presented in Figure 11.4. Each of the 10 bars (the second and tenth bars have no values in them) contains cover values within a range of 0.1. The third bar contains 11 values between 0.2 and 0.3, but we don't know if this represents 10 values of 0.2, 10 values of 0.3, or any of the other possible combinations of values between 0.2 and 0.3.

2. The number and width of bars is arbitrary. Changing these alters the shape of the histogram without conveying any additional information. Figure 11.5 is a histogram of the same data shown in Figure 11.4, but with 20 bars instead of 10.

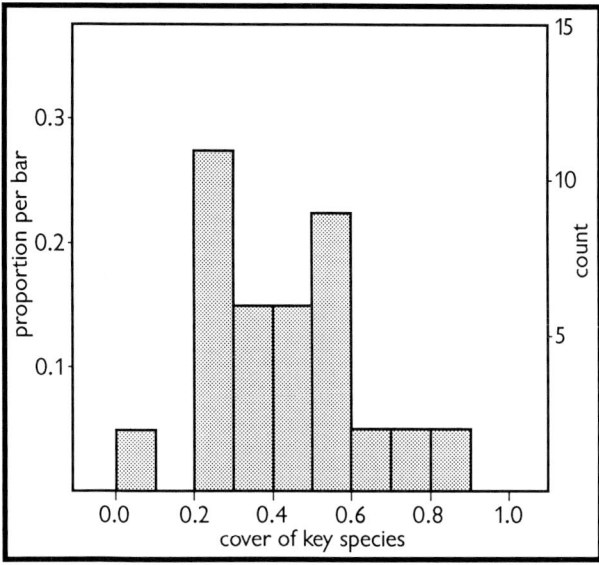

FIGURE 11.4. Histogram of cover data, with 10 bars chosen (2 of the bars contain no values). Notice individual data points cannot be distinguished.

3. Summary statistics (for example, means and medians) can't be computed from the data illustrated in a histogram.

One major advantage of histograms is that all the major statistical software programs produce them and, despite their disadvantages, they are an effective means of exploring your data. However, if your computer program can create one, you should also look at a dit plot of your data.

b. Dit plot

A dit plot, as illustrated in Figure 11.6, is a better type of density plot, because all the data points are presented, the underlying data structure is maintained, and the graph is easy to understand (Ellison 1993).

3. Box plot

A box plot, also called a box-and-whisker plot (Tukey 1977), is another good way to explore your data. As pointed out by Ellison (1993), a box plot provides more summary information without taking as much space or using as much ink as a histogram.

Figure 11.7 is a box plot of the same cover information used to construct Figures 11.4, 11.5, and 11.6. The following description of the box plot is adapted from Ellison (1993). The vertical line in the center of the box indicates the *sample median*. The median is the value that has an equal number of observations on either side,

FIGURE 11.5. Histogram of same cover values used to create Figure 11.4, but with 20 bars instead of 10 (6 of the 20 bars contain no values).

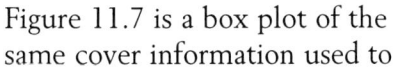

FIGURE 11.6. Dit plot of cover data used to create Figures 11.4 and 11.5. Note that each data point can be distinguished.

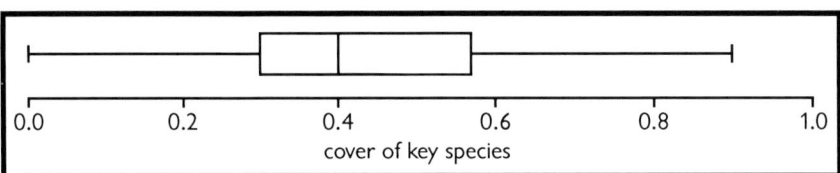

FIGURE 11.7. Box plot of same cover data used to create figures 11.4, 11.5, and 11.6. See text for explanation.

after the observations have been placed in order from smallest to largest. The left and right vertical sides of the box indicate the location of the 25th and 75th percentiles, respectively, of the data. This means that 25% of the data points lie to the left of the left vertical side of the box and 75% to the left of the right vertical side of the box. These 25th and 75th percentiles are also called lower and upper *quartiles* or *hinges*. The absolute value of the distance between the hinges (obtained by subtracting the value of the lower quartile from the value of the upper quartile) is the *hspread*. The whiskers on each side of the box extend to the last point between each hinge and its *inner fence*, a distance 1.5 *hspreads* from the hinge.

Outliers (data points lying farther from the rest of the data than one would usually expect, particularly if one were assuming the data came from an approximately normal distribution) are also indicated on box plots. There were no outliers in the data used to construct the box plot in Figure 11.7. Figure 11.8, a box plot of plant heights, illustrates the two kinds of outliers possible. Points occurring between 1.5 hspreads and 3 hspreads

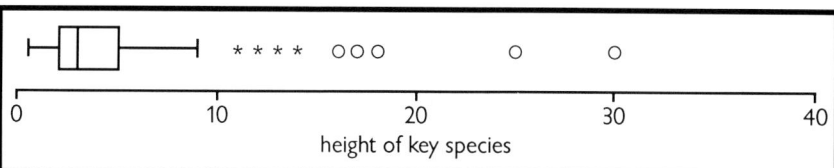

FIGURE 11.8. Box plot of plant heights illustrating the two types of outliers possible. See text for explanation.

(the outer fence) are indicated by an asterisk. Points occurring beyond the outer fence (*far outliers*) are indicated by open circles. These data are measurements of the heights of a plant species, after most of the plants had been grazed. There is, therefore, a preponderance of short plants (note the position of the median; half of all the plants measured are less than about 3 inches high), but some individual plants were ungrazed or not grazed as heavily, accounting for the outliers and far outliers shown in the box plot. These data follow a log-normal distribution and, as we learned when we examined a normal probability plot of these same data, we have to be careful when we use parametric statistics with data sets such as this (more on this subject later).

These box plots were constructed using the statistical package SYSTAT (Wilkinson 1991). You should be aware that other statistical packages may use different symbols to indicate outliers and far outliers. They may also define hspreads differently.

4. Combinations

Sometimes it is helpful to overlay different types of plots. Figure 11.9 overlays a symmetric dit plot onto a box plot. In addition to the information conveyed by the box plot

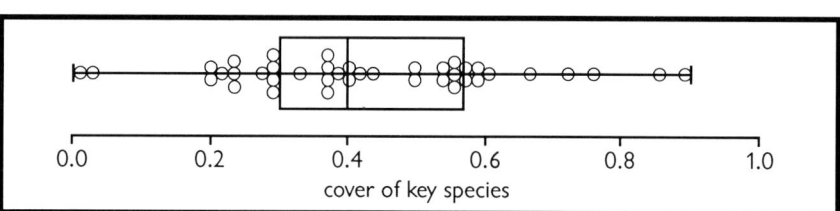

FIGURE 11.9. Overlay of symmetric dit plot on a box plot of cover data. The dits (actual data values) show the underlying data distribution of the box plot.

you can see how the individual data points are arrayed.

B. Parameter Estimation - Population is Sampled to Estimate a Population Mean, Proportion, or Total Population Size

The type of statistical analysis to which you intend to subject your data should be determined during the initial stages of your study. Two basic types of analysis can be identified based on the nature of your study's management and sampling objectives: parameter estimation (with confidence intervals) for target/threshold objectives and significance tests for change/trend objectives. We cover parameter estimation in this section.

If your management objective is a target or threshold objective, it is sufficient to estimate the parameter (mean, total, or proportion) and construct a confidence interval around the estimate. The analysis required is to calculate the sample statistic (mean, total, or proportion) and the confidence interval (the desired confidence level, α, should be specified in your sampling objective). Confidence intervals for estimates of population means and totals were introduced in Chapter 5. Appendix 8 gives directions on calculating confidence intervals around estimates of means and totals, as well as around estimates of proportions. You can calculate sample statistics and confidence intervals in each year of data collection and graph these using bar or point graphs with the confidence intervals as error bars (graphing results, including the use of bar and point graphs, is discussed in further detail in Section J below). The sample statistic and confidence interval of each sample would be compared to the target or threshold to determine if action is necessary or if the objective has been reached.

For example, your management objective is to maintain a population of at least 2000 individuals of *Lomatium bradshawii* at the Willow Creek Preserve over the next 5 years. Your sampling objective is to annually estimate the population size of *Lomatium bradshawii* at the Willow Creek Preserve and be 95% confident that the estimate is within 250 plants of the true population total. This is a threshold objective, because you are concerned with the population falling below the threshold. Therefore, data analysis consists of estimating the population size from the sample mean (by multiplying N, the total number of possible sampling units, by $\overline{\times}$, the sample mean) and calculating the confidence interval for this estimate (N times the standard error times the two-tailed, critical value of *t* for the number of degrees of freedom in the sample and the desired α level).

The estimated total and confidence interval are then compared to the threshold of 2000 plants. If both the estimated total and lower bound of the confidence interval are above the threshold, you can be confident (relative to the α level chosen) that you have met your objective. If both the estimated total and upper bound of the confidence interval are below 2000 plants, you can be confident (again relative to the selected α level) that you have failed to meet your objective. Less clear are situations where the threshold value is included within the confidence interval, with the estimated total either above or below the threshold (this is illustrated in Figure 11.22, later in this chapter). You should have prepared for this eventuality and prescribed the action you will take should this occur. There is further discussion on this in Section K, below.

A target/threshold objective can also be framed using a proportion. For example, your management objective is to decrease the frequency (in 1m² quadrats) of yellow star thistle to 30% or less at Key Area 1 in the Cache Creek Management Area by 2001 (the current frequency is 70%). This is the same thing as saying that, out of all the quadrats you could place in the sampled area (with no overlap), you want the proportion of quadrats containing yellow star thistle to be 30% or less. Your sampling objective is to annually estimate the percent frequency with 95% confidence intervals no wider than 10% of the estimated true percent frequency. Because you are dealing with a proportion, as opposed to a mean or total, the confidence interval width (10%) is expressed as an *absolute* rather than a *relative* value. So, for example, if your estimate of the true proportion is 40%, your target confidence interval width is from 30% to 50%.

For this example, data analysis entails estimating the percent frequency (by dividing the quadrats that contain yellow star thistle by the total number of quadrats sampled) and by calculating a confidence interval around this estimate (see Appendix 8 for instructions on doing this). The estimated frequency (proportion) and confidence interval are then compared to the target objective of 30%. If both the estimated proportion and upper bound of the confidence interval are below the target objective, you can be at least 95% confident that you have met your objective. If both the estimated proportion and lower bound of the confidence interval are above the target objective you can be at least 95% confident that you have failed to meet your objective. If the target objective falls within the confidence interval, your interpretation is more difficult. We discuss how to deal with this situation in Section K of this chapter.

C. Introduction to Significance Tests - Population is Sampled to Detect Changes in Some Average Value

If your management objective requires detecting change from one time period to another in some average value (such as a mean or proportion), then statistical analysis consists of a significance test, also called a hypothesis test. This situation often occurs in monitoring and involves analysis of two or more samples from the same monitoring site at different times (usually in different years). The major question asked is whether there has been change in the parameter of interest over a particular period of time. This parameter is often the mean, but we will also look at situations where the parameter is a proportion. Given a positive answer to this question a second question, usually (but not always) of equal importance, is the direction of this change. Significance tests are used to assess the probability of an observed difference being real or simply the result of the random variation that comes from taking different samples to estimate the parameter of interest.

1. Null hypothesis

A hypothesis is a prerequisite to the use of any significance test. In monitoring, this hypothesis is usually that no change has occurred in the parameter of interest. This hypothesis of no change is called the null hypothesis. If, through our significance test, we conclude that an observed change in a parameter between two or more years is not likely due to random variation, we reject the null hypothesis in favor of an alternative hypothesis: that there has been a change in the parameter of interest.

2. Example

The process can be illustrated by example. Let's say we've estimated the density of a rare species in a macroplot in two separate years. Each year we've taken a new random sample of forty 0.25 m x 5.0 m quadrats and counted the number of plants in each quadrat. The first year we obtain a mean of six plants per quadrat, and the second year we obtain a mean of four plants per quadrat. We wish to determine whether this change is statistically significant or simply due to random variation inherent in the population of all possible quadrats.

We start with the hypothesis that there is no real difference between the mean of six plants and the mean of four plants. What we are really saying is that the true population mean (unknown to us because we are sampling) has not changed, that these sample means could have been selected simply by chance from the same population.

a. *P* value

To test this null hypothesis we must first quantify the difference between these two sample means with a *test statistic* (Glantz 1992). When the test statistic is sufficiently large, we reject the null hypothesis of no difference between population means and conclude there is in fact a difference. However, we must specify in advance how large this test statistic must be for us to reject the null hypothesis. We do this by specifying a critical or threshold significance level, or *P* value. In this case we've specified a threshold *P* value of 20% or 0.20. This threshold *P* value is also called the α level.

The meaning of the *P* value can be described as follows (after Glantz 1992): The *P* value is the probability of obtaining a value of the test statistic as large as or larger than the one computed from the data when in reality there is no difference between the two populations.

Thus, if through our analysis we derive a *P* value of 0.18, and we therefore conclude that the true population mean has changed (because this is less than our threshold of 0.20), there is an 18% chance that we are wrong in that conclusion (that no true change has occurred). In other words, there is an 18% chance we have committed a false-change error. If, on the other hand, our analysis resulted in a calculated *P* value of 0.85, we would conclude the true population mean has not changed, because the calculated value is greater than our threshold *P* value of 0.20. In this case we cannot have committed a false-change error (since our conclusion is that no change has taken place), but we may have committed a missed-change error. The probability of a missed-change error (or its complement, power) must also be considered in analysis. We cover this in detail in Section K, below.

b. Significance test

Continuing with our example, we now enter our data into the computer. We can now conduct a significance test using a statistical software program (don't worry about which test at this point—we'll cover this in the following sections), which gives us a calculated *P* value of 0.125. Because this is smaller than the $P = 0.20$ (selected as our threshold level for determining significance), we conclude that the true population mean has changed. Our calculated *P* value of 0.125 tells us there is a 12.5% chance we are wrong, that there has been no real change at all.

Many scientific papers do not report actual *P* values. Instead, they report that an observed difference between samples "was not significant $(P > 0.05)$" or that the difference "was significant $(P < 0.05)$." This practice should be avoided. Actual *P* values calculated from your data should be reported, to enable the readers, who may have different thresholds of significance than you, to make up their own minds. In our example, a *P* value greater than 0.20 would indicate to us that no significant change occurred. But if the actual *P* value were 0.21, we'd be more concerned that we may have failed to detect a true change than would be the case if the actual *P* value were 0.85.

D. Significance Tests to Test for the Difference Between the Means or Proportions of Two or More Independent Samples

Two types of significance tests are commonly used to test for the difference between the means of two or more independent samples. Which type you apply to your monitoring data depends upon the nature of the data and how many samples (years) you wish to compare.

1. Independent-sample t test (for two samples)

The independent-sample *t* test is employed to test for difference in the means of *two* samples. This test is applicable to the analysis of density data, height data, and biomass data. It can be used to analyze cover data estimated in quadrats or along line intercepts. It is also appropriate for the analysis of cover data collected with points *if* the sampling unit is a group of points, such as points arranged along transects. If the points are treated as the sampling units, the chi-square test (discussed later in this section) is appropriate. The independent sample *t* test can also be used to analyze frequency data when quadrats are arranged along transects and the transects (*not* the quadrats) are treated as the sampling units. When the frequency quadrats are the sampling units then the chi-square test is the one to use.

The independent-sample *t* test can be easily carried out by many microcomputer software packages and some handheld calculators. The basic principle is that we examine the ratio (after Glantz 1992):

$$t = \frac{\text{difference of sample means}}{\text{standard error of difference of sample means}}$$

When this ratio is small, we do not reject the null hypothesis that there has been no change in the true population mean. If the ratio is large, we reject the null hypothesis and conclude there has been a change in the true population mean. How "large" the t value must be to reject the null hypothesis depends upon the P value we have previously chosen as our threshold of significance.

The fact that the t value is smaller than the value of t corresponding to our P value doesn't indicate there hasn't been a change in the true population mean. It only means we haven't demonstrated this change at a given level of significance through our monitoring study. To see how likely we would have been to detect a real change of a given magnitude, we can (and should) conduct a post hoc power analysis as discussed in Section K, below.

a. Two-tailed vs. one-tailed t test

Two types of t tests can be run on independent samples, a two-tailed test and a one-tailed test. The type of test selected depends on the type of null hypothesis being tested. If the null hypothesis is that there has been no change in the population mean, then a two-tailed t test would be used, because you need to detect change in either possible direction (smaller or larger values of the mean). If, however, the null hypothesis is, for example, that the population mean has not *increased*, then a one-tailed test would be used because you only need to detect change in one direction (an increase). Note, however, that an nonsignificant P value after a one-tailed test could mean either that the population mean has decreased or stayed the same; there is no way of testing which.

Although two-tailed tests are more commonly used in monitoring, in many cases, one-tailed tests are advantageous. If, for example, our management objective is to *increase* the density of a particular rare plant species, we may decide to frame our sampling objective in terms of detecting only whether an increase in density has occurred. If our monitoring study shows no increase between sampling periods then we institute a management change. The appropriate test would be a one-tailed test.

The advantage of a one-tailed test is that it is more powerful than a two-tailed test in detecting a true change in the population mean in the direction of interest. In many cases, this increase in power is considerable. The one-sided test, however, would only demonstrate significance in one direction—in the example given above, this is an increase.

b. Two-tailed example

Let's say that we have monitored the density of a rare plant species in each year over a 2-year period. We randomly place 50 quadrats, each 0.25 x 25m, in each of the years and calculate the mean and standard deviation for each of these two independent samples. In the first year our sample mean and standard deviation are 4.0 and 2.5, respectively (the units for both the mean and standard deviation are in plants/quadrat; the units are left out here for simplicity). In the second year our sample mean and standard deviation are 3.0 and 2.0, respectively. We now want to conduct a t test to determine if this observed difference is significant. Prior to sampling we have decided to set our false-change error rate (α) at 0.10. Thus, our threshold P value is 0.10.

Prior to testing, we must formulate a null hypothesis. In this instance we're interested in detecting change in either direction (either an increase or decrease in density). Our null (H_O) and alternative (H_A) hypotheses are therefore as follows:

H_O: The population mean has not changed between Year 1 and Year 2
H_A: The population mean has changed between Year 1 and Year 2

To test these hypotheses we calculate the t statistic as follows:

$$t = \frac{\overline{X}_1 - \overline{X}_2}{\sqrt{\frac{s^2}{n_1} + \frac{s^2}{n_2}}}$$

Where:

t = Test statistic.
$\overline{X}$ = Mean (subscripts denote samples 1 and 2, respectively).
n_1 = Sample size of sample 1.
n_2 = Sample size of sample 2.
s^2 = Pooled estimate of variance, calculated as follows:

$$s^2 = \frac{(s_1^2 + s_2^2)}{2}$$

Where:

s_1 = Standard deviation of sample 1.
s_2 = Standard deviation of sample 2.

Plugging our two sample standard deviation values into the pooled estimate formula we obtain:

$$s^2 = \frac{(2.5)^2 + (2.0)^2}{2} = 5.13$$

We now plug our pooled variance estimate into the formula for t and obtain:

$$t = \frac{4 - 3}{\sqrt{\frac{5.13}{50} + \frac{5.13}{50}}} = \frac{1}{\sqrt{0.1026 + 0.1026}} = 2.208$$

To determine the likelihood of H_O being true, we compare this calculated t statistic of 2.208[1] to the critical value of t in a t table for a α of 0.10 (remember we decided prior to testing that an α of 0.10 [P = 0.10] would be our threshold for significance) and the appropriate degrees of freedom (a t table can be found in Appendix 5). For an independent-sample t test like the one we're conducting here, degrees of freedom are determined by applying the formula 2(n-1), where n is the size of each sample. In our example the sample size is 50 in each year. The degrees of freedom are therefore 2(50-1) = 98.

The critical value of t from a t table (see table in Appendix 5) for $\alpha = 0.10$ (for a two-tailed test we use the α (2) row in the table, where the [2] stands for a two-tailed test) and 98 degrees of freedom (designated v in the t table) is 1.661. Since our calculated t value is greater than this critical value, we reject the null hypothesis of no change and conclude that there has been a downward change in the population mean (since the mean of the second year is less than the mean of the first year). We would also report our calculated P value, which we could interpolate from the t table, but could obtain more easily through a statistics program. For this example the P value is 0.0296, well below the threshold P value of 0.10. We can say there is about a 3% chance that we have committed a false-change error (concluding that there has been a change in the population mean when no true change has occurred).

---

[1] If we've sampled more than 5% of the population we should apply the finite population correction factor to the t test. This increases the t statistic and gives us greater power to detect change. See Section F of this chapter for instructions on how to do this.

c. One-tailed example

Using the same example we used for our two-tailed test, we will evaluate whether the population has decreased. We have decided to take action if the population decreases, but to take no action if the population remains the same or increases. In this situation we have a different set of hypotheses as follows:

H_O: The population has not decreased
H_A: The population has decreased

The first thing we do with the one-tailed test is look at the sample means. If the Year 2 sample mean is greater than the Year 1 sample mean, we won't bother to conduct the t test, since we already know we cannot reject the null hypothesis and say that the population has decreased (the population may have increased or it may have stayed the same—since we are conducting a one-tailed test, however, we will not be able to say which).

If the Year 2 sample mean is less than the Year 1 sample mean, we then conduct the t test, using the same formula as for the two-tailed test. The only difference is that we compare our calculated t value with the critical value for the one-tailed test (the row labeled $\alpha[1]$ in the t table of Appendix 5). The one-tailed critical t value for 98 degrees of freedom and $\alpha = 0.10$ is 1.290. Since this is less than our calculated t value of 2.208, we reject the null hypothesis in favor of the alternative hypothesis and conclude that the population has decreased. Using a statistical program we calculate the actual P value as 0.0148. Thus, we can state that there is about a 1.5% probability that we have committed a false-change error. Note that the P value for the one-tailed test is exactly one-half the P value for the two-tailed test. With the same data set this will always be the case. Thus, the one-tailed test is always more powerful than its two-tailed counterpart in detecting change in one direction.

2. Analysis of variance (for three or more independent samples)

The analysis of variance, often abbreviated as ANOVA, is used for testing for the difference between the means of three or more samples. All microcomputer statistics programs carry out this test.

Instead of t, ANOVA uses F as the test statistic. F is calculated as (from Glantz 1992):

$$F = \frac{s^2_{bet}}{s^2_{wit}}$$

where: s^2_{wit} = *within-groups variance*: population variance estimated from sample means.
s^2_{bet} = *between-groups variance*: population variance estimated as the average of sample variances.

The formulas for calculating the F test statistic are not given in this technical reference. These can be found in standard statistical text books such as Zar (1996). You will probably use some computer program to calculate F, so only the concept is presented here. The important thing to note is that under a null hypothesis of no difference between true population means, the two variances are estimates of the same population variance. Therefore, the closer this ratio is to 1, the less likely there is a difference between population means. How large the F statistic needs to be before you reject the null hypothesis and conclude there has been a change in the true population mean depends on the P value chosen.

ANOVA is a two-tailed test.[2] Also, a significant F statistic leads to the conclusion that at least one of the sample means tested comes from a different population. It does *not* tell you which means are different, although you can usually get a reasonable idea from your estimates.

As an example, assume we have collected three years of density data from the same macroplot in 1989, 1991, and 1993. Quadrats were randomly located in each year of measurement using different sets of random coordinates (their positions in any year are therefore independent of their positions in any previous year). Before sampling, we determined we would accept a false-change error rate of 0.05 and a missed-change error rate of 0.05. The summary statistics are as follows:

Year	Sample Size (n)	Mean	Standard Deviation	Standard Error
1989	30	21.467	10.136	1.851
1991	30	16.633	9.807	1.790
1993	30	14.800	9.539	1.742

The raw data are entered into a statistical computer program, and the analysis of variance option is chosen. The program creates an "ANOVA table," which gives the pertinent statistics for the analysis of variance test. This table may look slightly different from one computer program to another, but will have the same basic format as the one below. Following is an ANOVA table for our three years of data:

One-Way ANOVA Results

Source	DF	SS	MS	F	P
Between Groups	2	711.6667	355.8333	3.6822	0.0292
Within Groups	87	8407.2333	96.6349		
Total	89	9118.9000			

Alpha Level = 0.05
Critical F (0.0500,2,87) = 3.1013

The value of the test statistic, F, is 3.6822.[3] The P value, given in the last column, is 0.0292. Thus, there is about a 2.9% probability of obtaining an F value of 3.6822 or larger when in fact there is no difference between all three of the years. (The other values in the table are those used in the calculation of the F statistic. "DF," "SS," and "MS," stand for degrees of freedom, sum of squares, and mean squares, respectively. The MS value between groups divided by the MS value within groups yields the F statistic. The alpha level is the one we entered into the program, and the critical F value is the one corresponding to an alpha level of 0.05, with 2 and 87 degrees of freedom for the between and within group sources of variance, respectively.)

Our sampling objective specified a false-change error rate of 0.05. Since the P value is less than this, we conclude that one or more of the years is significantly different from the others.

---

[2] There are analysis of variance techniques that do not depend on the F statistic that can be used to test one-sided or directional hypotheses. However, few if any statistical programs can perform these techniques. See Rice and Gaines (1994) for an introduction to these techniques.

[3] If we've sampled more than 5% of the population, we should apply the finite population correction factor to the F statistic. This increases the F statistic and gives us greater power to detect change. See Section F of this chapter for instructions on how to do this.

To test statistically which of these three years is different, we can compare each of the pairs of means using two-sided t tests. However, we must modify the P value used for the ANOVA for each t test performed, by dividing the P value used for the overall ANOVA by the number of t tests to be performed. In this case, our overall P value is 0.05. If we want to compare all three mean values (mean 1 with mean 2, mean 2 with mean 3, and mean 1 with mean 3), we divide the overall P value by 3. Our new threshold P value for each of these tests is thus 0.05/3 = 0.0167.

When we do these pairwise t tests we come up with the following statistics:

Years Compared	DF	t-value	P
1989 vs. 1991	58	1.8771	0.0655
1989 vs. 1993	58	2.6234	0.0111
1991 vs. 1993	58	0.7340	0.4659

Only the P value of 0.0111 for the years 1989 vs. 1993 is less than our threshold of 0.0167. We therefore conclude that there has been a significant change between those two years (but not between any of the other pairs of years). This procedure is called the Bonferroni t test and works reasonably well when the number of comparisons are few (Glantz 1992). As the number of comparisons increases above 8 to 10, however, the value of t required to conclude a difference exists becomes much larger than it needs to be, and the method becomes overly conservative (Glantz 1992). Other multiple comparison tests are less conservative and preferable in these cases. Three such tests are the Student-Neuman-Keuls test, the Scheffe test, and the Tukey test, some or all of which are performed by many microcomputer statistical packages. There is debate over which of these is the preferable test; see Zar (1996:218) for a discussion of this. Another such test, the Duncan multiple-range test, is not conservative enough and should be avoided (Day and Quinn 1989).

3. Testing the difference between two proportions (independent samples): the chi-square test

The chi-square test is used to analyze frequency data when individual quadrats are the sampling units and point cover data when individual points are the sampling units. (Even though cover is expressed as a percentage, cover data are appropriately analyzed by calculating mean values, except when individual points are the sampling units.) If the frequency data are collected on more than one species, each species is usually analyzed separately. Another alternative is to lump species into functional groups, such as annual graminoids, and analyze each of the groups.

a. 2 x 2 contingency table to compare two years

To estimate the frequency of a plant species in two separate years, we've taken two independent random samples of 400 quadrats each. In each of these quadrats the species is either present or absent. For analysis we put these data into a 2 x 2 contingency table, as follows:

	1990	1994	Totals
Present	123 (0.31)	157 (0.39)	280 (0.35)
Absent	277 (0.69)	243 (0.61)	520 (0.65)
Totals	400 (1.00)	400 (1.00)	800 (1.00)

The numbers in parentheses are frequencies of occurrence in 1990 and 1994, and, in the last column, for both years combined. The chi-square test is conducted on actual numbers of quadrats, *not* percentages. The chi-square test is not appropriately applied to percentage data.

Just as for the *t* test and ANOVA, we must formulate a null hypothesis. Our null hypothesis states that the true proportion of the target plant species (the proportion we would get if we placed all of the quadrats of our particular size that could be placed in the sampled area) is the same in both years. This is equivalent to saying there has been no change in the proportion of the key species from 1990 to 1994.

Before we can calculate the chi-square statistic we must determine the values that would be expected in the event there was no difference between years. The total frequencies in the right hand column are used for this purpose. Thus, in both 1990 and 1994, 0.35 x 400 quadrats, or 140 quadrats, would be expected to contain the species, and in both 1990 and 1994, 0.65 x 400 quadrats, or 260 quadrats, would be expected to not contain the species. The following table shows these expected values:

	1990	1994	Totals
Present	140	140	280
Absent	260	260	520
Totals	400	400	800

Now we can compute the chi square statistic as follows:

$$\chi^2 = \sum \frac{(O - E)^2}{E}$$

Where: χ^2 is the chi square statistic.
Σ = summation symbol.
O = Number observed.
E = Number expected.

Applying this formula to our example we get:

$$\chi^2 = \frac{(123-140)^2}{140} + \frac{(277-260)^2}{260} + \frac{(157-140)^2}{140} + \frac{(243-260)^2}{260}$$

$$= 2.06 + 1.11 + 2.06 + 1.11 = 6.34$$

We then compare the chi-square value of 6.34 to a table of critical values of the chi-square statistic (see table in Appendix 5) to see if our chi-square value is sufficiently large to be significant.[4] The *P* value we have selected for our threshold before sampling began is 0.10. Now we need to determine the number of degrees of freedom. For a contingency table, the number of degrees of freedom, v, is given by:

$$v = (r - 1)(c - 1)$$

Where: r = number of rows in the contingency table.
c = number of columns in the contingency table.

---

[4] If we've sampled more than 5% of the population we should apply the finite population correction factor to the chi-square test. This increases the chi-square statistic and gives us greater power to detect change. See Section F of this chapter for instructions on how to do this.

For a 2 x 2 table $v = (2-1)(2-1) = 1$. Therefore, we enter the table at degrees of freedom = 1, and the P threshold of 0.10. The critical chi-square value from the table is 2.706. Since our value of 6.34 is larger than the critical value, we reject the null hypothesis of no difference in frequency of the plant species and conclude there has been an increase in its frequency. We would also report our calculated P value, which we could interpolate from the chi-square table, but could obtain more easily through a statistics program. For this example, the P value is 0.012.

Statistics texts differ on whether to use the chi-square statistic as calculated above in the special case of a 2 x 2 contingency table. Some authors (e.g., Zar 1996) state this value over-estimates the chi-square statistic and recommend that the Yates correction for continuity be applied to the formula as follows:

$$\chi^2 = \sum \frac{(|O - E| - \frac{1}{2})^2}{E}$$

Other authors (e.g., Steel and Torrie 1980; Sokal and Rohlf 1981) point out that the Yates correction is overly conservative and recommend against its use. Salzer (unpub. data) has shown through repeated sampling of simulated frequency data sets that the Yates correction is not needed. Munro and Page (1993) point out that the Yates correction is required only when the expected frequency of one of the cells in the table is less than 5. With the proper selection of quadrat size (see Chapters 7 and 8) this should rarely occur in plant frequency monitoring studies. Accordingly, we recommend calculating χ^2 without the Yates correction.

Statistical packages for personal computers calculate the chi-square statistic and give exact P values. For 2 x 2 tables, however, you should be aware of whether the program applies the Yates correction factor. Some programs, such as SYSTAT, give both the uncorrected and corrected chi-square values. Other programs such as STATMOST give only the corrected chi-square value. Because you want the uncorrected chi-square value, this presents a problem for 2 x 2 tables; no program applies the correction to larger tables.

b. Larger contingency tables for more than two years

When you have more than two years of data to compare, you can increase the size of the contingency table accordingly. For three years of data, you would use a 2 x 3 table; for four years, a 2 x 4 table; and so on. The chi-square statistic is computed according to the directions given above for a 2 x 2 table. Also, when using a table of critical values you need to calculate the degrees of freedom according to the directions given above. Because there will never be more than two rows (present and absent), the number of degrees of freedom will always be 1 fewer than the number of years. Thus, for a 2 x 3 table there, are 2 degrees of freedom; for a 2 x 4 table, there are 3 degrees of freedom; and so on.

It is important to realize that, just as for an ANOVA, a significant result in a chi-square table larger than 2 x 2 is an indication only that the frequency in at least one year is significantly different than expected. Which year(s) are different cannot be determined without further testing. This can be done by subdividing the larger contingency table into smaller 2 x 2 tables. Because this involves making multiple comparisons on the same set of data, however, the Bonferroni adjustment to the P value must be made before running these tests (directions on the use of the Bonferroni adjustment are given under Section D.2, above).

c. Contingency tables for analysis of point cover data

If you've collected cover data using a point intercept method and if the sampling units are the individual points (as opposed to transects or point frames), the data can be arrayed into a contingency table and analyzed using the chi-square statistic. The procedure is the same as for the frequency data described above (except you may wish to change "present" and "absent" to "hits" and "misses"). Just as for frequency data, analysis is done on a species-by-species basis or on functional groups of species. Total plant cover or any other type of cover (e.g., litter or bare ground) can also be analyzed this way.

E. Permanent Quadrats, Transects, and Points: the Use of Paired-Sample Significance Tests

1. Independent vs. paired samples

Thus far we've discussed significance tests for independent samples. Independent samples are ones in which different sets of sampling units are selected randomly (or systematically with random starts) in each year of measurement. Now we'll consider the case in which sampling units are randomly selected only in the first year of measurement. The sampling units are then permanently marked, and the same (or at least approximately the same) sampling units are measured in the subsequent monitoring year.

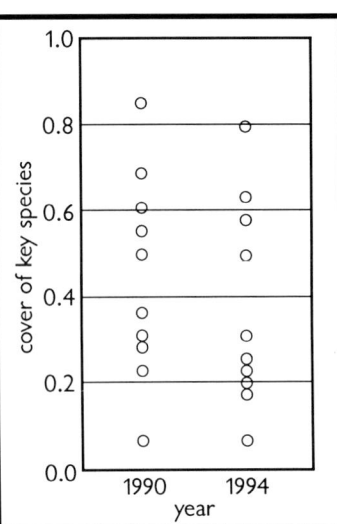

Because the two samples are no longer independent (the second sample is dependent upon the first), the use of the independent-sample significance tests discussed previously is not appropriate. Instead, a paired-sample significance test is used.

2. Paired t test: use it when you can

The appropriate significance test for two paired samples is the paired t test (unless the samples are proportions, in which case McNemar's test, discussed below, is the test to use). There is often a great advantage to testing change using a paired t test rather than an independent-sample t test. This is because the paired t test is often much more powerful in detecting change. To see why this is so, let's examine Figures 11.10 and 11.11 (adapted from Glantz 1992).

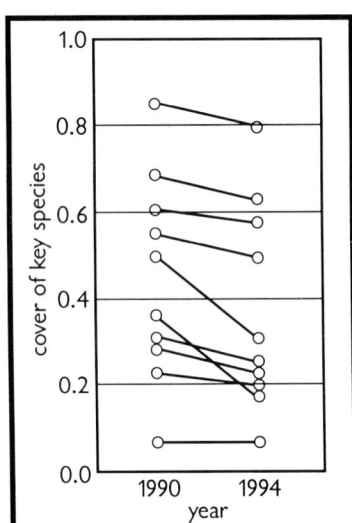

FIGURE 11.11. Cover estimates (in percent) for 1990 and 1994. Same data as in Figure 11.10 but by focusing on changes in each permanent transect, you can detect a change that was masked by the variability between transects obvious in Figure 11.10.

FIGURE 11.10. Cover estimates (in percent) for 1990 and 1994. Data from 10 permanent transects of 50 points each.

The data depicted in Figure 11.10 are cover estimates (in percent) for 10 transects in 1990 and 1994. The estimates were derived by placing 50 points at systematic intervals along a line (transect), recording whether the target plant species was present or absent, and reporting a total cover for the species on the transect. For example, if 16 out of the 50 points on the transect were "hits" on the target species, the cover for that transect is 16 divided by 50, or 0.32.

The spread of cover estimates for both years is great, ranging from 0.06 to 0.86 in 1990 and from 0.06 to 0.80 in 1994. As might be expected from this variability, the estimates of the mean—0.44 for 1990 and 0.38 for 1994—are not very precise: the 95% confidence interval for 1990 is 0.27 to 0.61 and for 1994 is 0.21 to 0.55. Not surprisingly, an independent-sample t test run on these samples results in a conclusion of no change. (The calculated t value is 0.617 and the actual P value is 0.55. This is not statistically significant at all.)

Consider now, however, Figure 11.11. These are the same data as shown in Figure 11.10, but now we can see that the *same* transects were measured in 1994 as in 1990. (The transect beginning and ending points were permanently marked in 1990, a measuring tape laid between the two points and 50 cover points read at systematic intervals along the tape. In 1994 the same procedure was used, with the same transect locations and the same systematic interval of cover points read.) Thus, an independent-sample t test is not appropriately applied to these data, because 1994's sample is not independent of 1990's sample. Each of the 10 transects read in 1990 is paired with one read in 1994.

Even if we could conduct an independent-sample t test, we wouldn't want to. To see why, notice that the cover values in 9 of the 10 paired transects have gone down between 1990 and 1994. A paired t test ignores the between-transect variability in both years and looks at only the *differences* between the 1990 and 1994 values for each of the transects. Conducting a paired t test on these same data results in a highly significant difference between years (the calculated t value is 3.34 and the actual P value associated with this is 0.009).

The message is clear: if you are interested only in documenting change, as is often the case in monitoring studies, paired t tests are more powerful than independent-sample t tests as long as the pairs of sampling units are correlated (i.e., a sampling unit with a large value the first year is likely to have a large value the second year, while a sampling unit with a small value the first year is likely to have a small value the second year; Zar 1996). The degree of correlation is measured by means of a correlation coefficient (see Zar 1996 for instructions on calculating a correlation coefficient; all statistical programs will perform the necessary calculations). The closer the correlation coefficient is to 1.0, the higher the correlation between samples (a value of 1.0 represents perfect correlation; a value of 0 represents no correlation). The paired samples illustrated in Figure 11.11 have a correlation coefficient of 0.96. The higher the degree of correlation the more powerful the test. Even if you don't know the degree of pairwise correlation, a paired t test is still valid (Snedecor and Cochran 1980). The lower the correlation, however, the less the advantage of the paired t test over the independent-sample t test (the latter test is invalid for a design that measures the same sampling units in both years of measurement).

Just as for other significance tests, you should apply the finite population correction factor to the paired t statistic if you've sampled more than 5% of the population. Section F of this chapter provides instructions on how to do this.

3. Repeated-measures analysis of variance

For 3 or more years of measurements on the same sampling units, there is a test analogous to the independent-sample analysis of variance discussed above. The test is the repeated-measures analysis of variance. An excellent introduction to the procedure, as it is used in medical experiments, can be found in Glantz (1992). Most statistical programs perform this test.

The repeated-measures ANOVA may not be the best choice in monitoring studies. One problem is the series of statistical decisions that must be made before tests of significance are calculated (Krebs 1989; Barcikowski and Robey 1984). One of the important assumptions of the repeated measures ANOVA is that the correlations between pairs of data for all the years analyzed are the same (Zar 1996). In other words, the correlation between the data of Year 1 and Year 2 is the same as that between Year 2 and Year 3, as well as that between Year 1 and Year 3, and so on. This condition of equal correlations is known as sphericity. Depending on the type of permanent sampling unit employed, this can be a problem. If the sampling unit is a quadrat, and the boundaries of the quadrat are permanently marked, this is less likely to be a problem than if the sampling unit is a line of points. Even if the endpoints of this line are permanently marked, and one takes care to place the points in the same place in each year of measurement, because the points themselves are not permanently marked there is more room for error. Therefore, correlations may not be the same between each year of measurement and the assumption of sphericity could be violated.

Just like the situation with ANOVA for independent samples, a significant result from the repeated-measures ANOVA indicates only that one or more years differ from each other, not which of these years is different. Most multiple comparison tests for an independent-sample ANOVA, such as Tukey's test, are not valid for the repeated-measures ANOVA. Paired t tests can be used to compare pairs of years. The Bonferroni adjustment to the t statistic must be applied if more than one pair of years are compared (see next paragraph).

We recommend using paired t tests to compare pairs of years, instead of using the repeated measures analysis of variance. If, however, you compare more than two years you will need to apply the Bonferroni adjustment to your threshold P value. Let's assume we've decided on a threshold P value of 0.20, meaning that if the paired t test results in a calculated P value less than 0.20, we will conclude a change has taken place between the two years tested. We take measurements in permanent quadrats for three years. If we compare *only* Year 3 with Year 1 or *only* Year 3 with Year 2, then no correction to the P value of 0.20 is required: a calculated P value less than 0.20 would lead to a decision of significance. If, however, we compare Year 3 with Year 1 *and* Year 3 with Year 2, we need to adjust our threshold P value by dividing it by the number of comparisons we are making. In this case, we are making two comparisons so our threshold P value is 0.20/2 = 0.10. If either of these comparisons results in a calculated P value less than 0.10, we can declare a significant difference.

4. Paired-sample testing for proportions: McNemar's test

Frequency data, when individual quadrats are the sampling units, and cover data collected by points, when the points are the sampling units, may also be analyzed as paired data using McNemar's test. The data are arrayed in a 2 x 2 table, similar to the contingency table discussed previously.

McNemar's test is used instead of chi-square to test for a difference in proportion between years when the same sampling units are measured each year. Unlike the chi-square test, McNemar's test is useful for comparing only two years; it cannot be used for more than two years.

Pairing of quadrats or points can be accomplished by permanently marking quadrats or points the first year and resampling them the next year. This can be accomplished by positioning quadrats systematically (with a random start) along randomly positioned permanent transect lines. Care must be taken, however, to permanently mark not only both ends of each transect, but intermediate points in between, and to stretch the tape to approximately the same tension at each time of measurement. You must then ensure that quadrats are placed at the same position along each transect in each year of measurement. It helps if in the first year at least two corners of each quadrat are marked with inexpensive markers such as long nails. See Chapter 7 for more information on this issue.

Just as with the paired t test, McNemar's test can be applied regardless of the level of correlation between the pairs of measurement, but the power of the test increases with the degree of correlation (J. Baldwin, 1996). When the degree of correlation between sampling units is high, the use of McNemar's test can be much more powerful in detecting change than will be the case if the sampling units are randomly located in each year of measurement. The following example illustrates this.

Let's first look at the situation with temporary frequency quadrats, where we decide to measure change in frequency by randomly locating 100 quadrats in a macroplot in each of two years. We decide that our P value for significance is 0.10. In the first year, 60 of the quadrats have one or more individuals of Species X in them. In the second year, 50 of the quadrats have Species X in them. The analysis in this case is a typical 2 x 2 contingency table using the chi-square statistic. The null and alternative hypotheses are as follows:

H_O: The proportion of quadrats containing Species X is the same in both years of measurement.
H_A: The proportion of quadrats containing Species X is not the same in both years of measurement.

Here is the contingency table:

	Year 1	Year 2	Totals
Present	60	50	110
Absent	40	50	90
Totals	100	100	200

A chi-square analysis of these data gives the following:

Chi-square statistic = 2.020
P value = 0.155

The observed change of 10 fewer quadrats is not significant at P=0.10. We therefore do not reject the null hypothesis that the proportion of quadrats containing Species X is the same in both years of measurement.

If we decide to permanently mark 100 quadrats (we could either actually mark all 100 quadrats or mark the ends and intermediate locations of several transects and systematically place the quadrats at the same points along tapes in each year of measurement), our null and alternative hypotheses are set up exactly the same way they were in the case of temporary quadrats:

H_O: The proportion of quadrats containing Species X is the same in both years of measurement.
H_A: The proportion of quadrats containing Species X is not the same in both years of measurement.

Just as before, we decide on a P value of 0.10 as our threshold of significance. In this case, however, we are going to either accept or reject the null hypothesis based on what happens in *permanently* established quadrats.

In the first year we find that Species X is found in 60 of the quadrats. In the second year we measure the same 100 quadrats and find that 10 of the 60 quadrats that contained the species the first year no longer contain the species. We also find that the 40 quadrats that did not contain the species the first year still did not contain the species in the second year. A 2 x 2 table set up for a McNemar analysis is shown below. Note the difference between this table and the contingency table given above: (1) the cell values total only 100, instead of 200 as in the contingency table; and (2) the years are not independent of one another (consequently, the values in the cells represent quadrats that meet both row and column requirements: 50 quadrats had Species X present in both Year 1 and Year 2, 40 had Species X absent in both years, 10 had Species X present in the first year but absent in the second, and no quadrats with Species X absent in the first year had it present in the second).

		Year 1	
		Present	Absent
Year 2	Present	50	0
	Absent	10	40

McNemar's test ignores the quadrats that responded in the same way each year. Thus the 50 quadrats with Species X present in both years and the 40 quadrats with Species X absent in both years are ignored (see Zar 1996:171-173 for the formulas used to calculate the McNemar chi-square statistic).

Here are the results of McNemar's test on these data:

McNemar chi-square statistic = 8.1000
P value = 0.0044

The calculated P value is well below our threshold P value of 0.10. We therefore reject the null hypothesis of no change. Even though only 10 quadrats went from containing the plant to not containing it, we have determined this to be significant, something we would not have done if we measured temporary quadrats in each year.[5]

See Chapter 7 and Appendix 18 for advice on deciding when to use a permanent frequency sampling design as opposed to a temporary design.

F. Applying the Finite Population Correction Factor to the Results of a Significance Test

If you have sampled more than 5% of an entire population then you should apply the finite population correction factor (FPC) to the results of a significance test.[6] The formula for the

---

[5] If we've sampled more than 5% of the population, we should apply the finite population correction factor to the McNemar test. This increases the McNemar chi-square statistic and gives us greater power to detect change. Section F, below, describes how to do this.

[6] Population as used here refers to the statistical population. In the context of the types of monitoring addressed by this technical reference, the FPC would be applied only to significance tests on data collected using quadrats. This is because there is a finite population of quadrats that can be placed in the area to be sampled (assuming quadrats are positioned, as they should be, to avoid any overlap). The FPC should never be applied to significance tests on line or point intercept data because a population of lines and points is by definition infinite.

FPC is 1 - (n/N). The procedure for applying the FPC depends on the nature of the test statistic. For tests that use the t statistic, the procedure involves dividing the t statistic from a significance test by the square root of the FPC. For tests involving the chi-square (χ^2) and F statistics, the procedure entails dividing the χ^2 or F statistic from a significance test by the FPC itself (not by its square root). The following examples illustrate the procedure for significance tests that use the t, χ^2, and F statistics.

1. Tests that use the t statistic

Independent-sample and paired t tests calculate the t statistic, which is compared to the critical value of t from a t table (a t table can be found in Appendix 5) for the appropriate degrees of freedom and the threshold P (α) value. If the calculated t value is larger than the critical t value, the null hypothesis of no change is rejected in favor of the alternative hypothesis that a change has taken place. The formulas for independent-sample and paired t tests given earlier in this chapter and in Appendix 8 do not include the FPC. Computer programs also do not apply the FPC to their calculated t values. If you have sampled more than 5% of the population you should correct the calculated t statistic by applying the FPC as in the following example. This will increase the size of the t statistic, resulting in greater power to detect change.

Let's say that the t statistic from a t test (either an independent-sample or paired t test) is 1.645 and in each of two years you sampled n = 26 quadrats out of a total of N = 100 possible quadrats. The FPC is applied as follows:

$$t' = \frac{t}{\sqrt{1-(n/N)}} \qquad t' = \frac{1.645}{\sqrt{1-(26/100)}} = 1.912$$

Where: t = The t statistic from a t test.
  t' = The corrected t statistic using the FPC.
  n = The sample size (the number of quadrats sampled in each year; note that you do *not* add the number of quadrats sampled the first year to the number of quadrats sampled in the second year).
  N = The total number of possible quadrat locations in the population. To calculate N, determine the total area of the population and divide by the area of each individual quadrat.

Following this calculation, you need to look up the P value of $t' = 1.912$ in a t table at the appropriate degrees of freedom (a t table can be found in Appendix 5). If this is an independent-sample t test the appropriate number of degrees of freedom would be $(n_1 - 1) + (n_2 - 1)$ = (26 - 1) + (26 - 1) = 50. If this is a paired t test, the values analyzed are the observed changes in each permanent quadrat. Since there are 26 permanent quadrats, n = 26, and the appropriate number of degrees of freedom is n - 1 = 26 - 1 = 25.

Looking up P values in a t table is difficult and inexact because it requires you to interpolate between values in the table. A more exact and convenient method is to use the computer program, NCSS PROBABILITY CALCULATOR, which is available as shareware from NCSS Statistical Software. Directions on obtaining this program are given in Section L of this chapter. Appendix 19 gives directions on how to use NCSS PROBABILITY CALCULATOR to calculate the P value for a given value of t.

2. Tests that use the chi-square statistic

The chi-square (χ^2) statistic is used to test the difference between years in a proportion when using temporary sampling units. McNemar's test, which tests the difference between two years in a proportion using permanent sampling units, also makes use of the chi-square statistic. In both cases, the chi-square statistic calculated using standard formulas and computer programs should be corrected using the FPC if you have sampled more than 5% of the population.

For example, the χ^2 statistic from a particular test is 2.706 and you sampled n = 77 quadrats out of a total of N = 300 possible quadrats. The FPC would be applied as follows:

$$\chi^{2\prime} = \frac{\chi^2}{1 - (n / N)} \qquad \chi^{2\prime} = \frac{2.706}{1 - (77 / 300)} = 3.640$$

Where: χ^2 = The χ^2 statistic from a chi-square test or McNemar's test.
$\chi^{2\prime}$ = The corrected χ^2 statistic using the FPC.
n = The sample size (the number of quadrats sampled in each year; note that you do not add the number of quadrats sampled the first year to the number of quadrats sampled in subsequent years).
N = The total number of possible quadrat locations in the population. To calculate N, determine the total area of the population and divide by the area of each individual quadrat.

Following this calculation, you need to look up the P value of χ^2 = 3.640 in a χ^2 table at the appropriate degrees of freedom (a χ^2 table can be found in Appendix 5). For McNemar's test, which can be used only to test for a difference between two years, there is always 1 degree of freedom. For a chi-square test applied to a contingency table, the number of degrees of freedom is always one less than the number of years being compared. Thus, for a 2 x 2 table comparing 2 years there is 1 degree of freedom, for a 2 x 3 table comparing 3 years there are 2 degrees of freedom, and so on.

Looking up P values in a χ^2 table is difficult and inexact because it requires you to interpolate between values in the table. A more exact and convenient method is to use the computer program, NCSS PROBABILITY CALCULATOR, which is available as shareware from NCSS Statistical Software. Directions on obtaining this program are given in Section L of this chapter. Appendix 19 gives directions on how to use NCSS PROBABILITY CALCULATOR to calculate the P value for a given value of χ^2.

3. Tests that use the F statistic

The analysis of variance and the repeated-measures analysis of variance use the F statistic to determine if one or more of the years sampled is different from the other years. The F statistic can also be corrected by the FPC. This is accomplished as illustrated in the following example.

An analysis of variance calculated by a computer program yields an F statistic of 3.077. In each of three years you sampled n = 50 quadrats out of a total of N = 400 possible quadrats. The calculated F is corrected as follows:

$$F' = \frac{F}{1 - (n / N)} \qquad F' = \frac{3.077}{1 - (50 / 400)} = 3.517$$

Where: F = The F statistic from an analysis of variance or a repeated-measures analysis of variance.
F' = The corrected F statistic using the FPC.
n = The sample size (the number of quadrats sampled in each year; note that you do *not* add the number of quadrats sampled the first year to the number of quadrats sampled in subsequent years).

N = The total number of possible quadrat locations in the population. To calculate N, determine the total area of the population and divide by the area of each individual quadrat.

Following this calculation, you need to look up the P value of $F = 3.517$ in an F table at the appropriate degrees of freedom. This technical reference does not include an F table, but you can find one in Zar (1996). Looking up P values in an F table is difficult and inexact because it requires you to interpolate between values in the table. A more exact and convenient method is to use the computer program, NCSS PROBABILITY CALCULATOR, which is available as shareware from NCSS Statistical Software. Directions on obtaining this program are given in Section L of this chapter. Appendix 19 gives directions on how to use NCSS PROBABILITY CALCULATOR to calculate the P value for a given value of F.

G. Assumptions Regarding the Statistics Discussed Above

Most statistics discussed thus far are *parametric statistics*, so called because they are used to estimate population parameters such as means and totals.[7] The use of parametric statistics requires that several assumptions be met, at least approximately (no monitoring data will meet these assumptions exactly):

1. That the population being sampled follows a normal distribution. A normal distribution is the familiar bell-shaped curve illustrated in Figure 11.12. This assumption holds both for the calculation of confidence intervals and for the use of t tests and analyses of variance. (For paired t tests, the *differences* between sampling units should come from a population that follows a normal distribution.)

2. That the sampling units are drawn from populations in which the variances are the same even if the means change from the first year of measurement to the next. This assumption, called homogeneity of variances, applies to significance tests to detect changes in means.

3. That the sampling units are drawn in some random manner from the population. This assumption applies both for the calculation of confidence intervals and for significance tests.

Prior to data analysis you need to determine if your monitoring data meet these assumptions. Although some tests can be used to assess whether your sample data are normally

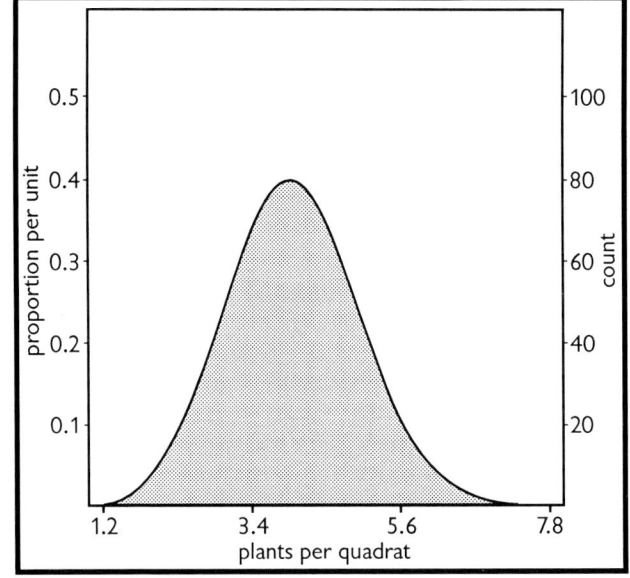

FIGURE 11.12. A normal distribution. The population is a set of quadrats; the variable measured is number of plants per quadrat. A distribution this close to normal is rare in practice.

---

[7] The chi-square and McNemar's chi-square are nonparametric statistics, as discussed below in Section H.

distributed,[8] it is often most effective to look at graphical analyses of your data. The use of probability plots, dit plots, and box plots to explore your data for normality was discussed at the beginning of this chapter.

Several tests are available to determine if the variances of two or more samples are equal, but none of these is very reliable. The most well-known, Bartlett's test, is not recommended because it is unduly sensitive to departures from normality (Sokal and Rohlf 1981). In fact, Zar (1996:206) recommends no test be used to assess whether the assumption of homogeneity of variances holds, because the analysis of variance is robust to departures from this assumption. The *t* test is similarly robust.

1. What happens if my data don't meet the assumptions of normality and homogeneity of variances?

Data on vegetation attributes will not meet either of these assumptions perfectly. As pointed out by Koch and Link (1970), few if any real data come from a population that is normal, or even quasi-normal. They go on to point out that the only consequences of failure to meet the assumption of normality are some distortion of the theoretical risk levels and a reduction in the efficiency of estimation. These problems, however, are far less serious than the failure to meet the assumption of randomness (Koch and Link 1970).

Fortunately, both *t* tests and analyses of variance are *robust* to moderate departures from either normality or homogeneity of variances (Zar 1996; the latter is true only if sample sizes are equal in each year of measurement). For severe departures from these assumptions there are several possible remedies:

a. Increase your sample size

According to Mattson (1981), a sample size of at least 100 sampling units will ensure against problems resulting from severe departures from normality (very skewed distributions). This is conservative; less severe departures from normality will not require as large a sample. We talk more about this below.

b. Transform your data

Transformations, whereby data in the original units are converted to another scale prior to analysis, are often applied to data prior to performing significance tests to make the data conform more closely to the assumptions of normality and homogeneity of variance. The use of transformations is covered in many text books, for example Zar (1996). They will not be covered further here, except to say that their utility for vegetation monitoring is limited because of several problems related to their use. These problems are as follows (Gilbert 1987):

1. Estimating quantities such as means, variances, and confidence intervals in the transformed scale typically leads to biased estimates when the data are transformed back into the original scale.

---

[8] Of these, the D'Agostino Omnibus test (D'Agostino et al. 1990) is probably the best (Hintze 1996). Some statistical programs, such as NCSS, conduct this test. Unfortunately, the test has small statistical power to detect departures from normality unless the sample sizes are large, say over 100 (Hintze 1996). What that means is if the test on a smaller sample size shows a departure from normality you can be reasonably sure the data are not normal. Less clear, however, is the situation when a smaller sample size does not indicate a departure from normality. In that case the data may be either approximately normal or nonnormal but the test simply failed to detect this.

2. It may be difficult to understand or apply the results of statistical analyses expressed in the transformed scale.

3. More calculations are required. As pointed out by Li (1964) the most common transformations are seldom helpful in practice.

Further useful information on transformations, including guidelines as to when they might be of use, can be found in Hoaglin et al. (1983).

c. Use nonparametric statistics

If you are greatly concerned whether your data meet the assumptions of normality and homogeneity of variance, you can use a class of statistics, called nonparametric statistics, that do not require these assumptions. Note, however, that nonparametric statistics, just like parametric statistics, require that data be collected in a random manner. Nonparametric statistics are discussed in more detail in Section H, below.

d. Use statistical analyses based on resampling

Resampling methods (also called computer-intensive methods) are becoming more and more popular with ecologists and other scientists. These methods can be used to calculate confidence intervals and to conduct significance testing. They are discussed further in Section I below.

2. When should I worry about using parametric statistics?

Glantz (1992) offers the following rules of thumb for deciding whether to use parametric statistics in significance testing. If the variances are within a factor of 2 to 3 of each other, then the assumption of homogeneity of variances can be considered to be met. If a density plot of the observations reveals they are not heavily skewed and there is no more than one peak, then you can assume the data are close enough to a normal distribution to use parametric statistics. Another "test" of normality is to compare the size of the mean with the standard deviation. When the standard deviation is about the same size or larger than the mean and the variable being measured can take on only positive values (which is true for most monitoring data discussed in this technical reference), this is an indication that the distribution is heavily skewed (Glantz 1992).

Hahn and Meeker (1991) point out that confidence intervals designed to include the population mean (as opposed to some other population parameter such as the variance) are relatively insensitive to the assumption of normality. Appendix 13 reports on a small experiment conducted by one of the authors (Willoughby), in which parametric statistics were used to construct confidence intervals around estimates of the population mean of an exponential distribution (a distribution very far from normal). The conclusion is that parametric procedures work well even for an exponential distribution, as long as sample sizes are reasonably large. Cochran's rule, following, gives guidance on when a sample is "reasonably large."

3. Cochran's rule for confidence intervals

Cochran (1977:42-43) offers what he terms a "crude rule" for determining how large the sample size must be to use the normal approximation in computing confidence intervals. This rule makes use of Fisher's measure of skewness, often designated G_1. Most statistical programs routinely calculate this measure, although many of them simply use the term "skewness" instead of G_1. The rule is designed so that a 95% confidence probability statement will be wrong no more than 6% of the time. The rule controls only for the *total* error rate and ignores the direction of the error of the estimate. Cochran's formula is:

$$n > 25G_1^2$$

Where: n = Sample size.

G_1 = Fisher's measure of skewness.

To illustrate how this formula works, let's use it on a simulated population of 4,000 observations that follow an exponential distribution (a histogram of this population is shown in Appendix 13, Figure 1). The computer program, STATMOST, gives a G_1 value (which it labels simply as "skewness") of 1.778.[9] Plugging this value into the above formula yields the value $(25)(1.778)^2 = 79$. We therefore know that we must have at least 79 sampling units to obtain a confidence interval that includes the true mean 94% or more of the time (this is consistent with the empirical results given in Appendix 13, Table 1). It still remains necessary, of course, to calculate the sample size necessary to obtain the level of precision needed to meet your sampling objective (see Chapter 7).

H. Nonparametric Statistics

The statistical analysis methods discussed so far (except for the chi-square McNemar's tests) are based upon estimates of two population parameters, the mean and the standard deviation. If our data approximate a normal distribution, these two parameters are all we need to know to fully characterize the population. For that reason, statistical analysis based on estimates of parameters and the assumption of normality are called parametric statistics.

When, however, the distribution is not normal, these two parameters decrease in their usefulness. This decrease is usually not sufficient to abandon the use of parametric statistics as long as sample sizes are reasonably large and, for significance testing, the samples being compared are the same (or nearly the same) size. If either or both these conditions is not satisfied, however, you will probably want to turn to methods based on nonparametric statistics (or on statistics based on resampling, discussed in the next section).

Distributions with high positive skews are rather common in biology. For example, we may measure the heights of a shrub species several years following a wildfire. A few plants of this shrub species, having survived the fire, might be very tall, while the rest of the plants, being new recruits, might be relatively short. Figure 11.13 shows how this distribution of heights might look. Depending on how severe this skew is, you might want to use nonparametric statistics in analyzing the data, unless your sample size is large enough to use parametric statistics.

Nonparametric statistics usually involve ordering (ranking) the data from the smallest value to the largest and using the ranks rather than the values themselves. For example, we might have a sample of 11 shrub heights shown in Table 11.1. Note how this ranking reduces the effect of the two large values, 4.5 m and 5.1 m, on the data set. Since the analysis is based on the ranks, not

---

[9] This skewness value, G_1, is the true value for the entire population. The value based on a sample of this population would be somewhat different. To ensure we have enough values to obtain a stable estimate of G_1 we could construct a sequential sampling graph of the mean and standard deviation, as shown in Chapter 7. When these values stabilize we know we have a large enough sample. We then calculate G_1 for this sample and plug it into Cochran's formula.

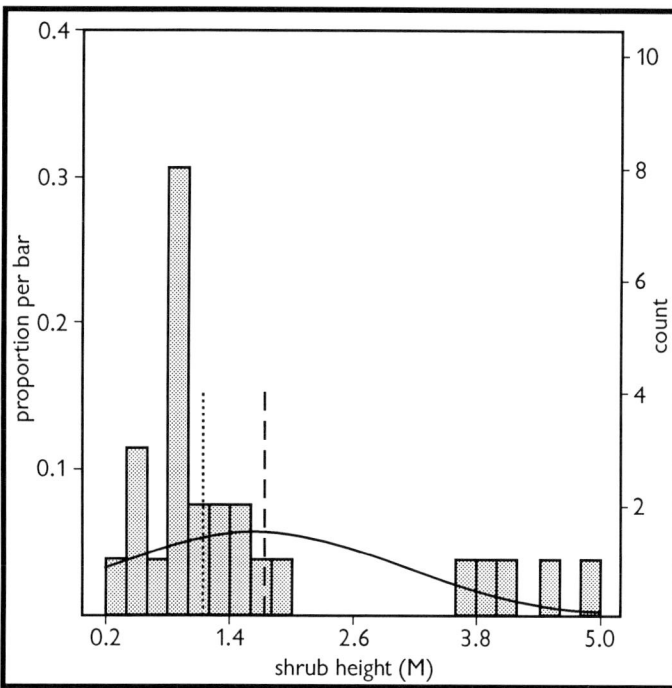

FIGURE 11.13. Histogram of shrub heights on which is superimposed a normal smoothing curve. Note positive skew, with tail to right. The dotted line is the median. The dashed line is the mean.

Height (m)	Rank
0.35	1
0.40	2
0.50	3
0.55	4
0.75	5
0.90	6
1.00	7
1.10	8
1.30	9
4.50	10
5.10	11

TABLE 11.1 Sample of 11 shrub. Heights ranked from smallest to largest.

the actual values, the difference between ranks 9 and 10 is only one unit, rather than 3.2 units in the original units.

1. Nonparametric confidence intervals

The sample mean is a parametric statistic. Because of this, there is no nonparametric way to estimate a confidence interval around the mean (except by the use of resampling, discussed in the next section) or, for that matter, around estimates of total population size because estimates of totals also depend on estimating mean values.

For distributions with a large positive skew, however, there are times that you may want to use the median instead of the mean as a measure of the central tendency in the distribution. The median is that value of the variable (after the values have been ranked) that has an equal number of values on either side of it. It divides the frequency distribution in half. Thus, the median for our sample of 11 shrub heights is 0.90, because there are five numbers above and five numbers below the value of 0.90. Notice the difference between this median and the mean of 1.50 for the same data set. The two large values of 4.50 and 5.10 have greatly affected the mean. (In a completely normal distribution the mean and median are equal).

Figure 11.13 shows another sample of shrub heights (n=26). Note that the distribution of values is positively skewed, with a long tail to the right. The median (indicated by the dotted line in the figure) for this sample is 1.10, while the mean (indicated by the dashed line) is 1.67. Again, the mean is affected by the few large heights, while the median is not.

If you are interested in inferring something about the height of the majority of the shrubs from which this sample was drawn, you may wish to estimate the median of the population. If so, you can calculate a confidence interval within which the true population median lies (with some confidence level).

The method for calculating a confidence interval for a median will not be elaborated here. Consult a text on statistics, for example Zar (1996), or the book by Hahn and Meeker (1991).

2. Nonparametric significance tests

There are nonparametric analogues to the parametric significance tests previously discussed. Before discussing these, however, it is important to point out that two of the significance tests we have already covered are nonparametric tests.

a. The chi-square test and McNemar's test are nonparametric tests

The chi-square test, discussed in Section D.3., above, is used to test for the difference between two proportions in a contingency table. It is a nonparametric test, requiring no assumptions regarding the distribution of data (although it requires that the sample be taken in a random fashion and that the samples are independent). Similarly, McNemar's test, discussed in Section E.4., to test for the difference in proportions between paired samples, is also a nonparametric test.

b. Nonparametric analogues to common parametric significance tests

There are nonparametric analogues to the parametric tests previously discussed. These are covered in Table 11.2. Refer to statistical texts such as Zar (1996) for descriptions of the tests. Most statistical computer packages perform these tests.

3. Why not use nonparametric statistics all the time?

Given the fact that nonparametric statistics require fewer assumptions than parametric statistics, you can ask the question, why bother with parametric statistics at all? The answer lies in the fact that, when the necessary assumptions are at least approximated, parametric statistical tests are more powerful than their nonparametric analogues. Also, other than the use of resampling techniques, discussed below, there is no nonparametric method available to calculate confidence intervals around means and totals, the two parameters often of the most interest in a monitoring study. If, however, the populations from which you sample are highly skewed, your sample size is small, and—in the case of significance tests—your sample sizes are very different at each time of measurement, you may want to use nonparametric methods or resampling techniques. Otherwise, you are perfectly justified in using parametric statistics.

Purpose of test	Parametric test	Nonparametric Test
Testing for change between two years; samples independent; not frequency data	Independent-sample t test	Mann-Whitney U test
Testing for change between two years; samples paired (permanent sampling units); not frequency data	Paired t test	Wilcoxin's signed rank test
Testing for change between two years; samples independent; frequency data		Chi-square test (2 x 2 contingency table)
Testing for change between two years; samples paired (permanent sampling units); frequency data		McNemar's test
Testing for change between three or more years; samples independent; not frequency data	Analysis of variance; independent-sample t-tests with Bonferroni correction	Kruskal-Wallis test; Mann-Whitney U tests with Bonferroni correction
Testing for change between three or more years; same samples measured each year (permanent sampling units); not frequency data	Repeated measures analysis of variance; paired t-tests with Bonferroni correction	Friedman's test; Wilcoxin's signed rank test with Bonferroni correction
Testing for change between three or more years; samples independent; frequency data		Chi-square test (2 x ≥ 3 contingency table)

TABLE 11.2. Matrix of statistical significance tests. Parametric and nonparametric significance tests corresponding to type of data and purpose of test. Note that for frequency (present-absent) data, only nonparametric tests are available.

I. Statistical Analysis Based on Resampling

With the advent of personal computers in the 1980s, statisticians began developing new theory and methods based on the power of electronic computation (Efron and Tibshirani 1991). In turn, many scientists interested in analyzing their data have begun to turn away from traditional parametric and nonparametric approaches. In their place they are substituting methods based on intensive resampling of the original data set.

These resampling methods (also called computer-intensive methods) can be used to calculate confidence intervals and to conduct significance testing. Two of the most commonly used methods are bootstrapping (which involves sampling the original data set with replacement) and randomization (also called permutation) testing (which involves sampling the original data set without replacement).

The only drawbacks to the use of these methods is their lack of familiarity to many scientists currently practicing and the fact that some of the theories behind them are relatively new and therefore little tested. The advantages of resampling methods, however, are many, including the fact that very few of the assumptions required for parametric statistics are needed (except, of course, for the assumption of random sampling) and they are apparently just as powerful (Manly 1991).

Resampling methods are also much easier than parametric statistics to intuitively understand, to the point that some (e.g., Bruce 1993) advocate teaching introductory courses in statistics using primarily resampling techniques. They also allow for estimation of parameters (including construction of confidence intervals) that would be difficult or virtually impossible to estimate using conventional statistics (Manly 1991; Good 1994).

Appendix 14 discusses the use of resampling methods in more detail and provides examples of how they can be used to calculate confidence intervals around means and percentiles and to conduct significance tests.

J. Graphing the Results of Data Analysis

Graphs are very important tools for displaying the results of data analysis and helping the investigator (as well as others) interpret the meaning of these data. When, as is usually the case, summary statistics such as a mean, total, or proportion are displayed, *error bars must be used to display the precision of the estimate*. Commonly encountered error bars are the sample standard deviation, the sample standard error, and *n* percent confidence interval (such as a 90 or 95 percent confidence interval). Because it is the true parameter (mean, total, or proportion) that is of interest, we recommend that you use only confidence intervals as error bars. You must clearly state what error bar you are using, as well as the sample size upon which the estimate and measure of error is based (Ellison 1993).

Following are some of the most important and commonly used graphs:

1. Bar charts with confidence intervals

Bar charts are commonly used to display the results of data analysis. They should not be confused with histograms. A histogram shows the density (or frequency) of the values occurring in the data set between the lower and upper bounds of each bar, whereas a bar chart is used to illustrate some summary measure (such as the mean, total, or percentage) of all the values within a given category, such as the year of measurement (Ellison 1993).

Figure 11.14 is an example of a bar chart showing the results of three years of monitoring. Mean density per quadrat of a hypothetical key species is displayed, along with error bars corresponding to 90% confidence intervals. When displaying information about more than one summary statistic per year (as, for example, data on the same species measured at two or more key areas per year), side-by-side bar charts (see Figure 11.15) can be used (again, with error bars for confidence intervals). Stacked bar charts should *not* be used; they are unintelligible and provide no way to display error bars (Ellison 1993).

2. Graphs of summary statistics plotted as points, with error bars

Graphs of points, with error bars corresponding to confidence intervals, can be used in lieu of bar graphs. An example of such a graph is given in Figure 11.16.

Sometimes lines are connected to each of the points as in Figure 11.17, although this is really unnecessary unless more than one summary statistic is presented in each year. Figure 11.18 illustrates means for two key areas in each year of measurement. Lines are appropriate here to clearly separate the two sets of means. Note, however, that the confidence intervals for the two key areas overlap in 1991, leading to some confusion. A side-by-side bar chart, as shown in Figure 11.15, would be more appropriate in this situation.

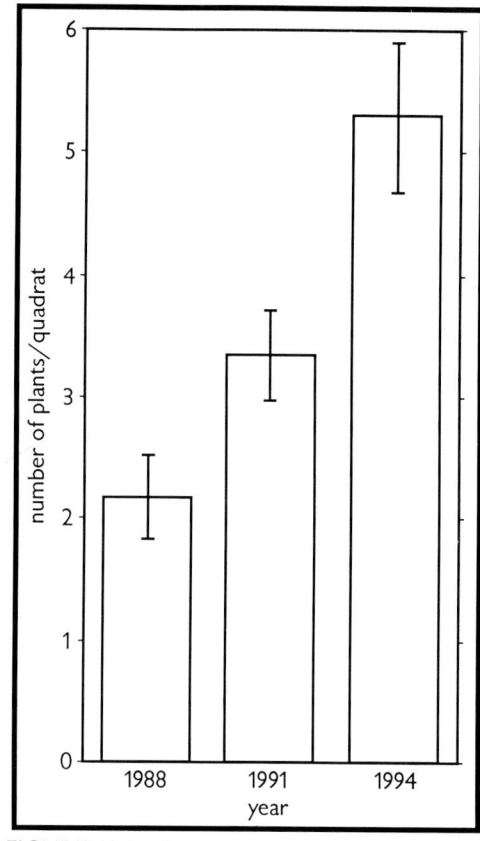

FIGURE 11.14. Bar chart of mean number of plants of the key species per 0.5m x 4.0m quadrat. Error bars are 90% confidence intervals. In each year n = 100.

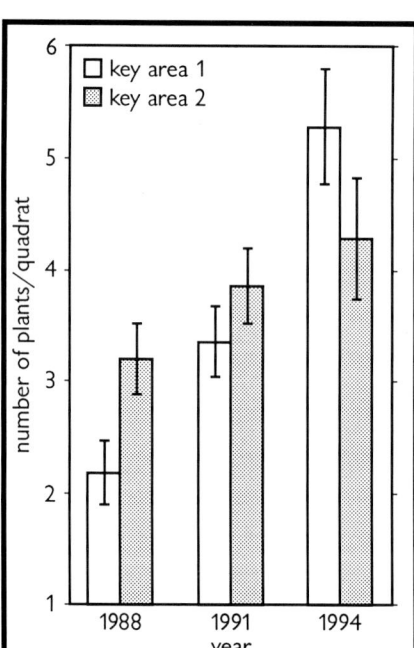

FIGURE 11.15. Side-by-side bar chart of mean number of plants of the key species per 0.5m x 4.0m quadrat, at key area 1 and key area 2. Error bars are 90% confidence intervals. All bars represent n = 100.

3. Box plots with "notches" for error bars

Another way of displaying summary statistics graphically is with box plots, already discussed above as a way of exploring your data prior to or during analysis. They can also be used to show the results of analysis, providing error bars for confidence intervals can be displayed. Some statistical packages offer the option to "notch" the box plots at a set confidence interval. Figure 11.19 shows such a notched box plot. These have the advantage of showing summary statistics (in this case the median and its 95% confidence interval), as well as other features relative to the distribution of data points. Note, however, that this option does not include the mean and that the confidence interval is one that includes the true *median* with 95% probability, not the true mean.

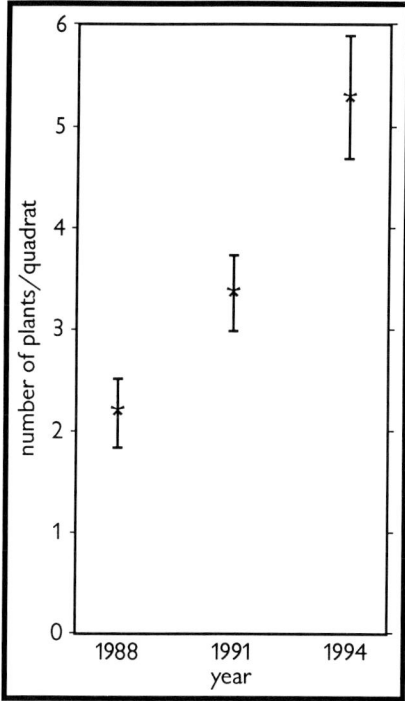

FIGURE 11.16. Point graph (also called category plot) of same data as shown in Figure 11.14. Error bars are 90% confidence intervals.

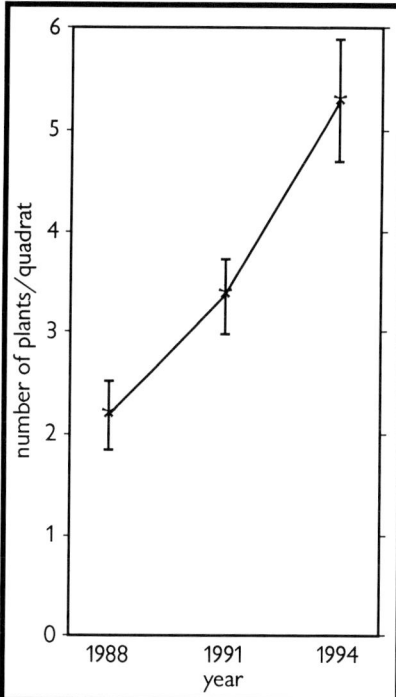

FIGURE 11.17. Point graph of same data as in Figure 11.16, but with lines connecting points.

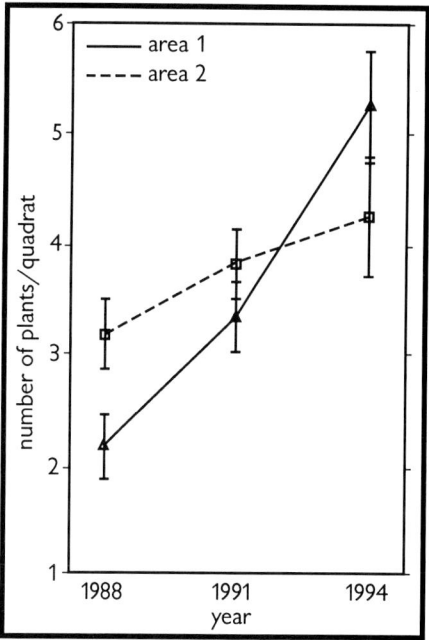

FIGURE 11.18. Point graph of same data as in Figure 11.15. Lines connect the means from each of the key areas. Error bars are 90% confidence intervals.

4. Graphing summary statistics when data are paired

Recall that significance tests are often much more powerful when data are collected in permanent plots or along permanent transects. The sampling units (plots or transects) in this case are said to be paired; that is, the data from the second year of measurement are dependent upon the data from the first year of measurement.

The use of paired methods is recommended, but graphical presentation is not as straightforward as with independent samples. Consider the data depicted in Figures 11.10 and 11.11. If one were to simply graph a summary statistic like the mean for each year of measurement, along with confidence intervals computed as if these data were independent, the graph would appear to illustrate no difference between the years 1990 and 1994. Figure 11.20 is a point graph that does just that.

Now consider the point graph shown in Figure 11.21. This graph is constructed with the same data used to produce Figure 11.20, but this time takes advantage of the fact that each of the 10 transects is paired. What is graphed in Figure 11.21 is the mean *difference* in cover between the paired transects. Because there was a decline in cover from 1990 to

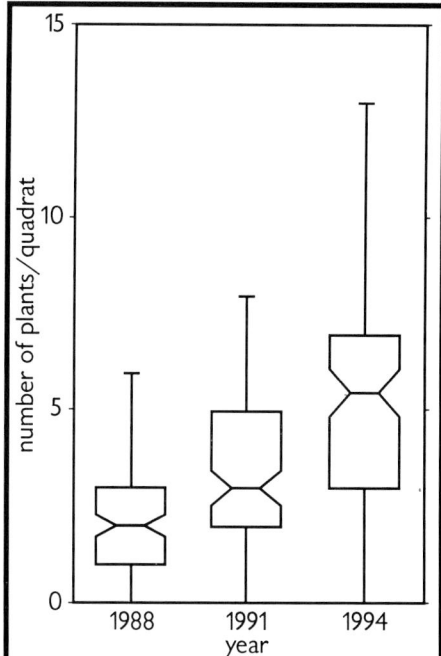

FIGURE 11.19. Notched box plots of the number of plants/quadrat in samples of one hundred 0.5m × 4.0m quadrats. The points at which the boxes reach full width on either side of the median represent the 95% confidence interval for the median.

1994 in all but one of the transects, the mean difference is negative. Also plotted is the 95% confidence interval around this mean difference. Because this interval does not include 0 (which would indicate the possibility of no change), this difference is significant at the 95% confidence level (i.e., $P < 0.05$).

Bar charts of the mean difference could be similarly constructed. If you are more interested in *median* differences you could use a notched box plot of these differences. It is also valuable to plot several mean differences on a single graph. For example, the graph could show points and confidence intervals for each of the differences between 1990-1994, 1994-1998, 1998-2002, and so on.

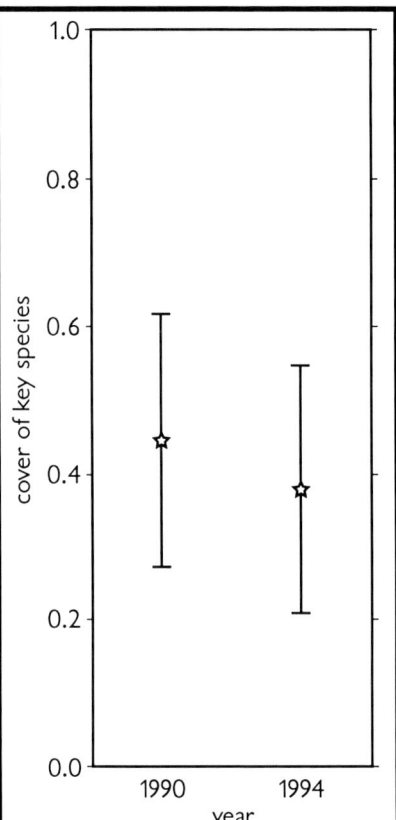

FIGURE 11.20. Point graph of cover data collected along permanent transects treated as if each year was independent. Error bars are 95% confidence intervals. See text for explanation.

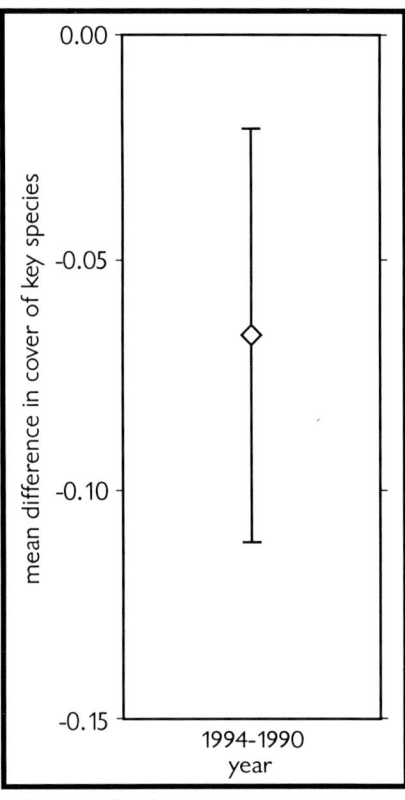

FIGURE 11.21. Point graph showing mean difference of cover in 10 paired transects of 50 points each. Error bar is 95% confidence interval.

5. Pie charts: don't use them

Pie charts are discussed only to state that they should not be used in presenting monitoring results. Though a favorite with news media, they do not offer much information for the amount of ink required, and it is not possible to present error bars on them. Instead of pie charts use the graphs discussed above.

K. Interpreting the Results of Monitoring

Following the analysis of monitoring data, it is necessary to explain the meaning of the results. This is the process of interpretation. We will address interpreting results from the two major types of data analysis discussed above: (1) parameter estimation, and (2) significance tests. The following discussion applies only to the situation in which sampling has taken place. If complete censuses have been conducted in each monitoring period, then any change observed is real; the only interpretation required is determining whether the change observed is biologically significant.

1. Interpreting the results of parameter estimation

Recall from our discussion in Section B of this chapter that one type of data analysis consists of calculating a sample mean, total, or proportion and constructing a confidence interval around that sample statistic in order to estimate the true parameter. This type of data analysis is done to facilitate a comparison with either a threshold or desired future condition management objective. Following are two examples of management responses to threshold and target management objectives.

1. Action X will occur if the mean density of rare species Y drops below value Z.

2. We will judge our restoration efforts to be successful if we have raised the mean density of species A to value B by the year 2000.

Because you have taken a sample (as opposed to conducting a complete census), you will not know the true population parameter (e.g., the true mean value). You will have only your estimate of the parameter (e.g., the sample mean) surrounded by a measure of precision such as a confidence interval. Interpretation then requires you to compare the parameter estimate and confidence interval to the threshold value. There are four possibilities, illustrated in Figure 11.22, and discussed below:[10]

1. Your threshold level has not been crossed by either the parameter estimate or the confidence interval (top arrow of Figure 11.22). Here the interpretation is relatively simple. You can be confident, at least to the degree of the confidence level you have selected for your confidence interval, that the true parameter has not crossed the threshold. For example, if your confidence interval is 95%, then you can be *at least* 95% confident that the true parameter is still below the threshold (the actual confidence may be greater than this if the upper bound of the 95% confidence is some distance from the threshold).

2. Your threshold level has been crossed by both the parameter estimate and the confidence interval (bottom arrow of Figure 11.22). Here again the interpretation is relatively simple. You can be confident, at least to the degree of the confidence level you have selected for your confidence interval, that the true parameter has crossed the threshold. If your confidence interval is 95%, then you can be *at least* 95% confident that the true parameter has crossed the threshold (the actual confidence may be greater than this if the lower bound of the 95% confidence interval is some distance from the threshold).

3 The parameter estimate does not exceed the threshold value, but the upper bound

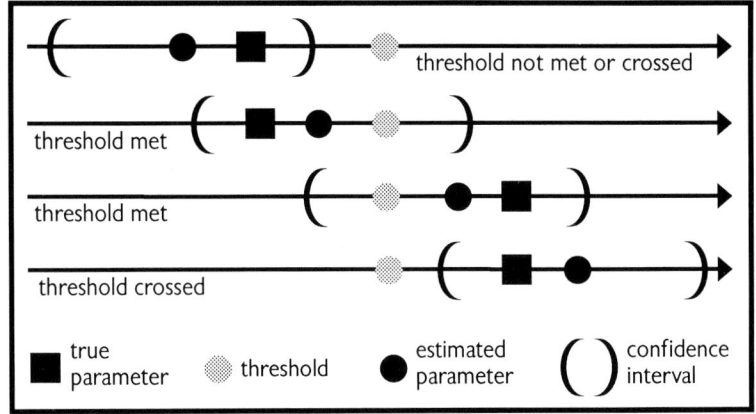

FIGURE 11.22. The four different possible outcomes when comparing a parameter estimate and confidence interval to a threshold level. The true parameter is shown only for illustrative purposes; we would never know it when conducting sampling. Adapted with permission from a figure prepared by Sylvia Mori, U.S. Forest Service, Pacific Southwest Research Station.

---

[10]The discussion here assumes that we want to remain *below* the threshold value, as would be the case, for example, if we were monitoring an invasive weed. Thus, it is the *upper* bound of the confidence interval that would cross this threshold first. If, on the other hand, our objective is to remain *above* a threshold, as would be the case if we wished to maintain a certain density of a rare plant, then it would be the *lower* bound of the confidence interval that would cross the threshold first.

of the confidence interval *does* exceed the threshold value (as in the second arrow of Figure 11.22). Now the interpretation is not nearly as clear cut. Because the true population parameter can be anywhere inside of the confidence interval, it is quite possible that the true population parameter has, in fact, crossed the threshold.

4. The fourth possibility is that both the parameter estimate *and* the upper bound of the confidence interval have crossed the threshold, but the lower bound of the confidence interval has not. This is illustrated by the third arrow of Figure 11.22. Again, because the true population parameter can be anywhere within the confidence interval, it may have crossed the threshold. In this case, because the midpoint of the confidence interval (the parameter estimate) has also crossed the threshold, it is more likely than in situation (3) that the true parameter has crossed the threshold.

How you will interpret situations like (3) and (4) should be determined prior to calculating your parameter estimate and confidence interval. In fact, you should have decided on this prior to even initiating sampling. One approach is to decide that if any part of the confidence interval crosses the threshold you will take action, based on the possibility that the true parameter has crossed the threshold. This minimizes the risk to the plant resource for which you are managing. Remember, however, that the size of the confidence interval depends on the confidence level you choose, the degree of variability in your sampling data (as expressed by the standard deviation), and your sample size. Thus, an inefficient sampling design and small sample size will result in much wider confidence intervals, which in turn will result in facing situations like (3) and (4) much more often. Good sampling design and reasonable sample sizes will therefore facilitate interpretation by making narrower confidence intervals and reducing the number of times you encounter the predicaments illustrated by (3) and (4).

2. Interpreting the results of significance tests.

A significance test is conducted when your management objective is to detect change from one time period to another in some average value (such as a mean or proportion). Once that test has been performed, you must now interpret the results from the test. Figure 11.23 is a flow chart to help you in your interpretation. Interpretation entails answering the following questions:

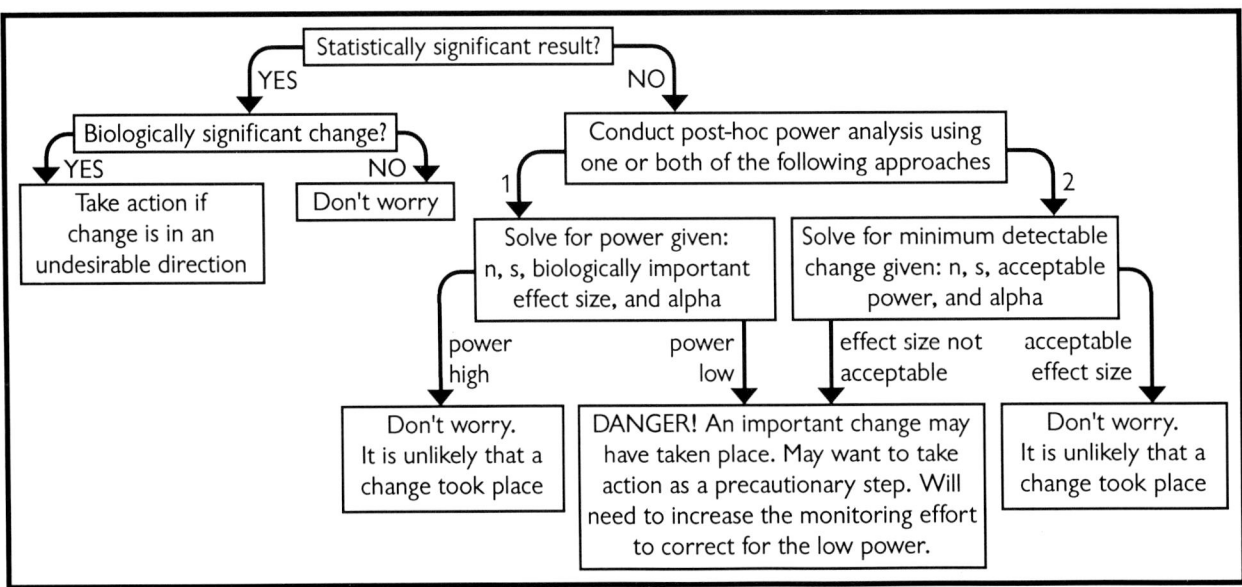

FIGURE 11.23. Interpreting the results from a statistical test comparing change over time.

a. Is there a statistically significant result? What is the likelihood that no true change occurred and that any observed difference is simply due to random chance?

The *P* value calculated from the significance test gives you the answers to these two questions. A threshold *P* value should be set prior to conducting the significance test so that the *P* value from the test can be assessed relative to the threshold. If the *P* value from the test is smaller than the threshold it is considered "significant" and the null hypothesis of no-change is rejected in favor of the hypothesis that a change did actually take place. If the *P* value from the test is larger than the threshold it is considered "non-significant" and the null hypothesis of no-change is not rejected. The *P* value calculated from the significance test is the likelihood that the observed difference is due to chance.

b. Does the observed magnitude of change have any biological significance?

Given a large enough sample size, a statistical test can find even an extremely small difference between two populations to be significant. It is unlikely that any two populations or the same population over any two time periods will ever be *exactly* the same. Therefore, it is important that you determine whether a statistically significant change has any biological significance. People often get stuck·on the idea of statistical significance. A helpful exercise is to pretend the difference observed through sampling is the true difference (i.e., pretend you conducted a complete census at each sampling period). Now ask yourself what action you will take if this observed difference is in fact the true difference. If your answer to this is that you would take no action, then the observed change, even though statistically significant, is not biologically significant.

c. If the test yields a non-significant result, what is the probability that a biologically important change actually occurred?

If your study results in a conclusion that an observed change is not significant, *your interpretation is not complete until you have conducted a post hoc power analysis.* The post hoc power analysis tells you the probability of your test failing to detect a true change (i.e., committing a missed-change error). Following are two approaches you can take in conducting this power analysis. Both of them are easy to do if you have a computer program developed for this purpose (we talk about computer programs for power analysis in Section L, below).

1. **Calculate a power value.** This is option 1 on the right side of Figure 11.23. Using this approach, you plug in your sample size, the sample standard deviation, the threshold significance level (α) you have chosen for the significance test, and an effect size you consider to be biologically important. A power value is then calculated. If the resulting power value is high, then it is unlikely that a change took place. If the resulting power value is low, then a biologically important change may have taken place. You need to improve your monitoring design immediately to ensure you can detect the level of change you believe is biologically important. If, based on ancillary information, you have reason to believe a deleterious change may have taken place, you may need to take action as a precautionary step until the monitoring design can be improved to address the low power issue.

   A threshold power value should be set in advance so that a decision can be made as to whether the power value calculated through the post hoc power analysis is considered high or low.

2. **Calculate the minimum detectable change (MDC)**. This is option 2 on the right side of Figure 11.23. This approach requires you to plug in the values for sample size, sample standard deviation, threshold significance level (α) for the test, and an acceptable level of power. The program then solves for the minimum detectable change (MDC) that can be detected. If the MDC is smaller than the size of change deemed to be biologically important then it is unlikely that the specified MDC actually occurred. If, however, the MDC is larger than this biologically important change, then an important change may have taken place. You need to improve your monitoring design immediately to ensure you can detect the level of change you believe is biologically important. If, based on ancillary information, you have reason to believe a deleterious change may have taken place, you may need to take action as a precautionary step until the monitoring design can be improved to address the inability to detect a change deemed to be biologically important.

Figure 11.24 exemplifies a post hoc power analysis, comparing two years of density data for *Lomatium cookii* at the Agate Desert Preserve in Oregon. Note that even though the significance test yielded a non-significant result, we cannot be confident that no change has taken place. This is because of the extremely low power (0.13) of the test to detect the 30% change we have determined to be biologically significant. Note also that the minimum detectable change is 155%; we could lose our entire population and not detect the loss! Although we may want to take action as a precautionary step, we very definitely want to improve the study design to reduce the standard deviation (in this case our standard deviation is more than twice the size of the mean, a very undesirable trait indeed).

Results of a statistical analysis comparing 1989 and 1990 data on <u>Lomatium cookii</u> from the Agate Desert Preserve. False-change threshold value = 0.10. Desired magnitude of change is 30% from the 1989 value.							
sample size	sample statistics				observed change (percent)	results of a statistical test (P)	calculated power $(1-\beta)$ to detect a 30% change from the 1989 mean	minimum detectable change size with a power of 0.9, $\alpha = 0.10$, (% change from 1989)
	1989		1990					
	mean	sd	mean	sd				
50	3.12	11.16	1.30	2.92	1.82(58%)	0.85	0.13	4.82 (155%)
INTERPRETATION: cannot conclude that a change took place (cannot reject the null hypothesis). Low confidence in the results due to low power and high minimum detectable change size. May want to take action as a precautionary step and make changes in the monitoring design to increase power.								

FIGURE 11.24. Example of a post hoc power analysis comparing two years of density data for *Lomatium cookii* at the Agate Desert Preserve in Oregon.

L. Statistical Software

1. For general statistical analysis

Several commercial software packages are available that perform all of the statistical procedures discussed above. Some of these are SAS, SPSS, SYSTAT, STATGRAPHICS, STATISTICA, JMP, NCSS, and STATMOST. No attempt is made here to review or recommend any of these, and any such review is quickly dated given the speed at which software manufacturers introduce new versions of their programs. Information on these packages can be found over the World Wide Web. The software companies have their own web pages, which can be accessed through "hotlinks" at the following two websites:

The Virtual Library of Statistics <http://www.stat.ufl.edu/vlib/statistics.html>
StatLib <http://lib.stat.cmu.edu/>

Both of these sites also have valuable links to other sources of information on statistics.

These major statistical packages are relatively expensive, in the neighborhood of several hundred to a thousand dollars or more (although special offers can sometimes be found for STATMOST, bringing its price down from $395 to about $150).

Inexpensive shareware programs are also available that will do many of the tests discussed in this chapter. Many of these programs, however, are DOS-based and tend to be rather "user unfriendly." One of these is the program EASISTAT (current version 2.1), available from several Internet addresses. One such address is:

http://oak.oakland.edu/simtel.net   (once at the homepage, click on "Dos Index," type in "estat21.zip" as the searchword, and download the file "estat21.zip")

Several other programs are also available from this source. The Illinois Natural History Survey also provides wildlife/ecology statistical software that can be downloaded from its website at:

http://nhsbig.inhs.uiuc.edu/www/index.html

A freeware version of the program NCSS ver. 6.0, which operates under Windows 3.1 and higher, was released in February 1996. Called NCSS JR., this is a fully functional statistical program. It doesn't have all the features of NCSS 97, which retails for $395, but is very usable nonetheless. It can be downloaded from the following website:

http://www.ncss.com

You can also download a fully functional version of NCSS 97 (which requires Windows 95, Windows NT 4.0, or higher) for a 30-day evaluation from this address.

SIMSTAT for Windows is a very reasonably priced ($129) statistical program that not only conducts the parametric and nonparametric tests discussed in this chapter, but also performs bootstrap analysis. A fully functional evaluation copy can be downloaded from the following website (it will function for 30 days):

http://ourworld.compuserve.com/homepages/Simstat/

STATIT is a UNIX-based statistical software program available to Bureau of Land Management personnel on UNIX workstations. It performs most of the parametric and nonparametric statistical analyses discussed in this chapter, but it's not particularly user friendly.

2. Software for calculating sample size and conducting power analysis

Knowledge of the power of significance tests to detect change is critical both in the planning stages of a monitoring study, where sample sizes must be calculated, and in interpreting the results of a monitoring study. Formulas for calculating sample sizes are given in Appendix 7, but these calculations are time-consuming when done on a calculator. An additional problem with these formulas is that they only calculate sample sizes; they do not solve for power or minimum detectable change as required in post hoc power analysis. Most of the general statistical programs discussed above do not calculate sample size or power, despite their rather hefty price tags. Fortunately, there are several good power analysis programs that will calculate both sample size and power.

Dr. Len Thomas maintains a website with information on all available power analysis and sample size programs. It can be accessed at the following address:

http://www.interchg.ubc.ca/cacb/power/

Dr. Thomas and Dr. Charles Krebs have also published a review of statistical power analysis software (Thomas and Krebs 1997; you can download a copy from the above internet address). For beginner to intermediate level use they recommend one of the following three commercial programs: PASS, NQUERY ADVISOR, or STAT POWER. The first one on this list, PASS, was the one most preferred by a graduate student class. A fully functional 30-day trial version of PASS can be downloaded from the address <http://www.ncss.com>. PASS retails for $249.95. Refer to Dr. Thomas' website for information on the cost of the other programs and how to order them. Thomas and Krebs also give relatively high marks to the program GPOWER, primarily because it is free. It is available from the following site:

http://www.psychologie.uni-trier.de:8000/projects/gpower.html

The documentation for GPOWER is extremely limited, and the user must have familiarity with Cohen's (1988) treatment of power analysis (Thomas and Krebs 1997). For these reasons we do not recommend the program for sample size determination and power analysis needed for the types of monitoring treated in this technical reference. Instead, we suggest you consider the following two programs (unless you have the money to purchase the commercial program PASS): STPLAN and PC SIZE: CONSULTANT. STPLAN (Brown et al. 1996), currently in version 4.1, is available free from the following website:

http://odin.mdacc.tmc.edu/ (once at the site, click on "Free computer code from the Section of Computer Science," click on "Software (Detailed List)," then go to "STPLAN" and follow directions).

The $15 program, PC SIZE: CONSULTANT, is shareware that can be downloaded from Dr. Len Thomas' website <http://www.interchg.ubc.ca/cacb/power/>. Go to "PC-SIZE and SIZE" and click on the link to download the program.

Both of these programs come with documentation files that can be printed. They are both DOS-based programs that are not very user friendly, but learning the basic procedures is not overly difficult. PC SIZE: CONSULTANT calculates sample sizes for all the types of significance tests discussed in this chapter (except for McNemar's test) *and* calculates sample sizes for estimating a population mean or total (but not a proportion). It also calculates power, but it does not calculate minimum detectable change.

STPLAN calculates sample sizes for all the types of significance tests discussed in this chapter (including McNemar's test), but it does not calculate sample sizes for estimating a single population mean, total, or proportion. It also calculates both power and minimum detectable change.

Appendix 16 provides instructions on how to use STPLAN and PC SIZE: CONSULTANT for the types of monitoring studies discussed in this technical reference.

3. Software for resampling

SIMSTAT for Windows, already mentioned above, conducts bootstrapping analyses on data sets. Two other programs warrant mention. RESAMPLING STATS is a command-based program that can perform both bootstrap analysis and randomization testing. It costs $225. Information on this package can be accessed at the following website:

> http://www.statistics.com/

RT (Manly 1996) is a DOS-based program that performs randomization tests, including one and two sample tests, linear regression, matrix randomization, time series, and multivariate analysis. It is marketed through WEST, Inc., 1402 South Greeley Highway, Cheyenne, Wyoming 82007. The phone number is 307-634-1756. The cost is about $100.

4. Software for calculating *P* values from test statistics

NCSS PROBABILITY CALCULATOR, which operates under Windows 3.1 and higher, is a freeware program that calculates *P* values for given test statistics, including t, χ^2, and F. This is valuable when you have adjusted the test statistic from a significance test with the finite population correction factor (as described in Section F of this chapter). You can download NCSS PROBABILITY CALCULATOR from the NCSS website <http://www.ncss.com>. Once at the NCSS homepage, click on "Shareware," then on "DOWNLOAD," to download the program. The program is provided as a "zip" file. Unzip it using PKUNZIP (or another program that handles zip files) into a directory of your choice. Then follow the instructions in the README.WRI file provided as part of the program.

The freeware program, NCSS JR., discussed previously in this section, also contains a version of the NCSS PROBABILITY CALCULATOR. One advantage of using the version in NCSS JR. is that this version contains a help function which is not available in the stand-alone program.

Appendix 19 gives directions on how to use NCSS PROBABILITY CALCULATOR to calculate *P* values for given values of t, χ^2, and F.

Literature Cited

Baldwin, J. 1996. [Personal communication]. U.S. Forest Service, Pacific Southwest Research Station, Albany, California.

Barcikowski, R. S.; Robey, R. R. 1984. Decisions in single group repeated measures analysis: statistical tests and three computer packages. American Statistician 38: 149-150.

Brown, B. W.; Brauner, C.; Chan, A.: Gutierrez, D.; Herson, J.; Lovato, J.; Polsley, J. 1996. STPLAN, Version 4.1. Houston, TX: University of Texas, M. D. Anderson Cancer Center, Department of Biomathematics.

Bruce, P. C. 1993. Resampling Stats user guide. Arlington, VA: Resampling Stats, Inc..

Cochran, W. G. 1977. Sampling techniques, 3rd ed. New York, NY: John Wiley & Sons.

Cohen, J. 1988. Statistical power analysis for the behavioral sciences. Hillsdale, NJ: Lawrence Erlbaum Associates.

D'Agostino, R. B.; Belanger, A.; D'Agostino, R. B., Jr. 1990. A suggestion for using powerful and informative tests of normality. American Statistician 44(4):316-321.

Day, R. W.; Quinn, G. P. 1989. Comparison of treatments after an analysis of variance in ecology. Ecological Monographs 59: 433-463.

Efron, B.; Tibshirani, R. 1991. Statistical data analysis in the computer age. Science 253:390-395.

Efron, B.; Tibshirani, R. 1993. An introduction to the bootstrap. New York, NY: Chapman and Hall.

Ellison, A. M. 1993. Exploratory data analysis and graphic display. In: Scheiner, S. M.; Gurevitch, J., eds. Design and Analysis of Ecological Experiments. New York, NY: Chapman and Hall.

Gilbert, R. O. 1987. Statistical methods for environmental pollution monitoring. New York, NY: Van Nostrand Reinhold.

Glantz, S. A. 1992. Primer of biostatistics, 3rd edition. New York, NY: McGraw-Hill.

Good, P. 1994. Permutation tests - a practical guide to resampling methods for testing hypotheses. New York, NY: Springer Verlag.

Hahn, G. J.; Meeker, W. Q. 1991. Statistical intervals: A guide for practitioners. New York, NY: John Wiley & Sons, Inc.

Hintze, J. 1996. NCSS 6.0.21 Jr., Help version 3.10.425. Kaysville, UT: NCSS, 329 North 1000 East.

Hoaglin, D. C.; Mosteller, F.; Tukey, J. W. 1983. Understanding robust and exploratory data analysis. New York, NY: John Wiley & Sons.

Krebs, C. J. 1989. Ecological methodology. New York, NY: Harper & Row.

Koch, G. S.; Link, R. F. 1970. Statistical analysis of geological data. New York, NY: John Wiley & Sons.

Li, J. C. R. 1964. Statistical inference, vol 1. Ann Arbor, MI: Edward Brothers.

Manly, B. F. J. 1991. Randomization and Monte Carlo methods in biology. New York, NY: Chapman and Hall.

Manly, B. F. J. 1996. RT: A program for randomization testing, Version 2.0. Available from Western Ecosystems Technology, 1402 South Greeley Highway, Cheyenne, WY 82007.

Mattson, D. E. 1981. Statistics: difficult concepts, understandable explanations. St. Louis, MO: C. V. Mosby Company.

Munro, B. H.; Page, E. P. 1993. Statistical methods for health care research, 2nd ed. Philadelphia, PA: J. B. Lippincott Company.

Rice, W. R.; Gaines, S. D. 1994. "Heads I win, tails you lose": testing directional alternative hypotheses in ecological and evolutionary research. Tree 9(6):235-237.

Snedecor, G. W.; Cochran, W. G. 1980. Statistical methods, 7th ed. Ames, IA: Iowa State University Press.

Sokal, R. R.; Rohlf, F. J. 1981. Biometry, 2nd ed. New York, NY: W. H. Freeman and Company.

Steel, R. G. D.; Torrie, J. H. 1980. Principles and procedures of statistics: a biometrical approach, 2nd ed. New York, NY: McGraw-Hill Book Company.

Thomas, L.; Krebs, C. J. 1997. A review of statistical power analysis software. Bulletin of the Ecological Society of America 78(2): 128-139.

Tukey, J. W. 1977. Exploratory data analysis. Reading, MA: Addison-Wesley.

Usher, M. B. 1991. Scientific requirements of a monitoring programme. In: Goldsmith, F. B., ed. Monitoring for Conservation and Ecology. New York, NY: Chapman and Hall.

Wilkinson, L. 1991. SYSTAT: The system for statistics, version 5.03. SYSTAT, Inc., Evanston, IL.

Zar, J. H. 1996. Biostatistical analysis, 3rd edition. Upper Saddle River, NJ: Prentice Hall.

CHAPTER 12
Demography

Leptochloa dubia
Green sprangletop
by Jennifer Shoemaker

CHAPTER 12. Demography

A. Introduction

Demographic methods for monitoring rare plants are popular, and a number of authors have emphasized these techniques over other monitoring approaches (Menges 1986, 1990; Pavlik and Barbour 1988; Kaye and Meinke 1991; Pavlik 1993). Of 98 recovery plans prepared for the U.S. Fish and Wildlife Service between 1980 and 1992, 84% proposed some form of demographic study or monitoring (Schemske et al. 1994). The methods are powerful, but also extremely time-consuming. They may be appropriate for some monitoring situations; for others they may be an over-allocation of precious and limited monitoring resources. The methods are most appropriate for species with certain life history patterns and morphologies.

This chapter is designed as an introduction to demographic methods for monitoring plant populations. The intent is to demonstrate both their power and usefulness and some of the difficulties in using them. Information provided here is insufficient to design a demographic monitoring project; an adequate treatment would require an additional technical reference. There are a number of good texts and papers on the subject, which are recommended for additional reading if you wish to learn more: population biology (Harper 1977; Solbrig 1980; Silvertown 1987; Hutchings 1986); matrix models (Menges 1986, 1990; Huenneke 1987; Manders 1987; Caswell 1989; Manly 1990); and comparisons under different management regimes (Hartnett and Richardson 1989; Eldridge et al. 1990; Charron and Gagnon 1991; Silva et al. 1991).

Consider the following descriptions of demography:

"Demography deals with the quantitative aspects of birth, growth, reproduction and death in a population." (Solbrig 1980)

"Ecology, genetics, evolution, development and physiology all converge on the study of the life cycle. The vital rates on which demography depend describe the development of individuals through the life cycle. The response of these rates to the environment determines population dynamics in ecological time and the evolution of life histories in evolutionary time." (Caswell 1989)

These two descriptions describe the unifying feature of demographic approaches: the measure of individuals and some measure of their success or fate (e.g., reproductive output, growth, mortality). At its simplest, a demographic approach may try to measure the number of flowers produced per individual annually, using changes in flower production as a measure of trend. In a more complex example, success at all phases of the life cycle may be measured and combined into a model of population dynamics.

A simple human illustration of demographic information for two Old-Time Fiddlers groups is shown in Figure 12.1. Most "recruits" for these musical groups are the kids of people who have played this type of music. A simple count of the two organizations suggest that they are equal. A comparison simply by sex suggests that there may be a problem. Apparently it is not cool to be a male fiddler in Picabo, Idaho. When age classes are added, it becomes clear that the Picabo group is headed for extinction. The interpretive power in the latter two groupings by sex and age lies in your knowledge of humans: their life-span, their age of marrying, and the importance of both sexes for continuation of the "population."

Demographic techniques for plants take advantage of similar understanding about the life cycle and ecology of plants. Information that includes the percentage of individuals within stage classes (e.g., 34 individuals are seedlings, 54 non-reproductive, 102 reproductive, and 21 senescent) and the likely probability of moving between stage classes (transitions), provides much more understanding and insight into the potential viability of a population than simply knowing the population size (211 individuals).

Demographic approach		Number of individuals			
		Old Time Fiddlers; Aspen, CO		Old Time Fiddlers; Picabo, ID	
none		150		150	
by sex		Male	Female	Male	Female
		70	80	5	145
by sex and by age	0 - 5	6	8		
	6 - 12	8	7		
	13 - 18	14	17		1
	19 - 27	16	22		2
	28 - 45	13	19		26
	46 - 60	9	6		45
	61 - 75	3	1	3	52
	> 75	1	0	2	19

FIGURE 12.1. Summary of two populations of Old Time Fiddlers. While the size of the groups is identical, the demographic distribution of the two groups differs dramatically. The likely fate of the two groups is also quite different.

B. Basic Concepts

1. Evaluating populations in stage or age classes

In animals, age classes are most often used to identify demographic stages, since life cycle events such as reproduction and death are fairly closely tied to certain age groups. Plants are more plastic, with life cycle events not always related to age. Trees, for example, can reproduce at a very young age on favorable sites, but may remain suppressed under the canopy for decades before release and reproduction. Plants are also extremely difficult to age, with the exception of trees, for which age classes can be determined by cores, and some herbaceous plants with annual abscision rings. For most plant populations, stage classes (such as seedling, reproductive, etc.) are more easily observed and interpreted than age classes. Most demographic studies use stage classes, but age classes may be appropriate in some situations, such as fire ecology studies. Dioecious plants may also require including sex as a class in the model.

Study objectives and the life history and morphology of the plant will dictate the number of classes you may wish to designate. Stage classes must be consistently recognizable in the field in order to be useful. Criteria for delineating classes can include phenology (reproductive or non-reproductive), size (rosette diameter, height, etc.), number of leaves, or other divisive characteristics. The best class divisions are those that reflect some ecological meaning. Reproductive plants, for example, obviously have a different function in the population than non-reproductive plants. Height may be ecologically important if sunlight is limiting, and a certain threshold height allows access to the canopy.

Gatsuk et al. (1980) recognize 10 stage classes in 4 main life stages, consistent across several life forms. Examples and illustrations of application of the classes are given for trees, shrubs, tussock grasses, and clonal species. Stages identified are latent (seed), pre-reproductive (seedling, juvenile, immature, virginile), reproductive (young, mature, old), and post-reproductive (subsenile, senile).

2. Environmental influences

Natural environmental factors affect individuals in different ways at different stages (refer to the ecological model section in Chapter 4—Management Objectives). Most plants, for example, have more specialized moisture requirements for the germination and seedling establishment stages compared to the adult stage. Native herbivores such as deer may browse flowering or fruiting plants more heavily than inconspicuous non-flowering plants. During drought, mortality may be highest among young plants, while older plants with larger root reserves may survive.

Human factors also may have different effects on different stages of the life-cycle. For example, domestic livestock will graze flowering stems of the biennial *Thelypodium repandum*, an east-central Idaho endemic that grows on unstable volcanic substrates. Germination within livestock exclosures, however, is nonexistent, compared with abundant germination around the periphery of exclosures and on adjacent unprotected slopes (Elzinga 1996). Domestic livestock reduce reproductive output, but also increase germination and establishment of seedlings. Is the overall effect of domestic livestock beneficial or detrimental? In another example, many populations of *Penstemon lemhiensis*, a species endemic to east-central Idaho and adjacent Montana, are found on disturbed road cuts and fills. The small populations on these sites, however, consist of mostly reproductive short-lived individuals. Populations in undisturbed habitats contain mostly non-flowering individuals (Elzinga 1997). Is *Penstemon lemhiensis* benefitted by road building disturbance? These types of complex interactions are difficult to address with standard vegetation measurements of cover, density, or frequency, but can be investigated with demographic techniques.

3. Three types of demographic approaches

Several different kinds of studies are termed "demographic" in the literature. The basic unifying theme for demographic studies is that the approach considers age or stage classes and rates of mortality, recruitment, or growth. Demographic studies can be generally classed into three types.

Population modeling and viability analysis. This type of approach is what many rare plant managers mean when they use the term "demographic study." Field work usually involves marking or mapping individual plants. The fate of individuals in all stages of the plant life cycle is measured and a model constructed that can be projected into the future. Based on the model, the future (viability) of the population can be assessed. Most of this chapter is devoted to this technique because it is the one most commonly used by rare plant managers, and because considerations pertinent to this technique are germane to demographic techniques in general.

Single age/stage class investigations. These studies focus on a single or few stages, such as the measure of reproductive output, survival of seedlings, or longevity of adults. Such investigations are also briefly discussed under "vigor measures" in Chapter 8—Field Techniques.

Demographic structure. The demographic structure of a population is the distribution of individuals in age or stage classes, e.g., the percentages of the population that are seedlings, juveniles, non-reproductive adults, reproductive, and senescent (old and dying). This is a point-in-time measure, and is the same approach as density counts by life form discussed in Chapter 8—Field Techniques. Monitoring can measure the change in demographic structure from year to year. Such monitoring may be more sensitive to downward changes in the population than simply measuring density (see Chapter 8—Field Techniques).

C. Population Modeling and Viability Analysis

1. Description

A key tool in population modeling and viability analysis of plants is the transition matrix, also called the Lefkovitch matrix (Lefkovitch 1965). Each cell in this table represents the probability that an individual will move to another class, a transition. A matrix of a species with a simple life history is illustrated in Figure 12.2. This plant has only three classes: seedling, rosette, and reproductive. It is a short-lived perennial. The values in the matrix represent the proportion of each stage making a transition. Seedlings in Year 1 are no longer seedlings in Year 2; they are, by definition, second year plants. They either become rosettes (10% of them) or flowering individuals (1%). The remaining 89% of the seedlings die in their first year. Some rosettes stay rosettes (25%); others flower the following year (15%).

	FROM:		
TO:	seedling	rosette	flowering
seedling	0.00	0.00	60.00
rosette	0.10	0.25	.20
flowering	0.01	0.15	.45

FIGURE 12.2. Simple life history and transition matrix for a plant species. The plant has no seed bank; seed production is moderate (an average of 60 effective seeds per flowering plant); and plants survive after flowering. The transitions between most stages are a percentage (less than 1.0), representing the percentage of the stage that moves to another stage. Some cells in the matrix are zero; these are transitions that do not occur. Seedlings, for example, are no longer seedlings by the following year.

The remaining 60% die. Of the reproductive individuals, 20% revert to rosettes the following year, and 45% flower again. The remaining 35% of the flowering individuals die before the next year. Flowering plants also produce seedlings (average of 60 seedlings per flowering plant). Seeds are not stored in the seed bank; thus, the number of seedlings produced the following year divided by the number of reproductive plants is the flowering to seedling transition. Note that this number is greater than 1.0 because the flowering plants are producing more than one seedling each.

This is an extremely simple life cycle and matrix. If size stages supplement the phenological stages (e.g., small and large non-reproductive and reproductive), the matrix will expand accordingly, and the number of transitions to be measured will also increase. Matrices can become extremely complex if many stages are identified. Nault and Gagnon (1993), for example, identified 15 stages in an onion, *Allium tricoccum*, based on the number of leaves and reproductive class, resulting in a matrix with 255 transitions.

2. Uses of matrices

a. Population parameters

Matrices can be used to calculate important population parameters. The most important for conservation purposes is lambda (λ), also called the finite rate of increase. Lambda values of > 1.0 represent a population that is increasing. The larger the lambda value, the more rapidly the population is increasing. Lambda values equal to 1.0 represent a population that is stable. Lambda values of < 1.0 are from populations experiencing decline.

Another useful table of values that can be calculated from a transition matrix is an elasticity matrix, which is the same size (same number of cells and form) as the transition matrix. Elasticity values are a measure of the sensitivity of the population growth rate to a change in the transition probability (deKroon et al. 1986; Caswell 1989). Changes in transition probabilities for transitions with high elasticity values will result in a greater change in the overall lambda value than a similar change in probability for a transition with a low elasticity value. Elasticity values can identify which stages and transitions should be managed to provide the largest overall population benefits. Note that for some species, the transition(s) with the highest natural variability may be the most limiting to population growth or survival (Schemske et al. 1994).

b. Projection

The matrix can be projected into the future by simple matrix algebra. Figure 12.3 demonstrates projection 1 year into the future. Each successive year can be projected by replacing the previous population values with the ones calculated. In this manner, populations can be projected many years into the future. The math can be done with a calculator (if you have time and persistence). You can also design a simple program in conjunction with a spreadsheet application. Of several commercial packages available, the one most appropriate for plant demography studies is RAMAS/stage (Ferson 1990), a matrix-based model that uses stage classes. Other matrix-based packages include RAMAS/space and RAMAS/age (Ferson and Akcakaya

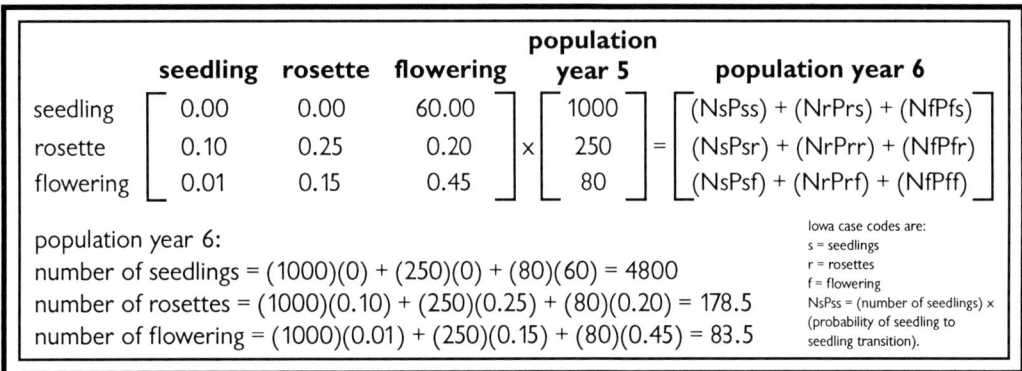

FIGURE 12.3. The matrix in Figure 12.2 is used here to project the population into the future. The population was measured in Year 1 and Year 2 to construct the matrix. The population was then projected into the future, year by year. Figure 12.3 shows the calculations from the projection of Year 5 to Year 6. The population of Year 7 can be calculated by replacing the values under "Population Year 5" with the Year 6 values of 4800, 178.5 and 83.5.

1990; Akcakaya and Ferson 1992). Individual-based models, some of which examine the reproduction and survival of different genotypes, are more often used for animal studies, but may be appropriate for some plant studies. Three examples are GAPPSII (Downer et al. 1992), VORTEX (Lacy and Kreeger 1992), and ALEX (Possingham et al. 1992).

Modeled projection is often confused with prediction. Projection describes what would happen if the measured conditions continue. Prediction attempts to describe what *will* happen (Caswell 1989). The difference is subtle, but important. A single transition projected into the future provides little predictive power because conditions occurring during the measurement period are unlikely to typify the projection period.

c. Sensitivity analysis

You can replace transition values within the matrix with other values and determine the effect on the population projection. For example, what happens to the population growth rate (λ) if reproduction is increased (perhaps by removing herbivores observed to consume inflorescences)? Simulations of the population response to changes in the transition values can identify areas where management can be most effective. The elasticity matrix shows the transitions on which to initially focus.

d. Viability analysis

Viability, a popular term in many Land Use Plans, is the persistence of a population or species into the future. Unfortunately, while viability is simple in concept, it is difficult to measure. Populations and species are never completely free from the risk of extinction; therefore, viability is a probabilistic concept, not an absolute one. The concept of viability also requires a consideration of time frame. For example, an incomplete objective would be: "maintain a viable population of Species X." A better objective is: "Population Y of Species X should have a 95% chance of persisting to the year 2100."

Demographic modeling is our only tool for measuring the viability of a population. Let's assume you have measured three transitions for a population (three transition matrices): a "bad" year, with a lambda value of 0.35 (lots of mortality from drought), an average year with a lambda value of 1.00, and a good year with a value of 1.76. A computer program can project these three matrices into the future for a specified number of years, perhaps 500. The computer can choose matrices at random each year, and calculate and project the population size through time to the year 500. If, by chance, the computer uses data representing many bad years in a row, the population may crash (go extinct) after only a few years. If the computer (again by chance) used transition data representing many good years, the population may survive to year 500, even grow. By doing a large number of these simulations, a frequency distribution is generated of the time to extinction for your population, based on random repeats of the three transitions you measured. Figure 12.4 illustrates a distribution of the probability of extinction. This frequency distribution can be presented in a different form, the cumulative extinction probability. This is the cumulative probability of extinction for that year and all the previous years. The cumulative probability that the species will become extinct within 100 years is 80.1% (Figure 12.4).

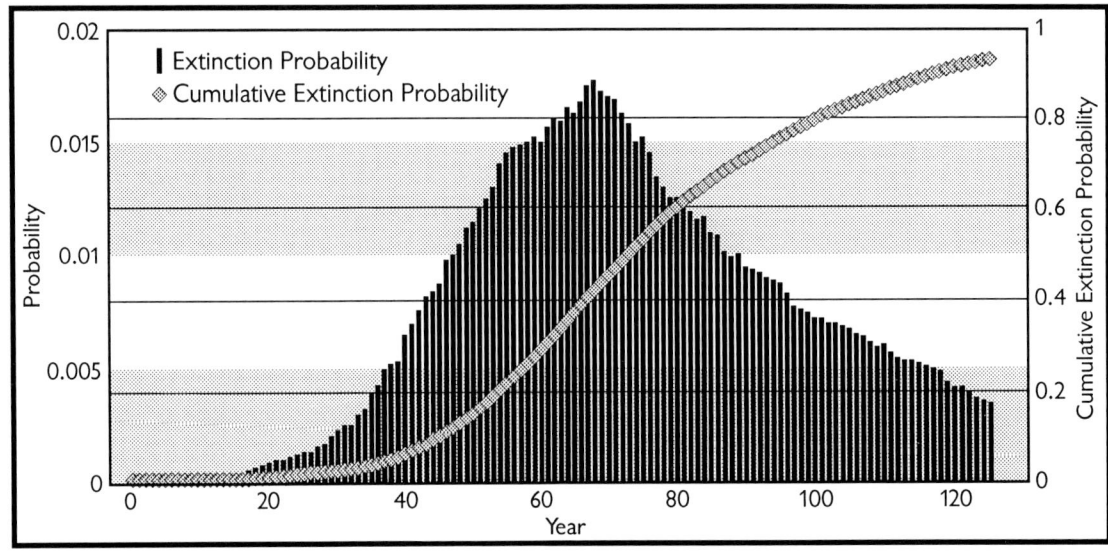

FIGURE 12.4. Extinction probabilities calculated from simulation using measured transitions. The long "tail" on the right indicates rare simulated successes, or survival, to the end of the simulation period. The majority of extinctions in this simulation, however, occur before 120 years have passed, as illustrated by the cumulative extinction probability.

You can run such simulations with either a random selection of complete matrices, or random selection of the transition probabilities for each stage (Menges 1986, 1987). Biologically, using complete matrices assumes that the transitions in each stage class are linked. Using a random choice of each stage assumes that the stage classes are not linked (conditions affecting the survival of seedlings are not related to those affecting survival of adults).

If annual variation is thought to be largely due to weather patterns, you can increase the biological realism of your simulations by matching transitions with long-term weather records. The number of times a given matrix or stage is used in a simulation can be dictated by the number of times a similar weather year arises in the weather record.

A final simulation tool is to alter transition probabilities, similar to sensitivity analysis, and then calculate the effect on extinction probabilities. For example, simulation can model the effect of excluding livestock from a population one year out of every three. If the percent utilization of inflorescences is known, as well as the percent success of surviving inflorescences, the probabilities for these transitions can be adjusted for the periodic rest option. The population response as measured by changes in modeled extinction probabilities can then be simulated. Fiedler (1987) provides a good example of this approach applied to a rare *Calochortus* species and Kalisz and McPeek (1992) for a winter annual.

e. Comparison of populations or treatments

Demographic parameters such as lambda or the extinction probability distribution can be sensitive tools for comparing treatments or populations. Treatments such as fire can often have variable effects on different stages (e.g., cause high mortality in the adult stage, but stimulate seedling establishment). If you only measure a single life stage (such as only the adult stage in the fire example), the results may suggest that a treatment is damaging, but if you consider all life stages together, the treatment may actually be beneficial. The advantage of demographic techniques is that they measure the integrated effects of a treatment on the population (i.e., the growth rate of the entire population), rather than on a specific attribute such as density, cover, or frequency, or on a specific life stage.

Simulation and sensitivity analysis can also be used to compare proposed treatments. In the grazing example above, simulation could be used to compare treatments of complete rest, every other year rest, and every third year rest.

3. Constructing a matrix

a. Identify stage classes

Stage classes should fit three criteria. First, they should be consistently recognizable in the field, and be so clearly defined that different observers will class individuals similarly. Second, the classes should have biological relevance so that the results can be interpreted ecologically. Third, the classes should complete the life cycle of the plant. No stage can be missed, including cryptic stages such as dormant seeds in the seed bank. Also include dormant plants if individuals "take time off" and remain below ground for a year or more (Gilbert and Lee 1980; Bierzychudek 1982).

Avoid splitting classes unnecessarily. Too many groups create classification difficulties in the field, and increase the complexity of the model. More important, unless you can measure all the individuals in the population, the transition probabilities will be estimates from a sample. The more classes, the more individuals you will need to measure to get a reasonably precise estimate for each transition probability.

b. Develop a model of transitions

Once you have determined the stage classes, you must identify the potential transitions between those classes and the transition period. Placing the stages on two lines, the top time t and the bottom time $t+1$, and connecting stages that represent transitions is an easy visual approach (Figure 12.5). These diagrams can sometimes help identify stages that are missing from your list (such as seed bank or dormant individuals). Another typical form is to put all the life cycle stages into a circle that represents a year, and join the stages of possible transitions with arrows.

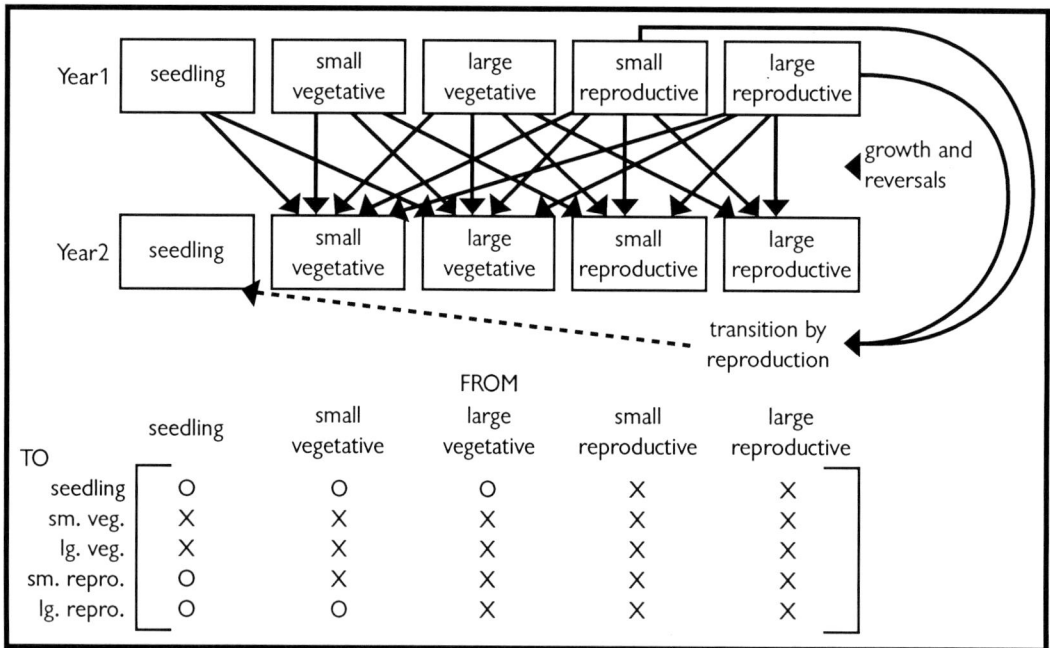

FIGURE 12.5. Life cycle diagram for a species with no seed bank, and a small and large class in both the vegetative and reproductive stage. Arrows represent possible transitions. Those possible transitions are identified in the transition matrix by an "X", while those that are biologically impossible (or known to not occur) are given a "zero" in the matrix.

The period between measurements is usually a year; thus, time t is Year 1 and time $t+1$ is Year 2 in Figure 12.5. Some species, such as those with a spring and fall germination "flush" or those with spring germination followed by late summer vegetative reproduction, may need to be measured more often. Including these dramatically increases the complexity of the model because the spring to fall transitions are so different from the fall to following spring transitions. However, this may be necessary to the accuracy and usefulness of the model (Caswell and Trevisan 1994; Van Groenendael et al. 1994). Rarely will the period between measures be greater than one year, unless seed germination is extremely rare (skipping years would miss the appearance of seedlings) and unless changes in adults are slow. It may be acceptable to monitor some tree species with periods of greater than one year.

c. Examples of transition matrices

Figures 12.5 through 12.9 illustrate five examples of transition matrices. Several types of species growth forms and life cycles are included.

D. Field Techniques

1. Marking/mapping individuals

The demographic analysis described above requires measurement of the stage class of individuals from year to year. You must mark or map individuals in such a manner that you can relocate each with confidence at the next measurement.

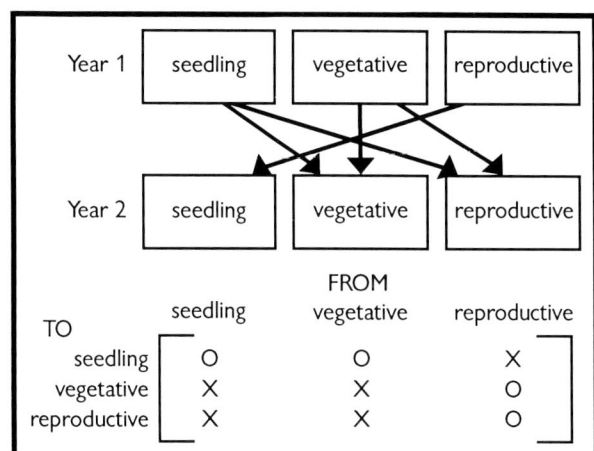

	FROM		
TO	seedling	vegetative	reproductive
seedling	O	O	X
vegetative	X	X	O
reproductive	X	X	O

FIGURE 12.6. Life cycle diagram of a short-lived perennial. Note that reproductive individuals can only transition to the seedling stage, because they die after flowering. The plant also lacks a seed bank. Potential transitions are shown by arrows in the diagram, and by an "X" in the transition matrix.

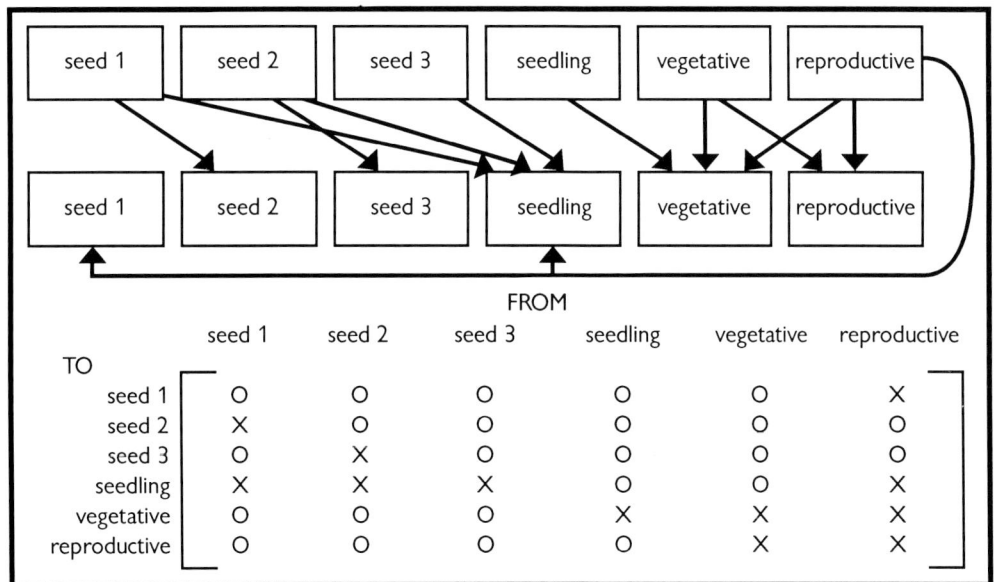

	FROM					
TO	seed 1	seed 2	seed 3	seedling	vegetative	reproductive
seed 1	O	O	O	O	O	X
seed 2	X	O	O	O	O	O
seed 3	O	X	O	O	O	O
seedling	X	X	X	O	O	X
vegetative	O	O	O	X	X	X
reproductive	O	O	O	O	X	X

FIGURE 12.7. Life cycle diagram of a perennial with a seed bank. Seeds in the seed bank are assumed to last only about 3 years. Note that seeds can only transition to the next year's seed class, or to seedlings. Some seeds have no dormancy and germinate the following spring (the reproductive to seedling transition). Possible transitions are shown by an "X" in the matrix. Some authors have split the seedling stage into classes that identify the age of the seed source (e.g., seed 1 to seedling 1, seed 2 to seedling 2, etc.) (Kalisz and McPeek 1992).

Most investigators use a combination of marking and mapping. Potential markers include swizzle sticks and coffee stirrers, popsicle sticks, pin flags, nails with tags attached, and tags that attach directly to the plant. Each marker must have a unique identifying number or label. These can be directly written on some objects (e.g., tick marks can be used on coffee stirrers). Ensure that the numbering tool or ink used will survive to the next measurement period.

All the markers listed above have disadvantages. Swizzle and popsicle sticks and coffee stirrers are easily uprooted by hoof action, frost heaving, or slope movement. Pin flags are noticeable, and may be removed or vandalized. Nails with tags are stable and inconspicuous, but the tags may blow in the wind and damage surrounding plants. Tags affixed directly to the plant can damage the plant, either by blowing around and damaging tissue, or by being too tight and restricting growth. Will the metal nails, for example, alter the micro-habitat near the plant (perhaps by adding minerals or by heating the ground)? Will brightly colored markers attract humans or animals? Will fence posts become perching spots for raptors or rub posts for livestock?

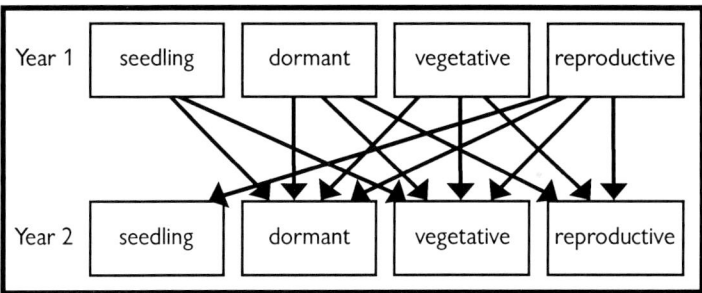

FIGURE 12.8. A perennial with a dormant phase. Note that plants in this model can be dormant for more than one year. The species has no seed bank, although reproductive individuals can flower more than once.

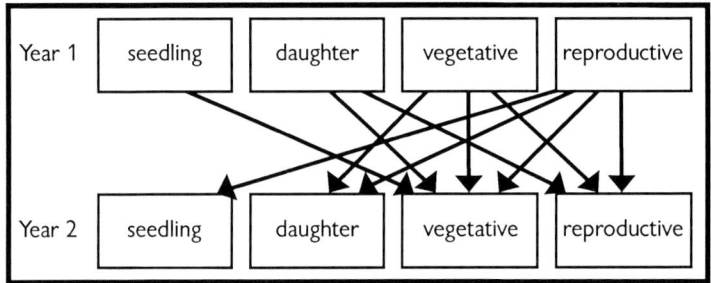

FIGURE 12.9. A life cycle with clonal growth. Daughter plants can be traced to vegetative or sexually reproductive plants.

Since markers are easily lost in the field, a back-up system of mapping is recommended. One method that works well for large sparsely distributed individuals is to use permanently staked baselines and to map each plant in terms of distance along and distance from the baseline (Figure 12.10). Other authors have used a transparent mapping table suspended over a permanently monumented plot, coordinate systems in small plots, or photoplot techniques (Cullen et al. 1978; Owens et al. 1985; Schwegman 1986; Windas 1986; Lesica 1987; Chambers and Brown 1988; Pyke 1990; Muir and McCune 1992).

Individuals of species found in high densities or associated with dense vegetation will be especially difficult to mark and relocate. Some species simply cannot be relocated with any degree of assurance. Before measuring many plots with a proposed method,

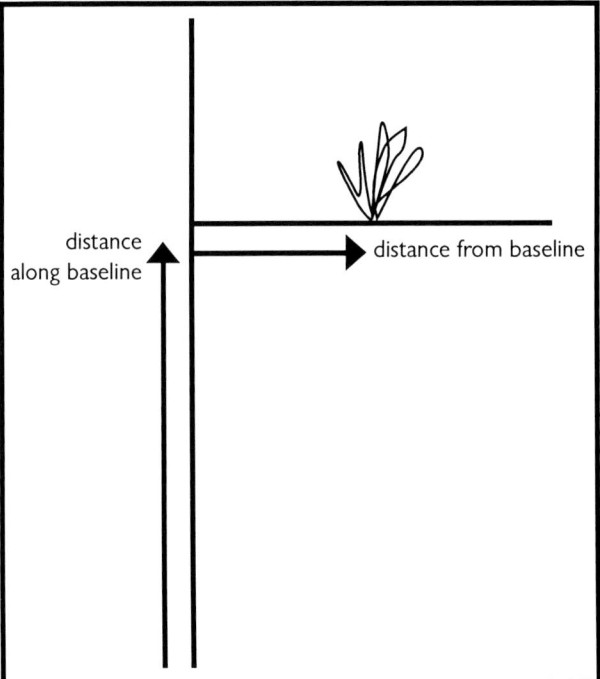

FIGURE 12.10. A potential method for mapping plants in populations of widely spaced, fairly large individuals.

try to measure and remeasure a few to determine if relocation will be possible on subsequent years. Using several trials measured by different people will test your method for consistency among observers.

2. Class delineation

Classes must be consistently recognizable and the boundaries of the class clearly articulated to communicate between different observers. Test classes with different observers before using a particular set of classes.

For seedlings, possible boundaries can be described by the number of true leaves and the presence and size of cotyledons. You can also identify all "new" plants as seedlings if you marked or mapped all plants the previous year and there is no vegetative reproduction or plant dormancy. If classes are separated based on reproduction, a rule for plants with few or aborted flowers/fruit will have to be made. Other potential divisions can be based on number of leaves, size of basal rosette, height of plant, height of flowering stem, number of flowering stems, etc.

3. Timing of monitoring

Schedule monitoring that uses classes such as seedlings or reproductive individuals, or that focuses on certain size classes, so that annual measurements are done at the same phenological time each year. If classes are based on reproduction, for example, the number of individuals included in each class will usually change as the season progresses. This can be even more problematic in populations that have various phenological stages expressed simultaneously. For example, a population situated on a gradually changing aspect may have plants that are fruiting on the most southern aspect and plants in bud on the least exposed aspect.

4. Sampling methods

Some populations are small enough that you can mark, map, and measure all the individuals in the population, and avoid the need to design a sampling strategy. Often, however, at least the seedling class, which may be quite dense even when adults are few and sparse, will need to be sampled rather than measured.

In populations that are too large to measure all individuals, you will likely estimate demographic parameters in quadrats. Density (such as number of seedlings) can be estimated from a sample of quadrats as sampling units. A cluster design (see Chapter 7) is appropriate for demographic sampling situations that involve estimating some parameter of individuals, say number of seeds produced per plant or plant height. You can also use a cluster sampling design to estimate the percentage of individuals moving to another stage class. Note that this is different than measuring a continuous variable such as height or number of seeds because the fate of the plant can be classified into one of several transitions (such as reproductive to dead, or reproductive to reproductive) and is thus estimated as a proportion. Cochran (1977; section 9.7) provides information on cluster sampling for proportions.

5. Measuring reproduction

Reproduction includes sexual (seed formation) and asexual (vegetative) reproduction. Both forms must be included in the model for species that form seeds and spread vegetatively. A number of studies have measured vegetative reproduction in the form of shoots or daughter plants (Silvertown et al. 1993). Counting rules must be developed for these. One option is to use a distance threshold from the parent plant; shoots exceeding that distance would be considered new individuals. Another approach is to use a threshold size for shoots. For some bulb species, the number of leaves can indicate when the bulblet has "split" from the parent plant (Prentice 1988, 1989; Nault and Gagnon 1993).

For species that do not spread asexually, reproduction is completely represented by the number of viable seeds produced. Counting or estimating large numbers of seeds produced per plant may be difficult. For plants with many seeds in few fruits, fruits can be counted and the average number of seed per fruit estimated by counting the seeds within a sample of fruits. For plants with dense spikelike inflorescences with many fruits, fruit number and seeds per fruit can both be estimated from a sample of inflorescences and correlated with inflorescence length. The type of sampling design (cluster, two-stage, or random) should be considered carefully to avoid bias caused by different sized individuals and different reproductive output (Pickart and Stauffer 1994). Chapter 7 discusses the difficulties in selecting a random sample of individual plants and offers some possible strategies. The same types of considerations apply to taking a random sample of inflorescences.

6. Measuring the reproductive to seedling transition

For plants without seed banks, the transition between reproductive and seedling classes is simply the number of seedlings that appear, averaged over the number of reproductive plants. Each reproductive plant is assigned a value for effective reproduction based on its number of seeds. For example, if Plant #1 produced 100 seeds and Plant #2 produced 200 seeds last fall, and this spring the number of seedlings was 30, the effective reproduction for Plant #1 is 10, and for Plant #2, 20. The value in the transition cell would be the average value (in this case 15) with some measure of the precision of the estimate (e.g., the standard error). The precision of the estimate is an important issue, and is addressed in more detail in Section E.4. If all the plants in the population are measured, the transition cell value is simply the average.

If plants form seed banks, the problem of estimating the reproductive transitions becomes more difficult. Of the seed produced, some portion is non-viable, some portion is lost to seed predators, some portion forms seedlings, and some portion is stored in the seed bank. You can lump the first two fates as mortality, but the last two fates, burial in the seed bank and germination, must be included in the model. As evident in the example shown in Figure 12.7, each year the seedlings that appear above ground may come from two sources: the seed bank or last year's seedcrop. Complicating the model is the problem that some seeds may be 2 years old, some 3, and so on. Reliable methods for estimating these transitions have not yet been developed. Most current approaches use seed burial of known quantities and observed germination rates (Kalisz 1991). See more on this issue in Section E.1.

7. Measuring the fate of seedlings

Measuring the fate of seedlings can be difficult if high seedling densities are impossible to mark and map as individuals. The fragility of seedlings adds to the difficulty of tagging them. One approach is to simply count the total number of seedlings. If all other individuals within the population or plot are accounted for, seedlings that survive will be evident as "new" young plants the following year, while those that die will disappear. The percentage transitioning to the next class is the number of "new" plants divided by the number of seedlings counted the year before.

E. Challenges

1. Certain life forms

Annuals, geophytes, and plants with a dormant phase all present problems in using demographic models because the hidden phases (underground seeds, bulbs, or roots) are difficult to measure. One approach used successfully for dormant plants and bulbs is to carefully excavate the root or bulb and determine if it is still alive (Bierzychudek 1982; Nault and Gagnon 1993). Another approach is to wait to declare mortality until the individual has been missing for more than 1 year.

Seed bank and germination ecology are especially important to annual plants, but information on the dynamics of seed banks and germination is extremely difficult and time-consuming to gather. In measuring seed dynamics, several factors need to be included: rate of seed mortality and aging, the amount of seed removed by predators, and the variability in germination events (weather related, exposure of buried seeds). The very fact the buried seeds (and even exposed ones) are not easily visible makes any investigation of them difficult.

Even if some data are available, seed bank dynamics are difficult to address in demographic models. Some authors have treated all seeds within the seed bank as a single class, ignoring age classes of seed (Schmidt and Lawlor 1983; Pyke 1995). While this might be appropriate for some species, Kalisz and McPeek (1992) demonstrated that an age-structured seed bank was critical to successful modeling of a winter annual. Mortality rates of stored seed and germination rates and timing of germination from the seed bank are often age-dependent (Leck et al. 1989; Kalisz 1991; Philippi 1993), and including these factors may be important to the predictive power of the demographic model. A further complication is that many species have "rescue" episodes from the seed bank, with large flushes appearing when germination conditions are suitable. These events may occur only once per decade, or once in several decades. Since the monitoring period will not likely include these rare events, models developed from the measured transitions will not provide a good mirror of the true population dynamics.

Rhizomatous growth forms are difficult to monitor using demographic techniques because of the problem of defining an "individual." While the "individual" in the model does not have to be an isolated genet, it does have to be consistently recognizable in the field. Studies have been done on shoots of rhizomatous species (Bernard 1976; Geber et al. 1992), but since shoots are often relatively short-lived, they must usually be measured more than once per year. An additional practical problem with rhizomatous species is marking and mapping the closely growing shoots.

Some species present no realistic recognizable unit. Rhizomatous matted plants and densely growing rhizomatous grass species (visualize the common lawn grass, Kentucky bluegrass) cannot be monitored using demographic techniques, simply because no consistent unit can be distinguished, marked, mapped, and measured.

A final problem with rhizomatous species is that of modeling vegetative reproduction and shoot mortality. First, because shoots are simply part (ramets) of a genet, the implications of the death of shoots for the viability of the population of genets are unclear in a matrix model. Second, because shoots arise from the roots rather than from another shoot, the transition from a reproductive class to some juvenile class cannot be modeled. Seedlings are often rare in rhizomatous species, thus shoots produced from the root are obviously important to the survival of the population. But the matrix models presented here do not adapt well to reproduction that cannot be traced to another stage. Some plants that reproduce vegetatively have been successfully modeled (Nault and Gagnon 1993), but for these species the vegetative daughter can be traced to an individual. A familiar example of a vegetatively reproducing plant that is amenable to matrix modeling is the garden strawberry, whose daughters can be traced to the parent plant by the runners.

2. Variability in time

Published matrices demonstrate significant variability from year to year in transition probabilities (Bierzychudyk 1982; Reinhartz 1984; Kalisz and McPeek 1992; Nault and Gagnon 1993). In other words, plants have "good" years and "bad" ones.

Because of this variability, lambda values can be misleading. Lambda values of less than 1.0 represent a population in decline, but this may not mean the population is in trouble. Declines can be natural downward adjustments caused by density-dependent factors, or short-term declines in a naturally fluctuating population (Figure 12.11). A continual decline is of conservation concern, but what constitutes "continual" over natural fluctuation varies by species, making conservation decisions difficult.

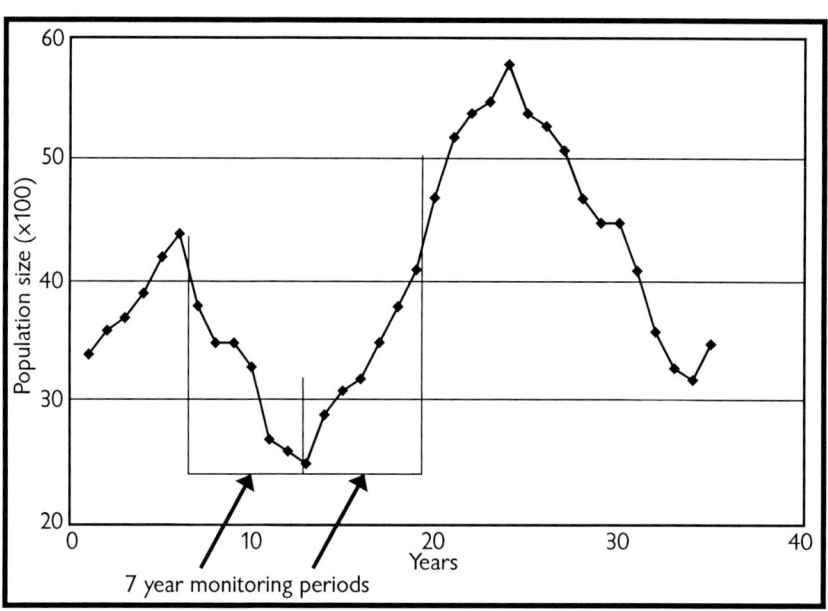

FIGURE 12.11. This diagram shows a population with naturally fluctuating numbers. Note how different projections and conclusions would be based on the first monitoring period compared to the second.

Lambda values can also be misleading if some measure of their annual variability is not reported along with the average value. Many published studies report only the average lambda value, calculated over the number of transitions measured. This provides only half the picture. An average lambda value of 0.82 (a population in decline) calculated from three transitions that are all less than 1.0 has different conservation implications compared to the same average value calculated from two good years (lambda values of greater than 1.0) and one very poor year. (Note that average lambda values cannot be calculated simply by adding up the values and dividing by the number of transitions. See more on this in Section E.4., below.)

Extinction probabilities, even those calculated from several transitions, are suspect for similar reasons. Some extinction risks are directional and caused by systematic factors, such as habitat degradation by humans and successional changes. Other natural risks of extinction for populations are stochastic, meaning they are random in time and space, and not predictable. Four types of stochastic risks can be recognized (Menges 1986, 1987). *Demographic stochasticity* involves random survival and reproductive occurrences. Consider, for example, a single pair of animals that colonizes a new habitat. By chance, it is possible for them to have a litter of all male offspring and for the female to die soon after reproduction. Such a chance event could be catastrophic for the population. Demographic stochasticity is most important in small populations. *Environmental stochasticity* results from random environmental influences, such as weather, habitat changes, and herbivores. *Genetic stochasticity* may affect populations by random changes in gene frequencies, especially in small populations (Lesica and Allendorf 1992). *Natural catastrophes* are environmental events with large impacts, such as fire and floods, which occur infrequently.

By their very nature, stochastic events are very difficult to include within a model because they are difficult to predict. They are, however, critical to estimates of population viability. Menges (1990) states: "Catastrophic mortality dominates estimates of population viability, causing the majority of extinctions when added to within-population environmental stochasticity." Unfortunately, few long-term demographic studies provide empirical insight into the impact of these stochastic events on population viability.

3. Variability in space

Plant populations not only vary in time, but the individuals within a population vary in space. Some botanists recommend placing permanent plots or lines in the area of the population containing the most plants (Lesica 1987). While this may be appropriate for some situations, such bias fails to incorporate the variability within a population. Population dynamics in the densest part of the population will probably differ from those on the periphery (Figure 12.12). In especially dense populations, there may be density-dependent interactions in the densest areas. High density portions of the population are also most likely to occur on the prime habitat of the population area; changes due to environmental stress will occur first on the less suitable but occupied fringe habitat.

If you wish to draw conclusions about an entire population, you must include the entire population in the monitoring. For large populations this will require sampling, most likely in randomly located plots (Chapter 7). Sampling the population, however, adds another level of variability in the analysis, and further complicates the calculations.

4. Dealing with variability

Assume a demographic study in which you monitor all individuals of a population rather than a sample, thus eliminating the variability in space. You monitor the population for 4 years (3 transitions). It would be tempting to simply average the 3 lambda values for an average lambda value over the 3 transitions. This approach would be incorrect because the lambda values are not linear, nor are the effects of the various transition values. Lambda values for declining populations can only range from greater than zero to less than 1.0. There is no upper limit for increasing populations, although the highest published lambda value is 11.8 (Silvertown et al. 1993).

When populations are sampled rather than completely monitored, there are two sources of variability for each cell within the matrix: time and space. Because the values within each cell are estimations, with a certain associated sampling error (Chapter 5), the lambda value calculated from the matrix also has an associated sampling error. Few published studies include an estimate of this error.

Methods to deal with variability (either in time, in space, or both) are basically of two types: analytical methods that assume either a normal or binomial distribution (depending on the type of data) and computer-intensive resampling methods (Chapter 11 and Appendix 14). The latter are probably the better approach because they require fewer assumptions. Several resampling methods have been developed, but all involve randomly choosing values within the range of variability and repeatedly recalculating the demographic parameter of interest. This can be done by estimating a confidence interval for each transition cell, then repeatedly drawing a value for each cell from that range, and calculating a lambda value. Alternatively, a sample of individuals could be randomly and repeatedly drawn from all the individuals measured, transitions calculated for each sample, and a distribution of lambda values generated. Extensive explanation of these methods is beyond the introductory nature of this chapter,

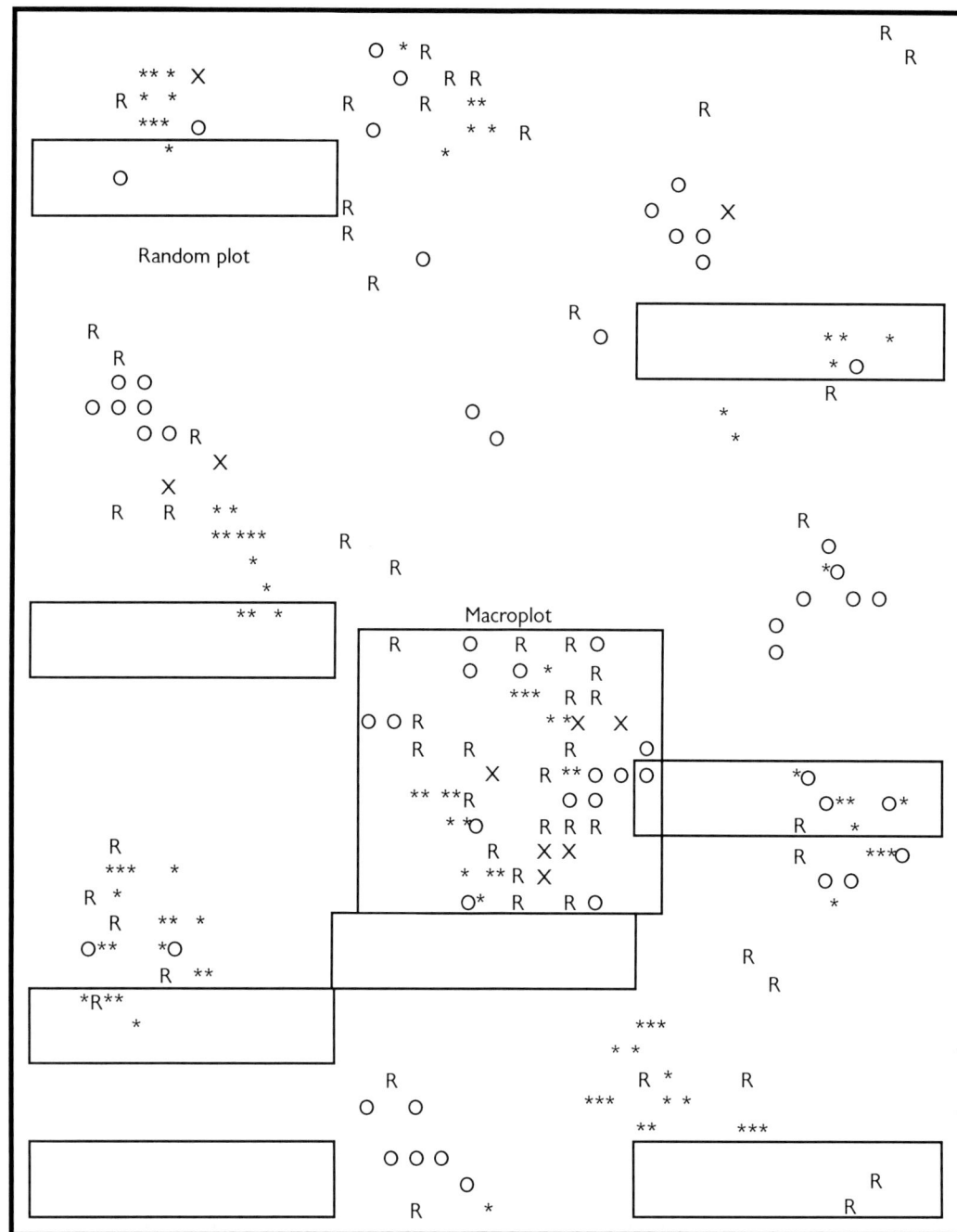

FIGURE 12.12. Population is represented by nonreproductive individuals (O), reproductive individuals (R), seedlings (*), and dead individuals (X). Population structure varies in space. A macroplot approach is positioned the plot in the densest portion of the population. An alternative approach is to use randomly located sampling units.

but can be found in Meyer et al (1986), Caswell (1989), Alvarez-Buylla and Slatkin (1991, 1993), Kalisz and McPeek (1992), and Stewart (1994).

Demographic techniques pose a sampling challenge because you cannot optimize the sampling strategy for all the elements to be measured. While the most efficient sampling unit could be specifically designed for each stage class, in practice all stage classes will likely be sampled within the same plot. You can use elasticity values from pilot data to identify the stage classes that have the most influence on lambda, and design the sampling to estimate most precisely those stages. In the absence of pilot data, an evaluation of 66 species suggests some guidelines regarding stages that are important based on elasticity values: (1) progression is more important than seedling recruitment in nearly all species studied; (2) clonal growth, when it occurs, is usually important; (3) growth and reproduction are most important for short-lived semalparous[1] herbs; (4) growth, reproduction, and survival are about equally important for iteroparous[2] herbs in open habitats; (5) fecundity is of low importance for iteroparous forest herbs; (6) survival is most important for woody species (Silvertown et al. 1993).

F. Demographic Monitoring and Modeling: General Cautions and Suggestions

1. Suitability

Demographic monitoring using transition matrices and modeling, while currently the most powerful of the monitoring methods available, is not always the most appropriate. Because of the time involved in implementing a good demographic monitoring program, other lower priority species will likely not be monitored at all. Whether a demographic approach is appropriate depends on several factors: the rarity of the species, the risk of losing it, the management sensitivity involved (conflicting uses), the suitability of the species for demographic techniques, and the availability of monitoring resources.

The amount of effort required for this type of demographic monitoring varies. Characteristics of species most suitable are summarized in Figure 12.13. Species with long-lived seed banks are especially problematic, and probably should not be monitored by demographic techniques, unless some extensive seed bank studies are done concurrently. Even with seed bank studies, however, modeling the episodic nature of seedling "flushes" is extremely difficult. A similar problem occurs with any species with episodic or occasional reproduction. Plants with short

Characteristics of:	
Species easily monitored by demographic techniques:	**Species not easily monitored by demographic techniques:**
• lack seedbank	• long-lived seedbank
• lack vegetative reproduction, or vegetative daughters easily traced to parents	• dense vegetative reproduction
• moderate life-span (3-7 years)	• very short (annual) or very long life-span
• regular reproduction	• episodic reproduction
• single-stem or trunk morphology	• multiple stem and mat-like morphology
• low densities	• high densities
• populations small enough to census	• large populations on heterogeneous habitats

FIGURE 12.13. Summary of characteristics of species that can be monitored by demographic techniques, and those species for which demographic methods may not be appropriate. These limitations are most pertinent to techniques that use modeling and viability analysis. Other demographic techniques, such as monitoring survival of a specific stage class or measuring reproductive output over time, are less affected by the characteristics listed here.

---

[1] Semalparous plants are those that reproduce only once. These are typically annuals or biennials, but also include monocarpic perennials.

[2] Iteroparous plants reproduce more than once.

and highly variable life cycles such as annuals and biennials are also difficult to model because of the large variation in lambda values from year to year. In addition, these species present logistical problems in capturing germination, mortality, and reproduction events— events that can occur weekly throughout the growing season.

Some plant morphologies are not amenable to demographic measurements. Mat-like species, species with dense vegetative reproduction, and small plants that occur at high densities are all difficult to monitor with demographic techniques.

2. Suggestions for success

Allocate adequate resources. If demographic monitoring is the tool of choice, allocate enough time to do the job well. Resources should be available for extensive planning, adequate sampling, and detailed analysis.

Solicit extensive review. Demographic techniques are relatively new, and few agency personnel have much experience with them. Because of this lack of internal expertise, external review from the academic and professional community, where some of these methods have been tried for research purposes, is critical.

Conduct a 2-year pilot study. A pilot study is absolutely necessary. Test field methods during the first field season to ensure individuals are relocatable. After measuring the first transition (Year 1 to Year 2), evaluate variability of estimated parameters, and compare sampling effort to elasticity matrices. Re-evaluate time commitments.

Administer contracts closely. Because of the technical sophistication of demographic monitoring, it is often designed and implemented under contract. While this can be an excellent way to acquire scarce skills, demographic monitoring projects must be administered closely to ensure that the data meet the management needs of the agency. Designs should be subject to the same pilot period requirements as an in-house study. Data summaries and reporting should be an annual requirement of the contract. Be wary of outside professionals/contractors attempting to "sell" demographic monitoring as the only appropriate methodology. Many of these people come from research experiences, where the best and most publishable methods are the ones of choice. Few have experience in the resource management arena. The decision to use demographic methods should be made by agency personnel after careful consideration of the biological and political situation and the availability and allocation of limited monitoring resources (Chapter 3).

G. Age/Stage Class Investigations

Monitoring that assesses a single or several stages often has increased biological interpretability over simple measures of cover, frequency, or density. Monitoring may also be easier to implement, if the focus is on obvious stages, such as flowering plants, while avoiding assessment of more cryptic stages such as seedlings. Such studies can often provide demographic insights into population dynamics without using a full-scale demographic modeling study.

The main drawback of the approach is that you may miss some important dynamics if only one or two age or stage classes in the population are monitored. If the focus is on reproductive individuals, but the decline in the population is caused by lack of recruitment because of some management activity, then the monitoring will not be very sensitive to changes. Woody riparian species provide a familiar example. A key problem in woody riparian species management is the lack of regeneration, caused either by heavy ungulate use of seedlings and sprouts, or by changes in hydrology (such as dams) that control or eliminate the floods that provide exposed mineral soils for seed germination. A monitoring program that focuses on the mortality of adult stems may not detect a decline until it becomes obvious that the total number of adults has declined because no new adults have grown into the monitored class. If the adults are long-lived, this may not become apparent for many years. Monitoring the seedling stage, however, would have exhibited a problem with regeneration earlier. Summarizing known information and using ecological models (Chapter 4) can help focus attention on the most sensitive stage class(es).

H. Demographic Structure and Changes in Demographic Structure

The snapshot-in-time measure of demographic structure can provide useful insights into the viability of a population, although it provides no measure of population viability. Populations with a large percentage of senescent individuals or a low percentage of seedlings or reproductive individuals are potentially declining populations. Monitoring the change in demographic structure is often more sensitive and more easily interpreted than changes in density (or cover or frequency). Density, for example, can remain constant in a population that is experiencing a negative change in demographic structure (more old or non-reproductive plants).

The drawbacks of measuring demographic structure are similar to those described for matrix models. First, the technique is not applicable to plants with a morphology that does not allow the identification of consistently classed counting units. Individuals (or stems) must not only be recognizable, one must be able to reliably place them in the correct stage class. The second problem is that this approach can be far more time-consuming compared to simple density counts, especially when classing the individual is difficult. Classes that are size-dependent, such as using the diameter of the basal rosette, require evaluating the diameter on every individual. Classes that use more obvious characteristics, such as reproductive and non-reproductive, can be more rapidly assessed.

The third problem is that the sampling design becomes more complex compared to simple density estimates. Instead of estimating the total number of individuals, you are estimating the number in each size class. Visualizing and designing a sampling approach to deal with variability in density throughout a population is difficult, but it is even more difficult to efficiently sample the spatial variability of all the different stage classes.

Literature Cited

Akcakaya, H. R.; Ferson, S. 1992. RAMAS/space user manual: spatially structured population models for conservation biology. Setauket, NY: Applied Biomathematics.

Alvarez-Buylla, E. R.; Slatkin, M. 1991. Finding confidence limits on population growth rates. Trends in Ecology and Evolution 6: 221-224.

Alvarez-Buylla, E. R.; Slatkin, M. 1993. Finding confidence limits on population growth rates: Monte Carlo test of a simple analytic method. Oikos 68: 273-282.

Bernard, J. M. 1976. The life history and population dynamics of shoots of *Carex rostrata*. Journal of Ecology 64: 1040-1045.

Bierzychudek, P. 1982. The demography of jack-in-the-pulpit, a forest perennial that changes sex. Ecological Monographs 52: 335-351.

Caswell, H. 1989. Matrix population models. Sutherland, MA: Sinauer Associates.

Caswell, H.; Trevisan, M. C. 1994. Sensitivity analysis of periodic matrix models. Ecology 75(5): 1299-1303.

Chambers, J. C.; Brown, R. W. 1988. A mapping table for obtaining plant population data. Journal of Range Management 41: 267-268.

Charron, D.; Gagnon, D. 1991. The demography of northern populations of *Panax quinquefolium* (American Ginseng). Journal of Ecology 79: 431-445.

Cochran, W. G. 1977. Sampling techniques. 3rd edition. New York, NY: John Wiley and Sons.

Cullen, J. M.; Weiss, P. W.; Wearne, G. R. 1978. A plotter for plant population studies. New Phytologist 81: 443-448.

deKroon, H.; Plaisier, A.; vanGroenendael, J.; Caswell, H. 1986. Elasticity: the relative contribution of demographic parameters to population growth rate. Ecology 67:1427-1431.

Downer, R. R.; Harris, B.; Metzgar, L. H. 1992. GAPPSII: Generalized animal population projection system, users manual. Setauket, NY: Applied Biomathematics.

Eldridge, D. J.; Westoby, M.; Stanley, R.J. 1990. Population dynamics of the perennial rangeland shrubs *Atriplex vesicaria, Maireana astrotricha* and *M. pyramidata* under grazing, 1980-1987. Journal of Applied Ecology 27: 502-512.

Elzinga, C. 1996. Habitat Conservation Assessment and Strategy for *Thelypodium repandum*. Unpublished report on file at: USDI-BLM, Salmon District; Salmon, ID.

Elzinga, C. 1997. Habitat Conservation Assessment and Strategy for Penstemon lemhiensis. Unpublished report on file at: USDA-FS, Region 1; Missoula, MT.

Ferson, S. 1990. RAMAS/stage user manual: general stage-based modeling for population dynamics. Setauket, NY: Applied Biomathematics.

Ferson, S.; Akcakaya, H. R. 1990. RAMAS/age user manual: modeling fluctuations in age-structured populations. Setauket, NY: Applied Biomathematics.

Fiedler, P. L. 1987. Life history and population dynamics of rare and common mariposa lilies (*Calochortus* Pursh: Liliaceae). Journal of Ecology 75: 977-995.

Gatsuk, L. E.; Smirnova, O. V.; Vorontzova, L. I.; Zaugolnova, L. B.; Zhukova, L. A. 1980. Age states of plants of various growth forms: a review. Journal of Ecology 68: 675-696.

Geber, M. A.; Watson, M. A.; Furnish, R. 1992. Genetic differences in clonal demography in *Eichhornia crassipes*. Journal of Ecology 80: 329-341.

Gilbert, N.; Lee, S. B. 1980. Two perils of plant population dynamics. Oecologia 46: 283-284.

Harper, J. L. 1977. Population biology of plants. London: Academic Press.

Hartnett, D. C.; Richardson, D. R. 1989. Population biology of *Bonamia grandiflora* (Convolvulaceae): effects of fire on plant and seed bank dynamics. American Journal of Botany 76: 361-369.

Huenneke, L. F. 1987. Stem dynamics of a clonal shrub: size class transition matrices. Ecology 68: 1234-1239.

Hutchings, M. J. 1986. Plant population biology. In: Moore, P. D.; Chapman, S. D., eds. Methods in Plant Ecology. Oxford: Blackwell Scientific Publications: 377-435.

Kadmon, R. 1993. Population dynamic consequences of habitat heterogeneity: an experimental study. Ecology 74: 816-825.

Kalisz, S. 1991. Experimental determination of seed bank age structure in the winter annual *Collinsia verna*. Ecology 72(2): 575-585.

Kalisz, S.; McPeek, M. A. 1992. Demography of an age-structured annual: resampled projection matrices, elasticity analyses, and seed bank effects. Ecology 73(3): 1082-1093.

Kaye, T.; Meinke, R. 1991. Long-term monitoring for *Mirabilis macfarleanei* in Hell's Canyon, Wallowa-Whitman National Forest. Unpublished report on file at: Conservation Biology Program, Natural Resources Division, Oregon Department of Agriculture, Salem, Oregon.

Lacy, R.; Kreeger, T. 1992. VORTEX users manual: a stochastic simulation of the extinction process. Chicago, IL: Brookfield Zoo.

Leck, M. A.; Parker, V. T.; Simpson, R. L., eds. 1989. Ecology of soil seed banks. San Diego, CA: Academic Press Inc.

Lefkovitch, L. P. 1965. The study of population growth in organisms grouped by stages. Biometrics 21: 1-18.

Lesica, P. 1987. A technique for monitoring nonrhizomatous perennial plant species in permanent belt transects. Natural Areas Journal 7: 65-68.

Lesica, P.; Allendorf, F. W. 1992. Are small populations of plants worth preserving? Conservation Biology 6: 135-139.

Manders, P. T. 1987. A transition matrix model of the population dynamics of the Clanwilliam cedar (*Widdringtonia cedarbergensis*) in natural stands subject to fire. Forest Ecological Management 20: 171-186.

Manly, B. F. 1990. Stage-structured populations; sampling, analysis and simulation. New York, NY: Chapman and Hall.

Menges, E. S. 1986. Predicting the future of rare plant populations: demographic monitoring and modeling. Natural Areas Journal 6: 13-25.

Menges, E. S. 1987. Stochastic modeling of population behavior in plants: effects of environmental and demographic stochasticity on extinction probability. Conservation Biology 1: 52-63.

Menges, E. S. 1990. Population viability analysis for an endangered plant. Conservation Biology 4: 52-62.

Meyer, J. S.; Ingersoll, C. G.; McDonald, L. L.; Boyce, M. S. 1986. Estimating uncertainty in population growth rates: jackknife vs. bootstrap techniques. Ecology 67: 1156-1166.

Muir, P. S.; McCune, B. 1992. A dial quadrat for mapping herbaceous plants. Natural Areas Journal 12: 136-138.

Nault, A.; Gagnon, D. 1993. Ramet demography of *Allium tricocum*, a spring ephemeral, perennial forest herb. Journal of Ecology 81: 101-119.

Owens, M. K.; Gardiner, H. G.; Norton, B. E. 1985. A photographic technique for repeated mapping of rangeland plant populations in permanent plots. Journal of Range Management 38: 231-232.

Pavlik, B. M. 1993. Demographic monitoring and the recovery of endangered plants. In: Bowles, M.; Whelan, C., eds. Recovery and restoration of endangered species. Cambridge: Cambridge University Press.

Pavlik, B. M.; Barbour, M. B. 1988. Demographic monitoring of endemic sand dune plants, Eureka Valley, California. Biological Conservation 46: 217-242.

Philippi, T. 1993. Bet-hedging germination of desert annuals: beyond the first year. The American Naturalist 142(3): 474-487.

Pickart, A. J.; Stauffer, H. B. 1994. The importance of selecting a sampling model before data collection: an example using the endangered Humboldt Milkvetch (*Astragalus agnicidus* Barneby). Natural Areas Journal 14: 90-98.

Possingham, H. P.; Davies, I.; Noble, I. R. 1992. ALEX: an operating manual. Adelaide, Australia: University of Adelaide.

Prentice C. 1988. 1988 Progress Report: A Study of the Life Cycle of *Allium aaseae* Ownby, Aase's onion. Unpublished report on file at: Bureau of Land Management, Idaho State Office, Boise, ID.

Prentice, C. 1989. 1989 Progress Report: A Study of the Life Cycle of *Allium aaseae* Ownby, Aase's Onion. Unpublished report on file at: Bureau of Land Management, Idaho State Office, Boise, ID.

Pyke, D. A. 1990. Comparative demography of co-occuring introduced and native tussock grasses: persistence and potential expansion. Oecologia 82: 537-543.

Pyke, D. A. 1995. Population diversity with special reference to rangeland plants. In West, N. E., ed. Biodiversity on rangelands. Natural resources and environmental issues, volume IV. Logan, UT: College of Natural Resources, Utah State University: 21-32.

Reinhartz, J. A. 1984 Life history variation of common mullein (*Verbascum thapsus*). III. Differences among sequential cohorts. Journal of Ecology 72: 927-936.

Schemske, D. W.; Husband, B. C.; Ruckelshaus, M. H.; Goodwillie, C.; Parker, I. M.; Bishop, J. G. 1994. Evaluating approaches to the conservation of rare and endangered plants. Ecology 75(3): 584-606.

Schmidt, K. P.; Lawlor, L. R. 1983. Growth rate projection and life history sensitivity for annual plants with a seed bank. American Naturalist 121: 525-539.

Schwegman, J. 1986. Two types of plots for monitoring individual herbaceous plants over time. Natural Areas Journal 6: 64-66.

Silva, J. F.; Ravetos, J.; Caswell, H.; Trevisan, M. C. 1991. Population responses to fire in a tropical savanna grass *Andropogon semiberbis*: a matrix model approach. Journal of Ecology. 79: 345-356.

Silvertown, J. W. 1987. Introduction to plant population ecology, second edition. New York: Longman.

Silvertown, J. W.; Franco, M.; Pisanty, I; Mendoza, A. 1993. Comparative plant demography—relative importance of life-cycle components to the finite rate of increase in woody and herbaceous perennials. Journal of Ecology 81: 465-476.

Stewart, S. C. 1994. Statistical analysis of cohort demographic data. American Midland Naturalist. 131: 238-247.

Solbrig, O. T. 1980. Demography and evolution in plant populations. Berkeley, California: University of California Press.

Van Groenendael, J.; de Kroon, H.; Kalisz, S.; Tuljapurkar, S. 1994. Loop analysis: evaluating life history pathways in population projection matrices. Ecology 75:2410-2415.

Windas, J. L. 1986. Photo-quadrat and compass-mapping tools. Natural Areas Journal 6: 66-67.

CHAPTER 13
Completing Monitoring and Reporting Results

Solidago missouriensis
Missouri goldenrod
by Jennifer Shoemaker

CHAPTER 13. Completing Monitoring and Reporting Results

A successful monitoring project is characterized by two traits. First, it is *implemented* as planned in spite of personnel changes, changes in funding, and changes in priorities. Successful implementation depends on good design and good communication and documentation over the life of the project. Second, the information from a successful monitoring program is *applied*, resulting in management changes or validation of existing management. A monitoring project that simply provides additional insights into the natural history of a species, or that languishes in a file read only by the specialist, does not meet the intent of monitoring. Successful application of monitoring results requires reporting them in a form accessible to all interested parties.

Monitoring projects that are implemented and applied will complete the adaptive management cycle described in Chapter 1. Successful monitoring affects management, either by suggesting a change or validating the continuation of current management (Gray and Jensen 1993).

A. Assessing Results at the End of the Pilot Period

In this technical reference, we have advocated the use of pilot studies to avoid the expense and waste of a monitoring project that yields inconclusive results. After the pilot period you should consider several issues before continuing the monitoring project:

1. Can the monitoring design be implemented as planned?

The pilot period should answer several questions about field design and implementation: If sampling units are permanent, can they be relocated? Are sampling units reasonably sized for the number of plants or do quadrats contain hundreds of individuals? Is it difficult to accurately position a tape because of dense growth? Are the investigator impacts from monitoring acceptable? Is the skill level of field personnel adequate for the field work, or is additional training needed?

Projects rarely work as smoothly in the field as anticipated in the office. Nearly all monitoring projects require some modification for effective field implementation. Occasionally you may find that the planned method does not work at all, and a major overhaul of the monitoring project is required.

2. Are the costs of monitoring within estimates?

The pilot period is important as a reality check on required resources: Does the monitoring take much longer than planned? Will the data entry, analysis, and reporting work take more time than allocated?

If the monitoring project as designed requires more resources than originally planned, either management must devote more resources to the project, or you will need to redesign the monitoring to be within budget.

3. Do the assumptions of the ecological model still seem valid?

Your understanding of the biology and ecology of a species may improve as you spend time on the site collecting data. Does new information suggest another vegetation attribute would be more sensitive or easier to measure (cover instead of density, for example)? Is the change you've targeted to monitor biologically significant, or is the natural annual variability due to weather conditions so extreme that it masks the target change? Does the frequency of monitoring still seem appropriate?

4. For sampling situations, does the monitoring meet the standards for precision and power that were set in the sampling objective?

After analyzing the pilot data, you may discover that you need many more sampling units than you planned to achieve the standards for precision, confidence, and power that you set in your sampling objective (Chapter 6). You have six alternatives:

1. **Reconsider the design.** The pilot study should improve your understanding of the population's spatial distribution. Will a different quadrat shape or size improve the efficiency and allow you to meet the sampling objective within the resources available for monitoring?

2. **Re-assess the scale.** Consider sampling only one or a few macroplots, rather than sampling the entire plant population.

3. **Lobby for additional resources to be devoted to this monitoring project.** Power curves such as those shown in Chapter 5 may help to graphically illustrate the tradeoffs of precision, power, and sampling costs for managers (Brady et al. 1995).

4. **Accept lower precision.** It may be prohibitively expensive, for example, to be 90% confident of being within 10% of the estimated true mean, but it may be possible to be 90% confident of being within 20% of the true mean using available monitoring resources.

5. **Accept higher error rates.** You may not, with the current design and expenditure of monitoring resources, be 90% certain of detecting a specified change, but you may be 80% certain. You may have to accept a 20% chance that you will make a false-change error, rather than the 10% level you set in your sampling objective. You may not be within 10% of the estimated true mean with a 95% confidence level, but your current design may allow you to be 90% confident. Look at the results from your pilot study, and consider whether the significance levels that can be achieved with the current design are acceptable, even though the levels may be less stringent than you originally set in your sampling objective.

6. **Start over.** Acknowledge that you cannot meet the sampling objective with reasonable precision or power within the budgetary constraints of the project.

5. Reporting results from a pilot project

The results from the pilot period should be reported even if your design and project require significant revision. Your audiences for this report would include all those who reviewed your initial project proposal or monitoring plan (Chapter 10). A report to managers is especially important to describe the recommended changes in design. Your report is also important to your successor and possibly other ecologists or botanists who work with similar situations or species. Reporting failures of techniques will help others avoid similar mistakes.

B. Assessing Results After the Pilot Period

Three possible conclusions result from a monitoring study: (1) objectives are (being) met; (2) objectives are not (being) met; (3) the data are inconclusive (see Chapter 11). The pilot period should eliminate the problem of inconclusive results caused by poor design, but such results can occur even with excellent design.

1. Objectives are met

Two management responses should result for objectives that have been met. First, the objective should be reevaluated and changed based on any new knowledge about a species and population. Second, both management and monitoring should be continued, although the latter perhaps less frequently or intensely.

It is important that monitoring does not cease when objectives have been met. Measured success may not be related to management, but simply a lucky correlation of an increasing population size or condition within the management period, caused by unknown factors. Fluctuations in population size caused by weather can give the appearance of success, especially with annuals and short-lived perennials. You should never assume that the resource is secured for the long-term. You may scale back the frequency and intensity of monitoring in a population that appears stable or increasing, but do not consider the job done and ignore the population or species permanently. Current management may in fact be detrimental, but its negative effects masked by fluctuations related to weather. In addition, conditions change— weeds invade, native ungulate populations increase, livestock use patterns change with the construction of a fence or water trough, and recreational pressure increases. All these things and more may pose new threats.

2. Objectives are not met

As described in Chapter 1, according to the adaptive management approach, failure to meet an objective should result in the change in management that was identified as the management response during the objective development phase (Chapter 4). Rarely, however, is resource management that simple. We need to remember that the inertia that resists changing management is very difficult to overcome. Managers will generally continue implementing existing management, the path of least resistance, unless monitoring or some other overriding reason clearly indicates a change.

Unfortunately, the data from most monitoring will not conclusively identify causes of failure to meet objectives or the corresponding corrective action (see the discussion on monitoring versus research, Chapter 1). The biologist monitoring the population may feel confident of the cause, but decision-makers may be uncomfortable making changes in management, especially unpopular ones, which have a basis only in the biologist's professional opinion.

Thus, the most common response in land management agencies is to first reevaluate the objective. Was the amount of change too optimistic and biologically unlikely? Was the rate of change too optimistic? While such assessment is necessary, it can result in changing the objective rather than implementing necessary management changes.

This scenario is extremely common, but may often be avoided by two techniques. The first is to articulate the management response along with the management objective (as suggested in Chapter 4). This clearly states the response to monitoring results before monitoring is even started. It represents a commitment by the agency to stand by its monitoring results and use them to adapt management. The second technique is to reach consensus among all interested parties concerning the monitoring and the management response before monitoring data are collected (Johnson 1993). This is discussed in detail in Chapter 10.

C. Reporting Results and Recommending Changes

1. Periodic summaries

You should analyze results of monitoring each year (or each year data are collected) and report them in a short summary. Analyzing data as soon as they are collected has several benefits. The most important is that analysis is completed while the field work is still fresh in your mind. Questions always arise during analysis, and the sooner analysis takes place after the field work the more likely you can answer those questions. You may also find after analysis that you would like supplementary information, but it may not be possible to collect this in the middle of the winter, or 5 years after the monitoring data were collected. You will have lost a valuable opportunity. Analysis after each data collection episode also means that you will assess the monitoring approach periodically. Although many problems will surface during the pilot period, some may not until after a few years of data collection. Periodic assessment ensures a long-term monitoring project against problems of inadequate precision and power, and problems of interpretation.

2. Final monitoring reports

At the end of the specified monitoring period, or when objectives are reached, you should summarize the results in a formal monitoring report (Box 1). Much of the information needed for the report can be lifted directly from the monitoring plan as described in Chapter 10, although deviations from the proposed approach and the reasons for them will need to be described. The final report should be a complete document so it can function as a communication tool, so you should include all pertinent elements from the monitoring plan. You can either cut and paste elec-tronically from the

---

BOX 1: Monitoring reports

Executive Summary
 I. Introduction.
 II. Description of ecological model.
 III. Management objective.
 IV. Monitoring design.
 V. Data sheet example.
 VI. Management implications of potential results.
 VII. Summary of results.
   Include tables and figures communicating the results as well as general natural history observations.
 VIII. Interpretation of results.
   Describe potential causes for the results observed, sources of uncertainty in the data, and implications of the results for the resource.
 IX. Assessment of the monitoring project.
   Describe time and resource requirements, efficiency of the methods, and suggestions for improvement.
 X. Management recommendations.
   A. Change in management.
     Recommended changes based on results and the management implications identified in Section VI.
   B. Change in monitoring.
     Analysis of costs vs information gain, effectiveness of current monitoring system, and recommended changes in monitoring.
 XI. References.
   Includes grey literature and personal communications.
 XII. Reviewers.
   List those who have reviewed drafts of the report.

monitoring report, or simply append the report to existing copies of the monitoring plan. The preparation of the report should not be a major task. If you've been completing annual data analysis and internal reporting (as you should), summarizing the entire monitoring project should be straightforward.

Completing the monitoring project with a final formal report is important. This report provides a complete document that describes the monitoring and its results for distribution to interested parties. It provides a complete summary of the monitoring activity for successors, avoiding needless repetition or misunderstanding of the work of the predecessor. Finally, a professional summary lends credibility to the recommended management changes by presenting all of the evidence in a single document.

3. Reporting results—other vehicles

If the results would be interesting to others, consider sharing those results through a technical paper or symposium proceedings. Much of the preparation work for a presentation has already been done with the completion of the monitoring plan and monitoring report documents. Sharing the results has three important benefits: (1) it increases the audience, possibly helping more people and improving other monitoring projects (similar problems, similar species, etc.); (2) it increases the professional credibility of the agency; and (3) it contributes to your professional growth.

Literature Cited

Brady, W. W.; Mitchell, J. E.; Bonham, C. D.; Cook, J. W. 1995. Assessing the power of the point-line transect to monitor changes in basal plant cover. Journal of Range Management 48: 187-190.

Gray, J. S.; Jensen, K. 1993. Feedback monitoring: a new way of protecting the environment. Trends in Ecology and Evolution 8: 267-268.

Johnson, R. 1993. What does it all mean? Environmental Management and Assessment 26: 307-312.

APPENDICES

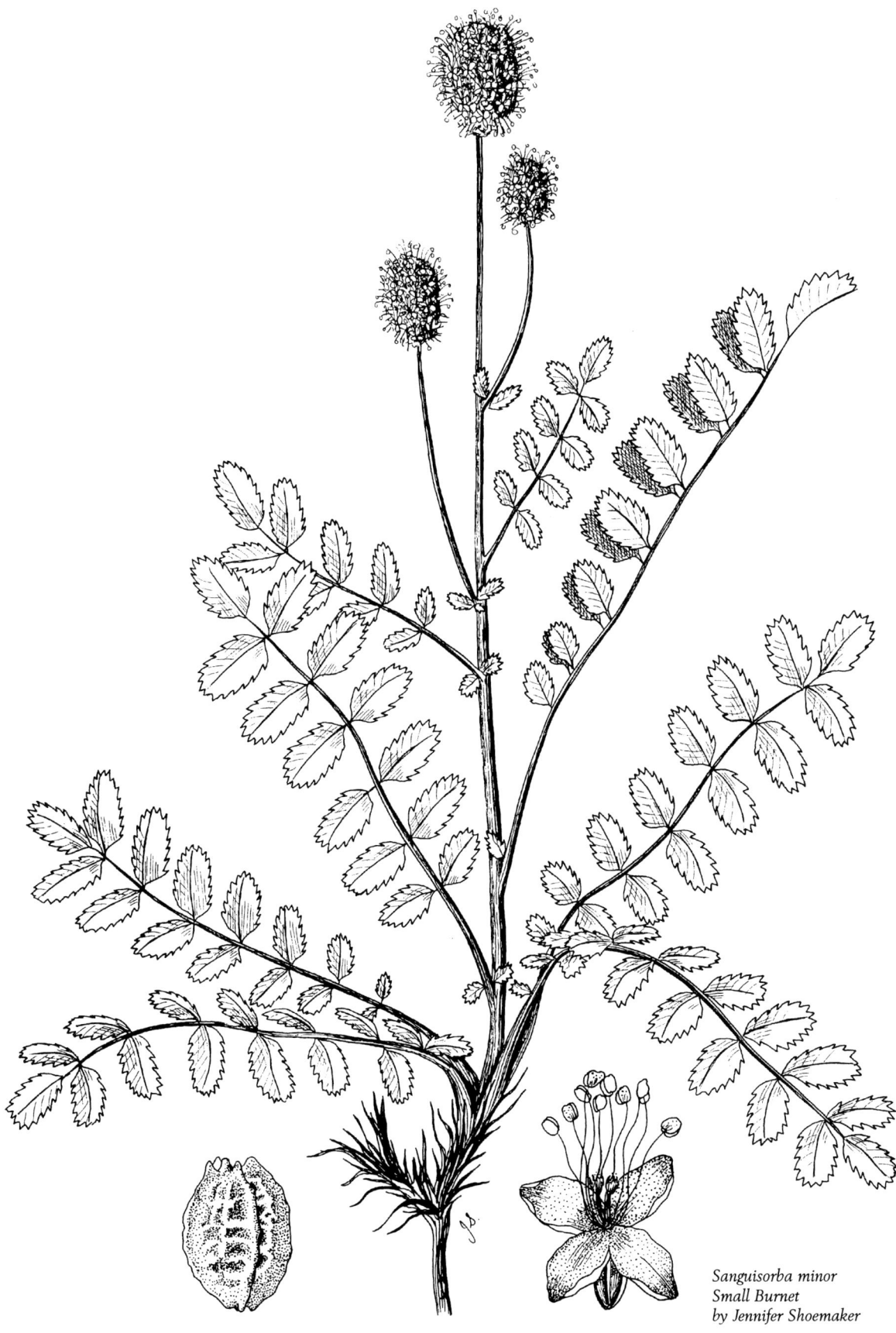

Sanguisorba minor
Small Burnet
by Jennifer Shoemaker

APPENDIX 1. Common Monitoring Problems

Monitoring projects often do not function as intended. The following are common scenarios and suggestions for avoiding problems.

A. Monitoring Not Implemented

1. Priorities changed and monitoring was not implemented after the first 2 years.

A signed monitoring plan (Chapter 10) represents a commitment by the agency to implement monitoring as designed. Although not a guarantee in the changing world of agency budgets and priorities, a monitoring plan provides some insurance that the monitoring will be implemented. If other parties outside the agency were part of the development of the monitoring plan, they may provide additional incentive to implement the monitoring as planned.

2. Data collection went as planned during the pilot period, but when we started using student interns for the field work after the pilot period, we found that they sometimes confused seedlings of a common shrub with the rare species.

The pilot period should function as a true test run of the monitoring. If technicians will be used for data collection over the life of the project, they should be used in the pilot period. Monitoring design needs to accommodate the skill levels of those doing the field work as well as those involved in analysis and interpretation.

3. The specialist in charge of the monitoring project was transferred to Washington and the monitoring project is faltering because of lack of an advocate.

Again, a monitoring plan may prove useful, especially if more than one person within the agency was involved in its development and can function as a replacement advocate, and if outside parties are actively involved (Chapter 10).

4. The specialist in charge of the monitoring project retired, and no one remaining knows where the transects are or what size quadrats were used.

Again, a monitoring plan can help. Not only are monitoring plans useful for communication, they also provide a link between predecessor and successors (Chapter 10). A cover sheet that describes monitoring methods provides further insurance that information such as transect locations is not lost (Chapter 9). Monitoring that has been poorly documented will not be continued once the originator leaves. Even worse, it is likely that all of the data already collected will be thrown out, since no one can interpret it.

B. Monitoring Data Not Analyzed

1. The field work was completed, but there is not enough time to analyze the data and report the results.

   When planning for monitoring, the time required for data entry, analysis, summary, and reporting are often forgotten, and only the field costs considered. Office work will likely require two to five times the field time and must be included in the budget. Commitment by decision-makers to allocate the time and resources required for the entire project, not just data gathering, should be part of the development of the monitoring plan (Chapter 10).

2. The field work was completed, but no one in the office knows how to analyze the data.

   Part of the monitoring design should be the identification of analysis methods (Chapter 11). If those can't be identified by available staff during the design stage, additional expertise should be brought in during design, not after the data are collected.

3. The field work was completed by student interns, who have since returned to college. We can't find some of the field notebooks, and no one in the office can decipher the notes in the ones we have.

   Field data sheets should be developed for each project, rather than using field notebooks for data recording (Chapter 9). Data collected by short-term employees or volunteers should be checked immediately, duplicated, and stored in a secure place.

C. Monitoring Yields Inconclusive Results

1. After 4 years of monitoring, the data were analyzed. The estimate of population size from the first year's data is 342 individuals, +/- 289 individuals at the 90% confidence level. Estimates of population size in subsequent years were no more precise.

   If the first year's data had been analyzed immediately as a pilot study, it would have been apparent that the methodology was not producing reliable estimates of population size (Chapters 5, 6, 7 and 11). As it is, four years of imprecise data have been collected.

2. During 10 years of monitoring, the population has exhibited an annual decline. It is still uncertain, however, whether the heavy livestock use in the area is responsible, and no decision to alter livestock management can be made.

   Developing a monitoring strategy of two phases—the first to identify an unacceptable decline and a second to determine reasons—would avoid this scenario (Chapter 4). Ten years is a long time to monitor a population decline and do nothing but watch.

3. After 12 years of monitoring, we've learned that the population size fluctuates up and down dramatically from year to year.

While this may be an interesting observation, it is not very useful for monitoring, and the annual fluctuations probably became apparent after 3-4 years of monitoring. Population size is not a sensitive measure to use for monitoring this species. You should have changed the measured attribute (here population size or density) have been changed after a few years, rather than continuing to measure it for 12 years. The potential for large annual variation in a chosen attribute should also be considered during the design phase (Chapter 4).

4. After 5 years of monitoring, we brought our data set to a statistician who said it was "nearly worthless."

Several mistakes were made here. During the design and pilot stages (Chapter 7), a statistician should have been consulted if the necessary skills were not available locally. Data should have been analyzed after the first year or two, so that changes in the monitoring could have been made before 5 years of time and effort were invested in the monitoring.

D. Monitoring Data Analyzed but not Presented

5. I don't have time to make fancy graphs and reports. I'm convinced of what the monitoring results say, and I'll use it to make better professional judgments concerning this species.

Such an attitude has two drawbacks. The first is that using the actual data is usually much more powerful than filtering it into "professional judgments," and the necessary changes will more likely be made if there are data to back them up. The improvement in the professional judgment of the specialist is important, but unless that translates into a management change, the monitoring really has not been successful. Second, failing to complete a report eliminates an important communication tool to describe results to successors, outside interested parties, and decision-makers.

6. The results are inconclusive. I don't have anything to report.

Inconclusive results need to be reported so others can avoid making the same mistakes.

E. Monitoring Results Encounter Antagonists

1. After 4 years of monitoring showing a significant decline in the population, the decision-maker refuses to change the grazing management because the range conservationist claims livestock never use the population area. I know I've seen herbivory and trampling in the population, but I don't have any data to prove it.

Other specialists may have information or concerns that need to be addressed when designing the monitoring (Chapter 10). Failing to include potential internal opposition during planning ensures their appearance after the data are collected.

2. We've monitored for 3 years, and have shown a statistically significant decline, but the timber company hired a consulting firm that has discredited our methodology.

Rare is the monitoring project that is not susceptible to criticism. Including the timber company during the development phase, and ensuring their support for the monitoring methodology and the potential results (Chapter 10), would have helped avoid this scenario.

APPENDIX 2. Legislation

Excerpts from National Environmental Policy Act of 1969

Declaration of National Environmental Policy
SEC. 101.

(a) The Congress, recognizing the profound impact of man's activity on the interrelations of all components of the natural environment, particularly the profound influences of population growth, high-density urbanization, industrial expansion, resource exploitation, and new and expanding technological advances and recognizing further the critical importance of restoring and maintaining environmental quality to the overall welfare and development of man, declares that it is the continuing policy of the Federal Government, in cooperation with State and local governments, and other concerned public and private organizations, to use all practicable means and measures, including financial and technical assistance, in a manner calculated to foster and promote the general welfare, to create and maintain conditions under which man and nature can exist in productive harmony, and fulfill the social, economic, and other requirements of present and future generations of Americans.

(b) In order to carry out the policy set forth in this Act, it is the continuing responsibility of the Federal Government to use all practicable means, consistent with other essential considerations of national policy, to improve and coordinate Federal plans, functions, programs, and resources to the end that the Nation may—

   (1) fulfill the responsibilities of each generation as trustee of the environment for succeeding generations;

   (2) assure for all Americans safe, healthful, productive, and aesthetically and culturally pleasing surroundings;

   (3) attain the widest range of beneficial uses of the environment without degradation, risk to health or safety, or other undesirable and unintended consequences;

   (4) preserve important historic, cultural, and natural aspects of our national heritage, and maintain, wherever possible, an environment which supports diversity and variety of individual choice;

   (5) achieve a balance between population and resource use which will permit high standards of living and a wide sharing of life's amenities; and

   (6) enhance the quality of renewable resources and approach the maximum attainable recycling of depletable resources.

(c) The Congress recognizes that each person should enjoy a healthful environment and that each person has a responsibility to contribute to the preservation and enhancement of the environment.

SEC. 102.

The Congress authorizes and directs that, to the fullest extent possible: (1) the policies, regulations, and public laws of the United States shall be interpreted and administered in accordance with the policies set forth in this Act, and (2) all agencies of the Federal Government shall—

(A) utilize a systematic, interdisciplinary approach which will insure the integrated use of the natural and social sciences and the environmental design arts in planning and in decision-making which may have an impact of man's environment.

(B) identify and develop methods and procedures, in consultation with the Council on Environmental Quality established by title II of this Act, which will insure that presently unquantified environmental amenities and values may be given appropriate consideration in decisionmaking along with the economic and technical considerations;

(C) include in every recommendation or report on proposals for legislation and other major Federal actions significantly affecting the quality of the human environment, a detailed statement by the responsible official on—

(i) the environmental impact of the proposed action,

(ii) any adverse environmental effects which cannot be avoided should the proposal be implemented,

(iii) alternatives to the proposed action,

(iv) the relationship between local short-term uses of man's environment and the maintenance and enhancement of long-term productivity, and

(v) any irreversible and irretrievable commitments of resources which would be involved in the proposed action should it be implemented.

Prior to making any detailed statement, the responsible Federal official shall consult with and obtain the comments of any Federal agency which has jurisdiction by law or special expertise with respect to any environmental impact involved. Copies of such statement and the comments and views of the appropriate Federal, State, and local agencies, which are authorized to develop and enforce environmental standards, shall be made available to the President, the Council on Environmental Quality and to the public as provided by section 552 of title 5, United States Code, and shall accompany the proposal through the existing agency review processes;

(D) study, develop, and describe appropriate alternatives to recommended courses of action in any proposal which involves unresolved conflicts concerning alternative uses of available resources;

Excerpts from The Endangered Species Act of 1973 (as amended through the 100th Congress, 1988)

SEC. 2. Findings, Purposes, and Policy

(a) FINDINGS.—The Congress finds and declares that—

(1) various species of fish, wildlife, and plants in the United States have been rendered extinct as a consequence of economic growth and development untempered by adequate concern and conservation;

(2) other species of fish, wildlife, and plants have been so depleted in numbers that they are in danger of or threatened with extinction;

(3) these species of fish, wildlife, and plants are of esthetic, ecological, educational, historical, recreational, and scientific value to the Nation and its people;

SEC. 3. Definitions

(3)   The terms "conserve," "conserving," and "conservation" mean to use and the use of all methods and procedures which are necessary to bring any endangered species or threatened species to the point at which the measures pursuant to this Act are no longer necessary. Such methods and procedures include, but are not limited to, all activities associated with scientific resources management such as research, census, law enforcement, habitat acquisition and maintenance, propagation, live trapping, and transplantation, and, in the extraordinary case where population pressures within a given ecosystem cannot be otherwise relieved, may include regulated taking.

SEC. 4. Determination of Endangered Species and Threatened Species

(b) Basis for Determinations.

(1)(A) The Secretary shall make determinations required by subsection (a)(1) solely on the basis of the best scientific and commercial data available to him after conducting a review of the status of the species and after taking into account those efforts, if any, being made by any State or foreign nation, or any political subdivision of a State or foreign nation, to protect such species, whether by predator control, protection of habitat and food supply, or other conservation practices, within any area under its jurisdiction or on the high seas.

(1)(B) In carrying out this section, the Secretary shall give consideration to species which have been—

(i) designated as requiring protection from unrestricted commerce by any foreign nation, or pursuant to any international agreement; or

(ii) identified as in danger of extinction, or likely to become so within the foreseeable future, by any State agency or by any agency of a foreign nation that is responsible for the conservation of fish or wildlife or plants

(2)   The Secretary shall designate critical habitat, and make revisions thereto, under subsection (a)(3) on the basis of the best scientific data available and after taking into consideration the economic impact, and any other relevant impact, of specifying any particular area as critical habitat. The Secretary may exclude any area from critical habitat if he determines that the benefits of such exclusion outweigh the benefits of specifying such area as part of the critical habitat, unless he determines, based on the best scientific and commercial data available, that the failure to designate such area as critical habitat will result in the extinction of the species concerned.

(3)(A) To the maximum extent practicable, within 90 days after receiving the petition of an interested person under section 553(e) of title 5, United States Code, to add a species to, or remove a species from, either of the lists published under subsection (c), the Secretary shall make a finding as to whether the petition presents substantial scientific or commercial information indicating that the petitioned action may be warranted. If such a petition is found to present such information, the Secretary shall promptly commence a review of the status of the species concerned. The Secretary shall promptly publish each finding made under this subparagraph in the Federal Register.

(3)(B) Within 12 months after receiving a petition that is found under subparagraph (A) to present substantial information indicating that the petitioned action may be warranted, the Secretary shall make one of the following findings:

(i) The petitioned action is not warranted, in which case the Secretary shall promptly publish such finding in the Federal Register.

(ii) The petitioned action is warranted, in which case the Secretary shall promptly publish in the Federal Register a general notice and the complete text of a proposed regulation to implement such action in accordance with paragraph (5).

(iii) The petitioned action is warranted but that—

(I) the immediate proposal and timely promulgation of a final regulation implementing the petitioned action in accordance with paragraphs (5) and (6) is precluded by pending proposals to determine whether any species is an endangered species or threatened species, and

(II) expeditious progress is being made to add qualified species to either of the lists published under subsection (c) and to remove from such lists species for which the protections of the Act are no longer necessary.

(C)(iii) The Secretary shall implement a system to monitor effectively the status of all species with respect to which a finding is made under subparagraph (B)(iii) and shall make prompt use of the authority under paragraph 7 to prevent a significant risk to the well being of any such species.

(f) Recovery Plans

(1) The Secretary shall develop and implement plans (hereinafter in this subsection referred to as 'recovery plans') for the conservation and survival of endangered species and threatened species listed pursuant to this section, unless he finds that such a plan will not promote the conservation of the species. The Secretary, in developing and implementing recovery plans, shall, to the maximum extent practicable—

(A) give priority to those endangered species or threatened species, without regard to taxonomic classification, that are most likely to benefit from such plan, particularly those species that are, or may be, in conflict with construction or other development projects or other forms of economic activity;

(B) incorporate in each plan—

(i) a description of such site-specific management actions as may be necessary to achieve the plan's goal for the conservation and survival of the species;

(ii) objective, measurable criteria which, when met, would result in a determination, in accordance with the provisions of this section, that the species be removed from the list; and

(iii) estimates of the time required and the cost to carry out those measures needed to achieve the plan's goal and to achieve intermediate steps toward that goal.

(g) Monitoring

(1) The Secretary shall implement a system in cooperation with the States to monitor effectively for not less than five years the status of all species which have recovered to the point at which the measures provided pursuant to this Act are no longer necessary and which, in accordance with the provisions of this section, have been removed from either of the lists published under subsection (c).

SEC. 7. Interagency Cooperation

(a) Federal Agency Actions and Consultations

(1) The Secretary shall review other programs administered by him and utilize such programs in furtherance of the purposes of this Act. All other Federal agencies shall, in consultation with and with the assistance of the Secretary, utilize their authorities in furtherance of the

purposes of this Act by carrying out programs of the conservation of endangered species and threatened species listed pursuant to Section 4 of this Act.

(2) Each Federal agency shall, in consultation with and with the assistance of the Secretary, insure that any action authorized, funded, or carried out by such agency (hereinafter in this section referred to as an 'agency action') is not likely to jeopardize the continued existence of any endangered species or threatened species or result in the destruction or adverse modification of habitat of such species which is determined by the Secretary, after consultation as appropriate with affected States, to be critical, unless such agency has been granted an exemption for such action by the Committee pursuant to subsection (h) of this section. In fulfilling the requirements of this paragraph each agency shall use the best scientific and commercial data available.

(3) Subject to such guidelines as the Secretary may establish, a Federal agency shall consult with the Secretary on any prospective agency action at the request of, and in cooperation with, the prospective permit or license applicant if the applicant has reason to believe that an endangered species or a threatened species may be present in the area affected by his project and that implementation of such action will likely affect such species.

(4) Each Federal agency shall confer with the Secretary on any agency action which is likely to jeopardize the continued existence of any species proposed to be listed under section 4 or result in the destruction or adverse modification of critical habitat proposed to be designated for such species. This paragraph does not require a limitation on the commitment of resources as described in subsection (d).

Excerpts from Federal Land Policy and Management Act of 1976

Declaration of Policy
SEC. 102.

(a) The Congress declares that it is the policy of the United States that—

(1) the public lands be retained in Federal ownership, unless as a result of the land use planning procedure provided for in this Act, it is determined that disposal of a particular parcel will serve the national interest;

(2) the national interest will be best realized if the public lands and their resources are periodically and systematically inventoried and their present and future use is projected through a land use planning process coordinated with other Federal and State planning efforts;

(3) public lands not previously designated for any specific use and all existing classifications of public lands that were effected by executive action or statute before the date of enactment of this Act be reviewed in accordance with the provisions of this Act;

(4) the Congress exercise its constitutional authority to withdraw or otherwise designate or dedicate Federal lands for specified purposes and that Congress delineate the extent to which the Executive may withdraw lands without legislative action;

(5) in administering public land statutes and exercising discretionary authority granted by them, the Secretary be required to establish comprehensive rules and regulations after considering the views of the general public; and to structure adjudication procedures to assure adequate third party participation, objective administrative review of initial decisions, and expeditious decisionmaking;

(6) judicial review of public land adjudication decisions be provided by law;

(7) goals and objectives be established by law as guidelines for public land use planning, and that management be on the basis of multiple use and sustained yield unless otherwise specified by law;

(8) the public lands be managed in a manner that will protect the quality of scientific, scenic, historical, ecological, environmental, air and atmospheric, water resource, and archeological values; that, where appropriate, will preserve and protect certain public lands in their natural condition; that will provide food and habitat for fish and wildlife and domestic animals; and that will provide for outdoor recreation and human occupancy and use;

(9) the United States receive fair market value of the use of the public lands and their resources unless otherwise provided for by statute;

(10) uniform procedures for any disposal of public land, acquisition of non-Federal land for public purposes, and the exchange of such lands be established by statute, requiring each disposal, acquisition, and exchange to be consistent with the prescribed mission of the department or agency involved, and reserving to the Congress review of disposals in excess of a specified acreage;

(11) regulations and plans for the protection of public land areas of critical environmental concern be promptly developed;

(12) the public lands be managed in a manner which recognizes the Nation's need for domestic sources of minerals, food, timber, and fiber from the public lands including implementation of the Mining and Minerals Policy Act of 1970 (84 Stat. 1876, 30 U.S.C. 21a) as it pertains to the public lands; and

(13) the Federal Government should, on a basis equitable to both the Federal and local taxpayer, provide for payments to compensate States and local governments for burdens created as a result of the immunity of Federal lands from State and local taxation.

(b) The policies of this Act shall become effective only as specific statutory authority for their implementation is enacted by this Act and by subsequent legislation and shall then be construed as supplemental to and not in derogation of the purposes for which public lands are administered under other provisions of law.

Definitions
SEC. 103.

Without altering in any way the meaning of the following terms as used in any other statute, whether or not such statute is referred to in, or amended by, this Act, as used in this Act—

(a) The term "areas of critical environmental concern" means areas within the public lands where special management attention is required (when such areas are developed or used or where no development is required) to protect and prevent irreparable damage to important historic, cultural, or scenic values, fish and wildlife resources or other natural systems or processes, or to protect life and safety from natural hazards.

(b) The term "holder" means any State or local governmental entity, individual, partnership, corporation, association, or other business entity receiving or using a right-of-way under title V of this Act.

(c) The term "multiple use" means the management of the public lands and their various resource values so that they are utilized in the combination that will best meet the present and future needs of the American people; making the most judicious use of the land for some or all of these resources or related services over areas large enough to provide sufficient latitude for periodic adjustments in use to conform to changing needs and conditions; the use of some land for less than all of the resources; a combination of balanced and diverse resource uses that takes into account the long-term needs of future generations for renewable and nonrenewable resources, including, but not limited to, recreation, range, timber, minerals, watershed, wildlife and fish, and natural scenic, scientific and historical values; and harmonious and coordinated management of the various resources without permanent impairment of the productivity of the land and the quality of the environment with consideration being given to the relative values of the resources and not necessarily to the combination of uses that will give the greatest economic return or the greatest unit output.

(d) The term "public involvement" means the opportunity for participation by affected citizens in rulemaking, decisionmaking, and planning with respect to the public lands, including public meetings or hearings held at locations near the affected lands, or advisory mechanisms, or such other procedures as may be necessary to provide public comment in a particular instance.

(e) The term "public lands" means any land and interest in land owned by the United States within the several States and administered by the Secretary of the Interior through the Bureau of Land Management, without regard to how the United States acquired ownership, except—

(1) lands located on the Outer Continental Shelf; and

(2) lands held for the benefit of Indians, Aleuts, and Eskimos.

(f) The term "right-of-way" includes an easement, lease, permit, or license to occupy, use, or traverse public lands granted for the purpose listed in title V of this Act.

(g) The term "Secretary", unless specifically designated otherwise, means the Secretary of the Interior.

(h) The term "sustained yield" means the achievement and maintenance in perpetuity of a high-level annual or regular periodic output of the various renewable resources of the public lands consistent with multiple use.

(i) The term "wilderness" as used in section 603 shall have the same meaning as it does in section 2(c) of the Wilderness Act (78 Stat. 890; 16 U.S.C. 1131-1136).

(j) The term "withdrawal" means withholding an area of Federal land from settlement, sale, location, or entry, under some or all of the general land laws, for the purpose of limiting activities under those laws in order to maintain other public values in the area or reserving the area for a particular public purpose or program; or transferring jurisdiction over an area of Federal land, other than "property" governed by the Federal Property and Administrative Services Act, as amended (40 U.S.C. 472) from one department, bureau or agency to another department, bureau or agency.

(k) An "allotment management plan" means a document prepared in consultation with the lessees or permittees involved, which applies to livestock operations on the public lands or on lands within National Forests in the eleven contiguous Western States and which:

(1) prescribes the manner in, and extent to, which livestock operations will be conducted in order to meet the multiple-use, sustained-yield, economic and other needs and objectives as determined for the lands by the Secretary concerned; and

(2) describes the type, location, ownership, and general specifications for the range improvements to be installed and maintained on the lands to meet the livestock grazing and other objectives of land management; and

(3) contains such other provisions relating to livestock grazing and other objectives found by the Secretary concerned to be consistent with the provisions of this Act and other applicable law.

(l) The term "principal or major uses" includes, and is limited to, domestic livestock grazing, fish and wildlife development and utilization, mineral exploration and production, rights-of-way, outdoor recreation, and timber production.

(m) The term "department" means a unit of the executive branch of the Federal Government which is headed by a member of the President's Cabinet and the term "agency" means a unit of the executive branch of the Federal Government which is not under the jurisdiction of a head of a department.

(n) The term "Bureau" means the Bureau of Land Management.

(o) The term "eleven contiguous Western States" means the States of Arizona, California, Colorado, Idaho, Montana, Nevada, New Mexico, Oregon, Utah, Washington, and Wyoming.

(p) The term "grazing permit and lease" means any document authorizing use of public lands or lands in National Forests in the eleven contiguous western States for the purpose of grazing domestic livestock.

Inventory and Identification
SEC. 201.

(a) The Secretary shall prepare and maintain on a continuing basis an inventory of all public lands and their resource and other values (including, but not limited to, outdoor recreation and scenic values), giving priority to areas of critical environmental concern. This inventory shall be kept current so as to reflect changes in conditions and to identify new and emerging resource and other values. The preparation and maintenance of such inventory or the identification of such areas shall not, of itself, change or prevent change of the management or use of public lands.

(b) As funds and manpower are made available, the Secretary shall ascertain the boundaries of the public lands; provide means of public identification thereof including, where appropriate, signs and maps; and provide State and local governments with data from the inventory for the purpose of planning and regulating the uses of non-Federal lands in proximity of such public lands.

APPENDIX 3. Examples of Management Objectives

This appendix contains 20 examples of management objectives, each paired with a management response. The examples are divided into two main categories: target/threshold and change/trend. Within each category, objectives are arranged in order approximating increasing intensity. Examples of desired condition and red flag types of objectives are included.

Many of the following management objectives illustrate examples where sampling is not occuring and therefore no sampling objective needs to be articulated. For management objectives where sampling is likely to occur, an example sampling objective is included.

Target/Threshold Objectives

Management Objective	Increase the estimated total cover of *Astragalus leptaleus* in Macroplot A at Birch Creek from Class 1 (1-10%) to Class 3 (21-30%) by 2010.
Management Response	Grazing will be changed to fall use only if an increase is not observed.
Management Objective	Eliminate OHV tracks in *Xanthoparmelia idahoensis* (illustrated on habitat areas Map 1) beginning in 1998.
Management Response	If OHV evidence is found, implement educational efforts to reduce OHV traffic in habitat areas. If these are unsuccessful, area closures will be effected and fences constructed.
Management Objective	Increase the number of population areas of *Penstemon lemhiensis* within the Iron Creek Drainage from 8 to 15 by 2010.
Management Response	If new populations fail to establish under current management, a transplant re-introduction program will be considered and, if approved, implemented by the year 2011.
Management Objective	Increase the number of acres of habitat for *Gymnosteris nudicaulis* that is protected from livestock grazing to 600 acres by 1998.
Management Response	Additional structural facilities (fences) will be constructed if cattle trespass occurs.
Management Objective	Maintain a minimum cover of 30% (plotless visual estimate) in at least 7 of the 10 macroplots established in the Bentonite Hills population area between 1998 and 2005.
Management Response	If, in any year, cover decreases below this threshold, reduce OHV use by effecting closures and erecting fences.
Management Objective	Maintain an estimated cover of at least 20% (plotless visual estimate) of *Xanthoparmelia idahoensis* in Macroplot A in the Warm Springs drainage between now and 2003.
Management Response	If cover declines below an estimated 20%, institute a more extensive, quantitative monitoring project that assesses the trend of the entire population in the Warm Springs drainage.

Management Objective	Increase the number of individuals of *Penstemon lemhiensis* in the Iron Creek Population to 160 individuals by the year 2000.
Management Response	Failure to detect an increase will result in more intensive monitoring to determine if the current population of 122 is stable and viable (demographic analysis), and the implementation of alternative management by 2005 if it is not.
Management Objective	Maintain at least 10 reproductive individuals of *Thelypodium repandum* at the Lime Creek population during mining operations.
Management Response	Collect seed the first year the population reaches 10 individuals, and for 3 years following.
Management Objective	Maintain a population of at least 200 individuals of *Thelypodium repandum* at the Malm Gulch site between 1998 and 2005.
Management Response	Failure to maintain a population of the minimum size will trigger additional monitoring and study to determine the reason for failure, and alternative management will be implemented by 2007.
Management Objective	Increase the mean density of *Viola adunca* in Macroplot A at the Clatsop Plains Preserves to 1.0 plants/m² by 1999.
Sampling Objective	Be 95% confident that estimates of density are within ± 30% of the estimated mean density.
Management Response	If the desired increase does not occur, additional monitoring of the population will be implemented, and alternative management implemented by 2003. If the mean density is equal to or greater than the target density, current management will continue and the population will be monitored again in 2003.
Management Objective	By 1999, decrease the percentage of woody first year leaders of *Chrysothamnus parryii* ssp. *montanus* that are grazed by ungulates to 30% at the Tick Creek population.
Sampling Objective	Obtain estimates of leader herbivory with 90% confidence intervals no wider than ± 10% of the estimated grazed percentage.
Management Response	Failure to meet this objective will result in fencing the populations to eliminate livestock use within population areas by 2001.
Management Objective	Allow herbivory of inflorescences on no more than 20% of the individuals of *Primula alcalina* at the Birch Creek population in any year.
Sampling Objective	Be 90% confident that estimates of inflorescence herbivory are within ± 8% of the estimated percent grazed.
Management Response	Exclude cattle use from the Birch Creek *Primula alcalina* site by constructing a buck and pole fence within 6 months of the time the threshold is exceeded.

Management Objective	Maintain a frequency of 20% (0.10m² square quadrats) or less of *Taeniatherum caput-medusae* in Macroplot A at the Agate Desert Preserve in any year between 1999 and 2005.
Sampling Objective	Be 95% confident that frequency estimates are within ± 5% of the estimated frequency values.
Management Response	Initiate chemical weed control the following field season if the frequency of *Taeniatherum caput-medusae* exceeds 20% in Macroplot A.
Management Objective	Allow no more than 30% of the population of 433 individuals of *Silene scaposa* var. *lobata* to be killed by logging operations at the Wood Creek site.
Sampling Objective	Obtain estimates of percent mortality with 95% confidence intervals that are no wider than ± 2% of the estimated percent mortality.
Management Response	Logging will not be allowed in population areas of *Silene scaposa* var. *lobata* if the mortality at this site exceeds the threshold.
Management Objective	Maintain a minimum population of 1000 clumps of *Sarracenia oreophila* at the Eller Seep Preserve between 1998 and 2010.
Sampling Objective	Estimate the number of *Sarracenia orephila* clumps with 95% confidence intervals no wider than ± 10% of the estimated number of total clumps.
Management Response	Additional monitoring will be initiated if the population falls below the threshold of 1000 clumps.

Change/Trend Objectives

Management Objective	Increase the mean density of *Penstemon lemhiensis* at the Warm Springs population by 30% between 1998 and 2006.
Sampling Objective	Be 90% certain of detecting a 30% increase in mean density with a false-change error rate of 0.10.
Management Response	Failure to meet the objective will result in more intensive monitoring to determine cause of failure, and the implementation of alternative management by the year 2009.
Management Objective	Increase the density of *Lomatium cookii* at the Agate Desert Preserve by 20% between 1998 and 2003.
Sampling Objective	Be 90% sure of detecting a 20% increase in density with a false-change error rate of 0.20.
Management Response	If the density fails to increase, additional research of potential management options will be initiated and alternate management implemented by 2003.

Management Objective	Maintain the mean density of *Primula alcalina* at the Summit Creek site within 20% of the 1998 density between 1998 and 2010.
Sampling Objective	Be 95% sure of detecting a 20% change in density with a false-change error rate of 0.10.
Management Response	Failure to maintain this minimum density will trigger a more intensive study of the interaction of livestock grazing and *Primula alcalina*, with the implementation of alternative management within 4 years after the first year the unacceptable level of decline is measured.
Management Objective	Allow a decrease of no more than 20% of the 1999 cover of *Astragalus diversifolius* at the Texas Creek population between 1999 and 2004.
Sampling Objective	Be 90% certain of detecting a 20% decrease in the percent cover with a false-change error rate of 0.10.
Management Response	Exceeding the decrease will trigger a change in grazing management to a fall-use only system, implemented the season after the 20% decrease is exceeded.
Management Objective	Allow a decrease of no more than 30% in the number of individuals of *Conradia glabra* at Apalachicola Bluffs and Ravines Preserve over a 2 year period after implementing prescribed fire.
Sampling Objective	Be 80% certain of detecting a 30% decrease in the total number of individuals with a false-change error rate of 0.20.
Management Response	If the reduction exceeds 30%, populations will be protected from subsequent burns at the Apalachicola Bluffs and Ravines Preserve. Prescribed fire in population areas of *Conradia glabra* at other preserves will be designed to affect 20% or less of the population and implemented only if resources are available to monitor species response.

APPENDIX 4. Selecting Random Samples

Following are two methods for using random numbers to select random samples. Either can be accomplished using a random number table or a random number generator on a computer or hand-held calculator. The first method is probably the most commonly used, but the second method is far more efficient, particularly with two and three digit numbers. The appendix concludes with a brief discussion of two additional ways to derive random numbers in the field when you've forgotten to bring along a random number table or a hand calculator.

Method 1: Treating Random Numbers as Whole Digits

Example 1: Selecting random pairs of coordinates

Let's say we have marked off a 10m x 20m macroplot within a key area and we wish to randomly place forty 0.25m x 4.0m quadrats within that macroplot (in actual practice the macroplot might be a lot larger, say 50m x 100m). We wish to place the quadrats so that the long side is parallel to the x-axis and the x-axis is one of the 20m sides of the macroplot. The total number of quadrats (N) that could be placed in that 10m x 20m macroplot without overlap comprises the sampled population. In this case N is equal to 200 quadrats. The total population of quadrats is shown in Figure 1.

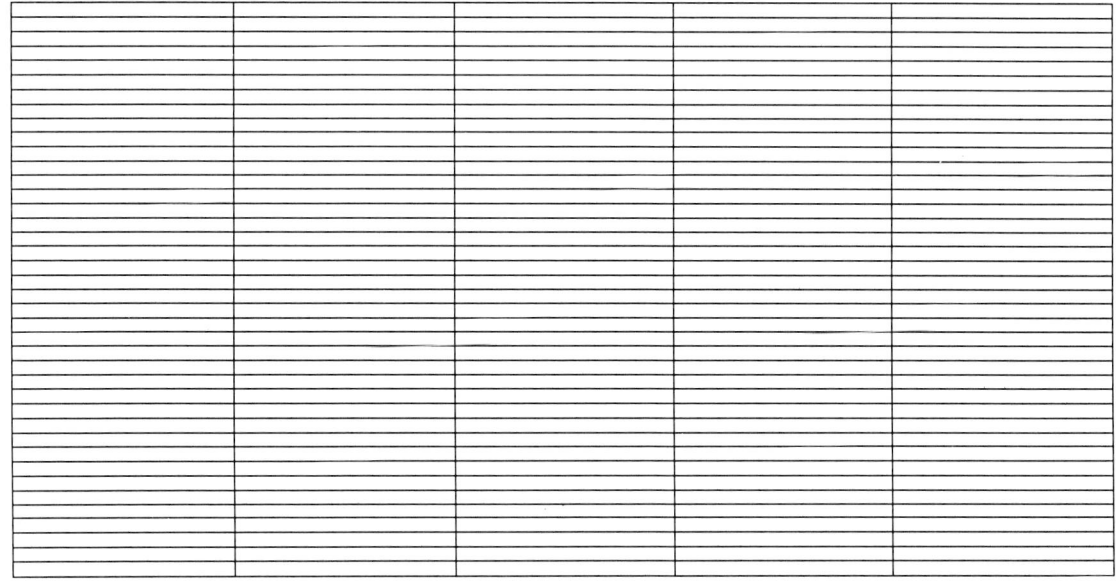

APPENDIX 4—FIG. 1. A 10m x 20m macroplot showing the 200 possible quadrats of size 0.25m x 4.0m that could be placed within it (assuming the long side of the quadrats is oriented in an east-west direction).

Along the x-axis there are 5 possible starting points for each 0.25m x 4.0m quadrat (at points 0, 4, 8, 12, and 16). Number these points 1 to 5 (in whole numbers) accordingly. Along the y-axis there are 40 possible starting points for each quadrat (at points 0, 0.25, 0.50, 0.75, 1.0, 1.25, and so on until point 9.75). Number these points 1 to 40 accordingly (again in whole numbers).

Using a table of random numbers. Consider Table 1, a table of random numbers (this table is presented only for instructional purposes; in practice you should use a much larger random number table, such as that found in Appendix 6). Using this random number table, you choose 40 numbers from 1 to 5 for the x-axis and 40 numbers from 1 to 40 for the y-axis. Because these numbers are in random order in the table, you can simply select these numbers in order, and they will be random. Arbitrarily begin at any 5 digit number in the table. To pick numbers along the x-axis read either across or down (it makes no difference which) and select the first 40 one digit numbers that correspond to the numbers 1 through 5. Reject any numbers that aren't 1 through 5. For example, if we start at column 2, row 3, and reads across the row, the first ten numbers are 5, 0, 3, 1, 2, 9, 2, 6, 8, and 3. From this list only the numbers 5, 3, 1, 2, 2, and 3 meet our criteria of being between 1 and 5. These become our first 6 randomly selected positions along the x-axis (note that we do not reject the second occurrences of the numbers 2 and 3 because the y-axis numbers we select to go with these repeat numbers may be different). We write these numbers under a column heading "x-axis," as follows:

x-axis	y-axis
5	
3	
1	
2	
2	
3	

	1	2	3	4	5
1	55457	60189	95970	71641	75935
2	37232	58802	85478	23088	48214
3	29229	50312	92683	27179	98501
4	13135	53586	20722	77003	93064
5	20387	52649	66532	26770	88003
6	66611	22679	69735	40297	66715
7	71488	93726	54025	56130	36901
8	99078	11154	69689	62223	74431
9	57171	73561	33584	40186	22910
10	55220	37500	60530	36185	56969

Table 1. A table of 250 random digits.

We continue until we have at least 40 numbers for the x-axis (in anticipation of ending up with at least some duplicate pairs of coordinates we may want to take 50 random numbers for the x-axis). We then do the same for the y-axis. To do this we enter the random number table at a different location (or simply continue from where we left off after obtaining the x-coordinates) and choose the first 40 (or more) two digit numbers that correspond to the numbers 1 through 40. We reject any numbers that aren't 1 through 40. For example, if one starts at column 1, row 9, and reads across the row, the first ten numbers are 57, 17, 17, 35, 61, 33, 58, 44, 01, and 86. From this list only the numbers 17, 17, 35, 33, and 01 meet our criteria of being between 1 and 40. We write these down in the column marked y-axis:

x-axis	y-axis
5	17
3	17
1	35
2	33
2	01
3	

We continue until we have at least 40 (or more) numbers under the y-axis. At the end of this process we will have 40 pairs of coordinates. If any pair of coordinates is repeated, we reject the second pair and pick another pair at random to replace it (because we are sampling without replacement). We continue until we have 40 unique pairs of coordinates.

Using a random number generator. Many hand-held calculators have random number generators, making their use in the field very easy. Several computer programs also have the ability to generate random numbers. For example, Lotus 1-2-3 will generate random numbers using the @RAND function. With both hand-held calculators and computer programs you need to consider whether you must reset the random number seed to generate different groups of random numbers. With some calculators and computer programs, failure to reset the seed will result in generation of the same set of random numbers (i.e., the numbers won't be "random" at all if you repeat the procedure more than once). Lotus 1-2-3 resets the random number seed automatically.

Random number generators yield numbers between 0 and 1 in decimals, usually to at least 5 places. In Method 1 the random numbers generated are used in the same way as numbers from the random number table, except that the decimals are ignored.

Example 2: Selecting random points along a baseline from which to run transects
In our second example, we have laid out a 200m baseline oriented in a north-south direction, and we wish to randomly select points along the baseline (see Figure 2). The 0 point is at the south end of the baseline. At each point we will run a 50m transect perpendicular to the baseline. We can go in either of two directions, east or west, so we also need to randomly select the direction in which to run each transect. We intend to treat the transects as our sampling units and have determined from pilot data that 20 transects are required.

Along each transect we intend to lay ten systematically spaced 1m x 1m quadrats, which will always be placed on the south side of the transect. Since we are sampling without replacement, we want to avoid the possibility that any two quadrats could overlap. Thus, we want to select values in whole meter increments beginning at the 1 meter point along the baseline (if the 1 meter point were chosen through our random process, quadrats placed along that transect—since they will always be on the south side of the transect—would reach the outer boundary, the 0 point, of our sampled population).

Using a table of random numbers. Using the same process that we used for selecting random coordinates, we enter into the random number table at some arbitrary point and begin reading numbers from left to right. In this case, however, we must look at groups of *three* digit numbers, since we are selecting points ranging from 1 to 200. Starting in column 2, row 2, and reading left to right, the first ten 3 digit numbers would be: 588, 028, 547, 823, 088, 482, 142, 922, 950, and 312. We accept the numbers 028, 088, and 142, because they meet our criterion of being between 1 and 200; we reject the others. We then continue on: 926, 832, 717, 985, 011, 313, 553, 586, 207, and 227. Of these numbers only 011 meets our criterion. We then continue on until we have 20 numbers from 1 to 200 (don't worry; there's a far more efficient means of deriving our random set of points discussed under Method 2 below).

Once we've generated a list of 20 random points along the baseline, we now need to determine in which direction we will run the transect. (Just as for the random coordinates we don't reject points that are the same; we only reject sets of points *and* directions that are the same. Thus, we can select both point 75 and direction E and point 75 and direction W. But if we select another point 75 and direction E we reject it and select another point and direction.) To determine direction we arbitrarily assign one digit numbers to E and W. For example E might be 0 and W

APPENDIX 4—FIG. 2. A 200m north-south baseline, showing 10 randomly positioned transects of 10 1m x 1m quadrats. We have determined through pilot sampling that 20 transects will be required to detect the level of change we want to be able to detect at a particular significance level and power. We therefore will need to randomly select an additional 10 transects.

might be 1. We then enter the random number table, read across and write down (next to the points we've already selected) the directions that correspond to every 0 and 1 we encounter (ignoring all numbers that aren't 0 or 1). Alternatively, we could flip a coin, assigning heads to one direction and tails to the other. Or, we could consider every even number to correspond to E and every odd number to correspond to W.

Using a random number generator. A random number generator would be used in the same way as the table of random numbers except that the decimal would be ignored.

Method 2: Treating Random Numbers as Decimals

This is by far the most efficient method of selecting random samples, particularly for two, three, and higher digit numbers. To use this method the random numbers must be treated as decimals. In our set of 250 random digits we would simply place a decimal point in front of every group of 5 digits and treat each group as one random number. Thus, if we entered the table at column 1, row 7, and read across, we would have the following six random numbers: 0.71488, 0.93726, 0.54025, 0.56130, 0.36901, and 0.99078. If we used a random number generator it would be even easier since these provide random numbers as decimals falling between 0 and 1.

The formula for using these decimal random numbers for selecting a sampling unit or point is:

$$[uN] + 1$$

Where: u = random number (expressed as decimal)
         N = total population size
         [ ] = used to indicate that only the integer part of the product is used in the calculation

To illustrate how this formula works, consider our baseline example. Here we need to select numbers between 1 and 200 as points along a baseline. Consider these points as a "population" of 200 possible points. Using the first of the six random numbers we came up with above, 0.71488, we calculate:

$$[0.71488 \times 200] + 1$$
$$= [142.976] + 1$$
$$= 142 + 1$$
$$= 143$$

Thus, 143 is our first point. Using the second random number we have:

$$[0.93726 \times 200] + 1$$
$$= [187.452] + 1$$
$$= 187 + 1$$
$$= 188$$

Now we have our second point, 188. We would continue in this manner until we had the 20 points we need. Although the formula may look difficult, a hand-held calculator or computer program with a random number generator makes it easy. With a hand-held calculator, for example, one could program the population size in memory, hit the button generating a random number, multiply it by the number in memory, and come up with the random point. Twenty such points could be produced in just a few minutes.

The reason for adding the 1 to the integer of the product of the random number and N may not be intuitively obvious. It is necessary because we are using only the integer of the product. Without adding 1 it would therefore not be possible to obtain the number 200. Consider the highest possible random number we could obtain, 0.99999. If we multiply this number by 200 we obtain 199.99800; taking the whole integer of this number yields the number 199. Adding 1 makes it 200. If, instead of choosing numbers from 1 to 200, you are choosing numbers between 0 to 199, there is no need to add the 1 to the integer of the product.

As a rule of thumb you should make sure the random numbers have more digits on the right side of the decimal point than the number of digits in N. In the example above, N is 200 and we are using random numbers with 5 digits to the right of the decimal point so we are okay.

Note that this process is much more efficient than Method 1, because we do not need to reject any numbers. When selecting random points along our 200 meter baseline using Method 1, we had to look at 20 three digit numbers just to come up with 4 numbers that met our criterion of being between 1 and 200. Given the fact that there is only a 1 in 5 chance of any three digit number falling between 1 and 200, this means we would—on the average—have to examine 100 three digit numbers to come up with 20 points. Using Method 2 we could use the first 20 random numbers to select the same 20 points. When we need to select 1 digit numbers (as, for example, to determine direction), it may be just as efficient (or even faster) to use Method 1.

Generating Random Numbers Without a Calculator or Random Number Table

You've left for a 5-day trip to the field, and realize, once you're 4 hours away from the office, that you've forgotten to bring either a random number table or your calculator with its random number generator. You are going to need to generate random numbers for the monitoring study you intend to design while in the field. How are you going to do it?[1] There are two options available to you. The best one involves the use of a digital watch. The other involves the use of a telephone directory.

Using a digital stopwatch to generate random numbers. If you have a digital watch with a stopwatch function, you can use the stopwatch to generate random numbers. Fulton (1996) describes how to do this and offers proof that the procedure is really random. The procedure is rather simple. Just start the stopwatch and let it run. Whenever a random digit is needed, simply stop the watch and read either or both of the two numbers in the tenths and hundredths of seconds place. For example, you stop the watch and it reads 27.45 seconds (ignore minutes and hours). You can use either the 4 in the .45 if you need only one random digit between 0 and 9, or you can use the 45 if you need two random digits between 0 and 99. If you need more than two random digits you can repeat the procedure as many times as necessary until you have the required number of digits. For example, we need three random digits between 0 and 999. After we stop the watch the first time, and get .45, we restart the watch and stop it again. This time, say, it stops at .82. Then we use the .8 as our random digit, and end up with a random number of 458. We can repeat this procedure until we get the required quantity of random numbers.

---

[1] Although you can certainly generate random numbers without a calculator, you're going to need a calculator to calculate means and standard deviations during plot sampling. This means you're probably going to have to stop somewhere and buy an inexpensive calculator. Some of the cheaper models may not have the capability of generating random numbers, so the methods discussed here may still be applicable.

Two things are important in this procedure: (1) to avoid bias, don't look at the watch before you stop it, and (2) wait long enough between starts and stops to allow a few seconds to elapse, making each reading independent and ensuring your selections are truly random. It's also important to note that you should use only the tenths and hundredths of seconds as your random digits or the digits you choose will likely not be truly random, unless you wait a long time (several tens of seconds) between starts and stops. Even then, however, you could only use numbers in the ones of seconds place, because the numbers in the tens of seconds place range only from 0 to 6. The safest bet is to use only the tenths and hundredths of seconds places.

Using a telephone directory as a source of random numbers. Another source of random numbers is a telephone directory. You can use the last 4 digits of the telephone numbers listed in the white pages as random numbers. Don't use the first 3 digits (prefixes), however, because these do not represent the full range of digits available between 0 and 999, they are not in random order, and they are not independent of one another.

Literature Cited

Fulton, M.R. 1996. The digital stopwatch as a source of random numbers. Bulletin of the Ecological Society of America 77:217-218.

APPENDIX 5. Tables of Critical Values for the t and Chi-square Distributions

This appendix includes tables of critical values for the t and chi-square distributions. The tables are reprinted from J. H. Zar, Biostatistical Analysis, 3rd ed., 1996, by permission of Prentice Hall, Inc., Upper Saddle River, New Jersey.

Critical Values of the *t* Distribution

ν	α(2): 0.50 α(1): 0.25	0.20 0.10	0.10 0.05	0.05 0.025	0.02 0.01	0.01 0.005	0.005 0.0025	0.002 0.001	0.001 0.0005
1	1.000	3.078	6.314	12.706	31.821	63.657	127.321	318.309	636.619
2	0.816	1.886	2.920	4.303	6.965	9.925	14.089	22.327	31.599
3	0.765	1.638	2.353	3.182	4.541	5.841	7.453	10.215	12.924
4	0.741	1.533	2.132	2.776	3.747	4.604	5.598	7.173	8.610
5	0.727	1.476	2.015	2.571	3.365	4.032	4.773	5.893	6.869
6	0.718	1.440	1.943	2.447	3.143	3.707	4.317	5.208	5.959
7	0.711	1.415	1.895	2.365	2.998	3.499	4.029	4.785	5.408
8	0.706	1.397	1.860	2.306	2.896	3.355	3.833	4.501	5.041
9	0.703	1.383	1.833	2.262	2.821	3.250	3.690	4.297	4.781
10	0.700	1.372	1.812	2.228	2.764	3.169	3.581	4.144	4.587
11	0.697	1.363	1.796	2.201	2.718	3.106	3.497	4.025	4.437
12	0.695	1.356	1.782	2.179	2.681	3.055	3.428	3.930	4.318
13	0.694	1.350	1.771	2.160	2.650	3.012	3.372	3.852	4.221
14	0.692	1.345	1.761	2.145	2.624	2.977	3.326	3.787	4.140
15	0.691	1.341	1.753	2.131	2.602	2.947	3.286	3.733	4.073
16	0.690	1.337	1.746	2.120	2.583	2.921	3.252	3.686	4.015
17	0.689	1.333	1.740	2.110	2.567	2.898	3.222	3.646	3.965
18	0.688	1.330	1.734	2.101	2.552	2.878	3.197	3.610	3.922
19	0.688	1.328	1.729	2.093	2.539	2.861	3.174	3.579	3.883
20	0.687	1.325	1.725	2.086	2.528	2.845	3.153	3.552	3.850
21	0.686	1.323	1.721	2.080	2.518	2.831	3.135	3.527	3.819
22	0.686	1.321	1.717	2.074	2.508	2.819	3.119	3.505	3.792
23	0.685	1.319	1.714	2.069	2.500	2.807	3.104	3.485	3.768
24	0.685	1.318	1.711	2.064	2.492	2.797	3.091	3.467	3.745
25	0.684	1.316	1.708	2.060	2.485	2.787	3.078	3.450	3.725
26	0.684	1.315	1.706	2.056	2.479	2.779	3.067	3.435	3.707
27	0.684	1.314	1.703	2.052	2.473	2.771	3.057	3.421	3.690
28	0.683	1.313	1.701	2.048	2.467	2.763	3.047	3.408	3.674
29	0.683	1.311	1.699	2.045	2.462	2.756	3.038	3.396	3.659
30	0.683	1.310	1.697	2.042	2.457	2.750	3.030	3.385	3.646
31	0.682	1.309	1.696	2.040	2.453	2.744	3.022	3.375	3.633
32	0.682	1.309	1.694	2.037	2.449	2.738	3.015	3.365	3.622
33	0.682	1.308	1.692	2.035	2.445	2.733	3.008	3.356	3.611
34	0.682	1.307	1.691	2.032	2.441	2.728	3.002	3.348	3.601
35	0.682	1.306	1.690	2.030	2.438	2.724	2.996	3.340	3.591
36	0.681	1.306	1.688	2.028	2.434	2.719	2.990	3.333	3.582
37	0.681	1.305	1.687	2.026	2.431	2.715	2.985	3.326	3.574
38	0.681	1.304	1.686	2.024	2.429	2.712	2.980	3.319	3.566
39	0.681	1.304	1.685	2.023	2.426	2.708	2.976	3.313	3.558
40	0.681	1.303	1.684	2.021	2.423	2.704	2.971	3.307	3.551
41	0.681	1.303	1.683	2.020	2.421	2.701	2.967	3.301	3.544
42	0.680	1.302	1.682	2.018	2.418	2.698	2.963	3.296	3.538
43	0.680	1.302	1.681	2.017	2.416	2.695	2.959	3.291	3.532
44	0.680	1.301	1.680	2.015	2.414	2.692	2.956	3.286	3.526
45	0.680	1.301	1.679	2.014	2.412	2.690	2.952	3.281	3.520
46	0.680	1.300	1.679	2.013	2.410	2.687	2.949	3.277	3.515
47	0.680	1.300	1.678	2.012	2.408	2.685	2.946	3.273	3.510
48	0.680	1.299	1.677	2.011	2.407	2.682	2.943	3.269	3.505
49	0.680	1.299	1.677	2.010	2.405	2.680	2.940	3.265	3.500
50	0.679	1.299	1.676	2.009	2.403	2.678	2.937	3.261	3.496

This table was prepared using Equations 26.7.3 and 26.7.4 of Zelen and Severo (1964), except for the values at infinity degrees of freedom, which are adapted from White (1970). Except for the values at infinity degrees of freedom, *t* was calculated to eight decimal places and then rounded to three decimal places.

Examples: $t_{0.05(2),13} = 2.160$ and $t_{0.01(1),19} = 2.539$

Critical Values of the t Distribution (cont.)

ν	$\alpha(2): 0.50$ $\alpha(1): 0.25$	0.20 0.10	0.10 0.05	0.05 0.025	0.02 0.01	0.01 0.005	0.005 0.0025	0.002 0.001	0.001 0.0005
52	0.679	1.298	1.675	2.007	2.400	2.674	2.932	3.255	3.488
54	0.679	1.297	1.674	2.005	2.397	2.670	2.927	3.248	3.480
56	0.679	1.297	1.673	2.003	2.395	2.667	2.923	3.242	3.473
58	0.679	1.296	1.672	2.002	2.392	2.663	2.918	3.237	3.466
60	0.679	1.296	1.671	2.000	2.390	2.660	2.915	3.232	3.460
62	0.678	1.295	1.670	1.999	2.388	2.657	2.911	3.227	3.454
64	0.678	1.295	1.669	1.998	2.386	2.655	2.908	3.223	3.449
66	0.678	1.295	1.668	1.997	2.384	2.652	2.904	3.218	3.444
68	0.678	1.294	1.668	1.995	2.382	2.650	2.902	3.214	3.439
70	0.678	1.294	1.667	1.994	2.381	2.648	2.899	3.211	3.435
72	0.678	1.293	1.666	1.993	2.379	2.646	2.896	3.207	3.431
74	0.678	1.293	1.666	1.993	2.378	2.644	2.894	3.204	3.427
76	0.678	1.293	1.665	1.992	2.376	2.642	2.891	3.201	3.423
78	0.678	1.292	1.665	1.991	2.375	2.640	2.889	3.198	3.420
80	0.678	1.292	1.664	1.990	2.374	2.639	2.887	3.195	3.416
82	0.677	1.292	1.664	1.989	2.373	2.637	2.885	3.193	3.413
84	0.677	1.292	1.663	1.989	2.372	2.636	2.883	3.190	3.410
86	0.677	1.291	1.663	1.988	2.370	2.634	2.881	3.188	3.407
88	0.677	1.291	1.662	1.987	2.369	2.633	2.880	3.185	3.405
90	0.677	1.291	1.662	1.987	2.368	2.632	2.878	3.183	3.402
92	0.677	1.291	1.662	1.986	2.368	2.630	2.876	3.181	3.399
94	0.677	1.291	1.661	1.986	2.367	2.629	2.875	3.179	3.397
96	0.677	1.290	1.661	1.985	2.366	2.628	2.873	3.177	3.395
98	0.677	1.290	1.661	1.984	2.365	2.627	2.872	3.175	3.393
100	0.677	1.290	1.660	1.984	2.364	2.626	2.871	3.174	3.390
105	0.677	1.290	1.659	1.983	2.362	2.623	2.868	3.170	3.386
110	0.677	1.289	1.659	1.982	2.361	2.621	2.865	3.166	3.381
115	0.677	1.289	1.658	1.981	2.359	2.619	2.862	3.163	3.377
120	0.677	1.289	1.658	1.980	2.358	2.617	2.860	3.160	3.373
125	0.676	1.288	1.657	1.979	2.357	2.616	2.858	3.157	3.370
130	0.676	1.288	1.657	1.978	2.355	2.614	2.856	3.154	3.367
135	0.676	1.288	1.656	1.978	2.354	2.613	2.854	3.152	3.364
140	0.676	1.288	1.656	1.977	2.353	2.611	2.852	3.149	3.361
145	0.676	1.287	1.655	1.976	2.352	2.610	2.851	3.147	3.359
150	0.676	1.287	1.655	1.976	2.351	2.609	2.849	3.145	3.357
160	0.676	1.287	1.654	1.975	2.350	2.607	2.846	3.142	3.352
170	0.676	1.287	1.654	1.974	2.348	2.605	2.844	3.139	3.349
180	0.676	1.286	1.653	1.973	2.347	2.603	2.842	3.136	3.345
190	0.676	1.286	1.653	1.973	2.346	2.602	2.840	3.134	3.342
200	0.676	1.286	1.653	1.972	2.345	2.601	2.839	3.131	3.340
250	0.675	1.285	1.651	1.969	2.341	2.596	2.832	3.123	3.330
300	0.675	1.284	1.650	1.968	2.339	2.592	2.828	3.118	3.323
350	0.675	1.284	1.649	1.967	2.337	2.590	2.825	3.114	3.319
400	0.675	1.284	1.649	1.966	2.336	2.588	2.823	3.111	3.315
450	0.675	1.283	1.648	1.965	2.335	2.587	2.821	3.108	3.312
500	0.675	1.283	1.648	1.965	2.334	2.586	2.820	3.107	3.310
600	0.675	1.283	1.647	1.964	2.333	2.584	2.817	3.104	3.307
700	0.675	1.283	1.647	1.963	2.332	2.583	2.816	3.102	3.304
800	0.675	1.283	1.647	1.963	2.331	2.582	2.815	3.100	3.303
900	0.675	1.282	1.647	1.963	2.330	2.581	2.814	3.099	3.301
1000	0.675	1.282	1.646	1.962	2.330	2.581	2.813	3.098	3.300
	0.6745	1.2816	1.6449	1.9600	2.3263	2.5758	2.8070	3.0902	3.2905

If a critical value is needed for degrees of freedom not on this table, one may conservatively employ the next smaller ν that is on the table. Or, the needed critical value, for $\nu < 1000$, may be calculated by linear interpolation, with an error of no more than 0.001. If a little more accuracy is desired, or if the needed ν is > 1000, then harmonic interpolation should be used.

Critical values of t for infinity degrees of freedom are related to critical values of Z and χ^2 as

$$t_{\alpha(1),\infty} = Z_{\alpha(1)} \qquad \text{and} \qquad t_{\alpha(1),\infty} = Z_{\alpha(2)} = \sqrt{\chi^2_{\alpha,1}}.$$

Critical Values of the Chi-Square Distribution

v	α = 0.999	0.995	0.99	0.975	0.95	0.90	0.75	0.50	0.25	0.10	0.05	0.025	0.01	0.005	0.001
1	0.000	0.000	0.000	0.001	0.004	0.016	0.102	0.455	1.323	2.706	3.841	5.024	6.635	7.879	10.828
2	0.002	0.010	0.020	0.051	0.103	0.211	0.575	1.386	2.773	4.605	5.991	7.378	9.210	10.597	13.816
3	0.024	0.072	0.115	0.216	0.352	0.584	1.213	2.366	4.108	6.251	7.815	9.348	11.345	12.838	16.266
4	0.091	0.207	0.297	0.484	0.711	1.064	1.923	3.357	5.385	7.779	9.488	11.143	13.277	14.860	18.467
5	0.210	0.412	0.554	0.831	1.145	1.610	2.675	4.351	6.626	9.236	11.070	12.833	15.086	16.750	20.515
6	0.381	0.676	0.872	1.237	1.635	2.204	3.455	5.348	7.841	10.645	12.592	14.449	16.812	18.548	22.458
7	0.599	0.989	1.239	1.690	2.167	2.833	4.255	6.346	9.037	12.017	14.067	16.013	18.475	20.278	24.322
8	0.857	1.344	1.646	2.180	2.733	3.490	5.071	7.344	10.219	13.362	15.507	17.535	20.090	21.955	26.124
9	1.152	1.735	2.088	2.700	3.325	4.168	5.899	8.343	11.389	14.684	16.919	19.023	21.666	23.589	27.877
10	1.479	2.156	2.558	3.247	3.940	4.865	6.737	9.342	12.549	15.987	18.307	20.483	23.209	25.188	29.588
11	1.834	2.603	3.053	3.816	4.575	5.578	7.584	10.341	13.701	17.275	19.675	21.920	24.725	26.757	31.264
12	2.214	3.074	3.571	4.404	5.226	6.304	8.438	11.340	14.845	18.549	21.026	23.337	26.217	28.300	32.909
13	2.617	3.565	4.107	5.009	5.892	7.042	9.299	12.340	15.984	19.812	22.362	24.736	27.688	29.819	34.528
14	3.041	4.075	4.660	5.629	6.571	7.790	10.165	13.339	17.117	21.064	23.685	26.119	29.141	31.319	36.123
15	3.483	4.601	5.229	6.262	7.261	8.547	11.037	14.339	18.245	22.307	24.996	27.488	30.578	32.801	37.697
16	3.942	5.142	5.812	6.908	7.962	9.312	11.912	15.338	19.369	23.542	26.296	28.845	32.000	34.267	39.252
17	4.416	5.697	6.408	7.564	8.672	10.085	12.792	16.338	20.489	24.769	27.587	30.191	33.409	35.718	40.790
18	4.905	6.265	7.015	8.231	9.390	10.865	13.675	17.338	21.605	25.989	28.869	31.526	34.805	37.156	42.312
19	5.407	6.844	7.633	8.907	10.117	11.651	14.562	18.338	22.718	27.204	30.144	32.852	36.191	38.582	43.820
20	5.921	7.434	8.260	9.591	10.851	12.443	15.452	19.337	23.828	28.412	31.410	34.170	37.566	39.997	45.315
21	6.447	8.034	8.897	10.283	11.591	13.240	16.344	20.337	24.935	29.615	32.671	35.479	38.932	41.401	46.797
22	6.983	8.643	9.542	10.982	12.338	14.041	17.240	21.337	26.039	30.813	33.924	36.781	40.289	42.796	48.268
23	7.529	9.260	10.196	11.689	13.091	14.848	18.137	22.337	27.141	32.007	35.172	38.076	41.638	44.181	49.728
24	8.085	9.886	10.856	12.401	13.848	15.659	19.037	23.337	28.241	33.196	36.415	39.364	42.980	45.559	51.179
25	8.649	10.520	11.524	13.120	14.611	16.473	19.939	24.337	29.339	34.382	37.652	40.646	44.314	46.928	52.620
26	9.222	11.160	12.198	13.844	15.379	17.292	20.843	25.336	30.435	35.563	38.885	41.923	45.642	48.290	54.052
27	9.803	11.808	12.879	14.573	16.151	18.114	21.749	26.336	31.528	36.741	40.113	43.195	46.963	49.645	55.476
28	10.391	12.461	13.565	15.308	16.928	18.939	22.657	27.336	32.620	37.916	41.337	44.461	48.278	50.993	56.892
29	10.986	13.121	14.256	16.047	17.708	19.768	23.567	28.336	33.711	39.087	42.557	45.722	49.588	52.336	58.301
30	11.588	13.787	14.953	16.791	18.493	20.599	24.478	29.336	34.800	40.256	43.773	46.979	50.892	53.672	59.703
31	12.196	14.458	15.655	17.539	19.281	21.434	25.390	30.336	35.887	41.422	44.985	48.232	52.191	55.003	61.098
32	12.811	15.134	16.362	18.291	20.072	22.271	26.304	31.336	36.973	42.585	46.194	49.480	53.486	56.328	62.487
33	13.431	15.815	17.074	19.047	20.867	23.110	27.219	32.336	38.058	43.745	47.400	50.725	54.776	57.648	63.870
34	14.057	16.501	17.789	19.806	21.664	23.952	28.136	33.336	39.141	44.903	48.602	51.966	56.061	58.964	65.247
35	14.688	17.192	18.509	20.569	22.465	24.797	29.054	34.336	40.223	46.059	49.802	53.203	57.342	60.275	66.619
36	15.324	17.887	19.233	21.336	23.269	25.643	29.973	35.336	41.304	47.212	50.998	54.437	58.619	61.581	67.985
37	15.965	18.586	19.960	22.106	24.075	26.492	30.893	36.336	42.383	48.363	52.192	55.668	59.893	62.883	69.346
38	16.611	19.289	20.691	22.878	24.884	27.343	31.815	37.335	43.462	49.513	53.384	56.896	61.162	64.181	70.703
39	17.262	19.996	21.426	23.654	25.695	28.196	32.737	38.335	44.539	50.660	54.572	58.120	62.428	65.476	72.055
40	17.916	20.707	22.164	24.433	26.509	29.051	33.660	39.335	45.616	51.805	55.758	59.342	63.691	66.766	73.402
41	18.576	21.421	22.906	25.215	27.326	29.907	34.585	40.335	46.692	52.949	56.942	60.561	64.950	68.053	74.745
42	19.239	22.138	23.650	25.999	28.144	30.765	35.510	41.335	47.766	54.090	58.124	61.777	66.206	69.336	76.084
43	19.906	22.859	24.398	26.785	28.965	31.625	36.436	42.335	48.840	55.230	59.304	62.990	67.459	70.616	77.419
44	20.576	23.584	25.148	27.575	29.787	32.487	37.363	43.335	49.913	56.369	60.481	64.201	68.710	71.893	78.750
45	21.251	24.311	25.901	28.366	30.612	33.350	38.291	44.335	50.985	57.505	61.656	65.410	69.957	73.166	80.077
46	21.929	25.041	26.657	29.160	31.439	34.215	39.220	45.335	52.056	58.641	62.830	66.617	71.201	74.437	81.400
47	22.610	25.775	27.416	29.956	32.268	35.081	40.149	46.335	53.127	59.774	64.001	67.821	72.443	75.704	82.720
48	23.295	26.511	28.177	30.755	33.098	35.949	41.079	47.335	54.196	60.907	65.171	69.023	73.683	76.969	84.037
49	23.983	27.249	28.941	31.555	33.930	36.818	42.010	48.335	55.265	62.038	66.339	70.222	74.919	78.231	85.351
50	24.674	27.991	29.707	32.357	34.764	37.689	42.942	49.335	56.334	63.167	67.505	71.420	76.154	79.490	86.661

Critical Values of the Chi-Square Distribution (cont.)

v \ α	0.999	0.995	0.99	0.975	0.95	0.90	0.75	0.50	0.25	0.10	0.05	0.025	0.01	0.005	0.001
51	25.368	28.735	30.475	33.162	35.600	38.560	43.874	50.335	57.401	64.295	68.669	72.616	77.386	80.747	87.968
52	26.065	29.481	31.246	33.968	36.437	39.433	44.808	51.335	58.468	65.422	69.832	73.810	78.616	82.001	89.272
53	26.765	30.230	32.019	34.776	37.276	40.308	45.741	52.335	59.534	66.548	70.993	75.002	79.843	83.253	90.573
54	27.468	30.981	32.793	35.586	38.116	41.183	46.676	53.335	60.600	67.673	72.153	76.192	81.069	84.502	91.872
55	28.173	31.735	33.570	36.398	38.958	42.060	47.610	54.335	61.665	68.796	73.311	77.380	82.292	85.749	93.168
56	28.881	32.491	34.350	37.212	39.801	42.937	48.546	55.335	62.729	69.919	74.468	78.567	83.513	86.994	94.461
57	29.592	33.248	35.131	38.027	40.646	43.816	49.482	56.335	63.793	71.040	75.624	79.752	84.733	88.236	95.751
58	30.305	34.008	35.913	38.844	41.492	44.696	50.419	57.335	64.857	72.160	76.778	80.936	85.950	89.477	97.039
59	31.021	34.770	36.698	39.662	42.339	45.577	51.356	58.335	65.919	73.279	77.931	82.117	87.166	90.715	98.324
60	31.738	35.535	37.485	40.482	43.188	46.459	52.294	59.335	66.981	74.397	79.082	83.298	88.379	91.952	99.607
61	32.459	36.301	38.273	41.303	44.038	47.342	53.232	60.335	68.043	75.514	80.232	84.476	89.591	93.186	100.888
62	33.181	37.068	39.063	42.126	44.889	48.226	54.171	61.335	69.104	76.630	81.381	85.654	90.802	94.419	102.166
63	33.906	37.838	39.855	42.950	45.741	49.111	55.110	62.335	70.165	77.745	82.529	86.830	92.010	95.649	103.442
64	34.633	38.610	40.649	43.776	46.595	49.996	56.050	63.335	71.225	78.860	83.675	88.004	93.217	96.878	104.716
65	35.362	39.383	41.444	44.603	47.450	50.883	56.990	64.335	72.285	79.973	84.821	89.177	94.422	98.105	105.988
66	36.093	40.158	42.240	45.431	48.305	51.770	57.931	65.335	73.344	81.085	85.965	90.349	95.626	99.330	107.258
67	36.826	40.935	43.038	46.261	49.162	52.659	58.872	66.335	74.403	82.197	87.108	91.519	96.828	100.554	108.526
68	37.561	41.713	43.838	47.092	50.020	53.548	59.814	67.335	75.461	83.308	88.250	92.689	98.028	101.776	109.791
69	38.298	42.494	44.639	47.924	50.879	54.438	60.756	68.334	76.519	84.418	89.391	93.856	99.228	102.996	111.055
70	39.036	43.275	45.442	48.758	51.739	55.329	61.698	69.334	77.577	85.527	90.531	95.023	100.425	104.215	112.317
71	39.777	44.058	46.246	49.592	52.600	56.221	62.641	70.334	78.634	86.635	91.670	96.189	101.621	105.432	113.577
72	40.520	44.843	47.051	50.428	53.462	57.113	63.585	71.334	79.690	87.743	92.808	97.353	102.816	106.648	114.835
73	41.264	45.629	47.858	51.265	54.325	58.006	64.528	72.334	80.747	88.850	93.945	98.516	104.010	107.862	116.092
74	42.010	46.417	48.666	52.103	55.189	58.900	65.472	73.334	81.803	89.956	95.081	99.678	105.202	109.074	117.346
75	42.757	47.206	49.475	52.942	56.054	59.795	66.417	74.334	82.858	91.061	96.217	100.839	106.393	110.286	118.599
76	43.507	47.997	50.286	53.782	56.920	60.690	67.362	75.334	83.913	92.166	97.351	101.999	107.583	111.495	119.850
77	44.258	48.788	51.097	54.623	57.786	61.586	68.307	76.334	84.968	93.270	98.484	103.158	108.771	112.704	121.100
78	45.010	49.582	51.910	55.466	58.654	62.483	69.252	77.334	86.022	94.374	99.617	104.316	109.958	113.911	122.348
79	45.764	50.376	52.725	56.309	59.522	63.380	70.198	78.334	87.077	95.476	100.749	105.473	111.144	115.117	123.594
80	46.520	51.172	53.540	57.153	60.391	64.278	71.145	79.334	88.130	96.578	101.879	106.629	112.329	116.321	124.839
81	47.277	51.969	54.357	57.998	61.261	65.176	72.091	80.334	89.184	97.680	103.010	107.783	113.512	117.524	126.083
82	48.036	52.767	55.174	58.845	62.132	66.076	73.038	81.334	90.237	98.780	104.139	108.937	114.695	118.726	127.324
83	48.796	53.567	55.993	59.692	63.004	66.976	73.985	82.334	91.289	99.880	105.267	110.090	115.876	119.927	128.565
84	49.557	54.368	56.813	60.540	63.876	67.876	74.933	83.334	92.342	100.980	106.395	111.242	117.057	121.126	129.804
85	50.320	55.170	57.634	61.389	64.749	68.777	75.881	84.334	93.394	102.079	107.522	112.393	118.236	122.325	131.041
86	51.085	55.973	58.456	62.239	65.623	69.679	76.829	85.334	94.446	103.177	108.648	113.544	119.414	123.522	132.277
87	51.850	56.777	59.279	63.089	66.498	70.581	77.777	86.334	95.497	104.275	109.773	114.693	120.591	124.718	133.512
88	52.617	57.582	60.103	63.941	67.373	71.484	78.726	87.334	96.548	105.372	110.898	115.841	121.767	125.913	134.745
89	53.386	58.389	60.928	64.793	68.249	72.387	79.675	88.334	97.599	106.469	112.022	116.989	122.942	127.106	135.978
90	54.155	59.196	61.754	65.647	69.126	73.291	80.625	89.334	98.650	107.565	113.145	118.136	124.116	128.299	137.208
91	54.926	60.005	62.581	66.501	70.003	74.196	81.574	90.334	99.700	108.661	114.268	119.282	125.289	129.491	138.438
92	55.698	60.815	63.409	67.356	70.882	75.100	82.524	91.334	100.750	109.756	115.390	120.427	126.462	130.681	139.666
93	56.472	61.625	64.238	68.211	71.760	76.006	83.474	92.334	101.800	110.850	116.511	121.571	127.633	131.871	140.893
94	57.246	62.437	65.068	69.068	72.640	76.912	84.425	93.334	102.850	111.944	117.632	122.715	128.803	133.059	142.119
95	58.022	63.250	65.898	69.925	73.520	77.818	85.376	94.334	103.899	113.038	118.752	123.858	129.973	134.247	143.344
96	58.799	64.063	66.730	70.783	74.401	78.725	86.327	95.334	104.948	114.131	119.871	125.000	131.141	135.433	144.567
97	59.577	64.878	67.562	71.642	75.282	79.633	87.278	96.334	105.997	115.223	120.990	126.141	132.309	136.619	145.789
98	60.356	65.694	68.396	72.501	76.164	80.541	88.229	97.334	107.045	116.315	122.108	127.282	133.476	137.803	147.010
99	61.137	66.510	69.230	73.361	77.046	81.449	89.181	98.334	108.093	117.407	123.225	128.422	134.642	138.987	148.230
100	61.918	67.328	70.065	74.222	77.929	82.358	90.133	99.334	109.141	118.498	124.342	129.561	135.807	140.169	149.449

Critical Values of the Chi-Square Distribution (cont.)

v	α = 0.999	0.995	0.99	0.975	0.95	0.90	0.75	0.50	0.25	0.10	0.05	0.025	0.01	0.005	0.001
101	62.701	68.146	70.901	75.083	78.813	83.267	91.085	100.334	110.189	119.589	125.458	130.700	136.971	141.351	150.667
102	63.484	68.965	71.737	75.946	79.697	84.177	92.038	101.334	111.236	120.679	126.574	131.838	138.134	142.532	151.884
103	64.269	69.785	72.575	76.809	80.582	85.088	92.991	102.334	112.284	121.769	127.689	132.975	139.297	143.712	153.099
104	65.055	70.608	73.413	77.672	81.468	85.998	93.944	103.334	113.331	122.858	128.804	134.111	140.459	144.891	154.314
105	65.841	71.428	74.252	78.536	82.354	86.909	94.897	104.334	114.378	123.947	129.918	135.247	141.620	146.070	155.528
106	66.629	72.251	75.092	79.401	83.240	87.821	95.850	105.334	115.424	125.035	131.031	136.382	142.780	147.247	156.740
107	67.418	73.075	75.932	80.267	84.127	88.753	96.804	106.334	116.471	126.123	132.144	137.517	143.940	148.424	157.952
108	68.207	73.899	76.774	81.133	85.015	89.645	97.758	107.334	117.517	127.211	133.257	138.651	145.099	149.599	159.162
109	68.998	74.724	77.616	82.000	85.903	90.558	98.712	108.334	118.563	128.298	134.369	139.784	146.257	150.774	160.372
110	69.790	75.550	78.458	82.867	86.792	91.471	99.666	109.334	119.608	129.385	135.480	140.917	147.414	151.948	161.581
111	70.582	76.377	79.302	83.735	87.681	92.385	100.620	110.334	120.654	130.472	136.591	142.049	148.571	153.122	162.738
112	71.376	77.204	80.146	84.604	88.570	93.299	101.575	111.334	121.699	131.558	137.701	143.180	149.727	154.294	163.995
113	72.170	78.033	80.991	85.473	89.461	94.213	102.530	112.334	122.744	132.643	138.811	144.310	150.882	155.466	165.201
114	72.965	78.862	81.836	86.342	90.351	95.128	103.485	113.334	123.789	133.729	139.921	145.441	152.037	156.637	166.406
115	73.761	79.692	82.682	87.213	91.242	96.043	104.440	114.334	124.834	134.813	141.030	146.571	153.191	157.808	167.610
116	74.558	80.522	83.529	88.084	92.134	96.958	105.396	115.334	125.878	135.898	142.138	147.700	154.344	158.977	168.813
117	75.356	81.353	84.377	88.955	93.026	97.874	106.352	116.334	126.923	136.982	143.246	148.829	155.496	160.146	170.016
118	76.155	82.185	85.225	89.827	93.918	98.790	107.307	117.334	127.967	138.066	144.354	149.957	156.648	161.314	171.217
119	76.955	83.018	86.074	90.700	94.811	99.707	108.263	118.334	129.011	139.149	145.461	151.084	157.800	162.481	172.418
120	77.755	83.852	86.923	91.573	95.705	100.624	109.220	119.334	130.055	140.233	146.567	152.211	158.950	163.648	173.617
121	78.557	84.686	87.773	92.446	96.598	101.541	110.176	120.334	131.098	141.315	147.674	153.338	160.100	164.814	174.816
122	79.359	85.521	88.624	93.320	97.493	102.458	111.133	121.334	132.142	142.398	148.779	154.464	161.250	165.980	176.014
123	80.162	86.356	89.475	94.195	98.387	103.376	112.089	122.334	133.185	143.480	149.885	155.589	162.398	167.144	177.212
124	80.965	87.192	90.327	95.070	99.283	104.295	113.046	123.334	134.228	144.562	150.989	156.714	163.546	168.308	178.408
125	81.770	88.029	91.180	95.946	100.178	105.213	114.004	124.334	135.271	145.643	152.094	157.839	164.694	169.471	179.604
126	82.575	88.866	92.033	96.822	101.074	106.132	114.961	125.334	136.313	146.724	153.198	158.962	165.841	170.634	180.799
127	83.381	89.704	92.887	97.698	101.971	107.051	115.918	126.334	137.356	147.805	154.302	160.086	166.987	171.796	181.993
128	84.188	90.543	93.741	98.576	102.867	107.971	116.876	127.334	138.398	148.885	155.405	161.209	168.133	172.957	183.186
129	84.996	91.383	94.596	99.453	103.765	108.891	117.834	128.334	139.440	149.965	156.508	162.331	169.278	174.118	184.379
130	85.804	92.223	95.451	100.331	104.662	109.811	118.792	129.334	140.482	151.045	157.610	163.453	170.423	175.278	185.571
131	86.613	93.063	96.307	101.210	105.560	110.732	119.750	130.334	141.524	152.125	158.712	164.575	171.567	176.438	186.762
132	87.423	93.904	97.163	102.089	106.459	111.652	120.708	131.334	142.566	153.204	159.814	165.696	172.711	177.597	187.953
133	88.233	94.746	98.021	102.968	107.357	112.573	121.667	132.334	143.608	154.302	160.915	166.816	173.854	178.755	189.142
134	89.044	95.588	98.878	103.848	108.257	113.495	122.625	133.334	144.649	155.361	162.016	167.936	174.996	179.913	190.331
135	89.856	96.431	99.736	104.729	109.156	114.417	123.584	134.334	145.690	156.440	163.116	169.056	176.138	181.070	191.520
136	90.669	97.275	100.595	105.609	110.056	115.338	124.543	135.334	146.731	157.518	164.216	170.175	177.280	182.226	192.707
137	91.482	98.119	101.454	106.491	110.956	116.261	125.502	136.334	147.772	158.595	165.316	171.294	178.421	183.382	193.894
138	92.296	98.964	102.314	107.372	111.857	117.183	126.461	137.334	148.813	159.673	166.415	172.412	179.561	184.538	195.080
139	93.111	99.809	103.174	108.254	112.758	118.106	127.421	138.334	149.854	160.750	167.514	173.530	180.701	185.693	196.266
140	93.926	100.655	104.034	109.137	113.659	119.029	128.380	139.334	150.894	161.827	168.613	174.648	181.840	186.847	197.451

Statistical Tables and Graphs

Table B.1 was prepared using Equation 26.4.6 of Zelen and Severo (1964). The chi-square values were calculated to six decimal places and then rounded to three decimal places.

Examples:

$$\chi^2_{0.05,12} = 21.026 \quad \text{and} \quad \chi^2_{0.0,138} = 61.162.$$

For large degrees of freedom (v), critical values of χ^2 can be approximated very well by

$$\chi^2_{\alpha,v} \cong v\left(1 - \frac{2}{9v} + Z_{\alpha(1)}\sqrt{\frac{2}{9v}}\right)^3$$

(Wilson and Hilferty, 1931). It is for this purpose that the values of $Z\alpha(1)$ are given below (from White, 1970).

$\alpha = 0.999$	0.995	0.99	0.975	0.95	0.90	0.75
$Z_{\alpha(1)} = -3.09023$	-2.57583	-2.32635	-1.95996	-1.64485	-1.28155	-0.67449

$\alpha = 0.50$	0.25	0.10	0.05	0.025	0.01	0.005	0.001
$Z_{\alpha(1)} = 0.00000$	0.67449	1.28155	1.64485	1.95996	2.32635	2.57583	3.09023

The percent error, i.e., (approximation − true value)/true value × 100%, resulting from the use of this approximation is as follows:

v	$\alpha = 0.999$	0.995	0.99	0.975	0.95	0.90	0.75	0.50	0.25	0.10	0.05	0.025	0.01	0.005	0.001
30	−0.7	−0.3	−0.2	−0.1	0.0*	0.0*	0.0*	0.0*	0.0*	0.0*	0.0*	0.0*	0.0*	0.1	0.2
100	−0.1	0.0*	0.0*	0.0*	0.0*	0.0*	0.0*	0.0*	0.0*	0.0*	0.0*	0.0*	0.0*	0.0*	0.0*
140	0.0*	0.0*	0.0*	0.0*	0.0*	0.0*	0.0*	0.0*	0.0*	0.0*	0.0*	0.0*	0.0*	0.0*	0.0*

where the asterisk indicates a percent error the absolute value of which is less than 0.05%. Zar (1978) discusses this and other approximations for $\chi^2_{\alpha,v}$.

For one degree of freedom, the χ^2 distribution is related to the normal distribution (Appendix Table B.2) and the t distribution (Appendix Table B.3) as

$$\chi^2_{\alpha,1} = \left(Z_{\alpha(2)}\right)^2 = \left(t_{\alpha(2),\infty}\right)^2.$$

For example, $\chi^2_{0.05,1} = 3.841$, and $(Z_{0.05(2)})^2 = (t_{0.05(2),\infty})^2 = (1.9600)^2 = 3.8416$.

The relationship between χ^2 and F (Appendix Table B.4) is

$$\chi^2_{\alpha,v} = v F_{\alpha(1),v,\infty}.$$

For example, $\chi^2_{0.05,9} = 16.919$, and $(9)(F_{0.05(1),9,\infty}) = (9)(1.88) = 16.92$.

APPENDIX 6. Table of Random Numbers

This appendix includes a table of 10,000 random digits, reprinted from J. H. Zar, Biostatistical Analysis, 3rd ed., 1996, by permission of Prentice-Hall, Inc., Upper Saddle River, New Jersey.

Ten Thousand Random Digits

	00-04	05-09	10-14	15-19	20-24	25-29	30-34	35-39	40-44	45-49
00	22808	04391	45529	53968	57136	98228	85485	13801	68194	56382
01	49305	36965	44849	64987	59501	35141	50159	57369	76913	75739
02	81934	19920	73316	69243	69605	17022	53264	83417	55193	92929
03	10840	13508	48120	22467	54505	70536	91206	81038	22418	34800
04	99555	73289	59605	37105	24621	44100	72832	12268	97089	68112
05	32677	45709	62337	35132	45128	96761	08745	53388	98353	46724
06	09401	75407	27704	11569	52842	83543	44750	03177	50511	15301
07	73424	31711	65519	74869	56744	40864	75315	89866	96563	75142
08	37075	81378	59472	71858	86903	66860	03757	32723	54273	45477
09	02060	37158	55244	44812	45369	78939	08048	28036	40946	03898
10	94719	43565	40028	79866	43137	28063	52513	66405	71511	66135
11	70234	48272	59621	88778	16536	36505	41724	24776	63971	01685
12	07972	71752	92745	86465	01845	27416	50519	48458	68460	63113
13	58521	64882	26993	48104	61307	73933	17214	44827	88306	78177
14	32580	45202	21148	09684	39411	04892	02055	75276	51831	85686
15	88796	30829	35009	22695	23694	11220	71006	26720	39476	60538
16	31525	82746	78935	82980	61236	28940	96341	13790	66247	33839
17	02747	35989	70387	89571	34570	17002	79223	96817	31681	15207
18	46651	28987	20625	61347	63981	41085	67412	29053	00724	14841
19	43598	14436	33521	55637	39789	26560	66404	71802	18763	80560
20	30596	92319	11474	64546	60030	73795	60809	24016	29166	36059
21	56198	64370	85771	62633	78240	05766	32419	35769	14057	80674
22	68266	67544	06464	84956	18431	04015	89049	15098	12018	89338
23	31107	28597	65102	75599	17496	87590	68848	33021	69855	54015
24	37555	05069	38680	87274	55152	21792	77219	48732	03377	01160
25	90463	27249	43845	94391	12145	36882	48906	52336	00780	74407
26	99189	88731	93531	52638	54989	04237	32978	59902	05463	09245
27	37631	74016	89072	59598	55356	27346	80856	80875	52850	36548
28	73829	21651	50141	76142	72303	06694	61697	76662	23745	96282
29	15634	89428	47090	12094	42134	62381	87236	90118	53463	46969
30	00571	45172	78532	63863	98597	15742	41967	11821	91389	07476
31	83374	10184	56384	27050	77700	13875	96607	76479	80535	17454
32	78666	85645	13181	08700	08289	62956	64439	39150	95690	18555
33	47890	88197	21368	65254	35917	54035	83028	84636	38186	50581
34	56238	13559	79344	83198	94642	35165	40188	21456	67024	62771
35	36369	32234	38129	59963	99237	72648	66504	99065	61161	16186
36	42934	34578	28968	74028	42164	56647	76806	61023	33099	48293
37	09010	15226	43474	30174	26727	39317	48508	55438	85336	40762
38	83897	90073	72941	85613	85569	24183	08247	15946	02957	68504
39	82206	01230	93252	89045	25141	91943	75531	87420	99012	80751
40	14175	32992	49046	41272	94040	44929	98531	27712	05106	35242
41	58968	88367	70927	74765	18635	85122	27722	95388	61523	91745
42	62601	04595	76926	11007	67631	64641	07994	04639	39314	83126
43	97030	71165	47032	85021	65554	66774	21560	04121	57297	85415
44	89074	31587	21360	41673	71192	85795	82757	52928	62586	02179
45	07806	81312	81215	99858	26762	28993	74951	64680	50934	32011
46	91540	86466	13229	76624	44092	96604	08590	89705	03424	48033
47	99279	27334	33804	77988	93592	90708	56780	70097	39907	51006
48	63224	05074	83941	25034	43516	22840	35230	66048	80754	46302
49	98361	97513	27529	66419	35328	19738	82366	38573	50967	72754

Ten Thousand Random Digits (cont.)

	00-04	05-09	10-14	15-19	20-24	25-29	30-34	35-39	40-44	45-49
50	27791	82504	33523	27623	16597	32089	81596	78429	14111	68245
51	33147	46058	92388	10150	63224	26003	56427	29945	44546	50233
52	67243	10454	40269	44324	46013	00061	21622	68213	47749	76398
53	78176	70368	95523	09134	31178	33857	26171	07063	41984	99310
54	70199	70547	94431	45423	48695	01370	68065	61982	20200	27066
55	19840	01143	18606	07622	77282	68422	70767	33026	15135	91212
56	32970	28267	17695	20571	50227	69447	45535	16845	68283	15919
57	43233	53872	68520	70013	31395	60361	39034	59444	17066	07418
58	08514	23921	16685	89184	71512	82239	72947	69523	75618	79826
59	28595	51196	96108	84384	80359	02346	60581	01488	63177	47496
60	83334	81552	88223	29934	68663	23726	18429	84855	26897	94782
61	66112	95787	84997	91207	67576	27496	01603	22395	41546	68178
62	25245	14749	30653	42355	88625	37412	87384	09392	11273	28116
63	21861	22185	41576	15238	92294	50643	69848	48020	19785	41518
64	74506	40569	90770	40812	57730	84150	91500	53850	52104	37988
65	23271	39549	33042	10661	37312	50914	73027	21010	76788	64037
66	08548	16021	64715	08275	50987	67327	11431	31492	86970	47335
67	14236	80869	90798	85659	10079	28535	35938	10710	67046	74021
68	55270	49583	86467	40633	27952	27187	35058	66628	94372	75665
69	02301	05524	91801	23647	51330	35677	05972	90729	26650	81684
70	72843	03767	62590	92077	91552	76853	45812	15503	93138	87788
71	49248	43346	29503	22494	08051	09035	75802	63967	74257	00046
72	62598	99092	87806	42727	30659	10118	83000	96198	47155	00361
73	27510	69457	98616	62172	07056	61015	22159	65590	51082	34912
74	84167	66640	69100	22944	19833	23961	80834	37418	42284	12951
75	14722	88488	54999	55244	03301	37344	01053	79305	94771	95215
76	46696	05477	32442	18738	43021	72933	14995	30408	64043	67834
77	13938	09867	28949	94761	38419	38695	90165	82841	75399	09932
78	48778	56434	42495	07050	35250	09660	56192	34793	36146	96806
79	00571	71281	01563	66448	94560	55920	31580	26640	91262	30863
80	96050	57641	21798	14917	21836	15053	33566	51177	91786	12610
81	30870	81575	14019	07831	81840	25506	29358	88668	42742	62048
82	59153	29135	00712	73025	14263	17253	95662	75535	26170	95240
83	78283	70379	54969	05821	26485	28990	40207	00434	38863	61892
84	12175	95800	41106	93962	06245	00883	65337	75506	66294	62241
85	14192	39242	17961	29448	84078	14545	39417	83649	26495	41672
86	69060	38669	00849	24991	84252	41611	62773	63024	57079	59283
87	46154	11705	29355	71523	21377	36745	00766	21549	51796	81340
88	93419	54353	41269	07014	28352	77594	57293	59219	26098	63041
89	13201	04017	68889	81388	60829	46231	46161	01360	25839	52380
90	62264	99963	98226	29972	95169	07546	01574	94986	06123	52804
91	58030	30054	27479	70354	12351	33761	94357	81081	74418	74297
92	81242	26739	92304	81425	29052	37708	49370	46749	59613	50749
93	16372	70531	92036	54496	50521	83872	30064	67555	40354	23671
94	54191	04574	58634	91370	40041	77649	42030	42547	47593	07435
95	15933	92602	19496	18703	63380	58017	14665	88867	84807	44672
96	21518	77770	53826	97114	82062	34592	87400	64938	75540	54751
97	34524	64627	92997	21198	14976	07071	91566	44335	83237	24335
98	46557	67780	59432	23250	63352	43890	07109	07911	85956	62699
99	31929	13996	05126	83561	03244	33635	26952	01638	22788	26393

Ten Thousand Random Digits (cont.)

	50-54	55-59	60-64	65-69	70-74	75-79	80-84	85-89	90-94	95-99
00	53330	26487	85005	06384	13822	83736	95876	71355	31226	56063
01	96990	62825	97110	73006	32661	63408	03893	10333	41902	69175
02	30385	16588	63609	09132	53081	14478	50813	22887	03746	10289
03	75252	66905	60536	13408	25158	35825	10447	47375	89249	91238
04	52615	66504	78496	90443	84414	31981	88768	49629	15174	99795
05	39992	51082	74547	31022	71980	40900	84729	34286	96944	49502
06	51788	87155	13272	92461	06466	25392	22330	17336	42528	78628
07	88569	35645	50602	94043	35316	66344	78064	89651	89025	12722
08	14513	34794	44976	71244	60548	03041	03300	46389	25340	23804
09	50257	53477	24546	01377	20292	85097	00660	39561	62367	61424
10	35170	69025	46214	27085	83416	48597	19494	49380	28469	77549
11	22225	83437	43912	30337	75784	77689	60425	85588	93438	61343
12	90103	12542	97828	85859	85859	64101	00924	89012	17889	01154
13	68240	89649	85705	18937	30114	89827	89460	01998	81745	31281
14	01589	18335	24024	39498	82052	07868	49486	25155	61730	08946
15	36375	61694	90654	16475	92703	59561	45517	90922	93357	00207
16	11237	60921	51162	74153	94774	84150	39274	10089	45020	09624
17	48667	68353	40567	79819	48551	26789	07281	14669	00576	17435
18	99286	42806	02956	73762	04419	21676	67533	50553	21115	26742
19	44651	48349	13003	39656	99757	74964	00141	21387	66777	68533
20	83251	70164	05732	66842	77717	25305	36218	85600	23736	06629
21	41551	54630	88759	10085	48806	08724	50685	95638	20829	37264
22	68990	51280	51368	73661	21764	71552	69654	17776	51935	53169
23	63393	76820	33106	23322	16783	35630	50938	90047	97577	27699
24	93317	87564	32371	04190	27608	40658	11517	19646	82335	60088
25	48546	41090	69890	58014	04093	39286	12253	55859	83853	15023
26	31435	57566	99741	77250	43165	31150	20735	57406	85891	04806
27	56405	29392	76998	66849	29175	11641	85284	89978	73169	62140
28	70102	50882	85960	85955	03828	69417	55854	63173	60485	00327
29	92746	32004	52242	94763	32955	39848	09724	30029	45196	67606
30	67737	34389	57920	47081	60714	04935	48278	90687	99290	18554
31	35606	76646	14813	51114	52492	46778	08156	22372	59999	43938
32	64836	28649	45759	45788	43183	25275	25300	21548	33941	66314
33	86319	92367	37873	48993	71443	22768	69124	65611	79267	49709
34	90632	32314	24446	60301	31376	13575	99663	81929	39343	17648
35	83752	51966	43895	03129	37539	72989	52393	45542	70344	96712
36	56755	21142	86355	33569	63096	66780	97539	75150	25718	33724
37	14100	28857	60648	86304	97397	97210	74842	87483	51558	52883
38	69227	24872	48057	29318	74385	02097	63266	26950	73173	53025
39	77718	56967	36560	87155	26021	70903	32086	11722	32053	63723
40	09550	38799	88929	80877	87779	99905	17122	25985	16866	76005
41	12404	42453	88609	89148	85892	96045	10310	45021	62023	70061
42	07985	27418	92734	80000	58969	99011	73815	49705	68076	69605
43	58124	53830	08705	20916	46048	30342	86530	72608	93074	80937
44	46173	77223	75661	57691	24055	27568	41227	58542	73196	44886
45	13476	72301	85793	80516	59479	66985	24801	84009	71317	87321
46	82472	98647	17053	94591	36790	42275	51154	77765	01115	09331
47	55370	63433	80653	30739	68821	46854	41939	38962	20703	69424
48	89274	74795	82231	69384	53605	67860	01309	27273	76316	54253
49	55242	74511	62992	17981	17323	79325	35238	21393	13114	70084

Ten Thousand Random Digits **(cont.)**

	50-54	55-59	60-64	65-69	70-74	75-79	80-84	85-89	90-94	95-99
50	03674	36059	46810	58367	82676	15051	57977	49410	02971	05797
51	26136	80623	96505	91089	02309	54743	15831	45538	96456	87272
52	61716	80405	84735	12997	86386	61606	75091	84996	76070	54923
53	67051	63246	99547	81223	52485	90333	24697	06266	07388	70389
54	17284	60347	87314	30218	87983	45426	84153	10569	64042	95618
55	12543	23999	95777	28105	66073	35174	67706	05181	35176	85558
56	45494	93037	29209	70724	86438	65354	71209	27969	85321	10216
57	39262	15415	93940	41615	43605	95675	53916	29580	07048	95838
58	29094	58703	92144	14287	50165	85661	95749	61118	36668	96852
59	77988	03222	57805	00725	91543	80021	16442	63360	33620	39324
60	02758	86823	52423	32355	96707	47448	06453	59430	43952	16775
61	46702	37467	66803	49344	59519	92717	97110	82087	36785	00880
62	61759	95153	80090	60626	55917	92812	63544	82295	50729	20116
63	82316	11402	28078	75325	43963	63105	99294	30285	61473	53613
64	92754	74241	14315	49697	61979	66711	61707	81589	53936	82115
65	37907	24080	31741	86653	81460	32304	99590	56644	41521	91172
66	10619	75264	12279	18996	16716	81959	65722	10058	91522	65410
67	66640	06195	84416	32836	53178	93810	36766	59778	26612	69017
68	45208	58525	07714	77126	67986	73140	12026	75550	84912	64691
69	00910	40237	91035	29125	03534	47246	64698	00608	39537	71755
70	19965	46945	59357	15551	20335	03145	21519	37882	99146	70161
71	37538	05747	54982	00494	51866	86172	82679	04152	56369	20356
72	38571	69663	03287	28101	46753	55715	93527	30508	19722	02072
73	76711	02864	00880	85518	25834	52317	48070	51582	03374	19540
74	07128	44400	48015	41449	21109	38948	21816	52089	64529	21510
75	00882	89357	80906	76476	58420	95793	34043	00991	38937	39859
76	96160	18580	40549	46562	45106	53768	76097	60504	85273	63076
77	13443	22235	46210	47755	05802	00311	15171	23818	83870	47578
78	99494	35395	71411	48281	92151	84465	63651	15969	61345	13324
79	90647	11809	96365	52409	17977	05971	35835	03889	43733	66100
80	33050	48785	92200	59319	36977	41111	28002	51580	10573	21763
81	21257	15066	72630	23206	03106	53140	50292	64012	83184	81304
82	45362	94324	81800	83980	97244	09691	08435	66723	06150	54972
83	93322	58684	95695	19096	98108	47678	98061	87193	99992	82870
84	20374	61803	62508	83696	54449	53649	86447	66115	90857	69114
85	00715	13209	17080	06890	38022	76469	27696	30778	31836	96676
86	85519	93677	90186	09579	98760	50320	98077	46048	79700	81431
87	71948	15871	84502	41330	46675	51342	93431	55566	90819	68923
88	43427	95500	02004	51802	59668	17806	87605	33010	20991	76269
89	64854	28815	74959	03531	77051	51807	89005	18898	23716	45862
90	62195	29095	23982	75883	41561	25897	43595	92703	86676	32038
91	61186	54041	60984	61602	18482	57941	59657	35924	21738	30646
92	88585	40218	69965	74354	62274	38948	44813	31558	40625	22477
93	15598	21389	79016	92151	21926	49901	16835	88055	30545	60306
94	27097	89653	21558	72731	66694	36703	92172	46129	32660	91356
95	40537	85697	78182	39711	59270	21934	78647	94801	78832	37287
96	74828	06544	13078	59528	31100	11132	91256	85899	72492	18200
97	43297	83195	66218	65838	63255	72093	38976	44892	96861	97848
98	32663	58127	73258	09220	49701	92357	43700	37214	56844	02048
99	45551	31330	08152	23712	23963	58274	94583	03761	73429	47328

APPENDIX 7. Sample Size Equations

Five different sample size equations are presented in this appendix for the following situations:

Each separate section is designed to stand alone from the others. Each section includes the sample size equation, a description of each term in the equation, a table of appropriate coefficients, and a worked out example based on a stated management and sampling objective.

The examples included in this appendix all refer to monitoring with a quadrat-based sampling procedure. The equations and calculations also work with other kinds of monitoring data such as measurements of plant height, number of flowers, or measures of cover.

The examples of management objectives included in this appendix for detecting changes between two means or two proportions could be evaluated with one-tailed significance tests (Chapter 11). The sampling objectives and worked-out examples show calculations for two-tailed significance tests. This implies an interest in being able to detect either *increases* or *decreases* over time, even though the management objectives specify a desire to achieve a change in only one direction or the other. If you are only interested in detecting changes in one direction, and you only plan on analyzing your monitoring results with one directional null hypotheses (e.g., H_o = density has not increased), then you should apply a simple modifiication to the simple size procedures. To change any sample size procedure to a one-tailed situation, simply double the false-change (Type I) error rate (α) and look up the new doubled-α value in the table of coefficients (e.g., use $\alpha = 0.20$ instead of $\alpha = 0.10$ for a one-tailed test with a false-change (Type I) error rate of $\alpha = 0.10$).

The coefficients used in all of the equations are from a standard normal distribution (Z_α and Z_β) instead of the t-distribution (t_α and t_β). These two distributions are nearly identical at large sample sizes but at small sample sizes (n < 30) the Z coefficients will slightly underestimate the number of sampling units needed. The correction procedure described for Equation #1 (using the sample size correction table) already adjusts the sample size using the appropriate *t*-value (see Appendix 5 for a copy of a *t*-table). For the other equations, t_α and t_β values can be obtained from a *t*-table and used in place of the Z_α and Z_β coefficients that are included with the sample size equations. The appropriate t_α-coefficient for the false-change (Type I) error rate can be taken directly from the $\alpha(2)$ column of a *t*-table at the appropriate degrees of freedom (v). For example, for a false-change error rate of 0.10 use the $\alpha(2) = 0.10$ column. The appropriate t_β coefficient for a specified missed-change error level can be looked up by calculating 2(1-power) and looking up that value in the appropriate $\alpha(2)$ column. For example, for a power of 0.90, the calculations for t_β would be 2(1-.90) = 0.20. Use the $\alpha(2) = 0.20$ column at the appropriate degrees of freedom (v) to obtain the appropriate t_β value.

Sample size equation #1: Determining the necessary sample size for estimating a single population mean or a population total with a specified level of precision.

Estimating a sample *mean vs. total population size*. The sample size needed to estimate confidence intervals that are within a given percentage of the estimated *total population size* is the same as the sample size needed to estimate confidence intervals that are within that percentage of the estimated *mean value*. The instructions below assume you are working with a sample mean.

Determining sample size for a single population mean or a single population total is a two- or three-step process.
(1) The first step is to use the equation provided below to calculate an uncorrected sample size estimate.
(2) The second step is to consult the Sample Size Correction Table (Table 1) appearing on pages 349-350 of these instructions to come up with the corrected sample size estimate. The use of the correction table is necessary because the equation below under-estimates the number of sampling units that will be needed to meet the specified level of precision. The use of the table to correct the underestimated sample size is simpler than using a more complex equation that does not require correction.
(3) The third step is to multiply the corrected sample size estimate by the finite population correction factor if more than 5% of the population area is being sampled.

1. Calculate an initial sample size using the following equation:

$$n = \frac{(Z_\alpha)^2 (s)^2}{(B)^2}$$

Where:
n = The uncorrected sample size estimate.
Z_α = The standard normal coefficient from the table below.
s = The standard deviation.
B = The desired precision level expressed as half of the maximum acceptable confidence interval width. This needs to be specified in absolute terms rather than as a percentage. For example, if you wanted your confidence interval width to be within 30% of your sample mean (i.e., $\bar{x} \pm 30\% * \bar{x}$) and your sample mean = 10 plants/quadrat then B = (0.30 x 10) = 3.0.

Table of standard normal deviates (Z_α) for various confidence levels		
Confidence level	Alpha (α) level	(Z_α)
80%	0.20	1.28
90%	0.10	1.64
95%	0.05	1.96
99%	0.01	2.58

2. To obtain the adjusted sample size estimate, consult Table 1 on page 349-350 of these instructions.

n = the uncorrected sample size value from the sample size equation.
n* = the corrected sample size value.

3. Additional correction for sampling finite populations.

The above formula assumes that the population is very large compared to the proportion of the population that is sampled. If you are sampling more than 5% of the whole population then you should apply a correction to the sample size estimate that incorporates the finite population correction (FPC) factor. This will reduce the sample size.

The formula for correcting the sample size estimate with the FPC for confidence intervals is:

$$n' = \frac{n^*}{(1 + (n^*/N))}$$

Where:

n' = The new FPC-corrected sample size.

n* = The corrected sample size from the sample size correction table (Table 1).

N = The total number of possible quadrat locations in the population. To calculate N, determine the total area of the population and divide by the size of one quadrat.

Example:
Management objective:
Restore the population of species Y in population Z to a density of at least 30 plants/quadrat by the year 2001.

Sampling objective:
Obtain estimates of the mean density and population size with 95% confidence intervals that are within 20% of the estimated true value.

Results of pilot sampling:
Mean ($\overline{x}$) = 25 plants/quadrat.
Standard deviation (s) = 7 plants.

Given:
The desired **confidence level** is 95% so the appropriate Z_α from the table above = 1.96.
The desired **confidence interval width** is 20% (0.20) of the estimated true value. Since the estimated true value is 25 plants/quadrat, the desired confidence interval (**B**) = 25 x 0.20 = 5 plants/quadrat.

Calculate an unadjusted estimate of the sample size needed by using the sample size formula:

$$n = \frac{(Z_\alpha)^2(s)^2}{(B)^2} \quad n = \frac{(1.96)^2(7)^2}{(5)^2} = 7.5$$

Round 7.5 plots up to 8 plots for the unadjusted sample size.

To adjust this preliminary estimate, go to Table 1 on pages 349-350 of these instructions and find n = 8 and the corresponding n* value in the 95% confidence level portion of the table. For n = 8, the corresponding n* value = 15.

The corrected estimated sample size needed to be 95% confident that the estimate of the population mean is within 20% (+/- 5 plants) of the true mean = **15 quadrats**.

If the pilot data described above was gathered using a 1m x 10m (10 m^2) quadrat and the total population being sampled was located within a 20m x 50m macroplot (1000 m^2) then N = 1000m^2/10m^2 = 100. The corrected sample size would then be:

$$n' = \frac{n^*}{(1 + (n^*/N))} \qquad n' = \frac{15}{(1 + (15/100))} = 13.0$$

The new, FPC-corrected, estimated sample size to be 95% confident that the estimate of the population mean is within 20% (+/- 5 plants) of the true mean = **13 quadrats**.

Sample size correction table for single parameter estimates, Part 1

80% confidence level						90% confidence level					
n	n*	n	n*	n	n*	n	n*	n	n*	n	n*
1	5	51	65	101	120	1	5	51	65	101	120
2	6	52	66	102	121	2	6	52	66	102	122
3	7	53	67	103	122	3	8	53	67	103	123
4	9	54	68	104	123	4	9	54	69	104	124
5	10	55	69	105	124	5	11	55	70	105	125
6	11	56	70	106	125	6	12	56	71	106	126
7	13	57	71	107	126	7	13	57	72	107	127
8	14	58	73	108	128	8	15	58	73	108	128
9	15	59	74	109	129	9	16	59	74	109	129
10	17	60	75	110	130	10	17	60	75	110	130
11	18	61	76	111	131	11	18	61	76	111	131
12	19	62	77	112	132	12	20	62	78	112	132
13	20	63	78	113	133	13	21	63	79	113	134
14	22	64	79	114	134	14	22	64	80	114	135
15	23	65	80	115	135	15	23	65	81	115	136
16	24	66	82	116	136	16	25	66	82	116	137
17	25	67	83	117	137	17	26	67	83	117	138
18	27	68	84	118	138	18	27	68	84	118	139
19	28	69	85	119	140	19	28	69	85	119	140
20	29	70	86	120	141	20	29	70	86	120	141
21	30	71	87	121	142	21	31	71	88	121	142
22	31	72	88	122	143	22	32	72	89	122	143
23	33	73	89	123	144	23	33	73	90	123	144
24	34	74	90	124	145	24	34	74	91	124	145
25	35	75	91	125	146	25	35	75	92	125	147
26	36	76	93	126	147	26	37	76	93	126	148
27	37	77	94	127	148	27	38	77	94	127	149
28	38	78	95	128	149	28	39	78	95	128	150
29	40	79	96	129	150	29	40	79	96	129	151
30	41	80	97	130	151	30	41	80	97	130	152
31	42	81	98	131	152	31	42	81	99	131	153
32	43	82	99	132	154	32	44	82	100	132	154
33	44	83	100	133	155	33	45	83	101	133	155
34	45	84	101	134	156	34	46	84	102	134	156
35	47	85	102	135	157	35	47	85	103	135	157
36	48	86	104	136	158	36	48	86	104	136	158
37	49	87	105	137	159	37	49	87	105	137	159
38	50	88	106	138	160	38	50	88	106	138	161
39	51	89	107	139	161	39	52	89	107	139	162
40	52	90	108	140	162	40	53	90	108	140	163
41	53	91	109	141	163	41	54	91	110	141	164
42	55	92	110	142	164	42	55	92	111	142	165
43	56	93	111	143	165	43	56	93	112	143	166
44	57	94	112	144	166	44	57	94	113	144	167
45	58	95	113	145	168	45	58	95	114	145	168
46	59	96	115	146	169	46	60	96	115	146	169
47	60	97	116	147	170	47	61	97	116	147	170
48	61	98	117	148	171	48	62	98	117	148	171
49	62	99	118	149	172	49	63	99	118	149	172
50	64	100	119	150	173	50	64	100	119	150	173

APPENDIX 7—TABLE 1. Sample size correction table for adjusting "point-in-time" parameter estimates. n = the uncorrected sample size value from the sample size equation. n* = the corrected sample size value. This table was created using the algorithm reported by Kupper and Hafner (1989) for a one-sample tolerance probability of 0.90. For more information consult Kupper and Hafner (1989).

Sample size correction table for single parameters, Part 2

95% confidence level						99% confidence level					
n	n*	n	n*	n	n*	n	n*	n	n*	n	n*
1	5	51	66	101	121	1	6	51	67	101	122
2	7	52	67	102	122	2	8	52	68	102	123
3	8	53	68	103	123	3	9	53	69	103	124
4	10	54	69	104	124	4	11	54	70	104	126
5	11	55	70	105	125	5	12	55	72	105	127
6	12	56	71	106	126	6	14	56	73	106	128
7	14	57	72	107	128	7	15	57	74	107	129
8	15	58	74	108	129	8	16	58	75	108	130
9	16	59	75	109	130	9	18	59	76	109	131
10	18	60	76	110	131	10	19	60	77	110	132
11	19	61	77	111	132	11	20	61	78	111	133
12	20	62	78	112	133	12	21	62	79	112	134
13	21	63	79	113	134	13	23	63	80	113	135
14	23	64	80	114	135	14	24	64	82	114	136
15	24	65	81	115	136	15	25	65	83	115	138
16	25	66	83	116	137	16	26	66	84	116	139
17	26	67	84	117	138	17	28	67	85	117	140
18	28	68	85	118	139	18	29	68	86	118	141
19	29	69	86	119	141	19	30	69	87	119	142
20	30	70	87	120	142	20	31	70	88	120	143
21	31	71	88	121	143	21	32	71	89	121	144
22	32	72	89	122	144	22	34	72	90	122	145
23	34	73	90	123	145	23	35	73	92	123	146
24	35	74	91	124	146	24	36	74	93	124	147
25	36	75	92	125	147	25	37	75	94	125	148
26	37	76	94	126	148	26	38	76	95	126	149
27	38	77	95	127	149	27	39	77	96	127	150
28	39	78	96	128	150	28	41	78	97	128	152
29	41	79	97	129	151	29	42	79	98	129	153
30	42	80	98	130	152	30	43	80	99	130	154
31	43	81	99	131	154	31	44	81	100	131	155
32	44	82	100	132	155	32	45	82	101	132	156
33	45	83	101	133	156	33	46	83	103	133	157
34	46	84	102	134	157	34	48	84	104	134	158
35	48	85	103	135	158	35	49	85	105	135	159
36	49	86	105	136	159	36	50	86	106	136	160
37	50	87	106	137	160	37	51	87	107	137	161
38	51	88	107	138	161	38	52	88	108	138	162
39	52	89	108	139	162	39	53	89	109	139	163
40	53	90	109	140	163	40	55	90	110	140	165
41	54	91	110	141	164	41	56	91	111	141	166
42	56	92	111	142	165	42	57	92	112	142	167
43	57	93	112	143	166	43	58	93	114	143	168
44	58	94	113	144	168	44	59	94	115	144	169
45	59	95	114	145	169	45	60	95	116	145	170
46	60	96	116	146	170	46	61	96	117	146	171
47	61	97	117	147	171	47	62	97	118	147	172
48	62	98	118	148	172	48	64	98	119	148	173
49	63	99	119	149	173	49	65	99	120	149	174
50	65	100	120	150	174	50	66	100	121	150	175

APPENDIX 7—TABLE 1. Sample size correction table for adjusting "point-in-time" parameter estimates. n = the uncorrected sample size value from the sample size equation. n* = the corrected sample size value. This table was created using the algorithm reported by Kupper and Hafner (1989) for a one-sample tolerance probability of 0.90. For more information consult Kupper and Hafner (1989).
(continued)

Sample size equation #2: Determining the necessary sample size for detecting differences between two means with temporary sampling units.

The equation for determining the number of samples necessary to detect some "true" difference between two sample means is:

$$n = \frac{2(s)^2(Z_\alpha + Z_\beta)^2}{(MDC)^2}$$

Where:

s = sample standard deviation.

Z_α = Z-coefficient for the false-change (Type I) error rate from the table below.

Z_β = Z-coefficient for the missed-change (Type II) error rate from the table below.

MDC = Minimum detectable change size. This needs to be specified in absolute terms rather than as a percentage. For example, if you wanted to detect a 20% change in the sample mean from one year to the next and your first year sample mean = 10 plants/quadrat then MDC = (0.20 x 10) = 2 plants/quadrat.

Table of standard normal deviates for Z_α		Table of standard normal deviates for Z_β		
False-change (Type I) error rate (α)	Z_α	Missed-change (Type II) error rate (β)	Power	Z_β
0.40	0.84	0.40	0.60	0.25
0.20	1.28	0.20	0.80	0.84
0.10	1.64	0.10	0.90	1.28
0.05	1.96	0.05	0.95	1.64
0.01	2.58	0.01	0.99	2.33

Example:

Management objective:
Increase the density of species F at Site Y by 20% between 1999 and 2004.

Sampling objective
I want to be 90% certain of detecting a 20% change in mean plant density and I am willing to accept a 10% chance that I will make a false-change error (conclude that a change took place when it really did not).

Results from pilot sampling:
Mean ($\bar{x}$) = 25 plants/quadrat.
Standard deviation (s) = 7 plants.

Given:
The acceptable **False-change error rate (α) = 0.10** so the appropriate Z_α from the table = 1.64.

The desired **Power** is 90% (0.90) so the **Missed-change error rate (β) = 0.10** and the appropriate Z_β coefficient from the table = 1.28.

The **Minimum Detectable Change (MDC)** is 20% of the 1999 value or (0.20)(25) = 5 plants/quadrat.

Calculate the estimated necessary sample size using the equation provided on page 351:

$$n = \frac{2(s)^2(Z_\alpha + Z_\beta)^2}{(MDC)^2} \qquad n = \frac{2(7)^2(1.64 + 1.28)^2}{(5)^2} = 33.4$$

Round up 33.4 to 34 plots.

Final estimated sample size needed to be 90% confident of detecting a change of 5 plants between 1999 and 2004 with a false-change error rate of 0.10 = **34 quadrats.** The sample size correction table is not needed for estimating sample sizes for detecting differences between two population means.

Correction for sampling finite populations:

The above formula assumes that the population is very large compared to the proportion of the population that is sampled. If you are sampling more than 5% of the whole population area then you should apply a correction to the sample size estimate that incorporates the finite population correction factor (FPC). This will reduce the sample size. The formula for correcting the sample size estimate is as follows:

$$n' = \frac{n}{(1 + (n/N))}$$

Where:

n' = The new sample size based upon inclusion of the finite population correction factor.

n = The sample size from the equation above.

N = The total number of possible quadrat locations in the population. To calculate N, determine the total area of the population and divide by the size of each individual sampling unit.

Example:

If the pilot data described above was gathered using a 1m x 10m (10 m²) quadrat and the total population being sampled was located within a 20m x 50m macroplot (1000 m²) then N = 1000m²/10m² = 100. The corrected sample size would then be:

$$n' = \frac{n}{(1 + (n/N))} \qquad n' = \frac{34}{(1 + (34/100))} = 25.3$$

Round up 25.3 to 26.

The new, FPC-corrected estimated sample size needed to be 90% certain of detecting a change of 5 plants between 1999 and 2004 with a false-change error rate of 0.10 = **26 quadrats.**

Note on the statistical analysis for two sample tests from finite populations.

If you have sampled more than 5% of an entire population then you should also apply the finite population correction factor to the results of the statistical test. This procedure involves dividing the test statistic by the square root of the finite population factor (1-n/N). For example, if your t-statistic from a particular test turned out to be 1.645 and you sampled n = 26 quadrats out of a total N=100 possible quadrats, then your correction procedure would look like the following:

$$t' = \frac{t}{\sqrt{1-(n/N)}} \qquad t' = \frac{1.645}{\sqrt{1-(26/100)}} = 1.912$$

Where:

t = The t-statistic from a t-test.

t' = The corrected t-statistic using the FPC.

n = The sample size from the equation above.

N = The total number of possible quadrat locations in the population. To calculate N, determine the total area of the population and divide by the size of each individual sampling unit.

You would need to look up the p-value of $t' = 1.912$ in a t-table at the appropriate degrees of freedom to obtain the correct p-value for this statistical test (a t-table can be found in Appendix 5).

Sample size equation #3: Determining the necessary sample size for detecting differences between two means when using paired or permanent sampling units.

When paired sampling units are being compared or when data from permanent quadrats are being compared between two time periods, then sample size determination requires a different procedure than if samples are independent of one another. The equation for determining the number of samples necessary to detect some "true" difference between two sample means is:

$$n = \frac{(s)^2(Z_\alpha + Z_\beta)^2}{(MDC)^2}$$

Where:

s = Standard deviation of the differences between paired samples (see examples below).

Z_α = Z-coefficient for the false-change (Type I) error rate from the table below.

Z_β = Z-coefficient for the missed-change (Type II) error rate from the table below.

MDC = Minimum detectable change size. This needs to be specified in absolute terms rather than as a percentage. For example, if you wanted to detect a 20% change in the sample mean from one year to the next and your first year sample mean = 10 plants/quadrat then MDC = (0.20 x 10) = 2 plants/quadrat.

Table of standard normal deviates for Z_α		Table of standard normal deviates for Z_β		
False-change (Type I) error rate (α)	Z_α	Missed-change (Type II) error rate (β)	Power	Z_β
0.40	0.84	0.40	0.60	0.25
0.20	1.28	0.20	0.80	0.84
0.10	1.64	0.10	0.90	1.28
0.05	1.96	0.05	0.95	1.64
0.01	2.58	0.01	0.99	2.33

If the objective is to track changes over time with permanent sampling units and only a single year of data is available, then you will not have a standard deviation of differences between the paired samples. If you have an estimate of the likely degree of correlation between the two years of data, and you assume that the among sampling units standard deviation is going to be the same in the second time period, then you can use the equation below to estimate the standard deviation of differences.

$$s_{diff} = (s_1)\left(\sqrt{(2\,(1 - corr_{diff}))}\right)$$

Where:

s_{diff} = Estimated standard deviation of the differences between paired samples.

s_1 = Sample standard deviation among sampling units at the first time period.

$corr_{diff}$ = Correlation coefficient between sampling unit values in the first time period and sampling unit values in the second time period.

Example #1:

Management objective:

Achieve at least a 20% higher density of species F at site Y in areas excluded from grazing as compared to grazed areas in 1999.

Sampling objective:

I want to be able to detect a 20% difference in mean plant density in areas excluded from grazing and adjacent paired grazed areas. I want to be 90% certain of detecting that difference, if it occurs, and I am willing to accept a 10% chance that I will make a false-change error (conclude that a difference exists when it really did not).

Results from pilot sampling:

Five paired quadrats were sampled where one member of the pair was excluded from grazing (with a small exclosure) and the other member of the pair was open to grazing.

Quadrat number	# of plants/quadrat		Difference between grazed and ungrazed	
	grazed	ungrazed		
1	2	3	1	
2	5	8	3	
3	4	9	5	
4	7	12	5	
5	3	7	4	
	$\bar{x} = 4.20$ s $=1.92$	$\bar{x} = 7.80$ s$=3.27$		
Summary statistics for the differences between the two sets of quadrats			$\bar{x}$ 3.60	s 1.67

Given:

The sampling objective specified a desired minimum detectable difference (i.e., equivalent to the MDC) of 20%. Taking the larger of the two mean values and multiplying by 20% leads to: (7.80) x (0.20) = **MDC = 1.56** plants quadrat

The appropriate **standard deviation** to use is **1.67**, the standard deviation of the differences between the pairs.

The acceptable **False-change error rate (α) = 0.10**, so the appropriate Z_α from the table = 1.64.

The desired Power is 90% (0.90), so the **Missed-change error rate (β) = 0.10** and the appropriate Z_β coefficient from the table = 1.28.

Calculate the estimated necessary sample size using the equation provided above:

$$n = \frac{(s)^2(Z_\alpha + Z_\beta)^2}{(MDC)^2} \qquad n = \frac{(1.67)^2(1.64 + 1.28)^2}{(1.56)^2} = 9.7$$

Round up 9.7 to 10 plots.

Final estimated sample size needed to be 90% certain of detecting a true difference of 1.56 plants/quadrat between the grazed and ungrazed quadrats with a false-change error rate of 0.10 = **10 quadrats**.

Example #2:
Management objective:
Increase the density of species F at Site Q by 20% between 1999 and 2002.

Sampling objective:
I want to be able to detect a 20% difference in mean plant density of species F at Site Q between 1999 and 2001. I want to be 90% certain of detecting that change, if it occurs, and I am willing to accept a 10% chance that I will make a false-change error (conclude that a change took place when it really did not).

The procedure for determining the necessary sample size for this example would be very similar to the previous example. Just replace "grazed" and "ungrazed" in the data table with "1999" and "2002" and the rest of the calculations would be the same. Because the sample size determination procedure needs the standard deviation of the difference between two samples, you will not have the necessary standard deviation term to plug into the equation until you have two years of data. The standard deviation of the difference can be estimated in the first year if some estimate of the correlation coefficient between sampling unit values in the first time period and the sampling unit values in the second time period is available (see the s_{diff} equation above).

Correction for sampling finite populations:
The above formula assumes that the population is very large compared to the proportion of the population that is sampled. If you are sampling more than 5% of the whole population area then you should apply a correction to the sample size estimate that incorporates the finite population correction factor (FPC). This will reduce the sample size. The formula for correcting the sample size estimate is as follows:

$$n' = \frac{n}{(1 + (n/N))}$$

Where:

n' = The new sample size based upon inclusion of the finite population correction factor.

n = The sample size from the equation above.

N = The total number of possible quadrat locations in the population. To calculate N, determine the total area of the population and divide by the size of each individual sampling unit.

Example:
If the pilot data described above were gathered using a 1m x 10m (10 m²) quadrat and the total population being sampled was located within a 10m x 50m macroplot (500 m²) then N = 500m²/10m² = 50. The corrected sample size would then be:

$$n' = \frac{n}{(1 + (n/N))} \qquad n' = \frac{10}{(1 + (10/50))} = 8.3$$

Round up 8.3 to 9.

The new, FPC-corrected estimated sample size needed to be 90% confident of detecting a true difference of 1.56 plants/quadrat between the two years with a false-change error rate of 0.10 = **9 quadrats.**

Note on the statistical analysis for two sample tests from finite populations.
If you have sampled more than 5% of an entire population then you should also apply the finite population correction factor to the results of the statistical test. This procedure involves dividing

the test statistic by the square root of (1-n/N). For example, if your t-statistic from a particular test turned out to be 1.782 and you sampled n=9 quadrats out of a total N=50 possible quadrats, then your correction procedure would look like the following:

$$t' = \frac{t}{\sqrt{1-(n/N)}} \qquad t' = \frac{1.782}{\sqrt{1-(9/50)}} = 1.968$$

Where:

t = The t-statistic from a t-test.

t' = The corrected t-statistic using the FPC.

n = The sample size from the equation above.

N = The total number of possible quadrat locations in the population. To calculate N, determine the total area of the population and divide by the size of each individual sampling unit.

You would need to look up the p-value of $t' = 1.968$ in a t-table for the appropriate degrees of freedom to obtain the correct p-value for this statistical test.

Sample size equation #4: Determining the necessary sample size for estimating a single population proportion with a specified level of precision.

The equation for determining the sample size for estimating a single proportion is:

$$n = \frac{(Z_\alpha)^2(p)(q)}{d^2}$$

Where:

n = Estimated necessary sample size.

Z_α = The coefficient from the table of standard normal deviates below.

p = The value of the proportion as a decimal percent (e.g., 0.45). If you don't have an estimate of the current proportion, use 0.50 as a conservative estimate.

q = 1 - p.

d = The desired precision level expressed as half of the maximum acceptable confidence interval width. This is also expressed as a decimal percent (e.g., 0.15) and this represents an *absolute* rather than a *relative* value. For example, if your proportion value is 30% and you want a precision level of ±10% this means you are targeting an interval width from 20% to 40%. Use 0.10 for the d-value and *not* 0.30 x 0.10 = 0.03.

Table of standard normal deviates (Z_α) for various confidence levels		
Confidence level	Alpha (α) level	(Z_α)
80%	0.20	1.28
90%	0.10	1.64
95%	0.05	1.96
99%	0.01	2.58

Example:

Management objective:

Maintain at least a 40% frequency (in 1m² quadrats) of species Y in population Z over the next 5 years.

Sampling objective:

Estimate percent frequency with 95% confidence intervals no wider than ± 10% of the estimated true value.

Results of pilot sampling:

The proportion of quadrats with species Z is estimated to be p = 65% (0.65). Because q = (1-p), q = (1-0.65) = 0.35.

Given:

The desired **confidence level** is 95% so the appropriate Z_α from the table above = 1.96.

The desired **confidence interval width (d)** is specified as 10% (0.10).

Using the equation provided above:

$$n = \frac{(Z_\alpha)^2(p)(q)}{d^2} \qquad n = \frac{(1.96)^2(0.65)(0.35)}{0.10^2} = 87.4$$

Round up 87.4 to 88.

The estimated sample size needed to be 95% confident that the estimate of the population percent frequency is within 10% (+/- 0.10) of the true percent frequency = **88 quadrats**.

This sample size formula works well as long as the proportion is more than 0.20 and less than 0.80 (Zar 1996). If you suspect the population proportion is less than 0.20 or greater than 0.80, use 0.20 or 0.80, respectively, as a conservative estimate of the proportion.

Correction for sampling finite populations:
The above formula assumes that the population is very large compared to the proportion of the population that is sampled. If you are sampling more than 5% of the whole population area then you should apply a correction for your sample size estimate that incorporates the finite population correction factor (FPC). This will reduce the sample size estimate. The formula for correcting the sample size estimate is as follows:

$$n' = \frac{n}{(1 + (n/N))}$$

Where:

n' = The new sample size with the inclusion of the FPC factor.

n = The sample size estimate from the above equation.

N = The total number of possible quadrat locations in the population. To calculate N, divide the total population area by the size of the sampling unit.

Example:
If the pilot data described above was gathered using a 1m x 1m (1 m²) quadrat and the total population being sampled was located within a 25m x 25m macroplot (625 m²) then N = 625m²/1m² = 625. The corrected sample size would then be:

$$n' = \frac{n}{(1 + (n/N))} \qquad n' = \frac{88}{(1 + (88/625))} = 77.1$$

Round up 77.1 to 78.

The new, FPC-corrected, estimated sample size needed to be 95% confident that the estimate of the population percent frequency is within 10% (+/- 0.10) of the true percent frequency = 78 quadrats.

Sample size equation #5: Determining the necessary sample size for detecting differences between two proportions with temporary sampling units.

The equation for determining the number of samples necessary to detect some "true" difference between two sample proportions is:

$$n = \frac{(Z_\alpha + Z_\beta)^2 (p_1 q_1 + p_2 q_2)}{(p_2 - p_1)^2}$$

Where:

n = Estimated necessary sample size.

Z_α = Z-coefficient for the false-change (Type I) error rate from the table below.

Z_β = Z-coefficient for the missed-change (Type II) error rate from the table below.

p_1 = The value of the proportion for the first sample as a decimal (e.g., 0.65). If you don't have an estimate of the current proportion, use 0.50 as a conservative estimate.

q_1 = 1 - p1.

p_2 = The value of the proportion for the second sample as a decimal (e.g., 0.45). This is determined based on the magnitude of change you wish to detect (see example, below).

q_2 = 1 - p2.

Table of standard normal deviates for Z_α		Table of standard normal deviates for Z_β		
False-change (Type I) error rate (α)	Z_α	Missed-change (Type II) error rate (β)	Power	Z_β
0.40	0.84	0.40	0.60	0.25
0.20	1.28	0.20	0.80	0.84
0.10	1.64	0.10	0.90	1.28
0.05	1.96	0.05	0.95	1.64
0.01	2.58	0.01	0.99	2.33

Example:
Management objective:
Decrease the frequency of invasive weed F at Site G by 20% between 1999 and 2001.

Sampling objective:
I want to be 90% certain of detecting an absolute change of 20% frequency and I am willing to accept a 10% chance that I will make a false-change error (conclude that a change took place when it really did not).

Note that the magnitude of change for detecting change over time for proportion data is expressed in absolute terms rather than in relative terms (relative terms were used in earlier examples that dealt with sample means values). The reason absolute terms are used instead of relative terms relates to the type of data being gathered (percent frequency is already expressed as a relative measure). Think of taking your population area and dividing it into a grid where the size of each grid cell equals your quadrat size. When you estimate a percent frequency, you are estimating the proportion of these grid cells occupied by a particular species. If 45% of all the grid cells in the population are occupied by a particular species then you hope that your sample values will be close to 45%. If over time the population changes so that now 65% of all the grid cells are occupied, then the true percent frequency has changed from 45% to 65%, representing a 20% absolute change.

Results from pilot sampling:

The proportion of quadrats with species Z in 1999 is estimated to be p_1 = 65% **(0.65)**.

Because $q_1 = (1-p_1)$, $q_1 = (1-0.65)$ = **0.35**.

Because we are interested in detecting a 20% shift in percent frequency, we will assign $p_2 = 0.45$. This represents a shift of 20% frequency from 1999 to 2001. A decline was selected instead of an increase (e.g., from 65% frequency to 85% frequency) because sample size requirements are higher at the mid-range of frequency values (i.e., closer to 50%) than they are closer to 0 or 100. Sticking closer to the mid-range gives us a more conservative sample size estimate.

Because $q_2 = (1-p_2)$, $q_2 = (1-0.45)$ = **0.55**.

Given:

The acceptable **False-change error rate** (α) = **0.10** so the appropriate Z_α from the table = 1.64.

The desired **Power is 90% (0.90)** so the **Missed-change error rate** (β) = **0.10** and the appropriate Z_β coefficient from the table = 1.28.

Using the equation provided above:

$$n = \frac{(Z_\alpha + Z_\beta)^2(p_1 q_1 + p_2 q_2)}{(p_2 - p_1)^2} \qquad n = \frac{(1.64 + 1.28)^2((0.65)(0.35)+(0.45)(0.55))}{(0.45 - 0.65)^2} = 101.3$$

Round up 101.3 to 102.

The estimated sample size needed to be 90% sure of detecting a shift of 20% frequency with a starting frequency of 65% and a false-change error rate of 0.10 = 102 quadrats.

Correction for sampling finite populations:

The above formula assumes that the population is very large compared to the proportion of the population that is sampled. If you are sampling more than 5% of the whole population area then you should apply a correction to the sample size estimate that incorporates the finite population correction factor (FPC). This will reduce the sample size. The formula for correcting the sample size estimate is as follows:

$$n' = \frac{n}{(1 + (n/N))}$$

Where:

n' = The new sample size based upon inclusion of the finite population correction factor.

n = The sample size from the equation above.

N = The total number of possible quadrat locations in the population. To calculate N, determine the total area of the population and divide by the size of each individual sampling unit.

Example:

If the pilot data described above was gathered using a 1m x 1m (1 m²) quadrat and the total population being sampled was located within a 10m x 30m macroplot (300 m²) then N = 300m²/1m² = 300. The corrected sample size would then be:

$$n' = \frac{n}{(1 + (n/N))} \qquad n' = \frac{102}{(1 + (102/300))} = 76.1$$

Round up 76.1 to 77.

The new, FPC-corrected estimated sample size needed to be 90% sure of detecting an absolute shift of 20% frequency with a starting frequency of 65% and a false-change error rate of 0.10 = **77 quadrats**.

Note on the statistical analysis for two sample tests from finite populations.

If you have sampled more than 5% of an entire population then you should also apply the finite population correction factor to the results of the statistical test. For proportion data, this procedure involves dividing the test statistic by (1-n/N). For example, if your χ^2-statistic from a particular test turned out to be 2.706 and you sampled n = 77 quadrats out of a total N = 300 possible quadrats, then your correction procedure would look like the following:

$$\chi^{2'} = \frac{\chi^2}{1-(n/N)} \qquad \chi^{2'} = \frac{2.706}{1-(77/300)} = 3.640$$

Where:

χ^2 = The χ^2-statistic from a χ^2-test.
$\chi^{2'}$ = The corrected χ^2-statistic using the FPC.
n = The sample size from the equation above.
N = The total number of possible quadrat locations in the population. To calculate N, determine the total area of the population and divide by the size of each individual sampling unit.

You would need to look up the *p*-value of $\chi^{2'}$ = 3.640 in a χ^2-table for the appropriate degrees of freedom to obtain the correct *p*-value for this statistical test (a χ^2- table can be found in Appendix 5).

Literature Cited

Kupper, L. L.; Hafner, K. B. 1989. How appropriate are popular sample size formulas? The American Statistician (43): 101-105.

Zar, J. H. 1996. Biostatistical analysis, 3rd edition. Upper Saddle River, New Jersey: Prentice Hall.

APPENDIX 8. Terms and Formulas Commonly Used in Statistics

Population (N): The entire collection of measurements about which one wishes to draw conclusions. There will almost always be a difference between the target population and the sampled population.

Parameter: A quantity that describes or characterizes a population. Examples of parameters are population means, population variances, population standard deviations, and population coefficients of variation. By convention, population parameters are designated by Greek letters.

Statistic: An estimate of a population parameter. By convention Latin letters are used to represent sample statistics.

Population Mean (μ) = $\dfrac{\text{sum of values[1] for each member of the population}}{\text{number of population members}}$

Mathematically this is given by:

$$\mu = \frac{X_1 + X_2 + \ldots X_N}{N}$$

Where:

X_1 = Value of the first member of the population.

X_2 = Value of the second member of the population.

X_N = Value of the last member of the population.

Or more concisely by:

$$\mu = \frac{\Sigma X}{N}$$

Population variance (σ^2) = $\dfrac{\text{sum of (value associated with member of population - population mean)}^2}{\text{number of population members}}$

Mathematically this is given by:

$$\sigma^2 = \frac{(X_1\text{-}\mu)^2 + (X_2\text{-}\mu)^2 + \ldots + (X_N\text{-}\mu)^2}{N}$$

Or more concisely by:

$$\sigma^2 = \frac{\Sigma(X\text{-}\mu)^2}{N}$$

Population standard deviation (σ)

$$= \sqrt{\text{population variance}}$$

Mathematically this is given by:

$$\sigma = \sqrt{\sigma^2} = \sqrt{\frac{\Sigma(X\text{-}\mu)^2}{N}}$$

---

[1] These values can be heights, counts, cover values, etc.

Sample (n): A subset of a population selected to estimate something about the whole population. A sample consists of n sampling units.

Sampling unit: One of the units comprising a sample. Sampling units can be quadrats, transects, points, individual plants, etc. Also called "observation."

Sample mean $(\overline{X})$ = $\dfrac{\text{sum of values, e.g., heights, of each observation in sample}}{\text{number of observations in sample}}$

The equivalent mathematical statement is:

$$\overline{X} = \frac{\Sigma X}{n}$$

Sample standard deviation:

Equivalent to the population standard deviation except that μ is replaced by its estimator $\overline{X}$ and N in the denominator is replaced by n - 1.

Mathematically this is given by:

$$s = \sqrt{\frac{(X_1-\overline{X})^2+(X_2-\overline{X})^2+...+(X_n-\overline{X})^2}{n-1}}$$

Or more concisely by:

$$s = \sqrt{\frac{\Sigma(X-\overline{X})^2}{n-1}}$$

Standard error of the mean (SE):

Usually referred to simply as "standard error", and abbreviated "SE." It is the standard deviation of all possible means of samples of size n from a population.

We estimate the standard error from a random sample taken from the population. The best estimate of the population standard error is:

$$SE = \sqrt{\frac{s^2}{n}} \quad also \quad SE = \frac{s}{\sqrt{n}}$$

This is called the sample standard error (or, more commonly, simply the "standard error," often abbreviated as SE).

The standard error quantifies the certainty with which the mean computed from a random sample estimates the true mean of the population from which the sample was drawn.

Confidence interval: The interval within which a true parameter value lies with known probability. It is a measure of the reliability of our sample estimate of the parameter value.

Confidence interval for a population mean:

We want to be able to specify the interval within which the true population mean most likely lies. In other words we want to be able to specify:

lower limit < μ < *upper limit*

We usually use a value from a table of the t distribution to determine a confidence interval. The formula for calculating a confidence interval is as follows:

$$\bar{X} - (t_{\alpha(2),v})(SE) < \mu < \bar{X} + (t_{\alpha(2),v})(SE)$$

Where $t_{\alpha(2),v}$ is the critical value from a t table for a given α value and v degrees of freedom (in determining a confidence interval for a mean $v = n - 1$). The (2) indicates that we are using both tails of the t distribution (which will always be the case for calculating a confidence interval of a mean).

The α value we choose depends upon how certain we wish to be that μ lies within our confidence interval. If we want to be 95% confident of this we choose $\alpha = .05$. If we want to be 80% confident we choose $\alpha = .20$ and so on.

Another, more concise, way of expressing the confidence interval is:

$$\bar{X} \pm (t_{\alpha(2),v})(SE)$$

If you are sampling from a finite population and you've sampled more than 5% of the population, you should apply the finite population correction factor (FPC) to your estimate of the SE. You do this as follows:

$$SE' = (SE)\left(\sqrt{1 - \frac{n}{N}}\right)$$

Where:
- SE' = Corrected standard error.
- SE = Uncorrected standard error.
- n = The sample size (the number of quadrats sampled).
- N = The total number of possible quadrats in the population. To calculate N, determine the total area of the population and divide by the area of each individual quadrat.

You then plug the corrected standard error (SE') into the equation for the confidence interval for the population mean.

Confidence interval for a population total:

To calculate a confidence interval for a population total you must know the size (N) of the population you have sampled from. You then calculate your estimate of the population total as follows:

$$\tau = (N)(\bar{x})$$

Where:
- τ = Estimate of population total.
- N = The total number of possible quadrats in the population. To calculate N, determine the total area of the population and divide by the area of each individual quadrat.
- $\bar{x}$ = Estimate of population mean.

The confidence interval around the estimate of the population total is then calculated as follows:

$$\tau \pm (N)(CI \text{ for population mean})$$

Where:

τ = Estimate of population total.

N = The total number of possible quadrats in the population. To calculate N, determine the total area of the population and divide by the area of each individual quadrat.

CI for = Confidence interval calculated for population mean as described above.
population
mean

Confidence interval for the difference between two population means (estimated from independent samples):

Several authors, particularly those in the behavioral sciences, have recently criticized the use of significance testing to determine whether two population means are different (see, for example, Cohen 1994). In its place they recommend calculating a confidence interval for the difference between two population means. This interval specifies:

lower limit $< \mu_1 - \mu_2 <$ *upper limit*

Just as for calculating the confidence interval around a single population mean, we resort to a value from the *t* distribution. Here is the formula:

$$\bar{X}_1 - \bar{X}_2 \pm (t_{\alpha(2),v}) \left(\sqrt{\frac{s_1^2}{n_1} + \frac{s_2^2}{n_2}} \right)$$

Where $t_{\alpha(2),v}$ is the critical value from a *t* table for a given α value and v degrees of freedom. In this case $v = n_1 + n_2 - 2$). The subscript (2) after the *t* indicates that we are using both tails of the *t* distribution.

Let's say we decide to calculate the 90 percent confidence interval for the difference in two population means. We take independent samples in two time periods and come up with the following information:

$\bar{X}_1 = 10$ plants
$\bar{X}_2 = 5$ plants
$s_1 = 3.5$ plants
$s_2 = 3$ plants
$n_1 = 40$ quadrats
$n_2 = 40$ quadrats

The 90 percent confidence interval for the difference between the means of the populations from which these two samples came is derived as follows:

$$10 - 5 \pm (1.665) \left(\sqrt{\frac{3.5^2}{40} + \frac{3^2}{40}} \right) = 5 \pm 1.21$$

Thus, we can be 90 percent confident that the true difference between the population means at times 1 and 2 fall within the interval 5 - 1.21 and 5 + 1.21 or between 3.79 and 6.21. Note that this interval does not include 0. If it did we would know that a significance test would yield a *P* value greater than 0.10 and we would conclude that the difference is not significant at the $\alpha = 0.10$ level. Because the interval is not even close to including 0 we can be very confident that the observed difference is real (this is not surprising since the difference between sample means is rather large and the estimates are rather precise). A significance test would yield a very low *P* value.

Consider the case, however, where either the difference between sample means is not so great and/or the estimates of the means are not very precise (s rather large compared to the means). Let's say we calculated a 90 percent confidence interval for the difference between two means and came up with the following interval:

-0.2 to 8

We can immediately determine two things from this. The first is that (because the interval contains 0) a significance test would yield a P value greater than 0.10. We would also note that the interval is rather large, meaning that our study design didn't have much power to detect change (the missed-change error rate would be high). The first thing we would have determined from a significance test (provided that the test provided exact P values). The second thing, however, is not as obvious. It is in providing this second important piece of information that the confidence interval between two population means is such a valuable statistical tool.

If you are sampling from a finite population and you've sampled more than 5% of the population, you should apply the finite population correction factor (FPC) to the formula for calculating the confidence interval as follows:

$$\bar{X}_1 - \bar{X}_2 \pm (t_{\alpha(2),v}) \left(\sqrt{\frac{s_1^2}{n_1} + \frac{s_2^2}{n_2}} \right) \left(\sqrt{1 - \frac{n}{N}} \right)$$

Where:
- n = The sample size (the number of quadrats sampled in each year; note that you do not add the number of quadrats sampled the first year to the number of quadrats sampled in the second year).
- N = The total number of possible quadrats in the population. To calculate N, determine the total area of the population and divide by the area of each individual quadrat.

Using our previous example, let's say that there were 500 possible quadrat locations in the population we sampled. Our sample size was 40 quadrats in each year. The confidence interval is therefore:

$$10 - 5 \pm (1.665) \left(\sqrt{\frac{3.5^2}{40} + \frac{3^2}{40}} \right) \left(\sqrt{1 - \frac{40}{500}} \right) = 5 \pm 1.16$$

Confidence interval for a population proportion

We want to be able to specify the interval within which the true population proportion most likely lies. In other words we want to be able to specify:

lower limit < p < upper limit

There are several ways of calculating a confidence interval around a proportion. Krebs (1989:21, Figure 2.2) provides a graph that can be used to estimate the confidence interval. Zar (1996:524-527) gives an "exact" method that uses a relationship between the F distribution and the binomial distribution. To use the method you need access to an F table, which can be also be found in Zar (1996); alternatively, you can use the program NCSS PROBABILITY CALCULATOR to calculate the F values needed for the procedure (see Appendix 19 for instructions on calculating F values using NCSS PROBABILITY CALCULATOR). Krebs (1989) provides the computer program, BINOM, which automates this exact procedure.

The following method, taken from Cochran (1977), approximates the confidence interval by using the normal distribution. It is accurate if the sample size is reasonably large, as shown in Table 1.

Sample Proportion $(\hat{p})$	Number of sampling units in the smaller class	Total sample size (n)
0.5	15	30
0.4	20	50
0.3	24	80
0.2	40	200
0.1	60	600
0.05	70	1400

TABLE 1. Sample sizes needed to use the normal approximation to calculate confidence intervals for proportions (Krebs 1989; Cochran 1977). Do not use the normal approximation unless you have a sample size this large or larger.

Calculate a confidence interval around the estimate of the population proportion $(\hat{p})$ obtained from your sample:

$$\hat{p} \pm \left[\left((Z_\alpha) \left(\sqrt{1 - (n/N)} \right) \left(\sqrt{\frac{\hat{p}\hat{q}}{n-1}} \right) \right) + \frac{1}{2n} \right]$$

Where:

$\hat{p}$ = Estimated proportion.
Z_α = Standard normal deviate from the table below.
$\hat{q}$ = 1 - $\hat{p}$.
n = Sample size.

Table of standard normal deviates (Z_α) for various confidence levels

Confidence level	Alpha (α) level	(Z_α)
80%	0.20	1.28
90%	0.10	1.64
95%	0.05	1.96
99%	0.01	2.58

The value, $1 - (n/N)$, in the above equation is the finite population correction factor (FPC). If your population is finite (i.e., you used quadrats and not points) and you've sampled more than 5% of the population you should use the above equation. Otherwise, you can leave the FPC out of the equation, in which case the equation reduces to:

$$\hat{p} \pm \left[\left((Z_\alpha) \left(\sqrt{\frac{\hat{p}\hat{q}}{n-1}} \right) \right) + \frac{1}{2n} \right]$$

For example, we sample 200 frequency quadrats and find species X in 75 of the 200 quadrats. Our estimate, $\hat{p}$, of the population proportion is $75/200 = 0.375$. There are 1000 possible quadrat positions in the population we sampled. The confidence interval around this estimate is therefore:

$$0.375 \pm \left[\left((1.96) \left(\sqrt{1 - (200/1000)} \right) \left(\sqrt{\frac{(0.375)(0.625)}{200 - 1}} \right) \right) + \frac{1}{2(200)} \right] = 0.375 \pm 0.063$$

For frequency sampling, you can (and should) adjust your quadrat size so the proportion of quadrats with the species of interest is close to 0.50. Doing this will ensure that the normal approximation method will give good estimates of the confidence interval at reasonable sample sizes (Table 1). When you are using the point-intercept method of estimating cover, however, the proportion of "hits" on the species of interest depends entirely on the amount of cover of the species. For a species with low cover values, you will end up with a small proportion of "hits." In this situation you must pay heed to the sample size requirements of Table 1. If your sample size is less than that given in Table 1, you should *not* use the normal approximation method. Instead, you should use the exact method given by Zar (1996) and automated by the program, BINOM (Krebs 1989).

Confidence intervals can also be calculated for the median, for a given percentile, and for many other statistics.

In addition to standard parametric and nonparametric procedures for calculating confidence intervals, methods based on resampling are commonly applied in constructing confidence intervals. See the discussion of this in Appendix 14.

Coefficient of variation:

The coefficient of variation, represented by CV, is defined as:

$$CV = \frac{s}{\overline{X}}$$

Often CV is multiplied by 100% in order to express CV as a percentage:

$$CV = \frac{s}{\overline{X}} (100\%)$$

The coefficient of variation is useful because, as a measure of variability, it does not depend upon the magnitude of the data. Contrast this with the standard deviation, which is dependent upon the magnitude and units of measurements of the data. This allows direct comparison of CV's from different studies. It also enables us to derive estimates of sample size when we do not have data from pilot studies but do have an idea of the likely magnitude of CV from similar studies and sites.

Parametric vs. nonparametric statistical methods:

Parametric statistical methods involve procedures that assume the population from which a sample is drawn can be completely described by population parameters such as means and standard deviations. These methods are valid only when the real population approximately follows the normal distribution, unless sample sizes are reasonably large, and—for significance testing—samples are approximately the same size in each sampling period.

Examples of parametric statistical methods:

- *t* test (both unpaired and paired)
- analysis of variance (both for independent samples and repeated measures)
- regression and correlation

Nonparametric statistical methods are based on frequencies, rates, or percentiles, and do not require the assumption that the population follows a normal distribution. These techniques, however, *still require random sampling.*

Examples of nonparametric statistical methods:

- Contingency table analysis (including chi-square)
- Mann-Whitney U test (analogous to unpaired *t* test)
- Wilcoxin signed-rank test (analogous to paired *t* test)
- Kruskal-Wallis test (analogous to analysis of variance of independent samples)
- Friedman test (analogous to repeated measures analysis of variance)

Resampling methods comprise a relatively new class of nonparametric techniques (see Appendix 14).

Central limit theorem:

- The distribution of sample means will be approximately normal regardless of the distribution of values in the original population from which the samples were drawn.

- The mean value of the collection of all possible sample means will equal the mean of the original population.

- The standard deviation of the collection of all possible means of samples of a given size, called the standard error of the mean, depends on both the standard deviation of the original population and the size of the sample.

- The central limit theorem allows us to apply parametric statistics to populations that are not normally distributed as long as sample sizes are reasonably large and—for significance testing—samples are approximately the same size at each sampling period.

ract *t* test (unpaired) to test hypotheses about two groups:

Start with two hypotheses:

H_o: The two means come from the same population (in a monitoring context this is equivalent to saying there has been no change in the population). This is called the *null hypothesis*.

H_a: The two means come from different populations (in a monitoring context this is equivalent to saying there has been a change in the population). This is called the *alternate hypothesis*.

Calculate the *t* statistic:

$$t = \frac{\text{difference of sample means}}{\text{standard error of difference of sample means}}$$

Mathematically *t* is calculated as follows:

$$t = \frac{\bar{X}_1 - \bar{X}_2}{\sqrt{\frac{s^2}{n_1} + \frac{s^2}{n_2}}}$$

Where:

t = Test statistic
$\bar{X}$ = Mean (subscripts denote samples 1 and 2, respectively)
n_1 = Sample size of sample 1

n_2 = Sample size of sample 2

s^2 = Pooled estimate of variance, calculated as follows:

$$s^2 = \frac{(s_1^2 + s_2^2)}{2}$$

Where:

s_1 = Standard deviation of sample 1

s_2 = Standard deviation of sample 2

To determine the likelihood of H_o being true we compare the t statistic we get using the above formula to the critical value of t in a t table for a given α and appropriate degrees of freedom. If t is sufficiently "big" we reject H_o.

Paired t test: In a paired t test we are interested in estimating the parameter δ, the average difference in response in each unit of the population. If we let d equal the observed change in each sampling unit, we can use d to estimate δ.

The standard deviation of the observed differences is:

$$s_d = \sqrt{\frac{\Sigma(d - \bar{d})^2}{n - 1}}$$

So the standard error of the differences is:

$$s_{\bar{d}} = \frac{s_d}{\sqrt{n}}$$

To test the null hypothesis (H_o) that there is, on the average, no difference, we calculate:

$$t = \frac{\bar{d}}{s_{\bar{d}}}$$

We then compare the resulting value of t with the critical value of t with $v = n - 1$ degrees of freedom and the α we have set as our threshold P value.

Chi square test: The chi square test is used to test for the difference between two or more proportions. It is used to analyze frequency data when individual quadrats are the sampling units and point cover data when individual points are the sampling units. If the frequency data are collected on more than one species, analysis is usually conducted separately on each species. Another alternative, however, is to lump species into functional groups such as annual graminoids, and conduct the analysis on that group.

> **2 x 2 contingency table to compare two years:**
> Chapter 11, Section D.3, gives an example of a chi-square test applied to a 2 x 2 contingency table, comparing frequency measurements collected in two separate years.

McNemar's test: McNemar's test is used to test for the difference between two proportions when sampling units are permanent. It is used to analyze frequency data when permanent quadrats are the sampling units and point cover data when permanent points are the sampling units. See Chapter 11, Section E.4, for further information.

Literature Cited

Cochran, W. G. 1977. Sampling techniques, 3rd ed. New York, NY: John Wiley & Sons.

Cohen, J. 1994. The earth is round (p<.05). American Psychologist 49:997-1003.

Krebs, C. J. 1989. Ecological methodology. New York, NY: Harper & Row.

Zar, J. H. 1996. Biostatistical analysis, 3rd ed. Upper Saddle River, NJ: Prentice Hall.

APPENDIX 9. Sampling Design Examples and Formulas from Platts et al. 1987

This appendix includes a reprint of the first 17 pages from the following publication:

Platts, W. S.; Armour, C.; Booth, G. D.; Bryant, M.; Bufford, J. L.; Cuplin, P.; Jensen, S.; Lienkaemper, G. W.; Minshall, G. W.; Monsen, S. B.; Nelson, R. L.; Sedell, J. R.; Tuhy, J. S. 1987. Methods for evaluating riparian habitats with applications to management. General Technical Report INT-221. Ogden, Utah: USDA Forest Service, Intermountain Research Station.

The material we have reprinted from this publication includes the following sections: (1) general field sampling, (2) concepts about populations and samples, (3) simple random sampling, (4) stratified random sampling, (5) cluster sampling, (6) two-stage sampling, and (7) monitoring.

We have included this material in this technical reference because it includes a clear discussion of how to calculate summary statistics and determine sample sizes for several sampling designs that are covered only briefly in Chapter 7. Clear, step-by-step instructions are provided using examples from ecological sampling situations.

Please note that the section titled "Simple Random Sampling" includes calculations for determining sample sizes that do not include a necessary correction that is discussed in Appendix 7—Sample Size Equations. When using simple random sampling procedures to determine sample sizes to estimate population means or population totals, use the section titled "Equation #1" in Appendix 7 instead of the calculations shown in the Platts et al. publication.

General Field Sampling

Information collection is necessary for inventory and monitoring activities associated with riparian management programs. Success for the programs is dependent upon the acquisition and use of information that must be appropriate for planning processes and the design of site-specific management. Unfortunately, widespread problems have resulted in inadequate, improper, or excessive information. This is usually attributed to a poorly thought-out approach to collecting information for specifically fulfilling resource management requirements. Therefore, the objective of this chapter is to present basic guidance for use when field sampling programs are being designed. We have presented information in a section pertaining to a general field sampling program and a second section in which considerations for monitoring approaches are discussed.

Six basic steps should be followed for a field sampling program (fig. 1) if useful information is to be obtained. Before sampling, justification for collecting the information (step 1) must be made. Considerations for establishing justifications include: (1) Is the information already available? (2) Is the acquisition of new information absolutely necessary for activities associated with riparian resource planning and management activities? (3) Would it be possible to measure a substitute condition to obtain essentially the same information at lower cost?

After specific information needs are defined, collection approaches must be determined (step 2). Considerations for this step must include evaluation of the suitability of a technique for achieving appropriate levels of accuracy and precision and the practicality of the technique based on ease of field application, costs, and other factors. Following step 2, pilot sampling (step 3) must be performed. Essentially, this step is a trial run designed to detect and correct problems that could seriously affect sampling. Additionally, this step is necessary for training of field crews and obtaining preliminary data for use in estimating the sample size for a predetermined level of statistical confidence. If problems are detected, which is usually the case (examples: sampling gear performs improperly, inadequate time was allocated for collecting and analyzing samples, more samples must be collected than originally planned), corrective measures must be taken. Step 3 is mandatory because serious flaws in the way sampling is conducted will adversely impact the quality of information that is collected.

When information is collected (step 4), it must be recorded accurately and assembled in a usable format for analysis (step 5). When the results are processed for use in planning and management procedures (step 6), careful thought must be given to the best way to present it to resource specialists and administrators. If the information is not presented with clarity and in a useful form, effort and costs expended for the work will be wasted.

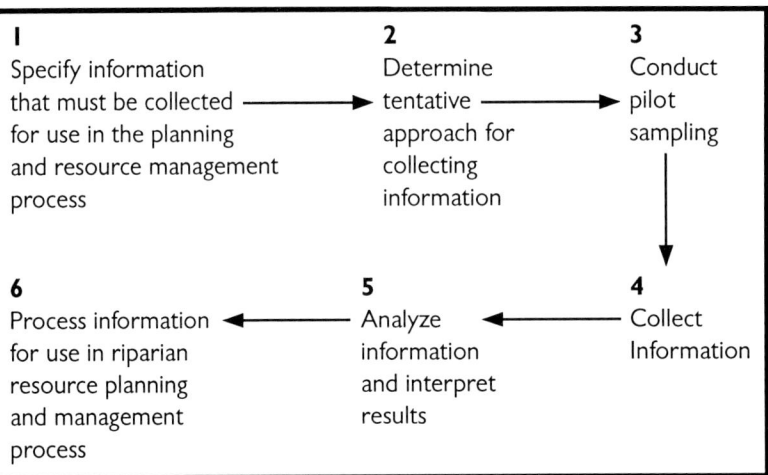

APPENDIX 9—FIG. 1. Steps for a field sampling program to obtain useful information for riparian resource planning and management processes.

Concepts About Populations and Samples

The entire collection of items in which we are interested is called the population. For example, the population might be a 100-ft. section of the stream to be divided into 100 cross sections of 1 ft. each. If we take measurements on only 20 of these cross sections, the cross sections we measure constitute the sample. The whole purpose of using sampling is to obtain information about the entire population when it is not possible or feasible to measure every element in it. We hope the items in the sample will give us accurate information about the whole population.

Populations can be either finite (with a fixed, countable number of elements) or infinite (with an infinite number of elements). Some populations are technically finite but with so many elements we could not reasonably count them. Such populations are considered to be infinite.

To illustrate, consider the example mentioned above. The 100-ft. stretch of stream is the population. We have arbitrarily divided it into 100 cross sections of 1 ft. each. Does this mean we have 100 elements in our population? Not necessarily. If we are interested in some characteristic that requirements measurement over the entire 1-ft. cross section, then the population could be considered finite with 100 elements in it. On the other hand, if we were interested in a characteristic that requires measurement at only a point along the stream (such as stream width, measured at a transect), it would be incorrect to consider the population as consisting of only 100 elements. In this case, the population should be dealt with as infinite. The methods that follow will often involve the finite population correction (*fpc*). It is defined as:

$$fpc = (1 - n/N)$$

where:

N = number of elements in the whole population

n = number of elements in the sample.

Notice that if N is large (essentially infinite), the *fpc* approaches 1. In the methods described later, if the population is infinite, we can ignore the *fpc* (that is, consider it equal to 1). This is true because the *fpc* is always used as a multiplier and multiplying by 1 has no effect.

We use "error of estimation" to denote the distance by which our estimate misses the true population value we are attempting to estimate. Although we cannot know the true error of estimation, it would be useful to be quite certain that after our sampling and estimating are complete, we have an error of estimation that is no greater than some upper boundary, say B. We will present some statistical methods designed to help us determine how large our sample must be to accomplish this.

Common field sampling procedures are simple random sampling, stratified random sampling, and cluster sampling (table 1). Most of the following computational examples for the procedures were adapted from Scheaffer and others (1979). The information presented here is expected to introduce field workers to some useful procedures; prior to application, a qualified statistician should be consulted.

Simple Random Sampling

A simple random sample (SRS) is, as its name implies, the sampling method that is simplest in concept. For its use, each element in the population (such as plots and transects) must be identifiable as individuals. Sampling must be performed in such a way that every element in the population has the same probability of being in the sample.

Using simple random sampling often results in samples that (1) are widely dispersed, causing considerable travel expense, and (2) leave some areas totally unsampled. Therefore, the most successful use of SRS is in relatively small geographical areas where a degree of homogeneity is known to exist. Simple random sampling could be used in other circumstances, but it would tend to be inefficient and more costly.

Simple random sampling should probably be within ecological types instead of across multiple types. This precaution will tend to reduce the variability and increase the precision of habitat parameter estimates. The precaution is reasonable, for example, when one considers the high variation that occurs between riparian habitat in meadows compared to headwater-timbered areas in an allotment that is heavily grazed.

TABLE 1. Comparison of simple, random, stratified random, and cluster sampling techniques

Sampling approach	Total number of elements or plots (potential samples) in population must be known in advance?	Key features	Application considerations	Appropriate field use
Simple random	Yes - identification of all elements or plots necessary for selection of random sample.	Through random sampling there is an equal chance for sampling of each element. This helps ensure that data representative of an overall population will be obtained.	Excessive costs can be incurred if elements are widely scattered through a large geographic areas.	Randomly distributed populations in relatively small geographic areas.
Stratified random	Yes - after strata are defined, elements or plots within each stratum are selected randomly for sampling.	Advantages over simple random sampling can be reduced and variance for parameter estimators and costs can be reduced substantially if sampling is restricted to a smaller geographic area. Additionally, conditions between strata can be compared statistically, that is, difference among means.	Within each stratum there must be relative homogeneity and heterogeneity must be maximized among strata. Homogeneity within helps to reduce sample variance.	Populations in homogeneous strata dissimilar from other strata. Recommended if sampling is conducted in recognizable homogeneous strata.
Cluster sampling	All elements are sampled for one-stage sampling. Two-stage sampling requires advance identification of elements for random selection for sampling.	Clusters to sample are selected randomly. Clusters must be alike (homogeneous between) with heterogeneous conditions within.	The sampling approach can be economical because heterogeneity within clusters helps to lower overall sampling costs because travel distance and time can be lessened when a representative sample is obtained. Clusters must have the same number of sampling units to avoid more complicated computations. Two-stage analysis is appropriate when there are too many elements per cluster to sample, or the elements are so similar that counting all of them is wasteful. Prior to using cluster sampling, a statistician should always be consulted.	Populations that are associated with heterogeneous conditions for which ordered, systematic sampling, simple random and stratified sampling is infeasible *and* there are an adequate number of clusters to sample.

APPENDIX 9. Sampling Design Examples and Formulas from Platts et al. 1987

Example 1. Twenty transects ($n = 20$) are placed along a stream in a meadow. They are selected randomly, and stream width is measured at each transect. What are the mean width, the upper bound on the error of estimation (in this case, B), and the 95 percent confidence interval on the population mean (μ)? Assuming that the information is preliminary, how many samples would have to be collected to be reasonably sure B does not exceed 1.07 ft.?

Step 1 - Calculate the sample mean and variance of the following 20 measurements on stream width: 10, 16, 11, 8, 9, 11, 3, 13, 10, 7, 5, 12, 9, 12, 11, 20, 11, 12, 14, 10.

NOTE: Almost any scientific calculator has the built-in capability of computing both the mean ($\overline{X}$) and the standard deviation (s) or the variance (s^2). If your calculator computes the standard deviation, the variance is obtained by squaring the standard deviation.

In this case we obtain $\overline{X} = 10.700$, $s^2 = 13.4843$.

Step 2 - Calculate the bound on the error of estimation (B)

$$B = 1.96 \sqrt{\frac{s^2}{n} \frac{N\text{-}n}{N}}$$

In this case, the population is infinite and the fpc = 1. Therefore:

$$B = 1.96 \sqrt{\frac{13.4843}{20}} = 1.96 \sqrt{0.6742} = 1.6094$$

where:

$\frac{N\text{-}n}{N}$ = the finite population correction (*fpc*)

1.96 = Z value from the normal distribution for the 95 percent level. If another level of confidence were used, the number 1.96 would be replaced by the appropriate value from the normal distribution.

Step 3 - Calculate the 95 percent confidence interval for the population mean (μ).
The interval is computed as:
Lower limit = $\overline{X}$ - B = 10.7000 - 1.6094 = 9.0906
Upper limit = $\overline{X}$ + B = 10.7000 + 1.6094 = 12.3094.
This means we are quite confident (95 percent) that the true population mean is between 9.0906 and 12.3094.

Step 4 - Calculate n' = estimated sample size if B is not to exceed 1.07 ft.

$$n' = \frac{(Z^2)\,(s^2)}{B^2}$$

$$= \frac{(1.96)^2\,(13.4843)}{(1.07)^2} = 45.2453$$

We always round to the next higher number. Therefore:
$n' = 46$

where:
Z = 1.96 at the 95 percent confidence level.

A sample size of $n = 46$ should give us a good chance of obtaining $B \leq 1.07$ ft.

Example 2. An inventory was conducted along a 60-mile stretch of a stream. Each 1-mile segment ($N = 60$) was designated as a possible sample site, and 20 sites ($n = 20$) were randomly selected for sampling along both sides of the stream to a distance of 200 ft. back from each bank. Snag trees in each sample site were counted. There was an average of 10 trees ($\overline{X}$) per site with a sample variance (s^2) of 8.3731. Estimate the total number of snags in the 60-mile stretch, the bound on the error of estimation (B), the 95 percent confidence interval for the total number of snags in the population, and the estimated sample size if our estimate is to be within 25 snags of the true total.

In this case, each 1-mile segment was a potential sample site and, if chosen for the sample, would be studied in its entirety—not a single point. This population can be considered finite with $N = 60$. (Of course, we might have chosen to use 120 segments of 0.5 mile each for a finite population of $N = 120$.)

Step 1 - Calculate $\hat{\tau}$, the estimate of the total number of snags in the 60-mile stretch
$$\hat{\tau} = N\,\overline{X} = (60)\,(10) = 600 \text{ snag trees}$$

Step 2 - Calculate the estimated variance of $\hat{\tau}$
$$\hat{V}(\hat{\tau}) = N^2 \left(\tfrac{s^2}{n}\right)\left(\tfrac{N\text{-}n}{N}\right) = 60^2 \left(\tfrac{8.3731}{20}\right)\left(\tfrac{60\text{-}20}{60}\right)$$
$$= (3{,}600)(0.4187)(0.6667) = 1{,}004.77$$

Step 3 - Calculate the bound on the error of estimation.
$$B = (1.96)\sqrt{\hat{V}(\hat{\tau})}$$
$$= (1.96)\sqrt{1{,}004.77} = 62.1284$$
Where:
    1.96 = Z for the 95 percent confidence level.

Step 4 - Calculate the 95 percent confidence interval for the total number of snag trees in the population.
The interval is computed as:
    Lower limit = $\hat{\tau}$ - B = 600 - 62.1284 = 537.9
    Upper limit = $\hat{\tau}$ + B = 600 + 62.1284 = 662.1

Step 5 - Calculate n', the estimated sample size for B not to exceed 25 snags
$$n' = \frac{NS^2}{(N\text{-}1)D + S^2}$$
where:
$$D = \frac{B^2}{Z^2 N^2}$$
$$= \frac{(25)^2}{(1.96)^2(60)^2} = 0.0452$$
$$n' = \frac{(60)(8.3731)}{(60\text{-}1)(0.0452) + 8.3731} = 45.5$$

Rounding up gives $n' = 46$.

Therefore, a sample of $n = 46$ should give us high probability of estimating the true number of snags within 25 trees.

Stratified Random Sampling

If the population of interest falls naturally into several subdivisions, or strata, stratified random sampling is found to be substantially more efficient than simple random sampling. For example, if the number of shrubs is a management concern in a riparian zone that extends through several homogeneous vegetation types (such as sagebrush, sagebrush-grass, and ponderosa pine-Idaho fescue), this method of sampling is suitable. This procedures requires that the investigator clearly identify each stratum in advance of sampling. Then a simple random sample (SRS) is taken independently within each stratum.

In addition to being more efficient in estimating the overall population mean or total, stratified random sampling provides separate estimates for each stratum. This feature alone might be reason enough for using this method over SRS.

Example 3. Assuming that the following information is collected from three strata, what are the mean number of shrubs per acres, the bound (B) on the error of estimation, and the 95 percent confidence interval for the population mean (μ)? Sample means and variances were calculated for each stratum. Approximately 13 percent of the acres were sampled in each stratum. This is a finite population with three strata such that $N_1 = 155$. $N_2 = 62$, and $N_3 = 93$.

Stratum	Total acres/ stratum (N_h)	Total acres sampled (n_h)	Sample statum mean $\bar{X}_h$	Total shrubs $N_h \bar{X}_h$	Straum variance s_h^2	$N_h s_h^2$
1 Sagebrush	155	20	33.900	5,254.500	35.358	5,480.49
2 Sagebrush-grass	62	8	25.125	1,557.750	232.411	14,409.48
3 Ponderosa pine- Idaho Fescue	93	12	19.000	1,767.000	87.636	8,150.15
	310	40		8,578.750		28,040.12

$$N = \Sigma N_h = 310 \qquad n = \Sigma n_h = 40 \qquad T = \Sigma N_h \bar{X}_h = 8,578.750 \qquad s^2 = \Sigma N_h s_h^2 = 28,040.12$$

Step 1 - Calculate sample mean

$$\bar{X}_{st} = \frac{T}{N}$$

$$= \frac{8,578.750}{310} = 27.673$$

$$= \text{sample estimate of } \mu, \text{ the population mean number of shrubs per acre}$$

Step 2 - Calculate an estimate of the variance of $\bar{X}_{st}$

$$\hat{V}(\bar{X}_{st}) = \frac{1}{N^2} \Sigma \left[N_h^2 \left(\frac{N_h - n_h}{N_h} \right) \left(\frac{s_h^2}{n_h} \right) \right]$$

$$= \frac{1}{(310)^2} \left[(155)^2 \left(\frac{155 - 20}{155} \right) \left(\frac{35.358}{20} \right) + (62)^2 \left(\frac{62 - 8}{62} \right) \left(\frac{232.411}{8} \right) + (93)^2 \left(\frac{93 - 12}{93} \right) \left(\frac{87.636}{12} \right) \right]$$

$$= \frac{1}{(310)^2} (36,993.308 + 97,264.004 + 55,013.499)$$

$$= \frac{189,270.81}{96,100} = 1.970$$

Step 3 - Calculate the bound on the error of estimation and the 95 percent confidence interval

$$B = (1.96)\sqrt{\hat{V}(\overline{X}_{st})} = (1.96)\sqrt{1.970} = 2.751$$

Step 4 - Calculate the 95 percent confidence interval for the population mean (μ) number of shrubs per acre.

The interval is calculated as:
Lower limit: $\overline{X}_{st} - B = 27.673 - 2.751 = 24.922$
Upper limit: $\overline{X}_{st} + B = 27.673 + 2.751 = 30.424$

Example 4 - What should the sample size be for each stratum if we want to be 95 percent confident that the error of estimation has a bound (B) no larger than 2.0?

Step 1 - Calculate the denominator for stratum weights
Denominator = $\sum N_h S_h$

$$= (155)\sqrt{35.358} + (62)\sqrt{232.411} + (93)\sqrt{87.636}$$

$$= 921.67 + 945.19 + 870.61$$

$$= 2,737.47$$

Step 2 - Calculate the stratum weights

$$w_h = \frac{N_h S_h}{\sum N_h S_h}$$

= the proportion of the total sample size, n, that will come from stratum h.

$$w_1 = \frac{921.67}{2,737.47} = 0.337$$

$$w_2 = \frac{945.19}{2,737.393} = 0.345$$

$$w_3 = \frac{870.573}{2,737.393} = 0.318$$

Notice that the weights over all three strata add up to 1.000. To determine the size of sample required from stratum h, multiply the total sample size by w_h. Therefore,

$$n_h = w_h n.$$

We still need to determine the overall sample size, n.

Step 3 - Calculate the numerator for the n' equation.

$$\text{Numerator} = \sum \frac{N_h^2 s_h^2}{w_h}$$

$$= \frac{(155)^2(35.358)}{0.337} + \frac{(62)^2(232.411)}{0.345} + \frac{(93)^2(87.636)}{0.318}$$

$$= 2,520,700.148 + 2,589,530.099 + 2,383,533.849$$

$$= 7,493,764.096$$

Step 4 Calculate n'

$$D = \frac{B^2}{Z^2} = \frac{(2.0)^2}{(1.96)^2}$$

   $= 1.041$, where $Z = 1.96$ comes from the normal distribution (appendix 1).

Finally

$$n' = \frac{\text{Numerator}}{N^2 D + s^2}$$

$$= \frac{7,493,764.096}{(310)^2(1.041) + 28,040.12} = \frac{7,493,764.096}{100,040.10 + 28,040.12}$$

$$= \frac{7,493,764.096}{128,080.22} = 58.508 \text{ or } 59$$

Therefore, an overall sample of $n = 59$ should give the investigator high probability of obtaining an estimate that is no more than 2.0 shrubs per acre from the population mean being estimated.

Step 5 - Calculate sample size for each stratum

$n_1 = w_1 n' = (0.337)(59) = 19.883$ or 20
$n_2 = w_2 n' = (0.345)(59) = 20.355$ or 20
$n_3 = w_3 n' = (0.318)(59) = 18.762$ or 19
$\qquad\qquad\qquad\qquad$ Total $\quad$ 59

NOTE: The weights, w_h, were determined in such a way that the variance of $\bar{X}_{st}$ is minimized for a fixed value of n. Therefore, once we determined an estimate of n, say n', we applied the weights to it to obtain the sample size in each stratum.

Example 5 - Using the results of example 4, what is the estimate of the total number of shrubs in the three strata, the bounds on the error of estimation (B), the 95 percent confidence interval for the estimate, and the estimated number of samples that would have to be collected for B not to exceed 400 shrubs?

Step 1 - Calculate the value for $\hat{\tau}$, the estimate of the population total number of shrubs
$\quad \hat{\tau} = N\bar{X}_{st}$

$\quad\quad = (310)(27.673)$

$\quad\quad = 8,578.630$ shrubs

Step 2 - Calculate the estimated variance of $\hat{\tau}$
$\quad \hat{V}(N\bar{X}_{st}) = N^2\hat{V}(\bar{X}_{st})$

$\quad\quad\quad = (310)^2(1.970)$

$\quad\quad\quad = 189,317$

Step 3 - Calculate the bounds on the error of estimation
$\quad B = 196\sqrt{\hat{V}(N\bar{X}_{st})} = 1.96\sqrt{189,317} = 852.81$

NOTE: Although the same symbol (B) is used in examples 4 and 5, its value is different for the mean (μ) than for the total (τ).

Step 4 - Calculate the 95 percent confidence interval for the total number of shrubs in the population.

The interval is computed as:
Lower limit: $\hat{\tau}_{st} - B = 8{,}578.63 - 852.81 = 7{,}725.82$
Upper limit: $\hat{\tau}_{st} + B = 8{,}578.63 + 852.81 = 9{,}431.44$

Step 5 - Calculate n', the estimated sample size for B not to exceed 400 shrubs.
The only difference between this case and the estimation of μ in example 4 is in the computation of D. We now have

$$D = \frac{B^2}{Z^2 N^2} = \frac{(400)^2}{(1.96)^2 (310)^2} = 0.433$$

Where Z is from a table of the normal distribution for 95 percent confidence.

$$n' = \frac{\text{Numerator}}{N^2 D + s^2} = \frac{7{,}493{,}764.096}{(310)^2 (0.433) + 28{,}040.12}$$

$$= \frac{7{,}509{,}992.786}{69{,}651.420}$$

$$= 107.59 \text{ or } 108 \text{ rounded up}$$

We can apply the weights from example 4 to obtain the sample sizes for each stratum. We get

$n_1 = (0.337)\ 108 = 36.40$ or 36
$n_2 = (0.345)\ 108 = 37.26$ or 37
$n_3 = (0.318)\ 108 = 34.34$ or 34

Cluster Sampling

Cluster sampling should not be confused with cluster analysis, which is a classification and taxonomic technique. Here, cluster sampling refers to a method of collecting a sample when the individual elements cannot be identified in advance. Instead, we are only able to identify groups or clusters of these elements. A sample of the clusters is then obtained, and every element in each cluster is measured.

For example, we may wish to take measurements on individual trees in a riparian area but are only able to identify 1-acre plots along the stream. Each plot can contain a different number of trees, and the individual trees cannot be identified before taking the sample. Cluster sampling allows us to select a sample of clusters, instead of individual trees. We would then measure every tree within each cluster.

Cluster sampling is convenient and inexpensive with regard to travel costs. To gain maximum advantage of this method, elements within a cluster should be close to each other geographically.

If we compare cluster sampling with either simple random sampling or stratified random sampling, we find one major advantage of the cluster method: the cost per element sampled is lower than for the other two methods. Unfortunately, two disadvantages of cluster sampling are: (1) the variance among elements sampled tends to be higher, and (2) the computations required to analyze the results of the sample are more extensive. Therefore, cluster sampling is preferable to the other methods if the cost benefits exceed the disadvantages.

If we have only a few clusters, each quite large, we minimize our costs—especially of travel. However, samples with only a few clusters produce estimates with low precision (that is, high variance). On the other hand, if we increase the number of clusters (making each cluster smaller), the variance is reduced while the cost is increased. The user must find a compromise.

Whether sampling 40 clusters of 0.5 acre each is better than 20 clusters of a full acre each is not clear, although approximately the same number of trees may be measured with either sample. There would be a larger number of the smaller clusters, and therefore they would be dispersed more evenly over the population. The estimates produced would have lower variability than those from fewer but larger clusters. However, the sampler would have to travel to twice as many sites, thus increasing costs. Knowledge of the variability and costs involved would be the key to planning such a study effectively.

Example 6 - Suppose that we have 30 clusters of 1 acre each ($N = 30$) in a riparian area. Calculate the average number of cavities per snag tree, the bound on the error of estimation (B), and the 95 percent confidence interval for the population mean (μ). Five clusters (n) are selected for sampling and data are collected for all snag trees in each cluster. Sampling data are tabulated below:

Cluster	Number of snag trees (m_i)	Total cavities (X_i)
1	8	5
2	9	7
3	4	8
4	5	9
5	6	10
	$\Sigma m_i = 32$	$\Sigma X_i = 39$

Step 1 - Calculate an estimate of μ, the population mean, for cavities per snag tree

$$\bar{X} = \frac{\Sigma X_i}{\Sigma m_i} = \frac{39}{32} = 1.22 \text{ cavities per snag tree}$$

Step 2 - Calculate $\bar{m}$, the average cluster size for the sample

$$\bar{m} = \frac{\Sigma m_i}{n} = \frac{32}{5} = 6.4 \text{ snag trees per cluster}$$

An estimate of the total number of snag trees in the 30 clusters is $N\bar{m} = (30)(6.4) = 192.0$ trees.

Step 3 - Calculate sum of squares

Cluster	m_i	X_i	$\bar{X}m_i$	$(X_i - \bar{X}m_i)^2$
1	8	5	9.76	22.66
2	9	7	10.98	15.84
3	4	8	4.88	9.73
4	5	9	6.10	8.41
5	6	10	7.32	7.18
			Total	63.82

where $\bar{X}$ came from step 1.

Step 4 - Calculate $\hat{V}(\overline{X})$ = estimated variance for $\overline{X}$

$$\hat{V}\overline{X} = \left(\frac{N-n}{(N)(n)(m)^2}\right)\left(\frac{\Sigma(X_i - \overline{X}m_i)^2}{n-1}\right)$$

$$= \left(\frac{30-5}{(30)(5)(6.4)^2}\right)\left(\frac{63.82}{4}\right)$$

$$= (0.004)(15.955) = 0.0649$$

Step 5 - Calculate the bound on the error of estimation

$$B = 1.96\sqrt{\hat{V}(\overline{X})} = 1.96\sqrt{0.064} = 0.4994$$

Step 6 - Calculate the 95 percent confidence interval for the population mean number of cavities per snag tree:

    Lower limit: 1.22 - 0.4994 = 0.7206

    Upper limit: 1.22 + 0.4994 = 1.7194.

Example 7 - Assuming that information for example 6 is preliminary, how can we determine the number of clusters to sample if we want the bound on the error of estimation (B) to be within 0.1?

Step 1 - Calculate s_c^2 = estimate of the population variance among clusters

$$s_c^2 = \frac{\Sigma(X_i - \overline{X}m_i)^2}{n-1}$$

$$= \frac{63.82}{4} = 15.955$$

Step 2 - Calculate

$$D = \frac{B^2\overline{m}^2}{Z^2} = \frac{(0.1)^2(6.4)^2}{(1.96)^2} = 0.1066$$

where:

    1.96 is the Z value from the normal distribution for 95 percent confidence.

Step 3 - Calculate n′ = total number of clusters to sample

$$n' = \frac{(N)(s_c^2)}{ND + s_c^2} = \frac{(30)(15.955)}{(30)(0.1066) + 15.955}$$

$$= \frac{(30)(15.955)}{19.153} = 24.99 \text{ or } 25 \text{ clusters rounded up}$$

Two-Stage Sampling

Suppose we have clusters with so many elements in them that it is prohibitive to measure all elements in the cluster. It is natural to think of sampling elements within each cluster—that is, to measure only part of the elements within each cluster. This situation is a common one and is referred to as two-stage sampling.

Another common use of two-stage sampling is when it is apparent that even though there are many elements within a cluster, all elements are so nearly the same that to sample all of them would provide little additional information. The reasonable thing to do might be to measure only a part of the elements available within the cluster.

Two-stage sampling introduces a high degree of flexibility in defining clusters and sampling within them. The give and take between the number of clusters and the number of elements to be sampled within each cluster has been studied in some detail. Unfortunately, the results are complicated and beyond the scope of this publication. Interested readers are referred to one of the more extensive books on sampling (Cochran 1963; Kish 1965).

The following examples serve to give the reader a brief introduction to the concepts of two-stage sampling.

Example 8 - Suppose that there are $N = 90$ clusters in a riparian zone and we can sample 10 clusters ($n = 10$) and 20 percent of the pools in each cluster. Estimate the mean depth of pools in the population, the bounds on the error of estimation (B), and the 95 percent confidence interval for the population mean (μ). Assume that there is a total of $M = 4,500$ pools in the 90 clusters. Data for each cluster have been used to calculate the cluster means ($\bar{X}_i$), and variances (s_i^2).

Step 1 - Tabulate data as follows:

cluster	total pools (M_i)	Pools sampled (m_i)	mean depth $\bar{X}_i$	$(M_i)(\bar{X}_i)$	$(M_i\bar{X}_i - \overline{MX})^{2*}$
1	50	10	5.40	270.00	900.00
2	65	13	4.00	260.00	400.00
3	45	9	5.67	255.15	229.52
4	48	10	4.80	230.40	92.16
5	52	10	4.30	223.60	268.96
6	58	12	3.83	222.14	318.98
7	42	8	5.00	210.00	900.00
8	66	13	3.85	254.10	198.81
9	40	8	4.88	195.20	2,007.04
10	56	11	5.00	280.00	1,600.00
	$\Sigma M_i = 522$			$\Sigma(M_i \bar{X}_i) = 2,400.59$	$\Sigma(M_i \bar{X}_i - \overline{MX})^2 = 6,915.47$

* Calculated $\overline{M}$ and $\overline{X}$ from Step 2 and Step 3 below

cluster	s_i^2	$M_i(M_i - m_i) = A_i$	$s_i^2/m_i = B_i$	$(A_i)(B_i)$
1	11.38	2,000	1.138	2,276.00
2	10.67	3,380	0.821	2,774.98
3	16.75	1,620	1.861	3,014.82
4	13.29	1,824	1.329	2,424.10
5	11.12	2,184	1.112	2,428.61
6	14.88	2,668	1.240	3,308.32
7	5.14	1,428	0.643	918.20
8	4.31	3,498	0.332	1,161.34
9	6.13	1,280	0.766	980.48
10	11.80	2,520	1.073	2,703.96
			$\Sigma M_i(M_i - m_i)\dfrac{s_i^2}{m_i} = 21,990.81$	

Step 2 - Calculate $\bar{M}$ = average number of elements (pools) in each cluster

$$\bar{M} = \frac{M}{N} = \frac{4,500}{90} = 50 \text{ pools}$$

Step 3 - Calculate $\bar{X}$ = the estimated population mean depth for pools

$$\bar{X} = \frac{N}{(M)(n)} \Sigma M_i \bar{X}_i$$

$$= \frac{90}{(4,500)(10)} (2,400.59) = 4.8012 \text{ ft deep}$$

Step 4 - Calculate the estimated variance for $\bar{X}$
A. Calculate:

$$s_b^2 = \frac{1}{n-1} \Sigma (M_i \bar{X}_i - \bar{M}\bar{X})^2 = \frac{1}{10-1} (6,915.47)$$

$$= \frac{6,915.47}{9} = 768.4;$$

B. and calculate:

$$\hat{V}(\bar{X}) = \left[\left(\frac{N-n}{N} \right) \left(\frac{1}{n\bar{M}^2} \right) (s_b^2) \right] + \left(\frac{1}{nN\bar{M}^2} \right) \left[\Sigma M_i (M_i - m_i) \left(\frac{s_i^2}{m_i} \right) \right]$$

$$= \left[\left(\frac{90-10}{90} \right) \left(\frac{1}{(10)(50)^2} \right) (768.4) \right] + \left[\frac{1}{(10)(90)(50)^2} \right] (21,990.81)$$

$$= 0.037095$$

Step 5 - Calculate bounds on the error of estimation

$$B = 1.96 \sqrt{\hat{V}(\bar{X})} = 1.96 \sqrt{0.037095} = 0.3775$$

Step 6 - Calculate the 95 percent confidence interval for the population mean pool depth (μ), which is:

Lower limit: $\bar{X} - B = 4.8012 - 0.3775 = 4.42$
Upper limit: $\bar{X} + B = 4.8012 + 0.3775 = 5.18$

Example 9 - If M is unknown in example 8, calculate the estimate of the population mean depth of pools, the bounds on the error of estimation (B), and the 95 percent confidence interval for the population mean depth of pools.

Step 1 - Estimate μ = ratio estimate of the population mean μ

$$\bar{X}_r = \frac{\Sigma M_i \bar{X}_i}{\Sigma M_i} = \frac{2,400.59}{522} = 4.599 \text{ ft}$$

Step 2 - Complete tabulations for extension of table for example 8

$M_i^2 \bar{X}_i$	$(M_i \bar{X}_i)^2$	M_i^2
13,500.00	72,900.00	2,500
16,900.00	67,600.00	4,225
11,481.75	65,101.52	2,025
11,059.20	53,084.16	2,304
11,627.20	49,996.96	2,704
12,884.12	49,346.18	3,364
8,820.00	44,100.00	1,764
16,770.60	64,566.81	4,356
7,808.00	38,103.04	1,600
15,680.00	78,400.00	3,136
$\Sigma M_i^2 \bar{X}_i = 126{,}530.87$	$\Sigma (M_i \bar{X}_i)^2 = 583{,}198.67$	$\Sigma M_i^2 = 27{,}978$

Step 3 - Calculate $\bar{M}$ = estimate of average number of pools per cluster

$$\bar{M} = \frac{\Sigma M_i}{n} = \frac{522}{10} = 52.2 \text{ pools per cluster}$$

Step 4 - Calculate estimated variance for μ

A. Calculate s_r^2:

$$s_r^2 = \frac{1}{n-1} \Sigma M_i^2 (X_i - \bar{X}_r)^2$$

$$= \frac{1}{n-1} \left[\Sigma (M_i \bar{X}_i)^2 - 2\bar{X}_r \Sigma M_i^2 \bar{X}_i + (\bar{X}_r)^2 \Sigma M_i^2 \right]$$

$$= \frac{583{,}198.67 - 2(4.599)(126{,}530.87) + (4.599)^2 (27{,}978)}{9}$$

$$= \frac{583{,}198.67 - 1{,}163{,}830.94 + 591{,}757.11}{9}$$

$$= \frac{11{,}124.84}{9} = 1{,}236.09;$$

B. and calculate $\hat{V}(\bar{X}_r)$, the estimated variance of $\bar{X}_r$

$$\hat{V}(\hat{\mu}) = \left(\frac{N-n}{N} \right) \left(\frac{1}{n\bar{M}^2} \right) (s_r^2) + \left(\frac{1}{nN\bar{M}^2} \right) \Sigma M_i (M_i - m_i) \left(\frac{s_i^2}{m_i} \right)$$

$$= \left(\frac{90-10}{90} \right) \left(\frac{1}{(10)(52.2)^2} \right) (1{,}236.09) + \left(\frac{1}{(10)(90)(52.2)^2} \right) (21{,}990.81)$$

$$= \left(\frac{80}{90} \right) \left(\frac{1}{(10)(2{,}724.84)} \right) (1{,}236.09) + \left(\frac{1}{(10)(90)(52.2)^2} \right) (21{,}990.81)$$

$$= \left(\frac{80}{2{,}452{,}356} \right) (1{,}236.09) + \left(\frac{1}{2{,}452{,}356} \right) (21{,}990.81)$$

$$= 0.0403 + 0.0090 = 0.0493$$

Step 5 - Calculate bounds on error of estimation

$$B = 1.96 \sqrt{\hat{V}(\hat{\mu})} = 1.96 \sqrt{0.0493} = 0.435$$

Step 6 - Calculate the 95 percent confidence interval for the population mean (μ) for pool depth, which is:

Lower limit: 4.599 - 0.435 = 4.164

Upper limit: 4.599 + 0.435 = 5.034

Monitoring

The purpose of monitoring is to obtain information for use in evaluating responses of land management practices. Specific steps (fig. 2) must be followed if meaningful results are to be obtained from a monitoring study. Step 1 is the documentation of baseline condition, management potential, and problems attributed to the mix of land use practices adversely affecting a riparian area. Management potential is the level of riparian habitat quality that could be achieved through application of improved management. Potential will vary between sites because of several variables, including rainfall patterns, landform, and history of use. If potential is evaluated to be higher than the response capability of a site, and an objective is made to achieve better conditions than are possible, a management failure will obviously occur. This emphasizes the importance of developing objectives that are compatible with site potential.

Documentation of problems from all land use practices that affect a site requires a thorough analysis. For example, if the objective is to improve habitat to increase numbers of trout, it is possible that complex problems (fig. 3) must be solved or controlled before trout will benefit.

Before completing the objectives for riparian habitat management (step 2, fig. 2) holistic planning by an interdisciplinary group will be necessary because most sites will be subjected to multiple-use management. Therefore, riparian habitat objectives will have to be compatible with those of the overall multiple-use plan. If dominant-use management is to be applied to solely benefit a riparian area, it is advisable to involve individuals in other disciplines to assess potential for response to management. Depending on site-specific problems, the disciplines could include hydrology, plant ecology, and perhaps engineering if structural physical changes (such as rechannelization or installation of stream improvements devices) are considered. When objectives are specified, they must be stated in quantifiable and measurable terms; this is of paramount importance. An example of an objective could be to increase the density of shrubs from 25 to 50 percent. This specifically requires that existing conditions be documented for comparison with future management results.

The design of site-specific management plans for achieving riparian area objectives (step 3, fig 2) requires multiple-use planning and conflict resolution. For example, suppose that timber harvesting, recreation, and mining are contributing to a degraded riparian habitat. It will be difficult, if not impossible, to design a management plan strictly for application in the area to solve problems caused by outside influences. Key considerations (Armour and others 1983) for a properly designed monitoring program (step 4, fig. 2) include the following:

1. Measurement of response to management is possible to determine through hypothesis testing if objectives are met. This prerequisite depends upon a clearly stated hypothesis (for example, H_o: shrub density increased 100 percent vs. H_a: shrub density increased <100 percent) that tracks with a management objective, and the variable must be responsive to management that will be applied. Additionally, measurement of the response with appropriate accuracy and precision must be feasible. Designation of variables that are difficult to measure and ones for which good measurement techniques have not been perfected should be avoided.

2. Control areas that will not receive management treatments must be included in the study. One precaution that must be taken in selecting control and treatment sites is that they must have the same premanagement characteristics and the same potential for response to management. This precaution is necessary if changes attributable to management are to be detectable. For example, if the objective is to improve overhanging stream-side cover by 50 percent in a meadow, a control must be established in a similar meadow, not in an area with different landform features and response capabilities. The recommended approach for selecting control and treatment sites for comparison is to make the selections randomly in areas with similar premanagement conditions.

3. Resources must be available for monitoring through an adequate period to permit management responses to occur. This requirement is frequently neglected. If it is uncertain whether a monitoring program can be completed with adherence to the plan, the program should not be initiated.

4. Management must be consistent with the original plan throughout the study. Noncompliance with this condition is one of the most common problems thwarting studies. The problem occurs when changes are made in management, preventing accurate interpretations of data. An example of the problem could be when the establishment of easier access by fishermen to study sites in a stream has resulted in depletion of fish in treatment and control sites, masking influences of improved habitat conditions. Another example that happens frequently is the trespass of livestock and subsequent overgrazing and habitat change in control sites.

5. Confounding factors that can adversely affect the study must be controlled. These factors are defined as unplanned events or influences that adversely affect results of a study. Factors in this category include institutional influences (such as when an agency changes emphasis away from monitoring and a study is stopped), political pressures (such as when a user group uses influence to stop a study because potential results are disliked), equipment failure problems, changes in personnel conducting the study and inability to find suitable replacements, and biological effects (such as when natural variation is excessive in time and space, and responses to management are masked). Although it is impossible to guarantee that confounding problems will not occur, individuals involved with monitoring should consider them in advance to eliminate as many as possible.

6. Statistical tests to analyze information are designated when the monitoring program is designed and assumptions for proper use of the tests are met. Unfortunately, there has been a tendency for the advance consideration of statistical tests to be neglected, resulting in the collection of data and the expectation that a statistician "can make something out of it" after completion of field work. When this happens, the result is usually a disappointing conclusion that the study was useless. To prevent problems, individuals involved with designing monitoring programs should always obtain assistance from a statistician during the design phase. This will help avoid serious problems that cannot be corrected. Essentially the pilot study (step 5, fig. 2) for a monitoring project is conducted for the same reasons discussed for step 3, fig. 1. To help ensure that meaningful statistical tests are feasible, assistance should be obtained from a statistician for this phase to refine approaches for the study. Once the pilot study is completed,

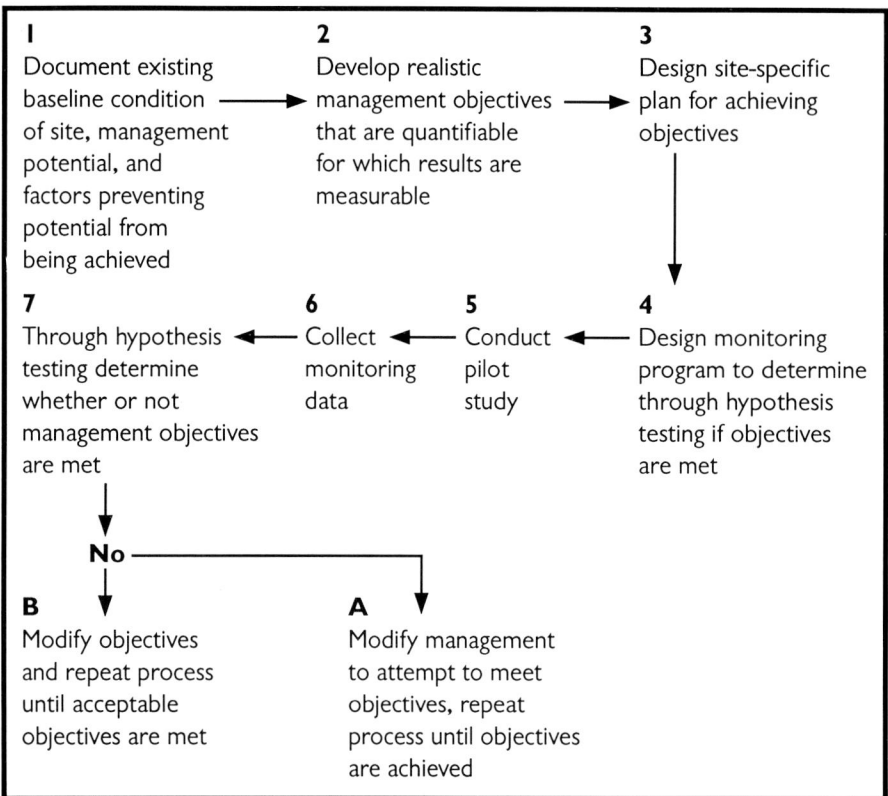

APPENDIX 9—FIG. 2. Steps for a monitoring program (modified from Armour and others 1983).

assuming that appropriate premanagement data for control and treatment sites have been collected, management can be applied and monitoring (step 6, fig. 2) can proceed with strict adherence to the design specifications. If appropriate premanagement data have not been collected, this requirement must be fulfilled before management is applied. Failure to obtain data from preconditions and postconditions will preclude evaluation if management resulted in the achievement of stipulated objectives. Special considerations for step 6 must include: (1) maintenance of accuracy and precision in collecting data, (2) the expending of equal levels of effort and adherence to the same technical standards in control and treatment sites to prevent bias from influencing results of the study, and (3) the recording and processing of data suitable for retrieval and use in statistical analyses.

Statistical tests are used in step 7 to evaluate with a predetermined level of statistical confidence whether objectives were met. This level might not have to be as high (say, 95 or 99 percent) as would be expected for research, but the price for a lower level is an increased chance for a type I error (claiming a difference when it does not exist). When tests are performed, the determined confidence level must not be arbitrarily altered (say, from 95 to 85 percent) if results do not conform with preconceived perceptions.

Common errors to avoid when using statistical tests include inaccurate data entry, errors in rounding numbers, use of incorrect degrees of freedom, and incorrectly reading statistical tables (such as tables of t and F values).

Based on results of hypothesis testing, it is possible to conclude with a stipulated level of statistical confidence whether objectives are met. If they are not met, there are two options: modify objectives and repeat the process in figure 2 until they are eventually met, or modify management and repeat the process until success is achieved.

One concept that must be emphasized is that monitoring should not result in a strict "pass" or "fail" conclusion. There cannot be a failure if, in the future, negative results contribute to avoidance of management practices that do not work. Therefore, it is equally important to document unsuitable practices to avoid if the art of riparian resource management is to progress.

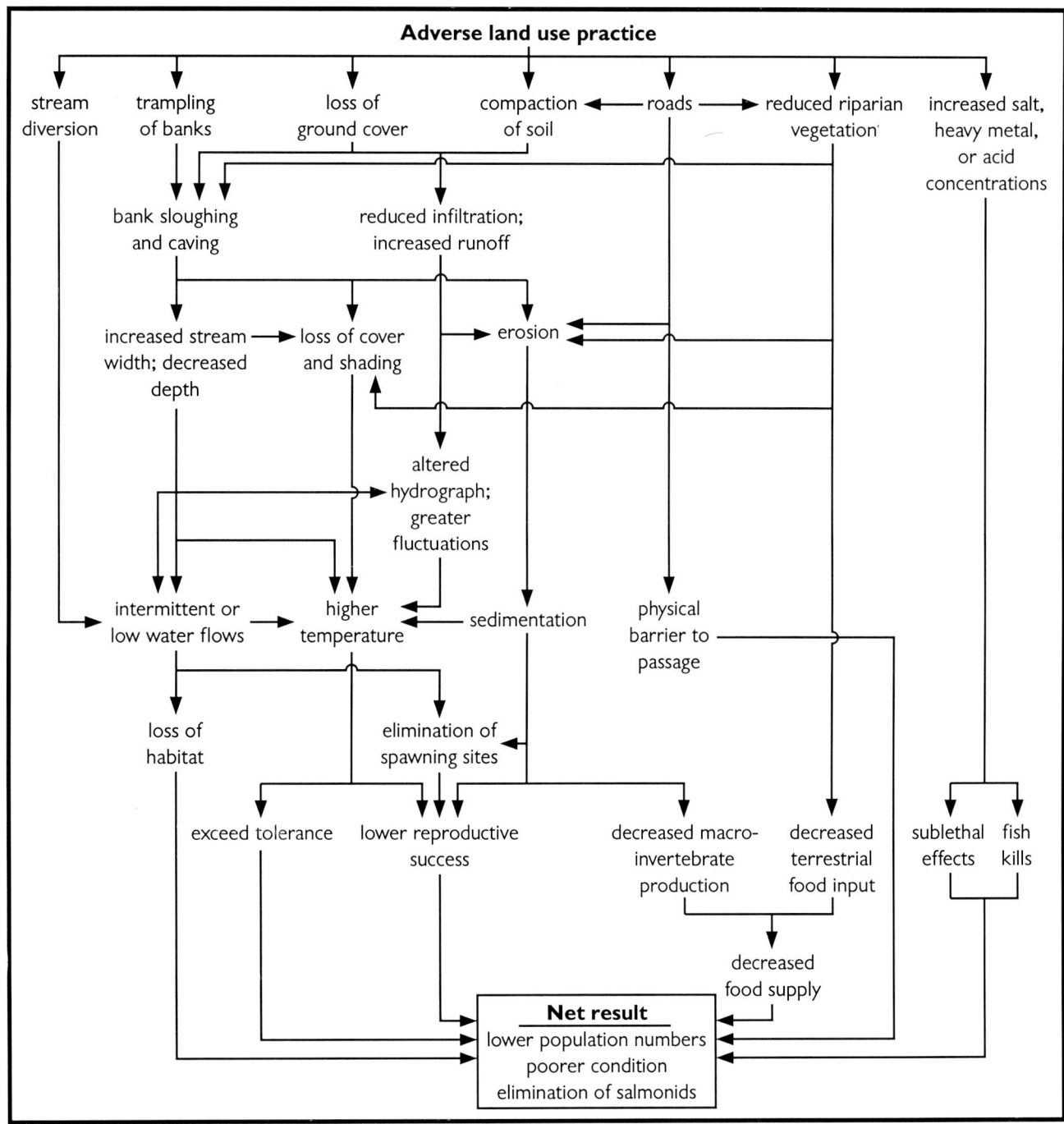

APPENDIX 9—FIG. 3. Some cause, effect, and impact relationships of adverse land use practices on salmonids (from Armour and others 1983).

Literature Cited

Armour, C. L.; Burnham, K. P.; Platts, W.S. 1983. Field methods and statistical analyses for monitoring small salmonid streams. FWS/OBS-83/33. Washington, D.C.: U.S. Fish and Wildlife Service.

Cochran, W. G. 1963. Sampling techniques, 2nd ed. New York, NY: John Wiley & Sons.

Kish, L. 1965. Survey sampling. New York, NY: John Wiley & Sons.

Scheaffer, R. L.; Mendenhall, W.; Ott, L. 1979. Elementary survey sampling, 2nd ed. Boston, MA: Duxbury Press.

APPENDIX 10. Qualitative Monitoring Examples

Two examples of qualitative monitoring worksheets are included in this appendix. The first was designed for assessing current conditions and management needs in Research Natural Areas by Forest Service personnel in the Intermountain Region (Evenden 1995). It has been tested throughout the region and functions well.

The second example is one developed for a rare species, *Penstemon lemhiensis*, found in east-central Idaho and adjacent Montana (Elzinga 1997). It is a large, showy species, usually confined to populations of 100 individuals or less. The monitoring data sheet was designed to be used by the plant specialist. The method is intended to detect large changes in habitat conditions and population size and structure.

Literature Cited

Elzinga, C. 1997. Habitat conservation assessment and conservation strategy for the Lemhi Penstemon, Penstemon lemhiensis. Unpublished paper on file at: U.S. Department of Agriculture, Forest Service, Northern Regional Office, Missoula, MT.

Evenden, A. G. 1995. Basic stewardship monitoring module for Research Natural Areas. Unpublished paper on file at: U.S. Department of Agriculture, Forest Service, Northern Region/Intermountain Region Research Natural Areas Program, Forestry Sciences Laboratory, Missoula, MT.

6/95

R1/R4/INT RESEARCH NATURAL AREAS
BASIC STEWARDSHIP MONITORING MODULE

This stewardship monitoring module is designed to provide a quick and standardized means of assessing current conditions and management needs in research natural areas (RNA).

One of the primary objectives of establishing research natural areas is to preserve a representative array of all significant natural ecosystems and their inherent processes as baseline areas. Protection and management standards for Research Natural Areas are identified in Forest Service Manual (FSM 4063.3). These standards indicate that RNAs should be protected against activities that directly or indirectly modify ecological processes. Certain activities are prohibited in RNAs and include: logging and wood gathering, livestock grazing (with certain exceptions), incompatible recreation use, and construction of new roads or trails.

A significant investment has been made in identifying and establishing the RNA network. Stewardship of the RNA network requires an ongoing management commitment. To date, management of Research Natural Areas has typically been a hands off approach, and in many cases, several years have lapsed since individual RNAs have been visited by RNA representatives. In some cases, this hands off approach has resulted in RNA values being compromised by incompatible uses. It is important to periodically visit each RNA to ensure that the values for which the RNA was established are being protected, and to make recommendations for management action when necessary.

This module is designed to meet the following objectives:

1. Document current uses of research natural areas.

2. Identify and document current or anticipated threats to condition (integrity) of research natural areas.

3. Identify management and monitoring needs.

Information from completed basic stewardship monitoring forms for each RNA will be entered into the R1/R4/INT Natural Areas Database (BCD System developed by The Nature Conservancy). The attached form is to be utilized for initial documentation visits, as well as return visits. This module is intended to monitor occurrence/current conditions only. Other RNA monitoring needs, such as monitoring trend of individual RNA elements, or monitoring vegetation response to a natural fire will require development of specific objectives and appropriate methods.

There are several benefits from completion of the stewardship monitoring module:

* Stewardship monitoring data will serve as a basis for communicating RNA management and monitoring needs to the responsible line officer.

* Provides a useful way for Forest/District RNA Coordinators and other interested individuals to participate in the RNA Program and become acquainted with individual RNAs.

* Provides an on-going effort to secure more active management interest in RNA stewardship.

* Provides a good opportunity to involve volunteers in the RNA program.

* Stewardship monitoring data will provide a basis for conducting general 'state of the RNA system assessments, identifying management and research needs and priorities.

6/95

BASIC STEWARDSHIP MONITORING DATA
USDA Forest Service - R1/R4/INT
RESEARCH NATURAL AREAS (RNAs)

INSTRUCTIONS: Complete Parts I and II of this form in the office, prior to going to the field. Parts III and IV are to be completed during a field visit, and Part V is to be completed following the field visit. Please utilize a copy of the RNA map and/or aerial photo to indicate location of documented observations. Please use complete sentences when recording information in individual fields.

Part I - Prework: (To Be Completed in the Office)

SITENAME: _____
Enter name of Research Natural Area.

STATUS: _____ Enter status of RNA: nominated, candidate, proposed or established.

NATIONAL FOREST: _____ **RANGER DISTRICT:** _____ **STATE:** _____
Enter name of National Forest, Ranger District, and State.

BACKGROUND/PREPARATION: Prior to initiating basic stewardship monitoring, obtain a copy of the Natural Areas Database Record, RNA Establishment Record, map and aerial photos for the site you will be visiting. Database records containing descriptive information on individual natural areas in R1 and R4 may be obtained from the Natural Areas Program, Intermountain Research Station, PO Box 8089, Missoula, MT 59807 and by calling Angela Evenden or Shannon Kimball (406) 329-3485 (A.Evenden:S22L01A). We would appreciate assistance in reviewing and improving the information in the natural areas records. Please submit additions or corrections to natural areas database to the address above.

PROPOSED ACTIONS: _____
_____
_____
_____

Enter information on projects and actions proposed for the RNA or vicinity. State how these proposed actions may affect the condition and management of the RNA. This information may be obtained by examining the quarterly list of proposed actions (available at the National Forest) and talking with district staff. Please indicate if the search turns up no scheduled projects or activities.

IMAGERY. COM: _____
_____

Please review the imagery field on the natural areas record. If new photos are available for the RNA, please provide dates and reference numbers.

PART II - Management and Monitoring Goals : (To Be Completed in the Office)

A general management goal for all RNAs is to protect and preserve featured ecosystem elements in their natural state for research and education. This means allowing natural, ecological, and physical processes to predominate, while preventing human-induced encroachment and activities which directly or indirectly modify ecological processes.

PART IV - Current Conditions (To Be Completed in the Field)

Use the following fields to identify and document current conditions and uses, as well as factors and problems which may compromise the integrity (natural condition) of the RNA. Use information from background and preparation section to assist with this assessment. Current conditions should be compared with identified management goals and ecological conditions described in the establishment record. Use a copy of the RNA map to note locations of specific observations or problems. PLEASE USE COMPLETE SENTENCES.

FACILITIES. Circle the appropriate items to indicate which facilities are present in the RNA.

>   trails   roads   culverts/ditches   bridges   fences/gates   signs

>   stream structures/water developments   other (specify):

FACILITIES.COM _____

_____

_____

_____

_____

_____

Enter comments on current condition of the identified facilities of concern.

Note: Please document existence of all roads and trails with reference numbers. Indicate if repairs are needed, and occurrence of vandalism.

RECREATION. Circle the appropriate items to indicate which recreation uses are occurring within the RNA.

>   hiking   camping   hunting/fishing   horse/stock use   rock climbing

>   off-road vehicle   snowmobiling   other (specify):

RECREATION.COM _____

_____

_____

_____

_____

_____

_____

Enter comments on current status of these recreation uses and associated conflicts within the RNA.

Note: Look for soil compaction or erosion resulting from concentrated camping, fishing, or hiking use. Look for unauthorized hiking, riding, and ORV trails and indicate locations on map.

ECOSYSTEM. Circle the appropriate items to indicate which ecological processes/parameters may require special management attention in order to maintain RNA integrity. If possible, the observer should compare existing ecological condition to conditions described in the RNA establishment record. Please note that exotic species and pests and pathogens are addressed in separate fields below.

fire regime/fuel loading **grazing/browsing** (native species)

hydrologic regime **pollution** **other (specify):**

ECOSYSTEM.COM _____

_____

_____

_____

_____

_____

Enter comments on current condition of the ecological parameters identified above.

Note: Completion of this field requires specialized ecological skills. Things to consider include: fire regimes which are outside the range of natural variability (e.g. former open stands of old-growth PIPO now with dense understories of PSME), native ungulate populations levels outside the range of natural variability, pollution, etc.. Please note observation of any evidence of recent wildfire (1-10 years).

EXOTIC.SPP

_____   _____

_____   _____

_____   _____

_____   _____

Enter names of exotic plants and animals observed within the RNA.

EXOTIC.COM _____

_____

_____

_____

_____

_____

Describe potentially damaging exotic flora and fauna on the site. Indicate their location (on map) and abundance, as well as their effect on the viability of natural communities.

Note: Completion of this field requires the ability to identify exotic plants and animals in the field.

CURRENT.COM: _____

_____
_____
_____
_____
_____
_____
_____

Enter a summary of the current overall condition of the RNA being monitored. Provide a very general summary here.

===

PART V - Recommendations

MANAGE.RECOMMEND: _____

_____
_____
_____
_____
_____

Enter a summary of any new management procedures you would recommend based on current condition of the RNA being monitored. This should address all of the concerns raised under specific categories above.

MONITOR.RECOMMEND: _____

_____
_____
_____
_____
_____

Enter a summary of any new monitoring procedures you would recommend based on current condition of the RNA being monitored.

INFO.NEEDS: _____

_____
_____
_____

Summarize the information that is still needed in order to effectively manage the RNA and elements on it. Include such items as the need for research on management techniques, a more detailed land use history, or baseline monitoring.

Lemhi Penstemon - Qualitative Observations

Species:_____ Date of Observation:_____ Observer(s) _____

Element Occurrence Record Number (Heritage or other State System Identifier):

Legal: T___ R___ S___ ¼___ ¼___ ¼___ T___ R___ S___ ¼___ ¼___ ¼___ County: _____

PHOTOGRAPHS

Roll/frame	From	Toward	Description

POPULATION DESCRIPTION

Population area (estimate size and draw on topographic map) ☐ 1m² ☐ 10m² ☐ 100m² ☐ 1000² ☐ 1ha

☐ count or ☐ estimate?   Give counts or estimates by age class:

____ juveniles (defined as: _____)

____ non-reproductive adults (defined as: _____)

____ reproductive (__flower, __fruit or __both)

____ senescent (defined as: _____)

Herbivory, insects or disease? Note livestock and wildlife use in area:_____

_____

_____

_____

_____

Collections: (give collection numbers): _____

Other notes:_____

_____

Disturbance (describe evidence of fire, surface disturbance, animal disturbance):_____

_____

_____

_____

Weeds (list any occurring on the site and describe extent of the infestation): _____

_____

ASSOCIATED SPECIES

Percent surface non-vegetated:_____   Percent surface cryptogams: _____   Associated species (cover class visual

estimates: t= trace, c=common, 1-5% 1=6-15%, 2=16-30%, 3=31-50%, 4=51-80%, 5= >80%; species list includes all

species recognizable during the survey period with cover of over 10% and any weedy species): _____

_____

_____

_____

_____

APPENDIX 11. Comparison of Vegetation Measures

This appendix compares density, cover, frequency, and vigor measures for plants with different life histories and morphological traits.

Life History Type	Density	Cover	Frequency	Vigor
(A) Annual (B) Biennial	**Field Notes:** Density counts in plots can be done. If germination is spatially variable from year to year, permanent plots may be of little value, depending on the type of spatial change. **Interpretation:** Biological interpretations of measurements on any annual are confounded by the yearly variability due to weather. Density is affected most by changes in weather patterns that affect germination and establishment.	**Field Notes:** Cover of annuals can be measured using any of the techniques discussed (quadrat estimation, line intercept, and point intercept). **Interpretation:** Cover measures are affected by changes in both density and vigor, thus weather patterns that affect germination, establishment, and growth will affect the annual variability of the cover measure. Cover may not be directly related to long term viability because seed production is not always directly related to cover.	**Field Notes:** Frequency based on rooted occurrence minimizes effects of vigor changes, although basal area changes will still have an effect. Note that changes can be dramatic from year to year, thus nested quadrats may be necessary. **Interpretation:** For most annuals, frequency measures are affected primarily by changes in spatial distribution.	**Field Notes:** Any of the techniques discussed in Chapter 8 can be used to measure the vigor of annuals. **Interpretation:** Differences from year to year, or between sites, are due to factors affecting that year's growth. For annuals, these factors will primarily be related to weather.
(C) Geophyte	**Field Notes:** Density counts are often difficult because vegetative reproduction obscures individuals. Density counts of leaves or flowering scapes (a vigor measure) may be a useful substitute for individuals. **Interpretation:** Changes in flowering scapes or leaf number may be vigor changes (and thus probably strongly influenced by annual weather patterns) rather than mortality or recruitment of individuals.	**Field Notes:** Many geophytes have a morphology that is not conducive to most methods of cover estimates: narrow, few leaves, lack of a defined "cylinder." Cover measurement techniques using points may work, if cover is large enough to be measured by points (over 10%). **Interpretation:** Changes in cover may be due to annual weather variation. Measures of cover do not identify whether the change is from density changes or vigor changes, or some combination.	**Field Notes:** Frequency is effective for most geophytes. **Interpretation:** Changes in frequency can be the result of dormancy, mortality, recruitment, and vegetative reproduction. Most geophytes demonstrate little annual spatial changes; thus, frequency changes can be primarily attributed to density changes.	**Field Notes:** Vigor measures such as number of leaves, flowering scapes, and flower scape height are all relatively easy to measure for most geophytes. **Interpretation:** Most vigor measures are strongly influenced by either current or previous year's weather.

Life History Type	Density	Cover	Frequency	Vigor
(D) Short-lived (3-5 yrs) perennial, individuals discernible	**Field Notes:** Density is easy to measure on plants of this type because the individual is recognizable. **Interpretation:** Because mortality and recruitment are high (each cohort is short-lived), stage classification may be very useful in interpretation. Establishment must be a common event, or the population would crash quickly (unless buffered by a seed bank). Counting individuals in classes helps to evaluate whether the density changes observed are a change in, for example, seedlings (which may mean just a bad year for establishment) or in all classes.	**Field Notes:** Cover values can be measured by any of the techniques mentioned. If individuals are solid cylinders, line transect would work well. If the canopy is lacy, points may be better unless overall cover is low. **Interpretation:** Because individuals are short-lived, population structure, and thus cover values (because younger plants are generally smaller), may change rapidly.	**Field Notes:** Frequency is effective for this type of plant. Since frequency may change dramatically from year to year, nested plots may be needed. **Interpretation:** Changes in frequency may be caused by changes in density, spatial pattern or both.	**Field Notes:** Any of the mentioned measures could be used on short-lived perennials. **Interpretation:** Since the plants are short-lived, changes in response to weather variation may be large.
(E) Short-lived perennial, individuals not discernible but stems can be identified	**Field Notes:** To use density a recognizable "individual" or counting unit must be identified. Options include ramets or stems. **Interpretation:** Changes in density of ramets or stems are related to both vigor changes and mortality of genets.	**Field Notes:** Cover is often used for these types of plants. Any of the methods can be used. **Interpretation:** Interpretation of cover changes is subject to all the problems discussed above.	**Field Notes:** Frequency can be used, but rules governing how to determine whether a plant is considered in or out of the plot become very important. **Interpretation:** See life history type D.	**Field Notes:** A consistently recognizable sampling unit will need to be identified (e.g., a recognizable ramet, stem, flowering scape, etc.). **Interpretation:** See life history type D.
(F) Short-lived perennial, neither stems nor individuals discernible (matted)	**Field Notes:** It is very difficult to use density for these types of plants. It may be possible to use clumps as a counting unit. **Interpretation:** Because of the influence of observer bias in the identification of counting unit, interpretation may be difficult.	**Field Notes:** Canopy cover is the most common measure for these types of plants. Basal cover is difficult to measure on species that hug the ground. If the mats are dense with clearly defined boundaries, line intercept is an especially efficient method. Photographic methods of cover measurement are most applicable to this growth form. **Interpretation:** Because these are short-lived species, they will probably demonstrate moderate to high response to annual weather variation.	**Field Notes:** Boundary rules are critical for this type of species because irregular plant mats will often intersect the boundary. If plants are tightly matted to the ground, the canopy boundary may have to be used rather than the typical "rooted rule." **Interpretation:** See life history type D.	**Field Notes:** Plant parts are often so tightly matted that they are hard to separate. Inflorescences may be measurable. **Interpretation:** See life history type D.

Life History Type	Density	Cover	Frequency	Vigor
(G) Long-lived (>5 yrs) perennial, individuals discernible	**Field Notes:** Density counts of these types of plants are easy. **Interpretation:** Changes in density may not occur quickly enough to answer a management question. As a measure of change in long-lived species, density is relatively insensitive. The more long-lived, the less sensitive (think of measuring the density of trees). Large or rapid changes in density are indicative of a major environmental change.	**Field Notes:** Cover can be measured by any of the methods discussed. **Interpretation:** Long-lived plants generally respond less dramatically to annual weather patterns than short-lived plants. Basal cover of these types of plants will probably be fairly insensitive to annual weather patterns.	**Field Notes:** Frequency can be used for this type of plant. **Interpretation:** Since changes in density are probably occurring slowly, frequency may not change much from year to year, similar to density. Spatial changes generally occur very slowly, because recruitment is a rare event.	**Field Notes:** Any of the methods can be used. **Interpretation:** Same comments as under cover.
(H) Long-lived perennial, individuals not discernible, but stems can be identified	**Field Notes:** If a counting unit can be identified, density can be used. **Interpretation:** Stem changes may be more rapid for a long lived plant than individual changes; thus, density in this case may be a sensitive measure.	See life history type G.	**Field Notes:** Frequency can be used for this type of species, but consistent units and boundary rules must be established. **Interpretation:** See life history type G.	**Field Notes:** A consistently recognizable sampling unit will need to be identified. **Interpretation:** See life history type G.
(I) Long-lived perennial, neither stems or individuals discernible	**Field Notes:** It is unlikely that a counting unit could be established.	**Field Notes:** Cover is a good measure to use for this type of species. Line intercept can be used when the canopy forms a fairly solid cylinder. If the canopy is lacy, point intercept would probably be better. **Interpretation:** These species are generally slow-growing, and often respond minimally to annual weather patterns. Most changes in cover are probably attributable to mortality of genets or ramets.	**Field Notes:** Boundary rules are critical. Matted plants may need to be assessed using the interception of the canopy with the boundary, rather than the standard "rooted rule." **Interpretation:** See life history type G.	See life history type G.

Life History Type	Density	Cover	Frequency	Vigor
(J) Woody Species (shrubs-multiple trunks)	**Field Notes:** Counts of stems are relatively easy, but it is often difficult to identify genets (individual plants). **Interpretation:** Changes in individuals may be very slow; changes in stems may be more rapid, and may be a more sensitive measure.	**Field Notes:** Line intercept has been commonly used for woody species. Point intercept will also work well if cover exceeds 10%. An optical sighting device should be used for canopies that are above the observer. **Interpretation:** Changes in cover may be very slow, or very rapid depending on the species. As always, cover is subject to influences of weather, but probably less than with herbaceous perennials.	**Field Notes:** Frequency has rarely been used for woody species except for seedlings. Plots would need to be large enough to achieve reasonable frequency. Boundary rules would need to be established for clumps that lie along the boundary of the plot. **Interpretation:** Similar problems as density.	**Field Notes:** Vigor measures such as flowering may be subject to masting (cyclic reproduction). Current year's annual vegetative production is a common measure. **Interpretation:** Because of the large root/above ground biomass ratio of most woody species, vigor responses to current or previous year's weather are often buffered by root reserves. Thus, vigor measures may be especially suitable for detecting real changes.
K Woody Species (trees-isolated trunks)	**Field Notes:** Density counts are easy. The large plot size needed may be difficult to establish. **Interpretation:** Because mortality and recruitment may be relatively rare events, density may be insensitive to changes that can be measured within the lifetime of the investigator. Conversely, measurable changes are likely important ecologically.	**Field Notes:** Line intercept measure of the canopy has often been used for trees. Basal area based on the diameter at 4.5ft (diameter breast height) is also common. **Interpretation:** Changes in cover that are measurable over a few years are likely important. A long-lived tree should not vary much from year to year in either basal or canopy cover, whatever the weather, because the canopy is defined by woody structures.	**Field Notes:** Similar to density. **Interpretation:** Similar to density.	**Field Notes:** Vigor measures such as flowering may be subject to masting (cyclic reproduction). Change in diameter is the most commonly measured vigor parameter. **Interpretation:** See life history type J.

APPENDIX 12. Field Equipment and Field Hints

Field Equipment

Following is a list of field equipment often used in rare plant monitoring. We recommend storing the equipment in a single box that can be taken to the field in its entirety.

100m tape (two to four)
10m tape (two to four)
30cm ruler
ball of string, brightly colored
binder clips, large (can be used to hold tapes to intermediate monuments)
binder clips, small (to hold field sheets together, and to your clipboard)
camera and three rolls of film
chaining pins (for holding tape ends)
clinometer
clipboard
compass
data sheets
diameter tape (if you work in wooded systems)
ensolite pad or knee pads (for cushioning)
field notebook
flagging (at least four rolls, two each of an unusual color or pattern)
graph paper (for creating impromptu field sheets)
hammer
hand lens
mechanical pencils (three)
meter stick
nails (large enough to serve as markers, or to hold down tapes)
newspaper for plant press
permanent markers, two each of two colors in both thick and thin-tipped (eight)
photo-ID clipboard sheets
pin flags
plant press
plastic bags (several ziploc and a few garbage bags for storing collections)
plumb bob
rebar of varying lengths, 20-50cm long (at least 30 pieces)
screwdrivers (2)
small hatchet (for work in forested areas)
spray marking paint (two colors)
T-posts (several)

Field Hints

◆ For comfort while you're measuring tedious plots, take an ensolite pad to kneel on, or a small gardening stool to sit on. This will also reduce trampling damage.

◆ Always paint stakes and monuments just before you leave the site.

◆ Paint stakes and monuments every year you monitor.

◆ Paint a stake that marks the corner of a permanent plot carefully to avoid spraying any plants.

◆ You may want to paint the handles of your field equipment bright orange so they can be easily found if dropped.

◆ Pin flags have all sorts of uses. They can be temporary markers of your population boundary. They can mark clusters of plants in the field so that you can get a better visual picture of the distribution of individuals in a population. They can mark plot boundaries. Pin flags are preferable to flagging because they are quickly placed, easily moved, and easily picked up when no longer needed.

◆ Screwdrivers are another multi-purpose tool. Use them to dig up plant specimens, hold down tapes, and secure temporary frames.

◆ To secure a long tape so that it is remeasured in the same spot at each measurement, place rebar stakes or large nails periodically along the line at establishment, and secure the tape to the rebar with binder clips or clothespins.

◆ To keep track of location when counting dense density quadrats, use two sticks the width of the plot to mark temporary counting sub-plots as you work the length of the plot.

◆ Two people can usually lay a tape in dense brush much more easily and accurately than one. The first person uses a compass to sight on an object past the end of the transect, then guides the second person with the tape over and around objects.

◆ Photocopy field notes after each day and store in a safe place (off-site). This will eliminate the chance of losing an entire field season's worth of data and observations.

◆ Photocopy monitoring data sheets and store off-site from the originals. This will reduce the chance of losing data to a catastrophe such as fire.

◆ Field vests or cruising vests can be purchased that come with a myriad of different-sized pockets. Compass, pens, pencils, field notebook, clipboard, camera, film, etc., can all be kept at your fingertips. Most are colored bright orange for visability.

◆ If you work in either a wet climate or a very hot arid one, there will be times while you are sampling vegetation when some form of shade or rain protection would be welcomed. If you are doing a field project that requires fairly long periods in a spot (perhaps mapping small plants in a small quadrat), a moveable gazebo may make you more comfortable. These are like a tent fly without the tent. Look for ones that are free-standing, requiring no stakes, so you can pick it up and move it to the next plot when you move on. You can also rig a large umbrella to a lightweight frame.

APPENDIX 13. A Test of the Effects of Using Parametric Statistics on a Very Non-normal Population

One of the questions that must be answered before analyzing quantitative data from a monitoring study is whether to use parametric statistics. We learned in Chapter 11—Statistical Analysis—that several assumptions must be approximately met before we can feel comfortable in using parametric statistics, but we also learned that no monitoring data will ever meet these assumptions perfectly and that the parametric procedures discussed in this technical reference are robust to departures from normality. Let's take a further look at how much a very serious departure from normality actually affects conclusions reached through the use of parametric statistics and why parametric methods can often be employed even when the population being sampled is very non-normal.

Using the computer program RESAMPLING STATS (Bruce 1993), we created a population of 4000 values following the exponential distribution, specifying a mean of 1. Technical descriptions of the exponential distribution are available (see, for example, Evans et al. 1993), but for our purposes it suffices to state that this population is extremely non-normal, with many values near zero and a long tail to the right (i.e., it has a high positive skew). All of the commonly employed procedures to test for normality, including the D'Agostino Omnibus Test, easily reject the hypothesis that this population follows a normal distribution, but Figures 1 and 2 show this most clearly.

Because we've created this population, we know the true population parameters. The true population mean is 0.9953. The true population standard deviation is 0.9624. We tested how well confidence intervals estimated from different sized samples taken from this population would perform. Using the program RESAMPLING STATS, we took 10,000 samples from this population for each of seven different sample sizes: n=10, n=20, n=30, n=50, n=100, n=150, and n=200. For each of these

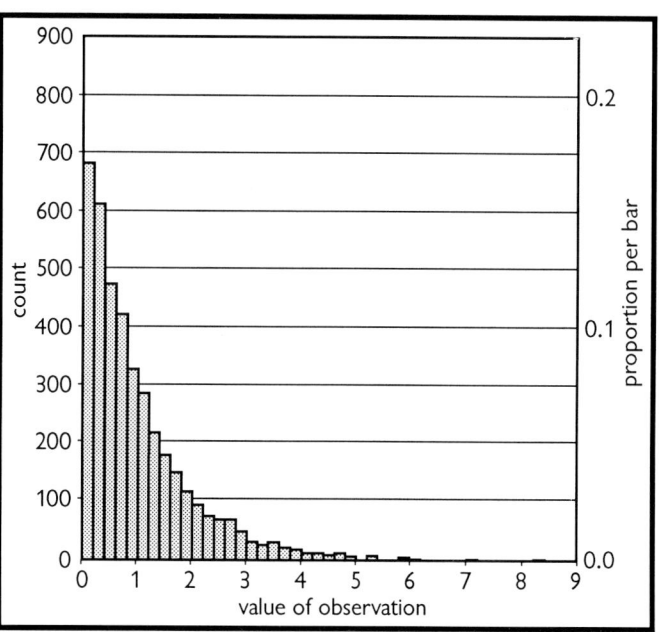

APPENDIX 13—FIG. 1. Histogram of a simulated population of 4000 observations. The population follows the exponential distribution. The mean of this population is 0.995, and the standard deviation is 0.962. Note the large number of small values and the very long tail to the right caused by a few very large values (though hard to see, there is a single value larger than 8, which is more than eight times the standard deviation).

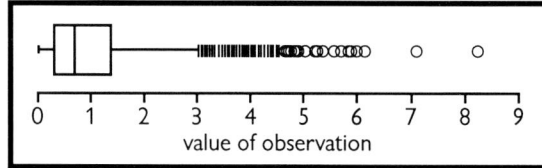

APPENDIX 13—FIG. 2. Box plot of the exponential distribution shown in Figure 1. There are so many outliers, it is impossible to tell the near outliers from the far outliers.

samples, we calculated a 95% confidence interval in the usual parametric manner: by multiplying the sample standard error by the appropriate (two-sided) critical value of *t*. We then kept track of the number of samples for which the confidence interval so calculated did not include the true mean of 0.9953, as well as whether those confidence intervals that missed the true mean missed it with the lower or upper confidence limit. The results are summarized in Table 1.

sample size	proportion of 10,000 samples missing true mean with lower confidence limit	proportion of 10,000 samples missing true mean with higher confidence limit	actual alpha level	actual confidence level (1-actual alpha level)
10	0.0039	0.0924	0.0963	0.9037
20	0.0098	0.0670	0.0768	0.9232
30	0.0084	0.0651	0.0735	0.9265
50	0.0106	0.0522	0.0628	0.9372
100	0.0121	0.0407	0.0528	0.9472
150	0.0133	0.0334	0.0467	0.9533
200	0.0142	0.0318	0.0460	0.9540

TABLE 1. Results from taking 10,000 random samples of a simulated population of 4,000 observations. The observations were created using the program RESAMPLING STATS (Bruce 1993). The population follows the exponential distribution. The true population mean is 0.99527. The true population standard deviation is 0.96243. Each repeated sample calculates a 95% confidence interval in the usual parametric manner: by multiplying the sample standard error by the critical value of **t**. Ten thousand samples were taken (without replacement) from the simulated population for sample sizes of 10, 20, 30, 50, 100, 150, and 200. The true mean fell outside the lower confidence limit with probability less (often much less) than the expected 0.025. The overall confidence level, however, was very close to the intended 95% level with a sample size of 100 and reasonably close with a sample size of 50. Sample sizes of 150 and 200 had empirical ("actual") overall confidence levels greater than the 95% target, although the upper confidence limit missed the true mean more than 2.5% of the time. These results are consistent with what one would expect when sampling a population with a very long tail to the right (Cochran 1977:39-44).

For a perfectly normal population, exactly 5% of the samples would result in confidence intervals that missed the true mean. Further, 2.5% of these confidence intervals would miss with the lower confidence limit (i.e., the true mean would fall below the lower limit) and 2.5% would miss with the upper confidence limit (the true mean would fall above the upper limit). With this highly skewed population, however, we see that the lower confidence limit misses the true mean with a probability less (often much less) than the expected 2.5%. For the smallest sample size tested, n = 10, only 0.39% of the lower confidence limits missed the true mean. With larger sample sizes, n = 50 and above, the lower confidence limit misses the true mean more than 1.0% of the time and begins to more closely approach the 2.5% level. Even at n = 200, however, the lower limit misses only 1.4% of the time. If we are most concerned with the lower limit, the confidence interval will actually perform better than expected. We might, for example, be managing to ensure that mean plant height doesn't fall below a certain threshold. If, based on the calculated confidence interval, the lower limit of the interval falls above the threshold value, we can be more confident with a positively skewed population that the population hasn't fallen below this threshold then would be the case if the population were normal.

Now consider the situation with the upper confidence limit. Here the probability is greater than 2.5% that the upper limit has missed the true mean. For the smallest sample size, n = 10, 9.2% of the upper confidence limits fail to include the true mean. This is clearly unacceptable for a 95%

confidence interval. As the sample size increases, however, the situation improves. Once you get to n = 50, the overall probability of missing the true mean (counting misses on both ends of the confidence interval) falls to 0.0628, at n = 100 it falls to 0.0528 (very close to the expected value of 0.05), and at n = 150 and n = 200 the confidence intervals actually perform better than expected. Note, however, that the upper limit still misses more often than the lower limit even at n = 100.

This small experiment alleviates a lot of concern with respect to the reliability of confidence intervals in plant monitoring, when sample sizes are reasonably large. Figure 3 illustrates why the parametric procedure still works for populations that are far from being normal. Even though the underlying population is highly skewed, a distribution of means of samples taken from this population will tend toward normality as the sample size increases (this phenomenon is known as the central limit theorem; see Zar 1996:75-76). The result is that parametric statistics work fine even for very skewed population distributions, as long as sample sizes are reasonably large. Cochran's rule, discussed in Chapter 11—Statistical Analysis, gives guidance on when a sample is "reasonably large."

Seldom will populations be as skewed as the exponential population used here. This is particularly true if you heed the advice in Chapter 7—Sampling Design—and design your study so that your sampling units do not have a lot of zeros or extreme values in them.

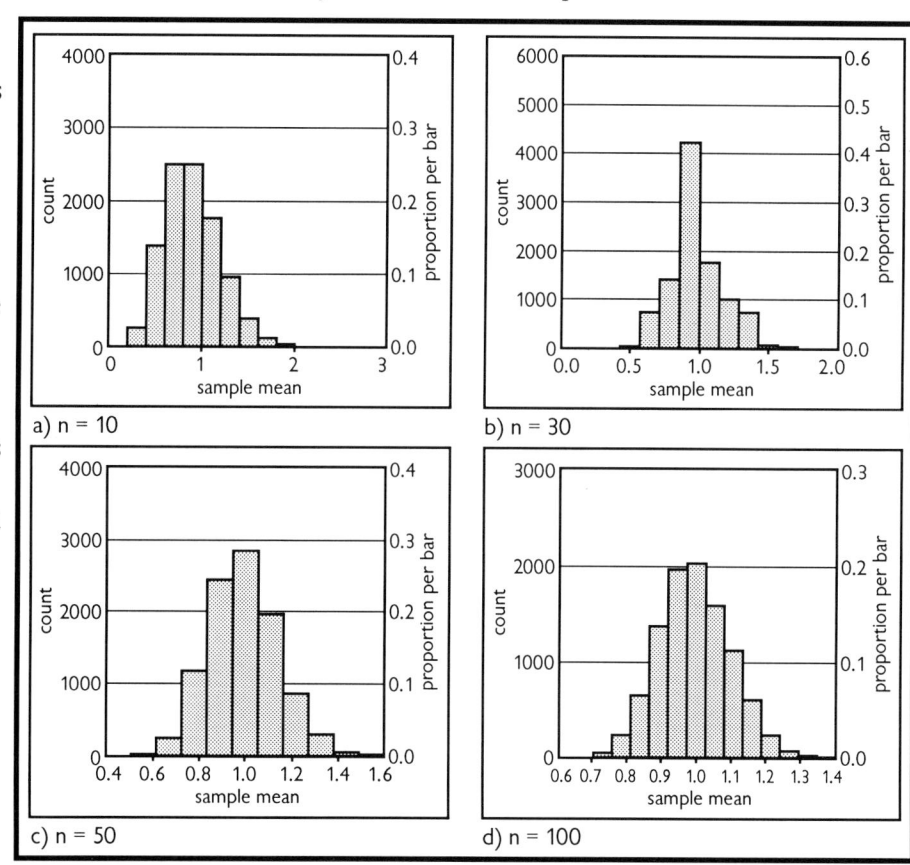

a) n = 10

b) n = 30

c) n = 50

d) n = 100

APPENDIX 13—FIG. 3. Histograms of means of 10,000 samples of four different sizes taken from the exponential population shown in Figures 1 and 2. Note how the sampling distribution more closely approximates a normal distribution as sample size increases.

Literature Cited

Bruce, P. C. 1993. Resampling Stats user guide. Resampling Stats, Inc., Arlington, Virginia.

Cochran, W. G. 1977. Sampling techniques, 3rd ed. John Wiley & Sons, New York.

Evans, M.; Hastings, N.; Peacock, B. 1993. Statistical distributions, 2nd ed. John Wiley & Sons, Inc., New York.

APPENDIX 14. Introduction to Statistical Analysis Using Resampling Methods

Some general comments were made in Chapter 11—Statistical Analysis—on the use of resampling methods (also called computer-intensive methods) to analyze data. These methods take advantage of the tremendous computing power of modern personal computers. They entail intensive resampling of the original data set.

Resampling methods can be used to calculate confidence intervals and to conduct significance testing. Two of the most commonly used methods are the bootstrap (which involves sampling the original data set *with* replacement) and randomization (also called permutation) testing (which involves sampling the original data set *without* replacement). The advantages of resampling methods are many, including the fact that very few of the assumptions required for parametric statistics are needed (except, of course, for the assumption of random sampling) and they are apparently just as powerful (Manly 1991a).

We start this introduction with a discussion on the use of bootstrapping to calculate confidence intervals and end with a description of randomization testing to determine if two means are significantly different.

A. Bootstrapping to Calculate Confidence Intervals

With bootstrapping, the calculation of confidence intervals around just about any statistic is easy. Chapter 5—Basic Principles of Sampling—showed how to calculate a confidence interval around a mean using the standard error and a *t* table. This is a relatively simply technique, and—despite the fact that it depends on the assumption of normality—it is robust to moderate violations of that assumption. If, however, we had reason to believe our data came from a very skewed distribution, we might choose to calculate bootstrap confidence intervals. We could then compare the parametric and bootstrap intervals to see if there is cause for concern. Calculation of confidence intervals around other statistics such as the median or another percentile is also possible using traditional methods (see Hahn and Meeker 1991), but the calculation is easier and more straightforward using a computer program that performs bootstrapping.

To illustrate the concept of bootstrapping, let's look at a simple example. Say we take 10 measurements of the height of individuals of a particular plant species and come up with the following heights (in inches): 25, 4, 30, 4.5, 4, 1.75, 2, 4, 2.5, and 4.5. Without even plotting these data we can see we have a problem: there are two large values, 25 and 30, and the remainder of the values are small relative to these. A quick analysis confirms this: the sample mean is 8.23 and the sample standard deviation, 10.27, is larger than the mean. In addition, our sample size is small (remember that if our sample size is reasonably large[1] we can use parametric statistics in most cases.)

---

[1] The G_1, or skewness value (obtained from a statistical program) for the set of 10 plant heights is 1.79. Using Cochran's (1977) formula $n > (25)(G_1)^2$, we see that an n greater than $(25)(1.79)^2 = 80$ will be sufficient to calculate a parametric 95% confidence interval and be sure that the interval will include the true mean at least 94% of the time.

1. Bootstrap confidence interval for the mean

If we calculate a 95% confidence interval in the usual way (by multiplying the standard error by the appropriate value of *t*) we come up with a 95% confidence interval of 0.874 to 15.576. We suspect, however, that we can't trust this confidence interval because of the large standard deviation (relative to the mean) and the rather small sample size. Now let's look at calculating a confidence interval for the mean using the bootstrap.

We resample the original sample of 10 a large number of times *with replacement*. These new samples are called bootstrap samples. This means that after we randomly select our first value "from the hat" we put that value back into the hat so that it can be selected again. So, for example, one of our bootstrap samples might be 25, 25, 1.75, 1.75, 4.5, 4.5, 4, 2, 4, 30 (here the single values of 25 and 1.75 in the original sample have been selected twice in this one bootstrap sample). We then take the mean of each of these bootstrap samples and keep track of it.

We end up with the distribution of bootstrap means shown in the histogram of Figure 1. To calculate a 95% confidence interval for the mean we record the value corresponding to the 2.5 percentile at the low end of the distribution and the 97.5 percentile at the high end of the distribution. In this case a bootstrap sample of size 10,000, conducted using the RESAMPLING STATS computer program (Bruce 1993),[2] yielded a 95% confidence interval of 3.175 to 15.175. This is far less conservative than the confidence interval calculated using parametric statistics, especially on the lower end. It is still, however, very wide. This low precision is a result of the large spread in the data and the relatively small sample size.

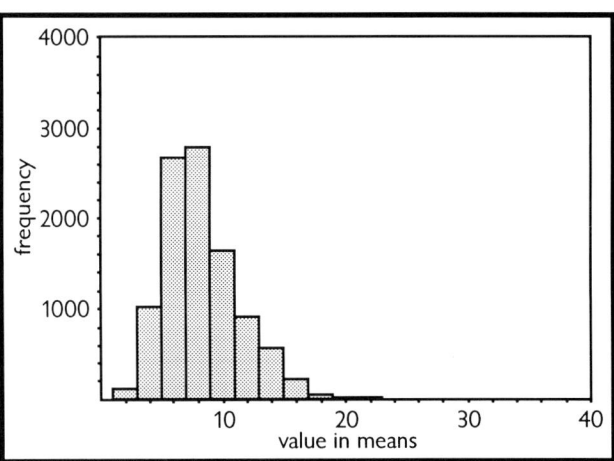

APPENDIX 14—FIG. 1. Histogram of the means of 10,000 bootstrap samples of size 10 taken from an original sample of 10 plant heights (see text for original sample values).

2. Bootstrap confidence interval for the median

Instead of the mean we may be interested in the median as a measure of central tendency, particularly if we're more concerned with *how many* plants fall below or above a certain threshold. In the above example the mean height is 8.23 inches, but the median height is only 4 inches. Thus, half of the plants in our sample are below 4 inches and half are above.

There are traditional methods available to *conservatively* estimate a confidence interval around a median. Zar (1996) shows how to do this making use of a table of critical values for the binomial test with *p* = 0.05. Use of this method (which will not be described here) results in a estimated confidence interval of 2.5 to 25.

We can also calculate a confidence interval around our estimate of the population median using bootstrapping. We do this the same way we did for the mean, except that instead of keeping score of the *mean* of each bootstrap sample we keep score of the *median* of each

---

[2] Instructions on obtaining this program are given in Chapter 11, Section L.

bootstrap sample. We then look at the distribution of medians and take the values corresponding to the 2.5 and 97.5 percentiles as the outer limits of our confidence interval. From 10,000 bootstrap samples we calculate a 95% confidence interval for the median of 2.5 to 14.75. Note that although the lower confidence limit obtained through bootstrapping is the same as that estimated through the method described by Zar (1996), the higher confidence limit obtained through bootstrapping is much less conservative.

A histogram of these samples is given in Figure 2. Note that this histogram looks considerably different than the one for the mean. The fact that this distribution of sample medians is not normal is not, however, of concern: there is no assumption of normality required when using statistics based on resampling.

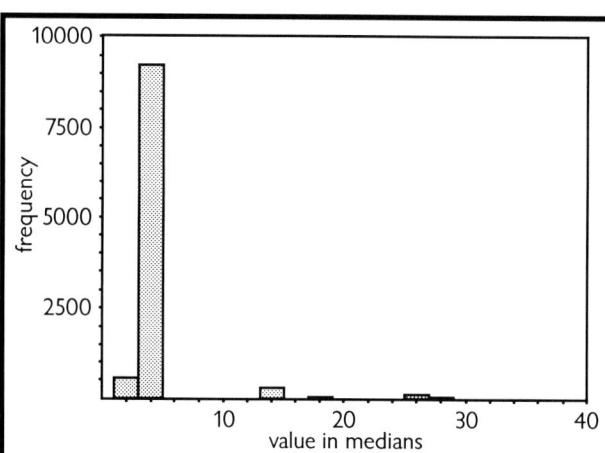

APPENDIX 14—FIG. 2. Histogram of the medians of 10,000 bootstrap samples of size 10 taken from an original sample of 10 plant heights (see text for original values). Note that more than 90% of the median values fall between 3 and 5.

3. Practical considerations in bootstrap sampling

In practice we wouldn't want to take an original sample as small as only 10 individuals. The bootstrap procedure for calculating the percentile confidence interval may not perform well with small sample sizes. According to Bryan Manly (1991b), bootstrap confidence intervals calculated from a sample of about size 40 work well even for an exponential distribution (a distribution with a high positive skew, very far from normal—refer to Figures 1 and 2 in Appendix 13).

When calculating bootstrap confidence intervals you need to bootstrap sample at least 1,000 times (Efron and Tibshirani 1993). With a relatively fast computer this is easy. A Pentium machine with a 60 MHZ chip takes less than 20 seconds to process 10,000 bootstrap samples.[3]

4. An example of calculating bootstrap confidence intervals for the 25th percentile

Here is an example of calculating bootstrap confidence intervals for the 25th percentile. This would be important to you if you set a threshold requiring a population estimate of something other than the mean or median. You may decide, for example, that no more than 25 percent of the plants in a certain population should be grazed lower than 4 inches. Using conventional statistics, construction of confidence intervals around percentiles other than the median is difficult and not covered by standard statistical texts (but Hahn and Meeker 1991, in a text devoted to statistical intervals, do provide tables that can be used to calculate intervals around particular percentiles). Using the bootstrap resampling technique, construction of a confidence interval around *any* percentile is easy.

---

[3] This is the time required to calculate a confidence interval around a mean or median using the DOS version of RESAMPLING STATS. The Windows 95 version takes about 70 seconds on the same computer.

The data set is 100 height measurements (in inches) on individual plants along a transect in a riparian key area in the Salmon, Idaho, BLM District. Plants are treated as the sampling units. Here are the measurements: 25, 4, 30, 4.5, 4, 1.75, 2, 4, 2.5, 4.5, 4.5, 4, 2.5, 2.5, 3, 4, 2, 4, 1.5, 0.75, 2.5, 1.5, 1, 1.5, 1.5, 2.5, 1, 7, 6, 3, 5, 3.5, 3, 2.5, 3.5, 3, 3, 2.5, 1.5, 0.5, 2, 2, 1.5, 3.5, 4, 3, 1, 1.5, 3, 4, 8, 5, 6, 3, 3, 5, 3, 5, 4, 3.5, 6, 7, 3, 6, 4, 2.5, 1.5, 3, 2, 3, 2, 3.5, 4, 2, 4, 12, 5, 7, 5, 4, 2, 2.75, 3, 6, 9, 5.5, 2.5, 17, 2.5, 13, 5, 8, 4.5, 4, 5.5, 5, 5.5, 8, 4, 1.

Let's say our management objective is to maintain 75% of the plants in our key area at a height of 4 inches or taller. We can determine whether we've achieved this objective by estimating the 25th percentile plant height. The true 25th percentile height is the height exceeded by 75% of the plants in the population. We again use the program RESAMPLING STATS. Here is the RESAMPLING STATS program to estimate the 25th percentile of the true plant height:

```
 1 : MAXSIZE B 10000 C 10000 D 10000
 2 : READ FILE "D:\DATA\STUBBLE.TXT" A
 3 : PERCENTILE A 25 PRCENT25
 4 : REPEAT 10000
 5 : SAMPLE 100 A B
 6 : PERCENTILE B 25 C
 7 : SCORE C D
 8 : END
 9 : HISTOGRAM D
10 : PERCENTILE D (2.5 97.5) CI
11 : PRINT PRCENT25
12 : PRINT CI
```

Here's the histogram of 25th percentiles from the 10,000 samples of size 100 taken from the original sample:

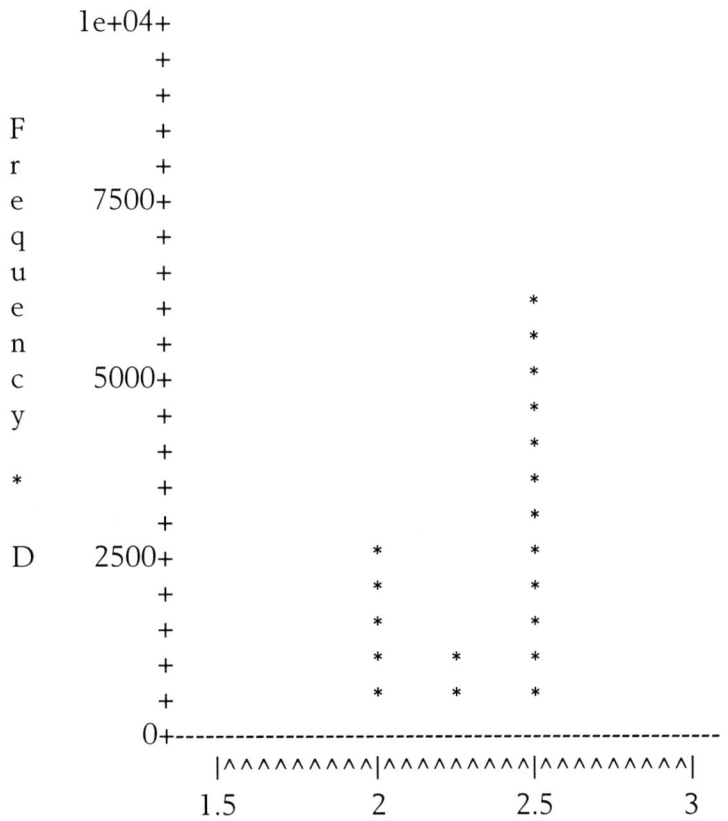

Here are the values used to construct the histogram:

Vector no. 1: D

Bin Center	Freq	Pct	Cum Pct
1.5	10	0.1	0.1
1.65	12	0.1	0.2
1.75	18	0.2	0.4
1.9	28	0.3	0.7
2	2690	26.9	27.6
2.25	804	8.0	35.6
2.5	5875	58.8	94.4
2.65	140	1.4	95.8
2.75	133	1.3	97.1
2.9	87	0.9	98.0
3	203	2.0	100.0

Note: Each bin covers all values within 0.025 of its center.

Here are the calculated values. The 25th percentile is calculated from our original sample. The 95% confidence interval was constructed from the 10,000 samples of the original data set.

PRCENT25 = 2.5

CI = 2 2.875

The bootstrap confidence interval for the 25th percentile is thus 2.00 inches to 2.88 inches. Although the tables in Hahn and Meeker (1991) do not include the 25th percentile, interpolation between the tables they give for the 20th and 30th percentiles yields a confidence interval close to 2.00 inches to 2.75 inches, which is a little less conservative than our bootstrap interval.

Interpretation of these data is easy. Because the entire 95% confidence interval is far below the threshold of 4 inches, we must conclude that we have failed to meet our objective.

B. Two-sample Randomization Significance Test

1. Example

To illustrate the principles of a randomization test we'll use a non-plant example. This example comes from Manly (1991a:51) and involves the sampling of the gut contents of two size classes of eastern horned lizards. Measurements are in units of milligrams of dry biomass found in the guts of these lizards.

There were 24 lizards in the first size class (adult males and yearling females). Here are the measurements: 256, 209, 0, 0, 0, 44, 49, 117, 6, 0, 0, 75, 34, 13, 0, 90, 0, 32, 0, 205, 332, 0, 31, 0.

Here are the measurements for the 21 lizards in the second size class (adult females): 0, 89, 0, 0, 0, 163, 286, 3, 843, 0, 158, 443, 311, 232, 179, 179, 19, 142, 100, 0, 432.

Notice the large differences between the values for both size classes and the fact that there are quite a few zeros. One would hesitate to use parametric significance tests, particularly since the samples are relatively small and of different sizes.

Recall that a significance test requires a null hypothesis. Our null hypothesis is that there is no difference in gut content between the two size classes of lizard. If this null hypothesis is true then both groups belong to a single population of lizards. That means we should be able to combine all the 45 values together, take a large number of random samples of size 24 (to correspond to the first age class) and size 21 (to correspond to the second age class), and then compare this collection of randomization samples to the original sample.

Here, then, is the randomization procedure for this example (after Manly 1991a):

1. Find the mean scores for each of the two size classes and the difference between these for the original sample. The mean difference between the two groups in the original sample is -108.6.

2. Randomly reallocate 24 of the sample weights to a "first age class" group and the remaining 21 sample weights to a "second age class" group. Calculate the difference in means between the "first age class" group and the "second age class" group.

3. Repeat step (2) a large number of times to come up with a distribution of mean differences between the "first age class" and the "second age class" group.

4. Compare the original mean difference to the set of randomization mean differences.

5. If the original mean difference looks like a typical value from the distribution of randomization mean differences then you conclude there is no difference between the two age classes.

6. If, however, the original mean difference is unusually large then this difference is unlikely to have arisen if the null hypothesis were true; you can then conclude that there is a difference between the two age classes.

For the data above the mean difference between the two means is -108.6. A t test for the significance of this difference gives a P value of 0.027. A Mann-Whitney non-parametric test gives a P value of 0.080. A randomization test, with significance estimated from 5000 randomizations, gives an estimated probability of 0.018 for a sample mean difference as large as that actually observed (Manly 1991a:51).

As Manly (1991a:51-52) points out:

> In this example the significance level from the t-table is not much different from the level obtained by randomization. However, using the t-table would certainly be questionable if the result of the randomization test was not known. The level of significance from the Mann-Whitney test was calculated using a standard statistical computer package. The large discrepancy between this level and the randomization level can be accounted for by the high proportion of tied zero observations affecting the Mann-Whitney probability calculation.

2. Using the program RT to conduct the above randomization test

An example of how the randomization test discussed in the above example would be conducted using Bryan Manly's computer program, RT (Manly 1993), is given below.

The program RT (Manly 1993) was used to carry out 5000 randomizations of the lizard data.[4] The program first calculates the difference between the means of the original samples. As previously noted, this difference is -108.6. This difference is then compared to the distribution of mean differences obtained by randomly allocating the 45 data values to groups of 24 and 21 a total of 5000 times.

The following printout from RT shows only the data for every 100th of these randomizations. (The program allows you to set any interval you want for the randomizations that will be printed; thus, you could see the results of every randomization, of every 10th randomization, or, as here, every 100th randomization. This allows you to inspect the results to see if the program is performing as you would expect.) The calculated results, shown following the table of randomizations, are based on all 5000 randomizations.

```
##############################################
##############################################
##            NEW RT ANALYSIS            ##
##############################################
##############################################



###############################################
# The RT module to carry out a two sample randomization  #
# test on the mean difference between two samples and the #
# ratio of the two sample variances (F).  A randomization    #
# confidence interval for the mean difference between the  #
# two source populations can also be determined.  To stop   #
# at any time, try <control>C.  Output is sent to RT.LOG.  #
#                                                          #
#            Version 1.02B (April, 1993)                  #
###############################################
```

Data: Lizard data from Manly   # Cases: 45   # Variables: 2   This case: 1

    V 1   1.000   V 2   256.0

Sample values are in variable 2 and sample numbers in variable 1.

Random number seed = 235.

---

[4] See Chapter 11—Statistical Analysis, Section K, for information on obtaining this program.

Mean difference (sample 1 - sample 2) = -108.2202
Variance ratio (sample 1/sample 2), F = .2036

Rand	Mean diff	F ratio
100	7.6726	.4812
200	-37.5952	.4324
300	-56.0774	1.4101
400	-58.8452	.4046
500	16.6905	2.5375
600	55.2619	2.7182
700	-94.1131	.1558
800	18.9226	2.5298
900	63.0298	2.5959
1000	-3.1310	2.0258
1100	19.1905	1.7742
1200	25.1726	2.1035
1300	-104.8274	.1870
1400	42.0476	1.8302
1500	-31.3452	.3898
1600	-100.9881	.3305
1700	2.6726	3.6767
1800	-30.5417	.4726
1900	-5.0060	1.5697
2000	-62.5952	.3227
2100	72.8512	4.1177
2200	30.0833	1.9253
2300	-35.7202	2.4955
2400	11.7798	1.5892
2500	-11.2560	.4316
2600	-8.8452	.5780
2700	-56.4345	.4611
2800	65.0833	2.5199
2900	11.9583	1.5655
3000	-15.4524	1.4419
3100	-15.8095	.3573
3200	49.1905	2.1676
3300	6.5119	.5013
3400	-7.4167	1.5742
3500	-25.6310	1.5719
3600	-50.0060	.3946
3700	55.0833	2.5045
3800	.7083	.5077
3900	-38.5774	.2642
4000	49.6369	1.9485
4100	11.9583	2.1546
4200	9.3690	.4599
4300	77.2262	2.7914
4400	-40.5417	.5260
4500	-71.2560	.3528
4600	-14.4702	.4673
4700	-13.0417	1.3222
4800	-49.7381	.3439
4900	25.9762	.6489
5000	-9.5595	.5148

Results of Randomization Testing of the Mean Difference
        Less than or equal to observed = 1.04%
        Greater than or equal to observed = 98.98%
        As far or further from zero than observed = 1.76%

Results of Randomization Testing of the F-Ratio
        Less than/equal to observed = 3.06%
        Greater than/equal to observed = 96.96%
        (Largest variance)/(Smallest variance) greater than or
        equal to observed = 6.12%

The first set of results is for the difference in means. Only 1.04% of the 5000 randomization samples had mean values less than or equal to the difference of -108.6 observed from the original samples. This value of 1.04% (= 0.0104) is the one-tailed randomization P value. The two-tailed randomization P value is 1.76% (= 0.0176), which represents the percentage of randomization mean differences that are as far or farther from zero than observed (thus, values of +108.6 or larger would be tallied along with values of -108.6 or smaller).

Note that none of the 50 mean differences shown were as far from zero as the observed absolute difference of 108.22. Of those shown, only randomization number 1300 (-104.83) is very close. Remember, however, that there are another 4,950 randomizations that are not shown in the above table, of which 88 (1.76% x 5000) had absolute mean differences of 108.22 or greater.

In addition to calculating the difference in means for each pair of randomization samples, the program also calculates the F-ratio of the variances associated with each pair of samples. The F-ratio of variances (called the variance ratio test by Zar 1996) is sometimes used to determine whether the variances of two samples are significantly different from one another (in which case the assumption of homogeneity of variances would be called into question). The traditional F-ratio test uses the original sample values and involves simply dividing the largest variance by the smallest variance and comparing the F value so calculated to the critical value of F for a given alpha (usually 0.05) and the appropriate degrees of freedom. Unfortunately, this test is known to be severely and adversely affected when sampling non-normal populations (Zar 1996), as is the case in this example. Randomization methods offer a remedy to this problem.

RT calculates a randomization P value for the F-ratio test. It does this by calculating an F-ratio for each of the 5000 pairs of randomization samples and comparing these values to the F-ratio from the original pair of samples. The percentage of randomization F-ratios that are equal to or larger than the F-ratio of the original pair of samples corresponds to the P value for declaring there to be a significant difference between the two original sample variances. In our example the P value calculated through randomization is 6.12% (= 0.0612). If we had set our threshold P value (alpha) at 0.05, we would conclude there was no significant difference between the variances. Contrast this to the P value of 0.0004 calculated using the traditional F-ratio test, and the effect of non-normal data on the traditional F-ratio test becomes obvious.

Literature Cited

Bruce, P. C. 1993. Resampling Stats user guide. Arlington, VA: Resampling Stats, Inc.

Cochran, W. G. 1977. Sampling techniques, 3rd ed. New York, NY: John Wiley & Sons.

Efron, B.; Tibshirani, R. 1993. An introduction to the bootstrap. New York, NY: Chapman and Hall.

Hahn, G. J.; Meeker, W. Q. 1991. Statistical intervals: A guide for practitioners. New York, NY: John Wiley & Sons, Inc.

Manly, B. F. J. 1991a. Randomization and Monte Carlo methods in biology. New York, NY: Chapman and Hall.

Manly, B. F. J. 1991b. On bootstapping for sample design. Report No. 1, Department of Mathematics and Statistics, University of Otago, New Zealand.

Manly, B. F. J. 1993. RT: A program for randomization testing, Version 1.02B. Available from Western Ecosystems Technology, 1402 South Greeley Highway, Cheyenne, WY 82007. [A newer version, Version 2.0, is now available.]

Zar, J. H. 1996. Biostatistical analysis, 3rd edition. Upper Saddle River, NJ: Prentice Hall, Upper.

APPENDIX 15. Data Forms

A. Introduction

This appendix contains a number of forms for recording data from monitoring studies. Data sheets are included for studies that use density, cover (point intercept, line intercept, quadrat estimation), and frequency. We also include data sheets for documenting studies and photopoints.

You are encouraged to use these as examples and develop your own data sheets specifically appropriate to your study. Most word-processing programs contain table features that facilitate construction of data sheets such as the ones illustrated here. We have also tried to anticipate the types of data sheets that are most often needed, and we've created examples that may be photo-copied and used for many of your studies.

B. Explanation of Forms

1. Common Fields

Several fields that occur on most or all data sheets are described below:

Study location. Detailed directions should be given in the monitoring plan and Field Monitoring Cover Sheet. Locational information included on the data sheet provides an identifying field for the study site in addition to the study ID number, and serves as back-up locational information if the more detailed directions are lost.

Study Notes. These will vary depending on the use of the data form. Codes used in the data form may include those referencing notes, heights of interceptions, phenology, etc.

Transect or Macroplot #. Each transect or macroplot should be given a unique identifying number. Study designs will vary in the positioning of sampling units within the area of interest. In some cases, sampling units will be randomly positioned within a particular site. In other cases, sampling units will be randomly positioned within a defined macroplot. In still other cases, sampling units will be arranged along transects randomly positioned within a site or macroplot area. Many of the sample data sheets included in this appendix show situations where data from sampling units associated with multiple transects can be recorded on the same data sheet.

Transect coordinates. You will likely sample a defined sampling area, locating transects using a coordinate system. Record here the location of the transect based on that coordinate system. This is critical for relocating transects if the transects are permanent. This field could be filled out in the office before the field day to avoid wasting valuable field time determining random transect locations.

Genus and species, species code. The complete genus name and specific epithet should be given. If the specific epithet is not known, use "Z1" for an unknown species that is consistently recognized, and "Z2" if there are several species of the genus that are lumped together (e.g., *Carex* spp.). You can also separate species into stage classes, such as: "*Penstemon lemhiensis,* reproductive." Codes are included for ease of computer data entry. Coded duplication of species names also ensures that a handwritten and indecipherable genus and species name is accompanied by a (hopefully) more legible code. Species names and codes can be pre-printed alphabetically in the office for rapid field entry.

2. Density

Two forms are included for recording counts of plants within quadrats. The first form is designed to record a single category of counting unit (e.g., all plants). The second form is designed to record up to three different types of counting units (seedlings, vegetative plants, flowering plants).

Mplot (=Macroplot number)

This field is for recording a location identification label. Often, density quadrats are randomly or systematically positioned within a defined macroplot, or along a transect line.

Quad (= quadrat number)

This field is for recording the quadrat number.

Seg (= quadrat segment number)

If quadrats have a long rectangular shape (these are sometimes called belt transects), we recommend that you record plant counts within segments of the quadrat to facilitate the counting process and to provide some information on the spatial distribution of the species being counted. The sampling unit is the quadrat, not the individual segments, so for density estimates, counts from all segments within a quadrat need to be summed together to obtain a single count for the whole quadrat. Individual segment lengths can be whatever length seems appropriate to evenly subdivide the quadrat area. Plant counts are often made within 1m long segments of the quadrats. The example of a completed density form for a single category of counting unit shows counts made within 0.25m x 50m quadrats where plants were counted within 1m long x 0.25m wide segments. Only segments that contained plants were recorded on the field data form.

3. Frequency

Transect or Macroplot

This field is for recording a location identification label. Frequency quadrats are often randomly positioned within a defined macroplot, or systematically positioned along randomly positioned transects.

Quad # (= quadrat number) columns 1-25

The square cells of this form are for noting the presence of a species within quadrats. For simple frequency, a check mark, slash, or some other character can be entered into the appropriate cell. For nested frequency, numerical codes are entered into these cells with the numbers corresponding to the smallest quadrat size containing a particular species. With a nested quadrat design, the smaller quadrats are located within the corners 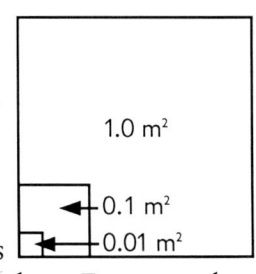 of larger quadrats and so they are, by definition, included within larger quadrats. For example, the figure shown to the right includes three quadrat sizes. A species occurring in the smallest (0.01 m²) quadrat size would receive a "1" on the field data form. A species occurring in the middle (0.1 m²) quadrat size but not in the smallest would receive a "2." A species occurring within the largest quadrat size (1.0 m²) but not present in the smaller two sizes would receive a "3." To summarize the frequency information for the smallest quadrat size, the number of quadrats with "1's" is divided by the total number of quadrats sampled and the number is expressed as a percentage. To summarize the frequency information for the middle quadrat size, all "1's" are tallied with all "2's" and the sum of these two counts is divided by the total number of quadrats sampled. For the largest quadrat size, all 1's, 2's, and 3's for a species are tallied together and divided by the total number of quadrats sampled.

4. Cover - Quadrats

Two forms are included for recording visual estimates of cover in plots. The form for single species sampling allows recording of several transects on a single sheet. The form for community sampling is intended for a single transect. A section is included on both sheets to record the cover percentage range for each cover class. For example, class 1 may be 1-3%, class 2 may be 4-9%, etc.

5. Cover - Points

The forms for recording point intercept data are designed to require an entry for each point. This approach is considered superior to simple dot tallies because it increases the flexibility in analysis for both permanent and temporary designs. Clearly, on the rare occasions that the points are considered a permanent sampling unit, data must be recorded by point, but even when the transect is considered the sampling unit, there are good reasons for the additional effort of recording the data by point.

Let's consider a temporary point intercept transect in which three species are recorded: A, B, and C. The data are recorded as a dot tally for the entire transect, with each species accorded a "hit" if it occurs under the point intercept. Now let's say you wish to combine species A and B for analysis. Because you recorded the data for all three species with a dot tally for the entire transect, you don't know how many of those point contacts represent a point where both species occurred. If you simply add the dot tallies for both species, you are including as two contacts those points where both species occurred, and thus overestimating the cover of the two species considered together.

Now consider a permanent design with the transect the permanent sampling unit. Consider an example of a permanent 50m long transect for which the point intercept estimate of cover of a rare species decreases by 50% between two measures. An inspection of the data shows that at the first measure, there were interceptions with the target species along the entire length of the transect; at the second measure, interceptions were only recorded along the first half of the transect. A field check finds that the density of cheatgrass (aggressive, exotic weedy grass of the Intermountain Region), appears especially high in the second half of the transect. While this design does not prove cheatgrass is the problem, the recording of data by point allows the evaluation of spatial patterns in the field, and the development of a possible (reasonable) hypothesis for the decline of the species that can now be tested.

A form is included for sampling communities and species using point intercept along transects. The species form provides room for recording several transects. The community form is designed for recording many species along a single transect.

Points will be located systematically along a transect for most point intercept studies. These data forms allow recording up to 50 points. If points are not located systematically, the sequential point numbers could be erased and replaced with the actual location along the transect, or another row added in which to record that information.

a. Single species

The data form can be used to record point interception of the canopy (c) and/or base (b) of a single species as follows:

Transect #/coordinates	Point number																			
	1	2	3	4	5	6	7	8	9	10	11	12	13	14	15	16	17	18	19	20
Transect ID01/15, 10	c	c	c/b	c/b	c					c	c					c	c/b	c		
Transect ID02/35, 5							c							c						
Transect ID03/55, 20							c	c	c/b	c	c/b	c	c	c						

When recording the canopy or basal cover of a single species in a few stage classes or layers, the data form can be adapted as shown:

Transect #	Point number																			
	1	2	3	4	5	6	7	8	9	10	11	12	13	14	15	16	17	18	19	20
Transect ID01																				
reproductive			c	c	c	c							c	c						
juvenile				c	c	c	c											c		
Transect ID02																				
reproductive	c	c																		
juvenile											c	c								c

Occasionally, a design may require the measurement of all interceptions rather than canopy or basal cover. This can be visualized as all plant matter touched by a sharp pin as it is lowered through vegetation layers. The same individual can be intercepted a number of times if it has a layered architecture. As discussed in Chapter 8, the recording of multiple interception is more closely related to biomass estimates than are canopy or basal cover measures. The data form can be used for recording multiple interception by using dot tallies. (•• = two interceptions; ⋮⋮ = 5 interceptions; ⊠ = 9 interceptions)

Transect #	Point number																			
	1	2	3	4	5	6	7	8	9	10	11	12	13	14	15	16	17	18	19	20
Transect ID01																				
reproductive	⁞•				••		•	⁞•	⋮⋮	⊠										
juvenile												•	••							
Transect ID02																				
reproductive			⁞⁞																	
juvenile					•	••														

b. Multiple species

When measuring the canopy cover of all species encountered at each point for community studies, each species is recorded once when it is first intersected. The data form can be used to track the top canopy species (in this example recorded as "1") as well as the ground cover (recorded as "G"). The order of intersection can also be recorded (at the first point, *B. tectorum* is the third species encountered as the point intercept is lowered through the vegetation). You could also record species by layer using codes, for example, for the upper layer (1-1.5m above the ground), middle layer (0.5-0.9m above the ground), and lower layer (<0.5m). Finally, the actual height of each intercept could be recorded on this form, if such detailed information is needed.

These examples are not meant to encourage you to record cover data in this manner. For most situations, recording occurrence of the species intersected by a particular point, and perhaps identifying the ground cover and the top canopy species, is adequate. Recording the order of intersection or the height of intersection is time-consuming, and you must have an important reason for recording these data to justify the time expense.

Genus species	Species code	Point number																			
		1	2	3	4	5	6	7	8	9	10	11	12	13	14	15	16	17	18	19	20
Artemisia tridentata	ARTTRI	1			1				1	1		1	1						1		
Agropyron spicatum	AGRSPI			2	1	1	2	2		1				1	1						1
Bromus tectorum	BROTEC	3		1		2	2	3		2				2	1	1	1		1		
Penstemon lemhiensis (reproductive)	PENLEMR	2						1			2								2		
Penstemon lemhiensis (non reproductive)	PENLEMN		1							3											2
Bare ground	BARGRO		G	G	G			G	G		G			G	G	G	G	G		G	G
Cryptogam	CRYPT	G				G	G			G		G	G						G		

The form can also be used to record multiple interception using dot tallies. At the first point in this example, *A. spicatum* is intercepted 6 times as the pin is lowered to the ground. Again, this example is provided to illustrate how the data sheet can be adapted for different situations, not to encourage you to always collect data in this manner.

Genus species	Species code	Point number																			
		1	2	3	4	5	6	7	8	9	10	11	12	13	14	15	16	17	18	19	20
Artemisia tridentata	ARTTRI																				
Agropyron spicatum	AGRSPI	⠶	⠔																		
Bromus tectorum	BROTEC				⠔																
Penstemon lemhiensis (reproductive)	PENLEMR						⠢			⠔	⠶										
Penstemon lemhiensis (non reproductive)	PENLEMN								⠔							⠢	⠄				
Bare ground	BARGRO		G	G	G			G	G		G			G	G	G	G	G		G	G
Cryptogam	CRYPT	G				G	G			G		G	G						G		

6. Cover - Line Intercept

A form is included for line intercept sampling of a single species and of communities. The "start" field is the first point on tape where the plant boundary intersects the transect, and the "stop" field is the end of the intersection length. The data form is arranged so that once the data are entered or scanned into a spreadsheet on the computer, rows can be inserted below each "stop" row for calculating the distance of each intercept and the total interception distance along the transect as shown in the example below:

Example of sampling a single species, and recording several transects on a single data sheet:

Transect		Distance on transect											
ID040	start	2.3	5.7	15.4	23.4								
	stop	4.5	5.8	18.2	25.0								
ID031	start	5.3	8.1	24.5									
	stop	6.1	10.3	25.0									
ID041	start	15.5											
	stop	21.3											

Same example with analysis rows added:

Transect Distance	on transect			total														
ID040	start	2.3	5.7	15.4	23.4													
	stop	4.5	5.8	18.2	25.0													
Distance total		2.2	0.1	2.8	1.6													6.7
ID031	start	5.3	8.1	24.5														
	stop	6.1	10.3	25.0														
Distance total		0.8	2.2	0.5														3.5
ID041	start	15.5																
	stop	21.3																
Distance total		5.8																5.8

Example of data entry for sampling communities by line intercept:

Genus/species/Code		Distance on transect															
Penstemon lemhiensis/	start	1.5	8.3	15.6	18.5												
PENLEM	stop	1.7	9.1	16.2	18.6												
Bromus tectorum/	start	1.9	6.8	9.9	19.2												
BROTEC	stop	6.2	9.1	15.1	19.3												
Artemisia tridentata/	start	6.5	17.3														
ARTTRI	stop	6.9	18.1														

7. Field Monitoring Cover Sheet

This data sheet should be completed for all monitoring studies, regardless of which other field data sheets are used. It is critical that the field monitoring cover sheet be completely filled out at the time of data collection to fully capture a description of the field methods. The information included on this form is necessary to insure that comparable data will be gathered during future monitoring intervals.

8. Photographs

Two data sheets are useful for documenting photographs. The photo log provides fields for listing the frame number, describing the position from which the photograph is taken, describing the direction and subject of the photograph, and providing comments. Maintaining a photo log sheet during a monitoring study will greatly facilitate sorting and labeling photographs after the film is returned from the lab.

The second sheet is used as photoboards for identifying film rolls or frames. You can use a completed photoboard as the first exposure of a roll of film to identify that roll. This is especially helpful if you are taking several rolls of film that will have similar subjects (such as riparian areas, or photo plots in grasslands). The photoboard can also be included in the photograph to identify each exposure. *Photocopies of this data sheet should be made on cream or colored paper to avoid glare.*

Density data sheet—single category of counting unit—rectangular quadrats divided into segments

Title/Description: Plant density monitoring in permanent quadrats				Date: 5/15/95—5/25/95		Page 1 of 4	
Location: Willow Creek Preserve (Eugene, OR)				Species: Lomatium bradshawii			
Treatment: Cleared in 1994		Field personnel: Ed Alverson		Quadrat dimensions	Width: 0.25m Length: 50m Segment length: 1m		
Notes:							

Mplot	Quad	Seg	# plants	Mplot	Quad	Seg	# plants	Mplot	Quad	Seg	# plants
3	1	27	1	4	1	4	0	4	6	23	4
3	2	20	1	4	1	5	15	4	6	24	1
3	2	25	1	4	1	6	5	4	6	25	13
3	3	48	10	4	1	7	1	4	6	26	17
3	4	3	1	4	1	8	2	4	6	27	9
3	4	19	4	4	1	22	5	4	6	28	19
3	4	20	0	4	1	23	5	4	6	29	28
3	4	24	5	4	1	24	0	4	6	30	2
3	4	36	3	4	2	46	0	4	6	31	8
3	4	37	1	4	2	47	0	4	6	32	9
3	4	49	0	4	4	40	2	4	6	33	3
3	5	7	3	4	4	41	1				
3	6	3	3	4	4	42	3				
3	6	4	1	4	4	43	3				
3	6	5	1	4	4	44	1				
3	7	5	4	4	3	6	0				
3	7	9	2	4	3	7	1				
3	7	10	1	4	3	8	1				
3	7	13	1	4	3	9	0				
3	7	21	1	4	3	10	0				
3	7	22	3	4	3	11	0				
3	7	23	6	4	3	12	20				
3	7	24	17	4	3	13	16				
3	8	5	4	4	3	14	4				
3	8	10	1	4	5	19	1				
3	8	18	2	4	5	28	2				
3	8	19	2	4	5	29	5				
3	9	16	3	4	5	30	10				
3	9	24	4	4	5	31	2				
3	9	25	27	4	5	32	1				
3	10	29	1	4	6	13	1				
3	10	39	1	4	6	15	1				

Density data sheet—single category of counting unit—rectangular quadrats divided into segments

Title/Description:								Date:		Page____ of ____	

Location:							Species:				

Treatment:				Field personnel:			Quadrat dimensions	Width:_____ Length:_____ Segment length:_____			

Notes:

Mplot	Quad	Seg	# plants	Mplot	Quad	Seg	# plants	Mplot	Quad	Seg	# plants

Density data sheet—3 categories of the counting unit—rectangular quadrats divided into segments

Title/Description: Plant density monitoring in permanent quadrats			Date: 5/15/95—5/25/95		Page 1 of 4	

Location: Agate Desert Preserve (SW Oregon)				Species: Lomatium cookii		

Treatment: To burn in 1996		Field personnel: Darren Borgias		Quadrat dimensions	Width: 0.25m Length: 90m Segment length: 1m	

Notes:

Mplot	Quad	Seg	Seedling	Veg	Flwring	Mplot	Quad	Seg	Seedling	Veg	Flwring
1	1	1	13	2	0	1	2	9	0	3	0
1	1	2	1	0	0	1	2	10	0	2	0
1	1	3	1	6	4	1	2	11	0	1	0
1	1	4	0	8	0	1	2	12	0	1	0
1	1	5	1	7	0	1	2	13	0	5	0
1	1	7	0	3	0	1	2	15	0	2	0
1	1	9	0	4	1	1	2	16	0	3	0
1	1	14	0	5	2	1	2	17	1	3	0
1	1	15	0	1	0	1	2	18	1	1	0
1	1	16	3	6	2	1	2	21	0	2	0
1	1	17	0	1	0	1	2	22	0	4	0
1	1	18	0	1	0	1	2	23	0	5	0
1	1	19	0	1	0	1	2	24	0	1	0
1	1	22	0	2	0	1	2	54	0	1	0
1	1	23	0	2	0	1	2	55	1	2	0
1	1	25	0	0	1	1	2	56	0	3	0
1	1	26	0	1	0						
1	1	27	0	2	0						
1	1	56	0	1	0						
1	1	58	0	1	0						
1	1	60	0	0	1						
1	1	61	0	3	0						
1	1	62	0	1	0						
1	1	63	1	0	0						
1	1	67	0	3	0						
1	1	77	1	2	0						
1	1	81	0	3	0						
1	1	82	5	8	7						
1	1	84	1	0	0						
1	1	87	0	1	0						
1	1	88	1	2	0						
1	2	1	2	0	0						
1	2	2	14	1	1						
1	2	3	16	2	0						

Density data sheet—3 categories of the counting unit—rectangular quadrats divided into segments

Title/Description:					Date:		Page____ of ____

Location:					Species:		

Treatment:		Field personnel:		Quadrat dimensions	Width:_____ Length:_____		Segment length:_____

Notes:

Mplot	Quad	Seg	Seedling	Veg	Flwring	Mplot	Quad	Seg	Seedling	Veg	Flwring

Nested Frequency Field Data Sheet Page _1_ of _1_

Title/Description:		Date: 7 July 98

Location: Kenny Creek Lemhi Penstemon population Species or community: Penstemon lemhiensis

Transect or Macroplot#: IDKØ1 Treatment: Burn Field personnel: Melanie Elzinga

Notes: Nested quadrat sizes: 1 = 10cm x 10cm 2 = 25cm x 25cm 3 = 25cm x 50cm 4 = 50cm x 50cm

CODE	Genus/species	Quad #	1	2	3	4	5	6	7	8	9	10	11	12	13	14	15	16	17	18	19	20
PENLEM 1	Penstemon lemhiensis seedlings		4	0	0	3	0	0	1	0	2	0	0	0	0	4	0	3	0	0	0	3
PENLEM 2	Penstemon lemhiensis repro		3	3	0	0	2	0	0	0	4	0	0	0	1	3	0	0	0	0	0	4
PENLEM 3	Penstemon lemhiensis nonrepro		0	2	0	0	4	0	0	0	4	0	0	4	0	2	0	0	0	0	4	0
BROTEC	Bromus tectorum		1	4	3	0	0	4	2	2	1	2	3	0	0	0	4	3	3	2	0	1
AGRSPI	Agropyron spicatum		2	4	3	3	2	0	0	0	0	3	4	0	0	3	2	0	0	0	0	0
ARTTRI	Artemisia tridentata		0	0	0	0	4	0	0	0	0	0	0	4	0	0	0	0	0	3	0	0
		Quad #	1	2	3	4	5	6	7	8	9	10	11	12	13	14	15	16	17	18	19	20

Nested Frequency Field Data Sheet

Page____ of ____

Title/Description:

Date:

Location:

Species or community:

Transect or Macroplot#:

Treatment:

Field personnel:

Notes:

Nested quadrat sizes: 1 =   2 =   3 =   4 =

CODE	Genus/species	Quad #	1	2	3	4	5	6	7	8	9	10	11	12	13	14	15	16	17	18	19	20
		Quad #	1	2	3	4	5	6	7	8	9	10	11	12	13	14	15	16	17	18	19	20

Cover—Visual Estimation in Plots for Single-Species Sampling Page 1 of 1

Study ID: IDCP 83.98

Species: *Penstemon lemhiensis*

Study Location: Kenny Creek population, approx. 50m east of Kenny Creek turn off. Transects aligned at 95° (true north), sampled area 50 x 100m, marked at corners w/sawed-off fence posts (no paint - look for study markers hanging on each post). Transects 42m long, plots begin at 1m and are positioned every 2m. Individual transects monumented at both ends w/rebar. See drawings & establishment photos in project binder.

Notes: Photo roll #3 (see photo - log sheet). Plants in full bloom. No obvious livestock use this season; a few cow pies from last year. Repainted transect rebars bright orange.

Cover classes: 1: >0 - 1 % 2: 1.1 - 3 % 3: 3.1 - 5 % 4: 5.1 - 15 % 5: 15.1 - 25 % 6: 25.1 - 45 % 7: 45.1 - 65 % 8: 65.1 - 85 %
9: 85.1 - 95 % 10: >95 - % 11: - % 12: - %

Date: 8/13/98 Observers: G & C Elzinga Plot size: 50cm x 50cm

Transect #	Transect coordinates	1	2	3	4	5	6	7	8	9	10	11	12	13	14	15	16	17	18	19	20
CP83.98 - 1	5,10	—	2	—	—	—	4	4	6	1	—	—	1	—	—	8	3	—	—	—	—
CP83.98 - 2	25, 40	—	—	—	—	—	—	—	2	3	1	1	3	—	3	—	—	—	—	—	—
CP83.98 - 3	30, 55	—	—	—	2	—	—	—	—	—	—	—	—	—	3	8	1	—	—	—	—

Plot number

Cover—Visual Estimation in Plots for Single-Species Sampling

Study ID:

Species:

Study Location:

Notes:

Cover classes: 1: _ - _ - _ % 2: _ - _ - _ % 3: _ - _ % 4: _ - _ % 5: _ - _ % 6: _ - _ % 7: _ - _ % 8: _ - _ %

9: _ - _ % 10: _ - _ % 11: _ - _ % 12: _ - _ %

Date:

Plot size:

Observers:

Transect #	Transect coordinates	Plot number																			
		1	2	3	4	5	6	7	8	9	10	11	12	13	14	15	16	17	18	19	20

Cover—Visual Estimation in Plots for Community Sampling

Study ID:		Community:	

Study Location:

Cover classes: 1: ____ - ____ % 2: ____ - ____ % 3: ____ - ____ % 4: ____ - ____ % 5: ____ - ____ % 6: ____ - ____ %
7: ____ - ____ % 8: ____ - ____ % 9: ____ - ____ % 10: ____ - ____ % 11: ____ - ____ % 12: ____ - ____ %

Notes:

Date:	Observers:		Plot size:
Transect #	Transect coordinates		

Genus/species	Species code	Plot number																			
		1	2	3	4	5	6	7	8	9	10	11	12	13	14	15	16	17	18	19	20
Plot #		1	2	3	4	5	6	7	8	9	10	11	12	13	14	15	16	17	18	19	20

Cover—Point Intercept for Species Sampling

Study ID: | Species: | Observers: | Date:

Study Location:

Notes:

Transect #/coordinates	Plot number

Plot number: 1 2 3 4 5 6 7 8 9 10 11 12 13 14 15 16 17 18 19 20 21 22 23 24 25 26 27 28 29 30 31 32 33 34 35 36 37 38 39 40 41 42 43 44 45 46 47 48 49 50

Cover—Point Intercept for Community Sampling

Study:

Study Location:

Study Notes:

Date:

Observer(s):

Community:

Transect coordinates:

Genus - Species	Species Code	Plot number																																																
		1	2	3	4	5	6	7	8	9	10	11	12	13	14	15	16	17	18	19	20	21	22	23	24	25	26	27	28	29	30	31	32	33	34	35	36	37	38	39	40	41	42	43	44	45	46	47	48	49	50

Cover—Line Intercept for Species Sampling																	Page____ of ____

Study ID:

Species:

Study Location:

Notes:

Date:

Observers:

Transect		Distance on transect																
	start																	
	stop																	
	start																	
	stop																	
	start																	
	stop																	
	start																	
	stop																	
	start																	
	stop																	
	start																	
	stop																	
	start																	
	stop																	
	start																	
	stop																	
	start																	
	stop																	
	start																	
	stop																	
	start																	
	stop																	
	start																	
	stop																	
	start																	
	stop																	
	start																	
	stop																	
	start																	
	stop																	
	start																	
	stop																	

Cover—Line Intercept for Community Sampling		Page____ of ____															
Study ID:		**Community:**															
Study Location:																	
Notes:																	
Date:		Observer(s):															
Transect #:		Transect coordinates:															
Genus species/Code																	
	start																
	stop																
	start																
	stop																
	start																
	stop																
	start																
	stop																
	start																
	stop																
	start																
	stop																
	start																
	stop																
	start																
	stop																
	start																
	stop																
	start																
	stop																
	start																
	stop																
	start																
	stop																
	start																
	stop																
	start																
	stop																
	start																
	stop																

Field Monitoring Cover Sheet—Page 1

1. Include information from the header portion of each of the filed data sheets.

Title or project description name:	
Location:	
Dates of data collection:	
Species or community name:	
Type of study (e.g., density, cover, frequency, etc.):	
Personnel:	
Treatment(s):	
Macroplots or transects or other location identifiers:	

2. Management Objective:

3. Sampling Objective:

4. Detailed description of data collection methods: This should include sufficient detail that someone unfamiliar with the project can understand how the data was gathered. Maps and diagrams should be included on the gridded pages at the end of this data form (and see item #5 below). Consider the following questions.

    a. What are the bounds of the population study area?

    b. If you are sampling within macroplots, what is the size and shape of the macroplots and how were they positioned?

Field Monitoring Cover Sheet—Page 2

c. What is the sampling unit (e.g., quadrats, lines, individual plants)?

d. What is the size and shape of the individual sampling units (quadrats, lines)?

e. How are the sampling units positioned in the population of interest?

f. How many sampling units were sampled? How was sample size determined?

g. Are sampling unit positions permanent or temporary? If permanent, what markers and methods were used to insure that positions will be accurately relocated?

h. Describe any boundary rules for plant counts or measurements that occur along the edge of sampling units.

Field Monitoring Cover Sheet—Page 3

i. For density measurements—describe the counting unit (e.g., genet, ramet, stem, flowering stem) and any rules that are used to discriminate among adjacent counting units.

j. For cover measurements—define whether basal or canopy cover is measured and define gap rules. If visual estimates of cover are made in cover classes describe the classes. For point-intercept cover measurements, describe the point diameter and type of tool being used.

k. Include a full description of any codes used on the field data sheets including species acronyms.

5. Location and layout of the study area: Use the gridded pages at the end of this data form to provide a diagram of the study area and study design. If the study area is quite large, you may need separate diagrams drawn at different scales (i.e., one diagram of the entire site, another diagram that shows how sampling units are arranged in a macroplot or along transects). Sketch location, including access. Denote key area, macroplot, or transect locations with macroplot numbers, names, and treatments as applicable and the approximate bounds of the population being studied. If the sampling units are placed along transect lines, show how they were placed. Provide approximate scale.

Field Monitoring Cover Sheet—Page 4

Field Monitoring Cover Sheet—Page 5

Field Monitoring Cover Sheet—Page 5

Photograph Log

| Site: | | Plot: | | Date of Observation: |
| Observer(s): | | Lens: | | Film Type: |

Roll/frame	From	Toward	Description

Site: _____

Plot/Point: _____

Date: _____

Film #: _____

Photographer: _____

APPENDIX 16. Instructions on Using the Programs STPLAN and PC SIZE: CONSULTANT to Estimate Sample Sizes and Conduct Post Hoc Power Analyses

This appendix gives instructions on using two easily obtained computer programs to calculate sample size and conduct post hoc power analyses. STPLAN, currently in version 4.1, is available free over the Internet. PC SIZE: CONSULTANT is available over the Internet as shareware (the author requests a payment of $15.00 if you find the program useful). Directions on obtaining these two programs are given in Chapter 11, Section K. Both programs run in DOS on IBM and compatible personal computers.

STPLAN will calculate sample size, power, and minimum detectable difference for all of the two-sample statistical tests discussed in this Technical Reference. It will not, however, calculate the sample size required to estimate a mean with a given level of precision. PC SIZE: CONSULTANT will calculate sample size and power (but not minimum detectable change) for most of the two-sample tests discussed in this Technical Reference. It also calculates the sample size required to estimate a mean with a given level of precision. Because STPLAN is free and more flexible (in that it allows you to calculate minimum detectable change), this appendix emphasizes that program. Thus, except for estimating a mean, where the instructions describe how to use PC SIZE: CONSULTANT, all of the instructions are for STPLAN only. Neither program allows you to calculate the sample size required to estimate a single proportion with a specified level of precision. For this you will have to use either Equation #4 of Appendix 7 or another program.

Remember that both programs assume the population from which you are sampling is infinite. If your population is finite (as it will be if you are using quadrats, and you position these using the grid-cell method or some other method that ensures quadrats do not overlap) and if you are sampling more than 5% of your population, you should apply the finite population correction factor to your estimates of sample size. See Appendix 7 for instructions on how to do this for each of the types of sample size calculations discussed below.

A. Calculating the Necessary Sample Size for Estimating a Single Population Mean or a Single Population Total with a Specified Level of Precision

The sample size needed to estimate confidence intervals that are within a given percentage of the estimated *total population size* is the same as the sample size needed to estimate confidence intervals that are within that percentage of the estimated *mean value*. The instructions below assume you are working with a sample mean.

1. Go to the DOS prompt.
2. Start PC SIZE: CONSULTANT (program is SIZE.EXE)
3. At the main menu select: (C) Confidence Interval Calculations
4. Select: (1) Estimating a single mean
5. At prompt "Enter the level of confidence for the interval" enter the confidence level you desire (e.g., 0.95 for a 95% confidence interval).
6. At prompt "Enter the required length of the confidence interval" enter the total length of the confidence interval between the upper and lower confidence limits. For example, for ± 5 plants you would enter 10. The number has to be in absolute terms, so, for example, if you want your confidence interval length to be within 30% of your sample mean and your sample mean is 10 plants/quadrat, then the absolute confidence interval is ± (0.30 x 10) = ± 3 plants. You would then enter 6 for the total length.
7. At prompt "Enter the probability that the interval be no longer than the required length" enter the same number as you entered for the level of confidence for the interval (e.g., enter 0.95 for a 95% confidence interval).[1]
8. At prompt "Would you rather provide the within group (S) standard deviation or an (I) interval that contains a specified percentage of the group's measurements?" enter (S).
9. At prompt "Enter the estimate of standard deviation" enter the estimate you have obtained from pilot sampling. For example, if your estimate of S was 7 plants/quadrat, enter 7.
10. The program then summarizes all the information you have entered and then prompts "Press enter to continue." Press enter for the results.
11. At prompt "Do you wish to generate a report? (Screen, File, No), press S, F, or N, accordingly. If you press "F" you will be prompted for a file name.
12. At prompt "Do you wish to continue?" press Y or N accordingly.

We'll use an example to make this procedure clearer. The example used here is the same as that used in Appendix 7, Sample Size Equation #1. Our sampling objective is to estimate the mean density of species Y in population Z in 1993 and to be 95% confident that this estimate is within 20% of the true mean density. From our pilot sampling data we have calculated a mean of 25 plants/quadrat and a standard deviation of 7 plants. Here is how we would enter these data into PC SIZE: CONSULTANT. The numbers below correspond to the steps listed above. Follow steps 1 through 4 as given above. We then enter the data from our example beginning with step 5:

5. Enter 0.95 for the 95% confidence interval.
6. We now need to enter the length of the confidence interval. Since we want our estimate to be within 20% of the true value, we multiply 20% times our estimate of the true mean density of 25 plants/quadrat: 0.20 x 25 = 5 plants/quadrat. Remember that this is 25 ± 5 plants/quadrat and that the program wants the total confidence interval length. In this case the lower confidence limit is 25 - 5 = 20, and the upper confidence limit is 25 + 5 = 30. The total confidence interval length is therefore 30 - 20 = 10. We enter 10 into the program.
7. Enter 0.95 for the probability that the interval be no longer than the required length.
8. Enter S to provide the standard deviation.
9. Enter 7 for the estimate of the standard deviation.
10. Review the summary of the data you have entered and press Enter.
11. The program displays the following:

---

[1] This is the "tolerance probability" of Kupper and Hafner (1989) that corrects for the underestimation of sample size resulting from the use of the usual sample size formula. Note that since this correction is factored into the PC SIZE: CONSULTANT program, you do *not* have to correct the program's sample size estimate like you do when you calculate sample size with Equation #1 of Appendix 7.

Sample size	Probability (interval is no longer than requested)
8	.35344
9	.45342
10	.56024
11	.66571
12	.76155
13	.84139
14	.90216
15	.94431
16	.97088

The sample size that meets your objective is 16 quadrats. This is 1 quadrat more than the 15 calculated using the procedure described in Appendix 7, Equation #1. Note, however, that the probability of 0.94431 that will be realized with 15 quadrats is very close to meeting the desired probability of 0.95. Using 16 quadrats is the conservative approach and will result in a probability of 0.97088.

B. Detecting Differences Between Two Means with Temporary Sampling Units

1. Go to the DOS prompt.
2. Start the program STPLAN (program is STPLAN.EXE).
3. You will be asked to enter a number corresponding to the following:
   (1) To print only answers to the screen
   (2) To print full problem report including answers to the screen
   (3) As (2) but also write to a file

   Choose (3) if you want to keep a record of your work. Once you've chosen (3) you will be prompted for a file name. If you choose the same file name you used last time you can append to the file or overwrite it.

4. You will now be at the main menu. Select (3) NORMALly distributed outcomes.
5. Select (2) Two-sample with equal variances (exact method).
6. Enter 1 or 2 for 1- or 2-sided test. (If your sampling objective calls for detecting change in either direction—an increase or decrease—choose the two sided test; if you're only interested in detecting change in one direction—e.g., a decrease—choose the one-sided option.)
7. Now you need to input the information necessary to perform the required operation. You are prompted to enter:

   1: Group 1 mean.
   2: Group 2 mean.
   3: Number in group 1.
   4: Number in group 2.
   5: Standard deviation of both groups.
   6: Significance level.
   7: Power.

You enter a "?" for the item you want calculated. Entering "?"s in positions 3 and 4 will calculate equal sample sizes for both groups. These values can be placed on a single line as long as they are separated by commas or spaces.

Sample size calculation. Let's say we want to calculate the sample size necessary to detect a 30% change with a significance level of 0.05 and a power of 0.95. Our pilot data yields a mean of 10 plants/quadrat and a standard deviation of 5 plants/quadrat. Because we want to detect a 30% change, the absolute change is 30% x 10 plants/quadrat = 3 plants/quadrat. We've decided that we're interested in a 30% change in either direction (up or down). Therefore we've already selected the two-sided option. Here are the values we enter into STPLAN (remember, they're in the same order shown in step 7 above):

10   13   ?   ?   5   0.05   0.95

Note that we've entered "?"s in spaces 3 and 4, in order to make the program calculate the equal sample sizes required at both time periods. Also note that we've entered "13" in space 2 for the Group 2 mean (this is the mean at the second time monitoring period, derived by adding 3 to the first year mean of 10; since we're interested in detecting changes in either direction we could have entered 7 here (10-3) and obtained the same answer).

STPLAN returns the answer 73.284 as the required sample size at both sampling times. We round this up to 74 quadrats.

Power analysis. If we've already collected two years' worth of data, we should conduct a post-hoc power analysis to look at how likely our study was to detect a change we deem to be biologically important. There are two ways of doing this. One is to specify the biologically important change and then look at the power value. The other is to specify the power value and look at the size of the minimum change that can be detected. We'll describe both methods below.

Solving for power. Let's say we've collected two years' worth of data. Our sample size in both years was 50 quadrats. We've determined that a change of 30% (in either direction) is biologically important. Since our first year's mean was 10 that means we want to detect a change of ±(10 x 0.30) = 3 plants. The standard deviation of the first year's data was 6 plants, and the standard deviation of the second year's data was 4 plants. We need to pool these two standard deviations to come up with the one we will plug into STPLAN. We do this by applying the following formula:

$$S_{pooled} = \sqrt{\frac{(s_1^2 + s_2^2)}{2}}$$

Where:

s_1 = Standard deviation of sample 1
s_2 = Standard deviation of sample 2

Applying this formula to our example we get:

$$S_{pooled} = \sqrt{\frac{(6^2 + 4^2)}{2}} = 5.1$$

Thus, we plug 5.1 into the position for standard deviation. Our sampling objective set the significance level at 0.05 and the power at 0.95. We want to see how close the actual power is to our objective of 0.95. Here is what we enter into STPLAN:

10   13   50   50   5.1   0.05   ?

STPLAN solves for power in this case and returns a value of 0.829. Since this is less than our objective power of 0.95 we decide to step up our monitoring by adding more quadrats in future years (we can use the sample size procedure described above to determine the number of quadrats necessary). Note an important thing about this example. Even though we've already estimated a mean for the second year, we do *not* enter the second year mean value into the program. Such a power analysis is meaningless (even though some text books and other power analysis programs do exactly that—see Thomas and Krebs 1997). Rather, you enter the value of the mean that you have determined will represent a biologically important change (in this case, since the biologically important change is ±3 plants we enter either 13 or 7 as the Group 2 mean—either value will result in the same answer).

Solving for minimum detectable change. Now let's take the same example, but in this case we specify the objective power value and ask STPLAN to solve for the minimum detectable change. Here are the values we enter:

10   ?   50   50   5   0.05   0.95

STPLAN returns the value 13.645. Thus, the minimum detectable change is 13.645 - 10 = 3.645 plants. Because this is greater than the 3 plants we set as the biologically important change we wanted to detect, we decide to step up our monitoring in future years to meet this objective.

C. Detecting differences between two means using paired or permanent sampling units

We'll use the program STPLAN to perform the necessary calculations for detecting change between two means using paired or permanent sampling units. To do this we will select the "one-sample" option in STPLAN, because a paired *t* test (the test we use to determine if there is a statistically significant difference between the means of paired or permanent sampling units) is the same as a one-sample *t* test comparing the results of a single sample to some predetermined value, as long as the mean corresponding to the null hypothesis (no difference) is 0. This should become clear in the examples below.

1. Go to the DOS prompt.
2. Start the program STPLAN (program is STPLAN.EXE).
3. You will be asked to enter a number corresponding to the following:
    (1) To print only answers to the screen
    (2) To print full problem report including answers to the screen
    (3) As (2) but also write to a file

Choose (3) if you want to keep a record of your work. Once you've chosen (3) you will be prompted for a file name. If you choose the same file name you used last time you can append to the file or overwrite it.

4. You will now be at the main menu. Select (3) NORMALly distributed outcomes.
5. Select (1) One-sample (exact method).
6. Enter 1 or 2 for 1- or 2-sided test. (If your sampling objective calls for detecting change in either direction—an increase or decrease—choose the two sided test; if you're only interested in detecting change in one direction—e.g., a decrease—choose the one-sided option.)
7. Now you need to input the information necessary to perform the required operation. You are prompted to enter:

    1. Null hypothesis mean. (For paired or permanent sampling units this will always be 0.)

2. Alternative hypothesis mean. (This will correspond to the absolute difference you wish to detect; for example, if you want to detect a change of 2 plants/quadrat, the number you enter is 2.)
3. Standard deviation. (This is the standard deviation of the set of *differences* between paired or permanent sampling units, not the standard deviation of the data sets at either time period.)
4. Sample size.
5. Significance level.
6. Power.

You enter a "?" for the value you want calculated. Following are directions on calculating sample sizes and conducting post-hoc power analyses using an example set of data. The example used is the same as that given in Appendix 7, Sample Size Equation #3. Because we are interested in change in either direction, we have already chosen the two-tailed option in STPLAN.

Sample size calculation. Let's say we're interested in detecting a change of 20% in the mean density of species F at Site Q between 1994 and 1995. We want to be 90% certain of detecting that change, if it occurs, and we are willing to accept a 10% chance of making a false-change error (concluding that a change has occurred when it really has not). We measure permanent quadrats in each of two years. The mean of the first year is 4.20 plants/quadrat, while the mean of the second year is 7.80 plants/quadrat. The standard deviation of the difference is 1.67 (see the table in Appendix 7, Sample Size Equation #3, for an example of how this standard deviation is calculated). Our sampling objective of detecting a 20% change must be framed in absolute terms. To derive this value we multiply the *larger* of the two year's means by 20%. In our example the second year's mean of 7.80 plants/quadrat is the larger, so the minimum detectable change we wish to detect is therefore 7.80 x 0.20 = 1.56 plants/quadrat. This is the value we enter as our alternative hypothesis mean. Following are the values we enter into STPLAN to calculate sample size:

0   1.56   1.67   ?   0.10   0.90

STPLAN returns the value 11.521. We round this up to 12 and conclude that we should establish 12 permanent quadrats.[2]

Power analysis. We've collected two years worth of data using 5 permanent quadrats, and we wish to know whether we've met the objective given under sample size calculation, above (a minimum detectable change of 20% at a significance level of 0.10 and a power of 0.90). The means and standard deviation of the difference are the same as given in the previous example.

Solving for Power. Here are the values we enter in STPLAN to calculate the power of our design:

0   1.56   1.67   5   0.10   ?

STPLAN returns the value 0.484, so it's clear that a biologically important change could have taken place that we have missed with our design. We decide, after conducting the sample size

---

[2] Note that this sample size of 12 is slightly larger than the 10 that was derived in the example using the sample size formula of Equation #3 of Appendix 7. The difference is due to the fact that Equation #3 uses a normal approximation, while STPLAN uses an exact (but much more complex) method. The STPLAN value is the more reliable of the two.

analysis described previously, to add another 7 permanent quadrats, bringing our total sample size to 12.

Solving for minimum detectable change. Here are the values we enter in STPLAN to calculate the minimum change we can detect given a significance level of 0.10 and a power of 0.90:

0   ?   1.67   5   0.10   0.90

Once we enter the data and press the Enter key, STPLAN asks us the following question: "Is null hypothesis mean less than the alternative hypothesis? (y,n)." What we enter here dictates whether STPLAN will give us the alternative hypothesis mean as a positive or negative number. If we enter "n" STPLAN returns the value 2.737; if we enter "y" it returns the value -2.737. Since we've chosen the two-tailed option, all we're interested in is the absolute value [2.737], which corresponds to the minimum detectable change so we can enter either "y" or "n". We compare the value of 2.737 plants/quadrat to our objective of 1.56 plants/quadrat and decide that our design is insufficient to meet our objective. We calculate the sample size needed (as in the example already given) and add a additional 7 permanent quadrats.

D. Detecting differences between two proportions with temporary sampling units

Here is how to calculate sample size and conduct post-hoc power analyses for frequency data and point cover data, when the frequency quadrats or cover points are the sampling units and these are positioned randomly in each year of measurement.

1. Go to the DOS prompt.
2. Start the program STPLAN (program is STPLAN.EXE).
3. You will be asked to enter a number corresponding to the following:
    (1) To print only answers to the screen
    (2) To print full problem report including answers to the screen
    (3) As (2) but also write to a file

Choose (3) if you want to keep a record of your work. Once you've chosen (3) you will be prompted for a file name. If you choose the same file name you used last time you can append to the file or overwrite it.

4. You will now be at the main menu. Select (1) BINOMIALly distributed outcomes.
5. Select (2) Two-sample (arcsin approximation).
6. Enter 1 or 2 for 1- or 2-sided test. (If your sampling objective calls for detecting change in either direction—an increase or decrease—choose the two sided test; if you're only interested in detecting change in one direction—e.g., a decrease—choose the one-sided option.)
7. Now you need to input the information necessary to perform the required operation. You are prompted to enter:

    1. Probability of an event in group 1.
    2. Probability of an event in group 2.
    3. Number of trials in group 3.
    4. Number of trials in group 4.
    5. Significance level.
    6. Power.

Let's say our sampling objective (the same as that given in the example of Appendix 7, Sample Size Equation # 5) is to detect an absolute change of 20% frequency of species F at Site G between 1993 and 1994. We want to be 90% certain of detecting that change, if it occurs, and we are willing to accept a 10% chance that we will make a false-change error (concluding that a change took place when it really didn't). We are interested in detecting a change in either direction so we have already chosen the two-tailed option in STPLAN.

We take a pilot sample, and based on this sample we estimate the proportion of quadrats with species Z in 1993 to be 0.65. Thus, the probability of an event in group 1 (the first number we must enter into the STPLAN program) is 0.65. Because we wish to detect an absolute change of 20% or 0.20, we subtract 0.20 from 0.65 and assign the probability of an event in group 2 the number 0.45. The reason we selected a decline instead of an increase (e.g., from 0.65 to 0.85) is because sample size requirements are higher at the mid-range of frequency values (i.e., closer to 0.50) than they are when closer to 0 or 1.00. Sticking closer to the mid-range givens us a more conservative sample size estimate.

Following are the values we enter into STPLAN to solve for sample size for this example:

0.65   0.45   ?   ?   0.10   0.90

STPLAN returns the value 104.493, which we round up to 105. Thus, we need to randomly place 105 quadrats in both years of measurement to meet our sampling objective.[3]

Power analysis. We're examining a study that has already been conducted to determine if it meets the same objectives used in our example above (capable of detecting an absolute change in proportion of 0.20, with a significance level of 0.10 and a power of 0.90). The study used 75 quadrats at each time of measurement. The estimated proportion in the first year was 0.65 and in the second year it was 0.70. What is the probability this design would detect a change of 0.20?

Solving for power. We use the estimated proportion of only one of the years in our calculation. Remember that the closer the proportion to the mid-range of frequency values (0.50) the more conservative the estimate of sample size. This holds true for power as well. Therefore we take the first year's estimate of 0.65 to plug into STPLAN (if the second year's estimate had been 0.55, we would have taken the second year's value). Here are the values we enter into STPLAN:

0.65   0.45   75   75   0.10   ?

Note that we use only one of the actual estimated proportions (the one closest to 0.50). We enter this as the Group 1 probability. We do *not* enter the second estimated proportion as the Group 2 probability, even though we have that value. Such a power analysis is meaningless (even though some text books and other power analysis programs do exactly that—see Thomas and Krebs 1997). Rather, we take the absolute value we have determined to represent a biologically important change (0.20) and either add it or subtract it from the Group 1 probability to come up with a Group 2 probability (whether we add or subtract the value depends on which action brings the Group 2 value closer to the mid-range value of 0.50).

---

[3] This is slightly higher than the estimate of 102 quadrats derived from Equation #5, Appendix 7. STPLAN uses a more complicated formula that corrects for values that are away from the mid-range of proportions. The STPLAN estimate is therefore more reliable than Equation #5.

STPLAN returns the value 0.798. Thus, the power of our test to detect the desired magnitude of change is less than our objective power of 0.90. Stated another way, there is about an 80% probability we would detect an absolute change in proportion of 0.20 at a significance level of 0.10. We therefore conclude that this design has not met the objective of 90% power. We can either add more quadrats or change the objective.

Solving for minimum detectable change. The alternative method of power analysis is to specify the objective power (0.90) and solve for the minimum change that can be detected. It would appear that we would simply enter the following values into STPLAN:

0.65   ?   75   75   0.10   0.90

If we do this, however, STPLAN returns the value 0.853. If we subtract the beginning proportion, 0.65, from this value we obtain 0.853 - 0.65 = 0.203, close enough to our value of 0.20 that it would appear we have met our objective. Remember, however, that sample size, power, and minimum detectable change values are less conservative the farther the proportions used to calculate these are from the mid-point value of 0.50. STPLAN returned a value of 0.853 for the Group 2 proportion. This value is quite a distance from 0.50. We'd like to tell STPLAN to give us a Group 2 proportion that is *less than* the Group 1 proportion of 0.65, so that we can be sure the minimum detectable change value is conservative. Unfortunately, there is no option in STPLAN to do this. There is a solution, however. We can take the complement of 0.65, which we derive by subtracting 0.65 from 1, and plug this value into STPLAN. That value is 1 - 0.65 = 0.35. For proportions, calculations performed on complements give the same results as calculations on the original values, except that using the complement of 0.35 in STPLAN forces the program to solve for a higher Group 2 proportion (and one that is therefore closer to 0.50). Here are the values we enter:

0.35   ?   75   75   0.10   0.90

This time, STPLAN returns the value 0.586. Thus the minimum detectable change is 0.586 - 0.35 = 0.236. Since this value is greater than our objective of 0.20, we conclude that this design does not meet our objective. We'll have to either add more quadrats or change our study objectives.

E. Detecting the difference between two proportions with permanent sampling units

Here is how to calculate sample size and conduct post-hoc power analyses for frequency data and point cover data, when the frequency quadrats or cover points are the sampling units and these are positioned randomly in the first year and the same quadrats or points are read in the second year of measurement.

1. Go to the DOS prompt.
2. Start the program STPLAN (program is STPLAN.EXE).
3. You will be asked to enter a number corresponding to the following:
    (1) To print only answers to the screen
    (2) To print full problem report including answers to the screen
    (3) As (2) but also write to a file

Choose (3) if you want to keep a record of your work. Once you've chosen (3) you will be prompted for a file name. If you choose the same file name you used last time you can append to the file or overwrite it.

4. You will now be at the main menu. Select (1) BINOMIALly distributed outcomes.
5. Select (5) Matched pairs test for equality of proportions (method of Miettinen).
6. Enter 1 or 2 for 1- or 2-sided test. (If your sampling objective calls for detecting change in either direction—an increase or decrease—choose the two sided test; if you're only interested in detecting change in one direction—e.g., a decrease—choose the one-sided option.)
7. Now you need to input the information necessary to perform the required operation. You are prompted to enter:

   1. Delta to be detected.
   2. Sample size.
   3. Significance level.
   4. Power.

Delta is the actual change in frequency you are interested in detecting specified *as a proportion* (e.g., plug in 0.20 for detecting a 20% change). Let's say our sampling objective calls for detecting a 20% change in either direction with $\alpha = 0.10$ and power = 0.90. We have therefore already selected the two-tailed option in STPLAN.

Sample size calculation. Here is what we enter into STPLAN to calculate sample size:

0.20   ?   0.1   0.9

8. The program then asks "Have you collected data from a preliminary sample? (y/n)." Specify "y."
9. The program then prompts you: "Enter Z11, Z10, Z01, Z00."

These refer to quadrat transitions and they can either be entered as actual counts *or* as proportions. The Z-code corresponds to the following matrix:

		Year 2	
		Present (1)	Absent (0)
Year 1	Present (1)	Z11	Z10
	Absent (0)	Z01	Z00

For example, let's say we've measured Species X in 100 permanent frequency quadrats in each of two years. We come up with the following data:

		Year 2	
		Present (1)	Absent (0)
Year 1	Present (1)	30	20
	Absent (0)	0	50

Thus, 30 of the 100 quadrats had Species X present in both the first year and second year, 20 of the quadrats that had Species X the first year no longer had it the second year, 0 quadrats that didn't have Species X the first year had it the second year, and 50 quadrats that had no Species X the first year still did not have it the second year.

Here are the Z values we enter into STPLAN:

30   20   0   50

(The same results will be obtained with: 0.30   0.20   0   0.50)

STPLAN then generates the sample size required and returns the number 24.597. We round this up to 25 quadrats.

Correction for sampling finite populations. Remember that if you sample more than 5 percent of the population, you should apply the finite population correction factor to the estimate of sample size you obtain from STPLAN. This is accomplished by use of the following formula:

$$n' = \frac{n}{(1 + (n/N))}$$

Where:

N = the total number of possible quadrat locations
n = the uncorrected sample size
n' = the FPC sample size.

Power analysis. We'll use the same objective and same example data set as given under sample size calculation, above. We've measured Species X in each of two years in 100 permanent frequency quadrats.

Solving for power. To solve for power we enter the following values into STPLAN:

0.20   100   0.1   ?

STPLAN then prompts you to enter Z11, Z10, Z01, and Z00, so we enter the values corresponding to our data:

30   20   0   50

STPLAN then returns a value of 1.000 meaning that we are 100% certain to detect a change of 0.20 with this design (i.e., our power is 100%). This is an indication that we are oversampling (remember that our objective power level was 90% and that the sample size necessary to meet that objective was calculated to be 25 permanent quadrats; we, on the other hand, have sampled with 100).

Solving for minimum detectable change. To solve for minimum detectable change we enter the following values into STPLAN:

?   100   .1   .9

We are then prompted for the Z11, Z10, Z01, and Z00 values and enter:

30   20   0   50

STPLAN returns the value 0.093, meaning that this design will detect an absolute change in frequency of this amount at a significance level of 0.10 and a power of 0.90. This is a much smaller change than our objective of 0.20.

Literature Cited

Thomas, L.; Krebs, C.J. 1997. A review of statistical power analysis software. Bulletin of the Ecological Society of America 78(2): 128-139.

APPENDIX 17. A Procedure to Compare the Efficiency of Different Quadrat Sizes and Shapes Using Pilot Sampling

One way to compare different quadrat sizes and shapes during pilot sampling is to subdivide a larger quadrat into separate sections and record counts separately for each section. For example, after examining the spatial distribution of a particular species, you can't decide whether to use a 0.25m width or a 0.5m width and whether the quadrat should be 25m or 50m long. Randomly locate some initial number (e.g., 10) of 0.5m x 50m quadrats in the population of interest.[1] Attach one end of a 50m tape to a pin or stake, pull it tight and treat one edge of the tape as the center of your quadrat. Count all plants that are within 25cm of either side of the tape edge (total width = 0.5m) and record separately, by side, on a field data sheet, similar to the one in Figure 1. You can also subdivide the long dimension of the quadrat and record plant counts separately within each segment (e.g., every meter) along your tape. This enables you to look at the performance of quadrats of different lengths.

You can save space by recording the segment number only if you have actual plant counts for that segment. For example, you've laid out your tape and started searching along both sides of the tape. You find your first plants (three of them on the left side and two of them on the right side in the third segment of the tape (between 2m and 3m along the tape). The next plants (two of them on the left side, none on the right) are found in the seventh segment (between 6m and 7m along the tape). The entries on the field data sheet would look like Table 1.

plot #	segment #	plant counts		
		left	right	total
1	3	3	2	5
1	7	2	0	2

TABLE 1. Entries on a field data sheet when plants are found in the third and seventh segments of plot number 1.

Continue this counting and recording procedure until all your preliminary quadrats have been sampled. Now you can use a hand calculator to calculate means and standard deviations for different size and shape quadrats. To compare quadrats of different sizes you should calculate the coefficient of variation (CV) for each quadrat size. The CV is calculated as follows:

$$CV = \frac{S}{\overline{X}}$$

Where: $\overline{X}$ = The sample mean
S = The sample standard deviation

---

[1] When randomly locating this initial number of quadrats, position the quadrats according to the design you plan to use for the study (designs are covered in Chapter 7, Section E). You should also make sure the quadrat lengths that you test are even multiples of the total length of the one dimension of the area you intend to sample. For example, if you plan to sample a 50m x 100m macroplot, with the long side of the quadrats parallel to the 50m side, you might test quadrat lengths of 1m, 2m, 5m, 10m, 25m, and 50m. If you follow these guidelines, you may be able to use the data from this initial test as part of your actual sample.

| Site: | | Date: | Page | of |

Observer: _____ Total quadrat length: _____ Segment length: _____

Total width of quadrat: _____ Width of left side: _____ Width of right side: _____

plot #	segment #	plant counts			plot #	segment #	plant counts		
		left	right	total			left	right	total

APPENDIX 17—FIG. 1. A field data sheet to help determine optimum quadrat size and shape.

Unlike the standard deviation, which has a magnitude dependent on the magnitude of the data, the coefficient of variation is a *relative* measure of variability. Thus, coefficients of variation from different sampling designs can be compared. The smaller the coefficient of variation the better. If two designs have similar coefficients of variation, choose the design that will be easiest to implement.

If, after evaluating the performance of different quadrat sizes, you select a size and shape that was some subcomponent of the larger quadrat sampled, you can still use the data as part of your first year's set of data. To do this you should randomly select the subcomponent from each of your pilot quadrats. Using the previous example, if you elected to use a 0.25m x 50m quadrat, you could randomly select one half of each of the 0.5m x 50m quadrats that you sampled as part of your pilot effort.

APPENDIX 18. Estimating the Sample Size Necessary to Detect Changes Between Two Time Periods in a Proportion When Using Permanent Sampling Units (Based on Data from Only the First Year)

Appendix 16 describes how to use the program STPLAN to calculate sample size when you are estimating a proportion using permanent sampling units, and you have two years of data. The following discussion applies to the situation where you want to estimate the sample size necessary under this sampling design, but you have data from only the first year. This would be the type of estimation you would need to do when planning a study using permanent frequency quadrats or permanent cover points, when the quadrats or the points are treated as the sampling units.

In a permanent frequency design, you track changes in presence/absence in each of the permanent quadrats. There are four possible quadrat transitions, as shown in the following matrix.

		Year 2	
		Present (1)	Absent (0)
Year 1	Present (1)	1:1	1:0
	Absent (0)	0:1	0:0

Without two years of permanent quadrat frequency data, you can confidently predict that the permanent quadrat sample size is likely to be less than the temporary quadrat sample size (which can be calculated with a single year of data), but how much less? This problem is further compounded in plant community monitoring projects where the sample size to detect some specified magnitude of change is likely to be quite different for different species.

One approach to calculating permanent frequency quadrat sample sizes is to determine the desired minimum detectable change size and then create a range of hypothetical population changes that could lead to such a magnitude of change. Sample sizes can be calculated for this range of population changes and used as a guide to guessing the "right" sample size to use.

Consider the following example where we specify a minimum detectable change of 10%, with a starting frequency of 50%. What are some different ways that this frequency change could occur? For simplicity's sake, let's work through our example assuming a sampling design with 100 quadrats (the actual numbers don't matter, only the ratio between them). In the first year, we have 50 quadrats with plants and 50 quadrats without plants. In the second year, we want to be able to detect a shift to either 40 quadrats now having plants or to 60 quadrats now having plants. Let's stick with the 40 quadrats with plants scenario. What are some different ways that a 50% to 40% frequency change could occur?

We could simply lose plants from 10 of our previously occupied quadrats and not have any new plants show up in previously unoccupied quadrats. This would create the quadrat transitions shown below. These quadrat transitions can be used along with the 0.10 minimum detectable change and some specified power and false-change error rate (power = 0.90 and false change error rate = 0.10 in the following examples) in the STPLAN program. (The finite population correction factor was not applied to any of the sample sizes in the examples listed below.)

1:1	1:0	0:1	0:0
40	10	0	50

Permanent quadrat sample size	51
Temporary quadrat sample size	423

Alternatively, we could obtain a 10% change in frequency by losing plants from 15 of the originally occupied quadrats and gaining plants in 5 quadrats that did not previously have plants. Here are the transitions and sample sizes.

1:1	1:0	0:1	0:0
35	15	5	45

Permanent quadrat sample size	156
Temporary quadrat sample size	423

We could look at any number of these hypothetical population changes and compare permanent vs. temporary quadrat sample sizes. Table 1 lists 9 different population changes (including the two listed above) that all lead to 10% declines in percent frequency. Each new row shows the result of 5 additional quadrats losing plants that had plants in the original population.

Quadrat transitions				Permanent quadrat sample size	Temporary quadrat sample size	Difference in sample size
1:1	1:0	0:1	0:0			
40	10	0	50	51	463	412
35	15	5	45	156	463	307
30	20	10	40	247	463	216
25	25	15	35	335	463	128
20	30	20	30	422	463	41
15	35	25	25	508	463	-45
10	40	30	20	595	463	-132
5	45	35	15	681	463	-218
0	50	40	10	767	463	-304

TABLE 1. Population changes that lead to a 50% to 40% change in frequency based upon 100 quadrats.

Table 2 shows the same type of information as Table 1 except now the magnitude of change is 20% and the frequency changes from 50% to 30%.

Quadrat transitions				Permanent quadrat sample size	Temporary quadrat sample size	Difference in sample size
1:1	1:0	0:1	0:0			
30	20	0	50	25	101	76
25	25	5	45	53	101	48
20	30	10	40	77	101	24
15	35	15	35	100	101	1
10	40	20	30	123	101	-22
5	45	25	25	145	101	-44
0	50	30	20	167	101	-66

Table 2. Population changes that lead to a 50% to 30% change in frequency based upon 100 quadrats.

With these sorts of tables in hand, we can evaluate the likelihood of various types of changes occurring and select a sample size that fits. For example, how likely is the following transition (taken from the middle line of Table 2)?

1:1	1:0	0:1	0:0
15	35	15	35

In this transition, the number of quadrats that had plants present in both years (15) is equal to the number of quadrats that were empty in the first year but gained at least one plant in the second year (15). This implies that quadrats having plants in them during the first year are no more likely to have plants during the second year than quadrats that did not have plants present in the first year. You could get this sort of result when the between-year correlation in quadrat counts is close to zero. How likely is this sort of result? For most plant species it is probably highly unlikely given typical patterns of sexual and asexual reproduction in plants (but perhaps it occurs in some situations where allelopathy is operating). Even many (most?) annual plants set a higher proportion of seed close to the parent plant location than at distances far from the parents.

The last transition listed in Table 2 (0-50-30-20) shows a permanent quadrat sample size penalty since it would take 167 permanent quadrats as compared to 101 temporary quadrats to detect this particular 20% change. In this transition, all 50 quadrats with plants in the first year lose their plants and 30 of the 50 quadrats that were previously empty gain plants. This would indicate a strongly negative correlation in plant counts between years and there would have to be a strong degree of allelopathy to create this sort of transition. Few, if any, plant species would be suspected of showing this type of response.

By developing an ecological model for your target plant species that describes the types of transitions likely to occur between years, you can therefore arrive at an initial estimate of the sample size necessary to detect a particular level of change based on the first year's data. You can then use the program STPLAN to calculate the sample size based on this model (see Chapter 11 —Statistical Analysis, Section K, for information on obtaining this program and Appendix 16 for instructions on how to use it). After you've collected the second year's data, you should use STPLAN to recalculate sample size based on the actual transitions.

APPENDIX 19. Instructions on Using the Program, NCSS PROBABILITY CALCULATOR, to Calculate P Values from Test Statistics that have been Corrected with the Finite Population Correction Factor

Chapter 11, Section F, gave directions on how to apply the finite population correction factor (FPC) to the test statistics (t, χ^2, and F) calculated by a computer program or the standard significance test formulas. Once these test statistics have been corrected by the FPC, it is necessary to calculate a P value corresponding to the new, corrected test statistic. Although tables (t, χ^2, and F) can be used for this purpose, looking up P values in these tables is difficult and inexact because it requires you to interpolate between values in the tables. A more exact and convenient method is to use the computer program, NCSS PROBABILITY CALCULATOR, which is available as freeware from NCSS Statistical Software. Directions on obtaining this program, which runs under Microsoft Windows 3.1 and higher, are given in Chapter 11, Section L. This appendix gives instructions on using NCSS PROBABILITY CALCULATOR to calculate P values for the t, χ^2, and F statistics. These instructions assume you have installed the NCSS PROBABILITY CALCULATOR on your computer according to the directions given by NCSS.

A. Calculating the P value for a given t statistic

Start up the program, NCSS PROBABILITY CALCULATOR. Select the Student's T Probability Distribution. (There are two buttons you can click on to go to the Student's T Probability Distribution. You should click on the right button, which brings up the "input" screen. Clicking on the left button brings up the "inverse input screen," which calculates the t value associated with a given P value.)

Once you have accessed the Student's T input screen, you enter values in three places on the left side of the screen. Under the heading "1<=DF," you enter the degrees of freedom appropriate to your test. If your t statistic resulted from an independent–sample t test, the appropriate number of degrees of freedom would be $(n_1 - 1) + (n_2 - 1)$, where n_1 is the first year's sample size and n_2 is the second year's sample size. If your t statistic resulted from a paired t test, the values analyzed are the observed changes in each permanent quadrat. The sample size, n, is the number of permanent quadrats, and the appropriate number of degrees of freedom is n – 1.

Under the heading "0<=NCP," you enter 0 ("NCP" standards for "noncentrality parameter;" for all of the calculations we will be doing here, this will always be 0).

Under the heading "T," you enter the corrected value of t that you calculated using the directions in Chapter 11, Section F.

Let's use the example we gave in Chapter 11, Section F. In that example our uncorrected t value was 1.645. After applying the PFC, the corrected t value, t', was 1.912. We sampled 26 quadrats in each of two years. If the sampling design used temporary quadrats, then the t statistic originated from an independent–sample t test. The appropriate degrees of freedom are therefore $(n_1 - 1) + (n_2 - 1) = (26 - 1) + (26 - 1) = 50$.

Under "1 <=DF" we enter 50, under "0 <=NCP" we enter 0, and under "T" we enter 1.912. We now click on the "Calculate" button and NCSS PROBABILITY CALCULATOR returns the following two values:

Prob(t <=T): 0.9691926480
Prob(t >=T): 0.0308073520

These two values are complements of each other. In other words Prob(t>=T) = 1 – (Prob(t<=T)). It is the second value, Prob(t>=T), that we use to determine the calculated P value from our data. "Prob(t>=T)" is the probability of obtaining a test statistic as large or larger than the one obtained from this test (1.912) when there is really no difference between the two populations.

The P value returned by NCSS PROBABILITY CALCULATOR, 0.0308, is the P value associated with a one–tailed test. If we had conducted a one–tailed t test this is the P value we would report. If, however, we conducted a two–tailed t test, we would need to multiply the one–tailed P value, 0.0308, by 2. That value, 0.0616, is the P value we would report for a two–tailed t test.

Using the same example data set, if our quadrats were permanent, then the t statistic originated from a paired t test. In this case we would enter 25 under "1<=DF," 0 under "0<=NCP," and 1.912 under "T." The program now returns the following two values:

Prob(t<=T): 0.9662990278
Prob(t>=T): 0.0337009722

If we had performed a one–tailed paired t test, we would report our P value of 0.0337. If, however, we had performed a two–tailed, paired t test, the P value we would report is 0.0337 x 2 – 0.0674.

B. Calculating the P value for a given X² statistic

Start up the program, NCSS PROBABILITY CALCULATOR. Select the Chi–Square Probability Distribution. (There are two buttons you can click on to go to the Chi–Square Probability Distribution. You should click on the right button, which brings up the "input" screen. Clicking on the left button brings up the "inverse input screen," which calculates the χ^2 value associated with a given P value.)

Once you have accessed the Chi–Square input screen, you enter values in three places on the left side of the screen. Under the heading "1<=DF" you enter the degrees of freedom appropriate to your test. For McNemar's test, which can be used only to test for a difference between two years, there is always 1 degree of freedom. For a chi–square test applied to a contingency table, the number of degrees of freedom is always one less than the number of years being compared.

Thus, for a 2 x 2 table comparing 2 years there is 1 degree of freedom, for a 2 x 3 table comparing 3 years there are 2 degrees of freedom, and so on.

Under "0<=NCP" you enter 0, and under "0<=X" you enter the corrected value of χ^2 that you calculated using the directions in Chapter 11, Section F.

Using the example we gave in Chapter 11, Section F, we sampled 77 quadrats in each year. Our uncorrected χ^2 was 2.706 and our FPC corrected χ^2 was 3.640. Now we want to calculate the P value associated with this corrected χ^2 of 3.640. Here is how we do this.

Under "1<=DF" we enter 1, under "0<=NCP" we enter 0, and under "0<=X" we enter 3.640. We now click on the "Calculate" button and NCSS PROBABILITY CALCULATOR returns the following two values:

Prob(0<=x<=X): 0.9435930657
Prob(x>=X): 0.0564069343

Unlike the situation with the t statistic, the P value returned by NCSS PROBABILITY CALCULATOR, 0.0524 (from the Prob(x>=X) value), is for the two–tailed test. If our test was two–tailed, we report our P value as 0.0564. If we conducted a one–tailed test we divide the P value of 0.0524 by 2. The P value for a one–tailed test is therefore 0.0262.

C. Calculating the P value for a given F statistic

Start up the program, NCSS PROBABILITY CALCULATOR. Select the F Probability Distribution. (There are two buttons you can click on to go to the F Probability Distribution. You should click on the right button, which brings up the "input" screen. Clicking on the left button brings up the "inverse input screen," which calculates the F value associated with a given P value.)

The following directions apply only to an analysis of variance (comparing three or more years with temporary sampling units used in each year), since we do not recommend use of the repeated–measures analysis of variance. Once you have accessed the F input screen, you enter values in four places on the left side of the screen. Under the heading "1<=DF1" you enter the degrees of freedom for the numerator of the test equation. This is the number of years of data minus 1. Under the heading "1<=DF2" you enter the degrees of freedom for the denominator of the test equation. This is the total number of sampling units for all years minus 1 minus the degrees of freedom for the numerator (this will become clearer with the example given below). Under the heading "0<=NCP" you enter 0. Under the heading "0<=F" you enter the corrected value of F that you calculated using the directions in Chapter 11, Section F.

Using the example given in Chapter 11, Section F, we sampled 50 quadrats each year for three years. The uncorrected F statistic was 3.077, and the FPC corrected F statistic was 3.517. To calculate the P value corresponding to the corrected F statistic of 3.517 we do the following.

Under the heading "1<=DF1" we enter 2 (the number of years, in this case 3, minus 1). Under the heading "1<=DF2" we enter 147 (the total number of quadrats for all years, 150, minus 1, minus the 2 degrees of freedom entered under "1<=DF1"). Under the heading "0<=NCP" we enter 0. Under the heading "0<=F" we enter 3.517. The program returns the following values.

Prob(0<=f<=F): 0.9677889794
Prob(f>=F): 0.0322110206

We report the *P* value of 0.0322 for our analysis of variance (remember that the analysis of variance is by its nature a two–tailed test).

INDEX

Page numbers in bold face type indicate figures and illustrations.